Evolution of the First Nervous Systems

NATO ASI Series

Advanced Science Institutes Series

A series presenting the results of activities sponsored by the NATO Science Committee, which aims at the dissemination of advanced scientific and technological knowledge, with a view to strengthening links between scientific communities.

The series is published by an international board of publishers in conjunction with the NATO Scientific Affairs Division

A	**Life Sciences**	Plenum Publishing Corporation
B	**Physics**	New York and London
C	**Mathematical and Physical Sciences**	Kluwer Academic Publishers Dordrecht, Boston, and London
D	**Behavioral and Social Sciences**	
E	**Applied Sciences**	
F	**Computer and Systems Sciences**	Springer-Verlag
G	**Ecological Sciences**	Berlin, Heidelberg, New York, London,
H	**Cell Biology**	Paris, and Tokyo

Recent Volumes in this Series

Volume 182—DNA Repair Mechanisms and their Biological Implications in Mammalian Cells
edited by Muriel W. Lambert and Jacques Laval

Volume 183—Protein Structure and Engineering
edited by Oleg Jardetzky

Volume 184—Bone Regulatory Factors: Morphology, Biochemistry, Physiology, and Pharmacology
edited by Antonio Pecile and Benedetto de Bernard

Volume 185—Modern Concepts in *Penicillium* and *Aspergillus* Classification
edited by Robert A. Samson and John I. Pitt

Volume 186—Plant Aging: Basic and Applied Approaches
edited by Roberto Rodríguez, R. Sánchez Tamés, and D. J. Durzan

Volume 187—Recent Advances in the Development and Germination of Seeds
edited by Raymond B. Taylorson

Volume 188—Evolution of the First Nervous Systems
edited by Peter A. V. Anderson

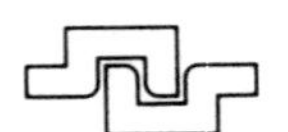

Series A: Life Sciences

Evolution of the First Nervous Systems

Edited by

Peter A. V. Anderson

University of Florida
St. Augustine, Florida

Plenum Press
New York and London
Published in cooperation with NATO Scientific Affairs Division

Proceedings of a NATO Advanced Research Workshop
on Evolution of the First Nervous Systems,
held July 2-5, 1989,
at St. Andrew's University,
Scotland, United Kingdom

Library of Congress Cataloging-in-Publication Data

NATO Advanced Research Workshop on Evolution of the First Nervous Systems (1989 : St. Andrew's University)
Evolution of the first nervous systems / edited by Peter A.V. Anderson.
p. cm. -- (NATO ASI series. Series A, Life sciences ; vol. 188)
"Proceedings of a NATO Advanced Research Workshop on Evolution of the First Nervous Systems, held July 2-5, 1989, at St. Andrew's University, Scotland, United Kingdom"--T.p. verso.
"Published in cooperation with NATO Scientific Affairs Division."
Includes bibliographical references.
ISBN 0-306-43529-2
1. Comparative neurobiology--Congresses. 2. Nervous system--Evolution--Congresses. I. Anderson, Peter A. V. II. Title. III. Series: NATO ASI series. Series A, Life sciences ; v. 188.
QP356.15.N38 1989
591.1'88--dc20 90-34441
CIP

A Division of Plenum Publishing Corporation
233 Spring Street, New York, N.Y. 10013

Printed in the United States of America

Contributors

Michel Anctil - Département de Sciences Biologiques et Centre de Recherche en Sciences Neurologiques, Université de Montréal, Québec, H3C 3J7, Canada

Peter A. V. Anderson - Whitney Laboratory and Departments of Physiology and Neurosciences, University of Florida, 9505 Ocean Shore Blvd., St. Augustine, Florida 32086, U.S.A.

Stuart A. Arkett - Department of Physiology, The Medical School, University of Bristol, Bristol BS8 1TD, U.K.

André Bilbaut - Laboratoire de Cytologie Expérimentale, Université de Nice, U.R.A. 651, Parc Valrose, 06034 Nice Cedex, France

Quentin Bone - The Marine Laboratory, Citadel Hill, Plymouth, PL1 2PB, U.K.

William E. S. Carr - Whitney Laboratory and Department of Zoology, University of Florida, 9505 Ocean Shore Blvd., St. Augustine, Florida 32086, U.S.A

James L. S. Cobb - Gatty Marine Laboratory, University of St. Andrews, St. Andrews, Fife KY16 8LB, U.K.

Charles N. David - Zoologisches Institut der Universität München, Luisenstrasse 14, 8000 München 2, F.R.G.

Joachim W. Deitmer - Abteilung für Allgemeine Zoologie, FB Biologie, Universität Kaiserslautern, Postfach 3049, D-6750, Kaiserslautern, F.R.G.

Stefan Dübel - German Cancer Research Center, Im Neuenheimger Feld 280, 6900 Heidelberg, F.R.G.

Jean Febvre - Laboratoire de Biologie Cellulaire Marine, U.R.A. 671, 06230, Villefranche-sur-Mer, France

Collette Febvre-Chevalier - Laboratoire de Biologie Cellulaire Marine, U.R.A. 671, 06230, Villefranche-sur-Mer, France

Christian Franke - Physiologisches Institut der Technischen Universität München, Beidersteiner, Strasse 29, 8000 München 40, F.R.G.

Doris Graff - Centre for Molecular Neurobiology, University of Hamburg, Martinistrasse 52, 2000 Hamburg 20, F.R.G.

Colin R. Green - Department of Anatomy and Developmental Biology, University College London, Gower Street, London WC1E 6BT, U.K.

Michael J. Greenberg - Whitney Laboratory, University of Florida, 9505 Ocean Shore Blvd., St. Augustine, Florida 32086, U.S.A.

Cornelius J. P. Grimmelikhuijzen - Centre for Molecular Neurobiology, University of Hamburg, Martinistrasse 52, 2000 Hamburg 20, F.R.G.

Hanns Hatt - Physiologisches Institut der Technischen Universität München, Beidersteiner Strasse 29, 8000 München 40, F.R.G.

Todd Hennessey - Department of Biological Sciences, State University of New York at Buffalo, Buffalo, New York 14260, U.S.A.

Mari-Luz Hernandez-Nicaise - Cytologie Expérimentale, Université de Nice, U.R.A. 651, Park Valrose, 06034 Nice Cedex, France

Bertil Hille - Dept. of Physiology and Biophysics, University of Washington, Seattle, Washington 98195, USA

Engelbert Hobmayer - Zoologisches Institut der Universitat München, Luisenstrasse 14, 8000 München 2, F.R.G.

Sabine A. H. Hoffmeister - Zentrum für Molekulare Biologie, Im Neuenheimer Feld 282, 6900 Heidelberg, F.R.G.

O. Koizumi - Physiological Laboratory, Fukuoka Women's University, 1-1-1 Kasumigaoka, Higashi-ku, Fukuoka 813, Japan

Harold Koopowitz - Ecology and Evolutionary Biology, University of California, Irvine, California 92717, U.S.A.

Ching Kung - Laboratory of Molecular Biology and Department of Genetics, University of Wisconsin, Madison, Wisconsin 53706, U.S.A.

Michael S. Laverack - Gatty Marine Laboratory, University of St. Andrews, St. Andrews, Fife KY16 8LB, U.K.

George O. Mackie - Biology Department, University of Victoria, Victoria, British Columbia V8W 2Y2, Canada

Ian D. McFarlane - Department of Applied Biology, University of Hull, Hull HU6 7RX, U.K.

Robert W. Meech - Department of Physiology, The Medical School, University of Bristol, University Walk, Bristol BS8 1TD, U.K.

Ghislain Nicaise - Cytologie Expérimentale, Université de Nice, U.R.A. 651, Park Valrose, 06034 Nice Cedex, France

C. Ladd Prosser - Department of Physiology and Biophysics, University of Illinois, Urbana, Illinois 61801, U.S.A.

Thomas S. Reese - Laboratory of Neurobiology, N.I.N.D.S., National Institutes of Health, Bethesda, Maryland 20892, U.S.A.

Eliana Scemes - Departamento de Fisiologia Geral, Instituto de Biociências, Universidade de São Paulo, SP, Brasil

Andrew N. Spencer - Department of Zoology, University of Alberta, Edmonton, Alberta, TG6 2E9, Canada

Judith Van Houten - Department of Zoology, University of Vermont, Burlington, Vermont 05405, U.S.A.

Jan van Marle - Department of Electron Microscopy, Faculty of Medicine, University of Amsterdam, Meibergdreef 15, 1105 AZ, Amsterdam, The Netherlands

J. A. Westfall - Department of Anatomy and Physiology, College of Veterinary Medicine, Kansas State University, Manhattan, Kansas 66506, U.S.A.

David C. Wood - Department of Behavioral Neurosciences, University of Pittsburgh, Pittsburgh, Pennsylvania 15260, U.S.A.

Preface

This book represents the proceedings of a NATO Advanced Research Workshop of the same name, held at St. Andrews University, Scotland in July of 1989. It was the first meeting of its kind and was convened as a forum to review and discuss the phylogeny of some of the cell biological functions that underlie nervous system function, such matters as intercellular communication in diverse, lower organisms, and the electrical excitability of protozoans and cnidarians, to mention but two. The rationale behind such work has not necessarily been to understand how the first nervous systems evolved; many of the animals in question provide excellent opportunities for examining general questions that are unapproachable in the more complex nervous systems of higher animals. Nevertheless, a curiosity about nervous system evolution has invariably pervaded much of the work.

The return on this effort has been mixed, depending to a large extent on the usefulness of the preparation under examination. For example, work on cnidarians, to many the keystone phylum in nervous system evolution simply because they possess the "first" nervous systems, lagged behind that carried out on protozoans, because the latter are large, single cells and, thus, far more amenable to microelectrode-based recording techniques. Furthermore, protozoans can be cultured easily and are more amenable to genetic and molecular analyses. Thus, our understanding of the properties and distribution, in lower animals, of the various factors that underlie nervous system function was very fragmented, so much so that realistic discussions of the evolution of the first nervous systems were not feasible simply because the comparative data base was not available.

The situation has changed dramatically in the past few years, with the introduction of new and powerful techniques, such as patch clamping, immunocytochemistry and peptide sequencing. As a result, we can now talk about such things as receptor potentials and ion channels in both the Protozoa and the Cnidaria, whereas until only a few years ago, those of us studying cnidarians were forced to rely on extracellular recordings. Thus, we have a far broader picture of the variety and distribution of many of the different "neural" mechanisms in several relevant organisms, and can now realistically compare the cellular properties in question across a range of appropriate organisms.

The NATO Advanced Research Workshop from which this book derives, was designed as the first real forum in which these various questions could be discussed.

Its aims were several. First, to bring together a group of experts working on relevant questions in diverse but appropriate, organisms ranging from unicellular protists through lower invertebrates, and by way of their presentations, to receive current reviews of the various fields. The diversity of speakers was such that many had not previously met, yet their interaction proved useful and important, and may be more so in the future through important new collaborations. A second aim was to highlight areas requiring particular attention and, finally, to provide a forum in which to take advantage of the experience of the participants in discussions of the whole question of nervous system evolution.

The bulk of this book represents the presentations by the main speakers. They have been broadly grouped into three general areas: intercellular communication, electrical excitability and sensory transduction. The latter section is rather sparse, but this is primarily a reflection of the fact that this area has received far less attention than it deserves, and was included in the Workshop to illustrate this point. In addition to the review presentations, we were treated to a Plenary lecture by George O. Mackie, who was invited to do this as an acknowledgment of his enormous contribution to the field over the years.

The discussions that followed each session of the Workshop were moderated by individuals who were chosen on the basis of their broad expertise in their general field, their interest in comparative questions, particularly as it pertains to the evolution of the nervous system and, finally, their ability to stimulate discussion. All were very effective in this role and the discussions were highly stimulating and enjoyable experiences. Each moderator has prepared a brief chapter in which he has attempted to summarize the content of the sessions he moderated, and any conclusions achieved during the respective sessions. This was an onerous task given the breadth of material covered and the sometimes boisterous discussions that ensued, but their summaries provide a good review of the proceedings and much food for thought.

The Workshop was, by all measures, very successful and enjoyable, and I would like to express my sincere thanks to all who made it such. Pride of place goes to Dr. Craig Sinclair and his staff at the Division of Scientific Affairs of NATO who provided generous financial support and considerable encouragement throughout the planning stages of the Workshop. Supplemental funds were supplied by the National Science Foundation (grant BNS 88-19742) and their assistance is gratefully acknowledged. I would also like to thank Mike Greenberg for the secretarial and other assistance provided by the Whitney Laboratory of the University of Florida. The Workshop was held at St. Andrews University, St. Andrews, Scotland, and I am very grateful for the facilities and assistance they provided through the services of Roger Smith and his staff. The local organizers, Jim Cobb and Glen Cottrell, provided invaluable advice and help. Finally, the success of any meeting is largely a reflection of the attitude and input of the participants, and I would like to thank everyone who participated for helping to make this such an enjoyable and stimulating Workshop.

This book was prepared in camera-ready format, using manuscripts supplied on floppy disks by the various authors. This process greatly accelerated publication while at the same time, I hope, minimized the introduction of errors etc. during retyping. I thank my wife, Jan Anderson, for reformatting the manuscripts and for her support during the preparation of this book.

Peter A. V. Anderson

Contents

I. INTERCELLULAR COMMUNICATION

Chapter 1

Cnidarian Gap Junctions: Structure, Function and Evolution
• C. R. Green

Chapter 2

Intercellular Junctions in Ctenophore Integument
• Mari-Luz Hernandez-Nicaise, Ghislain Nicaise and Thomas S. Reese

Chapter 3
Chemical and Electrical Synaptic Transmission in the Cnidaria • Andrew N. Spencer

Chapter 4
Control of Morphogenesis by Nervous System-derived Factors • S. A. H. Hoffmeister and S. Dübel

Chapter 5
Differentiation of a Nerve Cell-Battery Cell Complex in *Hydra* • Engelbert Hobmayer and Charles N. David

Chapter 6

Chemical Signaling Systems in Lower Organisms: A Prelude to the Evolution of Chemical Communication in the Nervous System

• William E. S. Carr

Chapter 7

Neurons and their Peptide Transmitters in Coelenterates

• C. J. P. Grimmelikhuijzen, D. Graff, O. Koizumi, J. A. Westfall and I. D. McFarlane

Chapter 8

Peptidergic Neurotransmitters in the Anthozoa

• I. D. McFarlane, D. Graff and C. J. P. Grimmelikhuijzen

Chapter 9

Catecholamines, Related Compounds and the Nervous System in the Tentacles of some Anthozoans

• J. Van Marle

Chapter 10

The Antiquity of Monaminergic Neurotransmitters: Evidence from Cnidaria

• Michel Anctil

Chapter 11

Rethinking the Role of Cholinergic Neurotransmitters in the Cnidaria

• Eliana Scemes

Chapter 12
Wide Range Transmitter Sensitivities of a Crustacean Chloride Channel
• Hanns Hatt and Ch. Franke

Chapter 13
Two Pathways of Evolution of Neurotransmitters-Modulators
• C. Ladd Prosser

Chapter 14
Summary of Session and Discussion on Intercellular Communication

II. ELECTRICAL EXCITABILITY

Chapter 15
Ion Channels of Unicellular Microbes
• Ching Kung

Chapter 16
Ion Currents of *Paramecium*: Effects of Mutations and Drugs • Todd M. Hennessey

Chapter 17
Membrane Excitability and Motile Responses in the Protozoa, with Particular Attention to the Heliozoan *Actinocoryne contractilis* • Colette Febvre-Chevalier, André Bilbaut, Jean Febvre and Quentin Bone

Chapter 20

The Electrophysiology of Swimming in the Jellyfish *Aglantha digitale* • Robert. W. Meech

Chapter 21

Ionic Currents in Ctenophore Muscle Cells • André Bilbaut, Marie-Lux Hernandez Nicaise and Robert W. Meech

Chapter 22

Polyclad Neurobiology and the Evolution of Central Nervous Systems

• Harold Koopowitz

Chapter 23

Enigmas of Echinoderm Nervous Systems

• James L. S. Cobb

Chapter 24

Summary of Session and Discussion of Electrical Excitability

III. SENSORY MECHANISMS

IV. PLENARY LECTURE

Part 1

INTERCELLULAR COMMUNICATION

Chapter 1

Cnidarian Gap Junctions: Structure, Function and Evolution

C. R. GREEN

1. Introduction

Gap junctions are plasma membrane specializations which allow cells to exchange ions and small molecules directly, without recourse to the extracellular space. They are found throughout the animal kingdom in virtually all tissues, including the nervous system where they form electrical synapses. Where gap junctions occur, the adjacent cell membranes are in close apposition, with protein channels spanning a 2-4 nm intercellular space. The junction usually consists of several channels forming a plaque-like structure; the number of channels varies considerably from junction to junction, and from tissue to tissue. It is possible that in some tissues only one or two channels are involved in coupling between cells, but in these cases there is no definitively recognizable structure. The gap junction channel, termed a connexon, has a central pore 2 nm in diameter, allowing the cell-to-cell passage of molecules with molecular weights up to about 1500, although the molecular weight cut-off depends on the shape and charge of the molecule involved, and upon the state of the channel itself. In the Arthropods, the channel is somewhat larger than in the other phyla, giving the upper size range quoted, but in many tissues the channels can be closed off to the extent that even very small tracer dyes are unable to pass. However, the cells may remain ionically coupled, and this is important when considering the role of gap junctions in the nervous system as opposed to their other putative roles in development, patterning, metabolic cooperation and the maintenance of homeostasis.

C. R. GREEN • Department of Anatomy and Developmental Biology, University College London, Gower Street, London WC1E 6BT, United Kingdom.

2. Gap Junction Structure

Gap junction structure varies slightly from tissue to tissue (for review see Larsen, 1983), but there has, so far, been no successful attempt to link structural variations with functional differences. In all cases, the junction is recognizable in thin section electron microscopy as a region where the adjacent cell membranes run parallel (Fig. 1), separated by a 2-4 nm gap (the visibility of which varies somewhat with the staining methods used) which gives the junction it's name. In freeze-fracture the connexons appear as particles on one-half of the lipid bilayer, with corresponding pits on the other. Particles are most commonly seen on the P-face (Fig. 1), but may occur on the E-face, or even be distributed between the two. This may reflect junction protein differences, but may equally reflect differences in the lipid bilayer or in the techniques used to prepare the sample (fixed or unfixed tissues, for example, will often fracture with particles remaining on opposite faces). The connexons often form a tightly packed hexagonal array within the membrane, but can also occur in random arrays. Once again, attempts to link connexon distribution with the functional state of the channels (open or closed) have been largely discredited (Green and Severs, 1984; Miller and Goodenough, 1985). In lanthanum-impregnated tissues, or, better yet, on negatively-stained isolated junctions, the channel structure becomes readily apparent (Fig. 1), with the central core delineated by a ring of the six subunits that make up the connexon. Each subunit is termed a connexin and consists of a single protein; six connexins make up a connexon, two connexons (one in each of the two contributing cell membranes) make up a single channel.

In *Hydra* the gap junctions are about 0.2 to 0.7 μm in diameter (Wood and Kuda, 1980), and occur predominantly at the laterobasal surfaces of the cells. In freeze-fracture, the connexon particles remain on the E-face (Filshie and Flower, 1977; Wood, 1977), a feature in common with the Arthropods (the majority of other phyla show P-face particles), and the junctional plaque is more clearly associated with a non-particulate matrix material in which the particles are embedded (Wood, 1977). Wood and Kuda (1980) suggest that this may indicate the presence of additional cell-cell adhesion components, since in *Hydra*, the fracture plane tends to move away from the junctions, rather than through them as is usual. Gap junctions link the epithelial cells within the endoderm and ectoderm, but also join heterotypic cells across the supporting mesoglea. The frequency of these junctions tends to vary with body region. In other Cnidarian classes and in Ctenophora, variations in structure are seen (see Hernandez-Nicaise et al. in this book), but there is little information available for the invertebrate groups as a whole. What there is suggests considerable variation in structure (see, for example, Berdan, 1987; Duvert et al., 1980; Flower, 1972; Flower and Green, 1982; Lane and Skaer, 1980; Wood and Hageman, 1982) within an overall common theme. In tissues where two structural variations of the junctions are seen, it is hard not to accept there must be differences in their roles, but there is, as yet, no firm evidence for this (for review see Larsen, 1983).

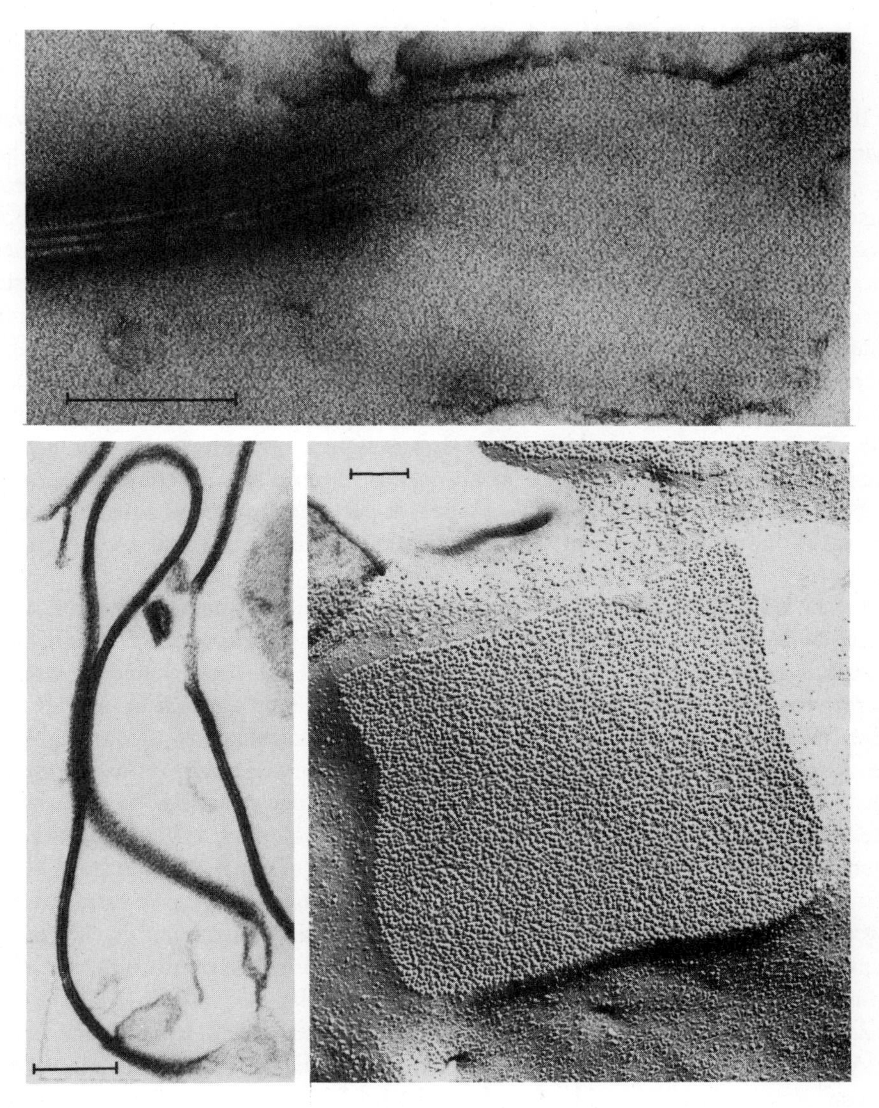

Figure 1. A negatively-stained gap junction (top) isolated from rat liver. The connexons are clearly outlined by the heavy metal stain. The central pores or channels have also filled with stain and appear as a dot within each connexon. The darker region to the left is where the junction has folded and the membranes and connexons are seen outlined in cross section. Two side-by-side views are seen, the gap between the two stain-excluding junctional membranes in each case appearing as a dark line with some cross-bridging of the connexons visible. At the bottom left is a thin section view of a gap junction, again isolated from rat liver, to show their structure more clearly. The typical pentalamina structure of thin sectioned junctions is apparent. The freeze-fracture image at the bottom right is of a gap junction in the rabbit interventricular septum. Each particle, here on the P-face, is a connexon, and the junctional plaque consists of several hundred of these. Scale bars represents 125 nm (from Green et al., 1988).

3. The Biochemistry of Gap Junctions

It is now clear that there is a family of gap junction proteins, their distribution varying with tissue and species type, in some cases two types occurring within the same cell. For some years gap junctions have been isolated, most commonly from rat or mouse liver, with the general consensus being that the major protein is a polypeptide of 26-29 kD MW. The situation was somewhat confused, however, by the presence of proteolytic breakdown products in some preparations, a tendency for the protein to form multimers, especially when heated (Hertzberg and Skibbens, 1984), and more recently, the demonstration that it runs anomalously on polyacrylamide gels; it's apparent molecular weight varies with gel concentration (Green et al., 1988). In addition, it is clear that the protein isolated from cardiac tissue is considerably larger, 44-47 kD (Manjunath et al., 1984), and that present in the vertebrate eye lens even more so, at around 70 kD (Kistler et al., 1985). Several gap junction proteins have now been fully sequenced from cDNA clones, or partially sequenced in from the amino terminal of the isolated protein. The major characteristics of these are summarized below.

1. 32 kD. This is the most commonly described gap junction protein, usually running at 26-29 kD on polyacrylamide gels, but it does run at 32 kD when the gel concentration is 15% or higher (Green et al., 1988). It has been cloned from human and rat liver cDNA libraries (Kumar and Gilula, 1986; Paul, 1986), and antibody studies (Western blotting and immunolocalization) indicate it is widespread in the vertebrates (Hertzberg and Skibbens, 1984). The human and rat liver forms are virtually identical, although in lower classes there is more variation. In *Xenopus*, the protein has a molecular weight of 30 kD, but it shows around 70% identity with that of human liver when aligned for optimum fit (Gimlich et al., 1988). Western blotting and immunolocalization experiments have shown that there is at least partial homology between this junction type and that in many invertebrate tissues and phyla, the main exception being the Arthropods, where the gap junction is structurally different and no homology has, so far, been found (C.R. Green, unpublished results).

2. 43 kD. First described in cardiac tissues, and cloned from a heart cDNA library (Beyer et al., 1987), this protein occurs in several tissues, but seems restricted to those of mesodermal origin. It is the protein that coordinates contraction of the myocytes in the heart and contractions of the uterus at parturition. It has also been reported in other organs such as the kidney, ovary and stomach (Ebihara et al., 1989; Dupont et al., 1988), but has not yet been located in the invertebrates.

3. 26 kD. This protein runs with an apparent molecular weight of 21 kD on polyacrylamide gels, and has, so far, been described only in rodents, where it occurs in the liver, along side the 32 kD protein (Nicholson et al., 1987; Traub et al., 1989).

4. 38 kD. This protein occurs in embryonic *Xenopus* tissues (Ebihara et al., 1989), being replaced later in development by the 30 kD variant (Gimlich et al., 1988). It appears to have more homology with the 43 kD form found in cardiac tissues than the

liver type junction protein, and has not yet been described in any other class of the vertebrates, or any other phylum.

5. 70 kD. This protein appears to be unique to the vertebrate eye lens. It has only been partially sequenced, but shows strong homology at the amino terminal end with all the other gap junction types (Kistler et al., 1988). It runs at 70 kD on gels, and breaks down, *in vivo*, to 38 kD, with older cells nearer the cortex of the lens having the latter form. This *in vivo* breakdown presumably reflects a functional change, or a change in some control requirements of the junction channel in the older cells. The only cDNA clone, so far, obtained that matches the known protein sequence data is for a 46 kD polypeptide; this anomaly has not yet been explained.

All of the proteins sequenced have strikingly similar hydropathy plots. These plots indicate that there are four transmembrane regions (Beyer et al., 1987; Paul, 1986; Milks et al., 1988), and models derived from protease digestion and antibody localization studies (Zimmer et al., 1987), and from electron image analysis and peptide antibody mapping (Milks et al., 1988) indicate that both the amino terminus and the carboxyl terminus are on the cytoplasmic surface of the membrane. The transmembrane regions are most probably alpha-helices; the third transmembrane region has a line of polar amino acids which are thought to line the hydrophilic gap junction channel formed when six of these proteins make a half connexon in the membrane. A corresponding six protein unit in the neighboring cell will join to complete the connexon. The major differences between the different protein types occurs in the cytoplasmic regions, especially at the C-terminal end, with most homology occurring in the transmembrane regions and the extracellular gap regions. Proteins with the higher molecular weights appear to have C-terminal extensions with the first 220 or so amino acids forming the basic structure described above. While it is tempting to ascribe functional differences to the different protein types, there is, as with the structural variations, no evidence for this, and the differences may reflect the tissue origin in which the junction is present, rather than having any practical significance.

Extrapolating to the Cnidaria, the major gap junction type is most likely the 32 kD form. Western blotting indicates that the gap junction protein in *Hydra* runs on gels with an apparent molecular weight equal to that of the rat liver protein from which the blotting antibodies were derived (Fraser et al., 1987). Although there will almost certainly be protein variation, especially in the cytoplasmic regions, there is, none the less, some homology here between the gap junctions found in rat liver and those found in *Hydra*, since the antibodies used for the immunoblotting experiments immunolocalize to the inner junction surface (Fraser et al., 1987; Zimmer et al., 1987). It is expected, however, that most homology will occur in the transmembrane and extracellular regions, and that the structural configuration of the protein in the membrane will be essentially the same as that in vertebrate tissues. However, attempts to clone the protein in *Hydra* using liver sequence-derived probes have not been successful, indicating considerable differences probably exist as well. There is, also, no reason to exclude the possibility that the different types of gap junction seen in the vertebrate phyla might have a common ancestor in the lower invertebrates, with those in the Cnidaria showing different regions of close homology with several of these.

4. Regulation of Gap Junction Communication

Gap junctional communication between cells is not necessarily permanent, and clearly at different times, especially during patterning or development, or following tissue damage, cells may uncouple from their neighbors. Unfortunately, there is little information available on the regulation of gap junction communication in the Cnidaria, but most of the factors which have been implicated in the control of junctional permeability in other phyla will almost certainly apply.

Turin and Warner (1977) found that an increase in intracellular pH caused a rapid (but reversible) uncoupling of *Xenopus* blastomeres. In this species, the main response occurs at pH 7.0; the normal cytoplasmic pH is generally close to pH 7.6. In liver (Spray and Bennett, 1985) and heart (Noma and Tsuboi, 1987) a pH level below 6.5 results in junction closure. An increase in intracellular calcium (Ca^{++}) levels has a similar effect (Rose and Loewenstein, 1976). Cardiac gap junctions are particularly sensitive to Ca^{++} concentrations at physiological levels, but this sensitivity falls with a drop in intracellular pH (Noma and Tsuboi, 1987). Any fall in pH during contractile activity (so long as it remained above 6.5) would, therefore, explain why cardiac cells do not appear to uncouple transiently with each beat, despite the increase in Ca^{++} levels required to sustain twitch tension (Warner, 1988). The Ca^{++} sensitivity of gap junctions, however, varies with tissue type; embryonic tissues, for example, appear to be much less sensitive (Spray et al., 1982). A major function of Ca^{++}-induced uncoupling may be in a cell's response to injury. While free Ca^{++} in the cytoplasm is normally sequestered and concentration changes are kept local, cell injury will lead to an influx of Ca^{++} ions, resulting in isolation of the injured cell from its neighbors and restricting the damage effect (Asada and Bennett, 1971). Gap junctions may also be voltage sensitive, in some cases asymmetrically so, as in the crayfish synapse, where transmission is in one direction only (Furshpan and Potter, 1959), in others symmetrically, as in amphibian embryos (Spray et al., 1981). In the latter, the channel acts as if it had two gates, one in each membrane, and it is sensitive to any difference in potential between the contributing cells and adjusts its permeability accordingly. Metabolic factors are also important, and to a large extent their importance has not been adequately considered. Hormones (Caveney et al., 1986), neurotransmitters (Piccolino et al., 1984), growth factors (Maldonado et al., 1988), calmodulin (Peracchia, 1988) and cyclic AMP (cAMP) (Azarnia et al., 1981; Saez et al., 1986) have all been shown to influence junctional communication. Furthermore, the rate at which junctions are formed or removed from the membrane could have a marked effect on communication levels, with control being possible at the transcription, translation or degradation level. In liver, protein half-life time may be as short as one and one-half hours (Traub et al., 1989), and a significant drop in the amount of junction protein present, with a subsequent rise back to normal levels, occurs following partial hepatectomy (Traub et al., 1983). Changes in junctional communication presumably occur concomitantly with tissue repair.

5. The Roles of Gap Junctions

Gap junctions have been attributed many roles, largely by correlating the presence of the gap junction structure with a physiological need. It is only with the development of gap junction protein-specific antibodies, which are able to block junctional communication, that more direct evidence has been obtained. Even so, the evidence available provides a comprehensive insight into gap junctional roles.

5.1. Metabolic Cooperation

Subak-Sharpe et al. (1966, 1969) carried out early experiments demonstrating a role of gap junctions in metabolic cooperation in cultured cells. They used a mutant BHK strain, which was unable to synthesize purines and was, therefore, resistant to culture in the presence of lethal base analogues which they were unable to incorporate. However, when cultured with normal cells, which did incorporate the lethal bases, both cell types died. This "Kiss of Death" was only administered to mutant cells in close contact with normal cells. The opposite, a "Kiss of Life," was demonstrated by Pitts (1971), using mutant cells which were unable to survive in HAT medium unless they were in contact with normal cells. In both cases, metabolites were clearly being transferred from cell to cell, in one case to the cells detriment, and in the other for its survival. Experiments using radiolabelled nucleotides show that the transfer is not via the extracellular medium, and electron microscopy shows the presence of gap junctions.

Metabolic cooperation has also been demonstrated in co-cultures of ovarian granulosa cells which respond to follicle stimulating hormone (FSH) by increasing synthesis of plasminogen activator, and myocardial cells, which respond to noradrenalin, by increasing their beat frequency (Lawrence et al., 1978). In these co-cultures, application of noradrenalin also caused higher plasminogen activator levels in the granulosa cells, and conversely, FSH caused an increase in myocardial cell beat rate. In each case, the cell with the required receptor molecules was, therefore, not only responding to the stimulus itself, but passing on a common mediator to its neighbors, which in these cultures could be heterotypic. The communicated mediator was thought to be the second messenger cAMP.

There are few proven cases of metabolic cooperation *in vivo*, but cells deeper within a tissue, and not immediately adjacent to a blood vessel, may well receive nutrients via gap junctions. The best example of this is in the avascular vertebrate eye lens. Here, gap junctions occur in large numbers, presumably to compensate for the lack of blood supply (Benedetti et al., 1976; Kistler et al., 1985).

5.2. Ionic Coupling

The above-mentioned experiments of Lawrence et al. (1978) also serve to illustrate an example of ionic coupling. When two myocytes were separated by a granulosa cell, not only was their beating coordinated, but the normally inexcitable granulosa cell showed rhythmic membrane depolarizations; ionic coupling of the myocytes occurred via the intermediary granulosa cell. The two classic examples of gap junctional ionic coupling *in vivo* are in cardiac myocytes, where junctional communication coordinates myocyte contraction so the heart beats as a syncytium (De Mello, 1982; Severs, 1985), and in the myometrium of the uterus at parturition. In the latter example, there are few gap junctions until about 24 hours before parturition, when there is a sudden increase in numbers. This increase correlates with an increase in electrical coupling between the smooth muscle cells, allowing coordination of their contractions (Garfield et al., 1977; Sims et al., 1982). Immediately after the fetus is expelled, there is a rapid decline in the number of junctions, with those present being internalized (Garfield et al.,1977). The increase in junction numbers is thought to be induced by a rise in levels of intracellular cAMP.

5.3. Transfer of Regulatory Molecules and Growth Control

In 1964, Stoker reported that the growth of transformed cells is arrested when they grow into contact with normal cells, and Loewenstein (1966) proposed that the formation of gap junctions might be involved in the passage of regulatory substances. Confirmation of the role of junctional communication in tumor cells and growth control has not, however, been clear. This is, in part, because junctional structures may remain in the membrane, but not necessarily with open channels, thereby limiting the reliability of structural studies. Furthermore, there may be a reduction in coupling, but not a complete block, making physiological experiments difficult to interpret. Despite these difficulties, virally-transformed cells have been shown to have reduced cell-cell communication, as have those treated with tumor-promoting agents (Chang et al., 1985). Furthermore, the actions of retinoids on cell growth correlates with their action on junctional communication (Mehta et al., 1989), growth inhibition of transformed cells correlates with the formation of gap junctions with normal cells (Mehta et al., 1986), and the over expression of the cellular SRC gene in 3T3 cells, results in reduced cell-cell communication. In the later case, regulation of communication during cell growth and development is possibly controlled by the level of expression of this gene (Azarnia et al., 1988).

5.4. Development and Patterning

Perhaps the best evidence for the role of gap junctions is available from patterning and developmental studies. Communication between cells of the insect larval epithelium has been shown to vary with location; those within a segment are well coupled, but those at the segment border that delineates a developmental compartment, have reduced coupling (Blennerhassett and Caveney, 1984; Warner and Lawrence, 1982). Coupling at the border is not totally abolished, however; gap junctions are present and small molecules are still able to pass. Similar restriction patterns have been proposed for other compartmental boundaries, for example in molluscan (Van den Biggelaar and Serras, 1988), *Drosophila*, and mammalian development (Lo, 1988). *Drosophila* wing disc mutants also have an abnormal gap junction distribution (Ryerse and Nagel, 1984). More recently, Buehr et al. (1987) have correlated reduced gap junctional communication with the lethal condition that occurs when DDK mouse eggs are fertilized by sperm from another strain. These embryos tend to decompact. A similar result is obtained when normal mouse embryos are injected with a gap junction protein-specific antibody which causes uncoupling of the junctions (Lee et al., 1987). The antibodies used were the same as those used by Warner et al. (1984) to demonstrate a role for gap junctions during the development of frog embryos. One particular cell of *Xenopus* embryos at the eight cell stage was injected with the affinity purified antibody, and subsequent tests at the 32 cell stage showed that both dye coupling and electrical coupling were abolished between the progeny of the injected cells. When the embryos were allowed to develop, some 63% of those injected with immune antibodies showed severe developmental defects on one side of the head; they often lacked one eye and had brain deformities. Control animals, and those injected with preimmune antibodies, had no change in their junctional coupling patterns or development.

6. Gap Junctions in the Nervous System

In the nervous system, gap junctions play an important role in providing low resistance pathways between excitable cells. In the Cnidaria and elsewhere, they are believed to be responsible for epithelial conduction properties (see refs. in Wood and Kuda, 1980), and electrical synapses have been studied extensively since their discovery by Furshpan and Potter (1959) in the crayfish giant motor neuron. However, the extent of gap junctional coupling in higher animals has only become apparent in more recent years, and it may be playing a far more significant role in neuronal communication and integration of the central nervous system than previously thought (Dudek et al., 1983; Llinas, 1985). Antibody localization studies have suggested that functional and structural compartmentalization of discrete central nervous system regions may occur (Nagy et al., 1988) and it has been suggested that gap junctions may play a role in modifying or regulating the behavior of discrete populations of neurons (Bennett et al., 1985). It is also becoming apparent that there may be several types of gap junction in the nervous system (Nagy et al., 1988), but that these might not necessarily fit into the known types described above (communication by Willecke at the 4th International Congress for Cell Biology, Montreal, 1988; Dupont et al., 1988).

Gap junctions occur between glial cells and between neurons, but heterologous junctions between the two are not seen (Berdan et al., 1987), even in the hippocampus, where gap junctions between the different cell types occur in close proximity (Nagy et al., 1988). The junctions between glial cells may facilitate the transfer of metabolites to more central regions, but in neurons they are believed to mediate communication between cells in, for example, escape responses, bidirectional signaling, and synchronization of neuronal activity. It is not clear how the neuron-neuron and glial-glial specificity is bought about, since different types of gap junction protein are considered able to interact to form heterologous junctions, such as those between endothelial cells and vascular smooth muscle. Cell adhesion molecules may be involved; antibodies to NCAM inhibit the formation of cell-cell communication in chicken neuroectoderm, by reducing adhesion (Keane at al, 1988), and during reaggregation of dissociated *Hydra* cells, heterotypic junctions do not form, even though they are present in the mature animal (Wood and Kuda, 1980). This lack of coupling is thought to be bought about by adhesive differences between the gastrodermal and epidermal cells and may assist in sorting of the epithelial layers. However, it may also have a role in determining the pattern of the nervous system in the reforming animals.

Gap junctions are also involved in establishing the nervous system during embryogenesis. The formation of neuromuscular junctions involves the cooperation of two morphologically distinct cells, the neuron and the muscle cell. Initial contact occurs via gap junctions, with subsequent development of the chemical synapse. This "sampling," by way of gap junctions, may be important in neuronal guidance and the specificity of nerve-muscle synapses, or may mediate the neuronal influence on the course of myoblast differentiation (Allen, 1987). Once the nervous system is formed, gap junctions appear to play a role in its maintenance. Acidosis, caused by anoxia and ischemia for example, leads to glial swelling and neuronal degeneration. It has been shown that one effect of the acidosis is a lactic acid-induced inhibition of junctional communication (Anders, 1988).

A single gap junction channel conductance is in the order of 150 - 300 pS (Neyton and Trautman, 1985; Young et al., 1987), and this poses problems for those investigating the nervous system. Clearly, ionic coupling can occur between cells in the absence of a recognizable gap junction structure. In the phylum Porifera, cell-cell communication with gap junction-like properties occurs when cells are pushed together (Loewenstein, 1967) and their free-swimming larvae have cilia which beat with a metachronal rhythm, even though each cilium arises from a separate cell (Green and Bergquist, 1978). However, while several types of particle arrangement have been observed in freeze-fractured sponge membranes, none form classical gap junctional arrangements (Lethias et al., 1983).

Gap junctions in the nervous system provide a more rapid means of transmission than the chemical synapse, although this advantage must be lost over long distances where neuronal conduction and chemical synapses play a role. They are also limited in their specificity. While it is clear that gap junctions can limit the flow of larger molecules, there is no evidence that they can select for ions. Thus, gap junctions are probably unable to impart the specificity possible with the different types of neurotransmitters and chemical synapses. This would explain the discrete compartmentalization seen in the nervous system. On the other hand, as noted earlier, it is not clear to what extent the different protein types and different gap junction structures seen in various tissues may reflect functional differences, and there are certainly different types of electrical synapses, the rectifying synapse in the crayfish, for example, compared with the symmetrical transmission seen in the hydroid epithelium.

7. Gap Junctions in the Cnidaria

Gap junctions in the Cnidaria have a major role in their nervous systems, especially in the simpler nerve nets, where they link cells, and in *Hydra*, they almost certainly coordinate contraction and relaxation of the animal in response to stimuli. This is supported by the subjective observation that animals loaded with gap junction-specific antibodies, which block junctional communication (see below), seem to take longer to contract when stimulated than those treated with preimmune antibodies. In *Obelia*, signaling through the gap junctions coordinates behavioral responses too, one of which is light emission from photocytes (Dunlap et al., 1987). In this hydrozoan, luminescence is emitted from the endogenous Ca^{++}-activated obelin. However, there are no inward Ca^{++} currents detectable in the photocytes, which are dependent upon neighboring support cells for activation. It is Ca^{++} entry into the support cells with subsequent chemical signalling through the gap junctions to the photocytes, which activates the luminescence. It isn't clear what the signalled substance is, although it seems most likely to be Ca^{++}, in which case the concentration must be low enough not to cause junction closure, or the junctions in this species must be relatively Ca^{++} insensitive.

The other major role for gap junctions in the Cnidaria is patterning. Of special interest is cell-cell coupling in *Hydra*, where well-defined models of patterning which require communication between cells, have been developed, largely on the basis of tissue regeneration experiments (Meinhardt and Gierer, 1974; MacWilliams, 1983a,b; Webster and Wolpert, 1966). A pair of gradients are thought to control the location of head formation. The first of these, termed the head activation gradient, is a local tissue property and is quite stable. An 11 amino acid peptide has been isolated from *Hydra*. It has activity consistent with head activation, in that it increases the rate of head regeneration and the number of tentacles formed when applied topologically (Schaller, 1973; Schaller and Bodenmuller, 1981). The second is the head inhibitor gradient, which is responsible for inhibiting head formation or regeneration (Berking, 1977, 1979). This molecule is small (less than 500 MW), hydrophilic, and labile, and is, therefore, consistent with the experiments of MacWilliams, which indicated that head inhibition is due to a diffusible substance. MacWilliams calculated that the rate of diffusion of the inhibitor molecule was 2×10^{-6} cm^2 s^{-1}, about ten times slower than might be expected if it were freely diffusing in an aqueous solution. Its passage down the body column is, therefore, most likely to be from cell to cell, via gap junctions, while that of the head activator is more likely to be passed extracellularly or within the membrane. Gap junction structures were first described in *Hydra* epithelium by Wood (1977) and Filshie and Flower (1977), and communication between cells was subsequently demonstrated by dye injection (Fraser and Bode, 1981). More recently, Fraser et al. (1987) have shown that the rate of diffusion down the body column of intracellularly injected molecules is consistent with that predicted for the head inhibitor molecule, and with the models proposed for patterning of *Hydra*. More direct evidence for the role of gap junctions has been possible by using gap junction-specific antibodies. These bring about patterning defects by blocking junctional communication (Bode et al., 1987; Fraser et al., 1987).

In this set of experiments, antibodies raised against eluted rat liver gap junction protein were shown to cross react with the gap junction protein in *Hydra*, using immunohistochemical localization and Western blotting experiments. The affinity-purified antibodies blocked cell-cell communication, as demonstrated with dye injection and electrical coupling experiments. The dye injection studies, using the small hydrophilic fluorescent dye lucifer yellow, were carried out at various times after antibody loading, and the junctions were found to be blocked for up to 18 hours. Antibody was loaded using a DMSO permeabilization technique. Control experiments indicated that this treatment did not adversely effect the normal patterning ability of the animal. Grafting experiments (Fig. 2) were then carried out, in which the apical eighth of a body column from one animal was grafted to the region just below the head of an animal that had been treated with immune or preimmune immunoglobulins. Control experiments were carried out on animals which were untreated or decapitated before grafting the donor tissue. In the control experiments, grafting of tissue to a normal animal had little effect; the donor tissue was suppressed by inhibitor released from the existing host head. In only a small percentage of cases did a secondary axis form (i.e. a second head grew from the grafted tissue as a result of its high activator level). When tissue is grafted to a decapitated animal, however, there is no source of inhibitor, and in the vast majority of cases, two heads are formed, one from the apical tip of the host animal, and one from the grafted donor tissue, since both regions have high activator levels. In theory, therefore, if the inhibitor moves from cell to cell down the body column via gap junctions, blocking gap junctional communication by loading the host animal with antibodies should result in a higher proportion of secondary axis formation; the restricted movement of the inhibitor, in effect, has the same result as decapitation, in that the source is effectively removed. This is exactly what happened; there was a significant rise in secondary axis formation in immune treated animals compared with those treated with preimmune antibodies. The difference was not as great as that obtained by decapitation of the host animal, but it is necessary to bear in mind that the antibody bulk loading technique will not give equal loading to all cells,

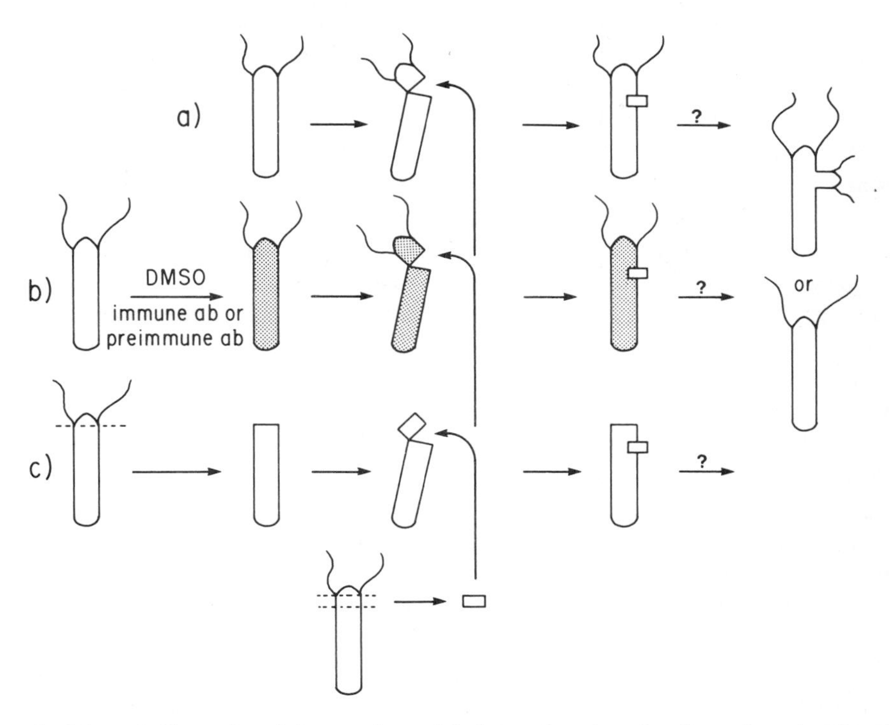

Figure 2. Schematic illustration of the experimental design used to show that distruption of gap junctional communication in *Hydra* interferes with a patterning process. In each experiment, the apical quarter of an untreated animal was grafted to the midgastric region of a normal animal (**a**), an antibody-treated animal (**b**) or a decaptitated animal (**c**), and the host screened 4-6 days later for the presence or absence of a secondary axis. For further details, see Fraser et al., 1987 (from Fraser et al., 1987. Copyright 1987 by the AAAS).

and at best, the communication block is only maintained for 18 hours. The return of communication as the antibody effect wears off can be mimicked by carrying out the decapitation control experiment, but regrafting a head on after 12-18 hours, in effect restoring the flow of inhibitor by replacing the source. In such experiments (described in Bode et al., 1987), the number of secondary axes formed fell to a value nearer that induced in the antibody experiments.

8. Gap Junctions and the Evolution of the First Nervous Systems

Electrical communication between cells seems to have been a feature of the earliest nervous systems. In the Porifera, which do not apparently possess nervous systems, ion communication channels probably act in conjunction with external signaling molecules (Green and Bergquist, 1978; Pavans de Ceccatty, 1974), and in the Cnidaria, where a true nervous system exits, we find nerve nets with fully developed gap junctional communication. In the vertebrates, gap junctions still play a major role in electrical coupling and are apparently more widespread in the central nervous system than has previously been appreciated, but there is also a trend toward compartmentalization of junctional communication within the nervous system, into discrete regions and between homologous cell types. This may reflect a need, in the higher organisms, to separate nervous function from the other roles in which the gap junction is involved, something *Hydra* is apparently unable to do.

The antibody experiments described above provide clear evidence for the role of gap junctions in pattern formation in *Hydra*, but these same gap junctions are also those involved in the electrical coupling between cells, and, thus, play a role in the Hydra's nervous system. Furthermore, experiments by Hoffmeister and Schaller (1987 and in this book) demonstrate that the same putative activator and inhibitor molecules involved in patterning also act as signals for nerve cell differentiation. Nerve cell precursors arrested before final differentiation will mature following topological application of activator, or after wounding, where there is loss of inhibitor by diffusion from the damaged tissue. Conversely, application of inhibitor to wounded tissue will prevent nerve cell differentiation. These regulatory substances are normally produced in nerve cells (Hoffmeister and Schaller, 1987), although both epithelial cells and nerve cells are involved in the head inhibitor system (Rubin and Bode, 1982), and nerve-free *Hydra*, which have only epithelial cells still exhibit normal development (Marcum and Campbell, 1978). The process of patterning and the control of nerve cell differentiation may, however, be quite separate, the former involving an internal gradient mediated by gap junctional communication, the latter involving an external messenger system. That the same molecule may have two separate modes of action is not unprecedented. Cyclic AMP enhances junctional communication (Mehta et al., 1986; Saez et al., 1986), is apparently a second messenger able to induce junction formation (Garfield et al., 1977), and can probably act as the communicated mediator during gap junctional communication (Lawrence et al., 1978). In the slime mold, however, it is an extracellular signal.

In the evolution of the first nervous systems, therefore, there seems to be two trends worthy of further consideration. First, there is the separation of gap junctional communication within the nervous system from other roles the gap junctions play, achieved by the compartmentalization of junctional communication and the development of chemical synapses. Secondly, there is the concept that molecules, which in earlier systems acted as external signals, may be used in higher organisms as the basis for internal transmitter systems, a point taken much further by Carr (this book) in relation to chemical communication in the nervous system.

ACKNOWLEDGEMENTS. I am grateful to Dr. Hans Bode for his comments during the formative stages of this chapter and to Dr. Nick Severs for the freeze-fracture image.

References

Allen, F., 1987, Gap junctions and Development, *Sci. Prog. Oxf.* **71**:275-292.

Anders, J. J., 1988, Lactic acid inhibition of gap junctional intercellular communication in in vitro astrocytes as measured by fluorescence recovery after laser photobleaching, *Glia* **1**:371-379.

Asada, Y., and Bennett, M. V. L., 1971, Experimental alteration of coupling resistance at an electronic synapse, *J. Cell Biol.* **49**:159-172.

Azarnia, R., Dahl, G., and Loewenstein, W. R., 1981, Cell junction and cyclic AMP: III Promotion of junctional membrane permeability and junctional membrane particles in a junction-deficient cell type, *J. Memb. Biol.* **63**:133-146.

Azarnia, R., Reddy, S., Kmiecik, T., Shalloway, D., and Loewenstein, W. R., 1988, The cellular src gene product regulates junctional cell-to-cell communication, *Science* **239**:398-401.

Benedetti, E. L., Dunia, I., Bentzel, C. J., Vermorken, A. J. M., Kibbelaar, M., and Blomendal, H., 1976, A portrait of plasma membrane specializations in eye lens epithelium and fibres, *Biochim. Biophys. Acta.* **457**:353-384.

Bennett, M. V. L., Zimering, M. B., Spira, M. E., and Spray, D. C., 1985, Interaction of electrical and chemical synapses, in: *Gap Junctions* (M. V. L. Bennett and D. C. Spray, eds.), Cold Spring Harbor, New York.

Berdan, R. C., 1987, Intercellular communication in Arthropods: biophysical, ultrastructural, and biochemical approaches, in: *Cell-to-Cell Communication* (W. C. De Mello, ed.), Plenum Press, New York.

Berdan, R. C., Shivers, R. R., and Bulloch, A. G. M., 1987, Chemical synapses, particle arrays, pseudo-gap junctions and gap junctions of neurons and glia in the buccal ganglion of Helisoma, *Synapse* **1**:304-323.

Berking, S., 1977, Bud formation in Hydra: inhibition by an endogenous morphogen, *Wilhelm Roux Arch.* **181**:215-225.

Berking, S., 1979, Analysis of head and foot formation in Hydra by means of an endogenous inhibitor, *Wilhelm Roux Arch.* **186**:189-210.

Beyer, E. C., Paul, D. L., and Goodenough, D. A., 1987, Molecular cloning of cDNA for connexin 43, a gap junction protein from rat heart, *J. Cell Biol.* **105**:2621-2629.

Blennerhassett, M. G., and Caveney, S., 1984, Separation of developmental compartments by a cell type with reduced junctional permeability, *Nature* **309**:361-364.

Bode, H. R., Fraser, S. E., Green, C. R., Bode, P. M., and Gilula, N. B., 1987, Gap junctions are involved in a patterning process in Hydra, in: *Genetic Regulation of Development* (W. F. Loomis, ed.), Alan. R. Liss, Inc., New York.

Buehr, M., Lee, S., McLaren, A., and Warner, A., 1987, Reduced gap junctional communication is associated with the lethal condition characteristic of DDK mouse eggs fertilized by foreign sperm, *Devel.* **101**:449-459.

Caveney, S., Berdan, R. C., Blennerhassett, M. G., and Safranyos, R. G. A., 1986, Cell-to-cell coupling via membrane junctions: methods that show its regulation by a developmental hormone in an insect epidermis, in: *Techniques in Cell Biology*, vol 2 (E. Kurstak, ed.), Elsevier, Amsterdam.

Chang, C. C., Trosko, J. E., Kung, H. J., Bombick, D., and Matsumura, F., 1985, Potential role of the src gene product in inhibition of gap junctional communication in NIH/3T3 cells, *Proc. Natl. Acad. Sci. USA* **82**:5360-5364.

De Mello, W. C., 1982, Intercellular communication in cardiac muscle, *Circ. Res.* **51**:1-9.

Dudek, F. E., Andrew, R. D., MacVicar, B. A., Snow, R. W., and Taylor, C. P., 1983, Recent evidence for and possible significance of gap junctions and electronic synapses in the mammalian brain, in: *Basic Mechanisms of Neuronal Hyperexcitability* (H. H. Jasper and N. M. van Gelder, eds.), Alan R. Liss, Inc., New York.

Dunlap, K., Takeda, K., and Brehm, P., 1987, Activation of a calcium-dependent photoprotein by chemical signalling through gap junctions, *Nature* **325**:60-62.

Dupont, E., El Aoumari, A., Roustiau-Severe, S., Briand, J. P. and Gros, D., 1988, Immunological characterization of rat cardiac gap junction: presence of common antigenic determinants in heart of other vertebrate species and in various organs, *J. Memb. Biol.* **104:**119-128.

Duvert, M., Gros, D., and Salat, C., 1980, Ultrastructural studies of the junctional complex in the musculature of the arrow-worm (*Sagitta setosa*) (Chaetognatha), *Tissue Cell* **12:**1-11.

Ebihara, L., Beyer, E. C., Swenson, K. I., Paul, D. L., and Goodenough, D. A., 1989, Cloning and expression of a Xenopus embryonic gap junction protein, *Science* **243:**1194-1195.

Filshie, B. K., and Flower, N. E., 1977, Junctional structures in Hydra, *J. Cell Sci.* **23:**151-172.

Flower, N. E., 1972, A junctional structure in the epithelia of insects of the order Dictyoptera, *J. Cell Sci.* **10:**683-691.

Flower, N. E., and Green, C. R., 1982, A new type of gap junction in the phylum Brachiopoda, *Cell Tissue Res.* **227:**231-234.

Fraser, S. E., and Bode, H. R., 1981, Epithelial cells of Hydra are dye-coupled, *Nature* **294:**356-358.

Fraser, S. E., Green, C. R., Bode, H. R., and Gilula, N. B., 1987, Selective disruption of gap junctional communication interferes with a patterning process in Hydra, *Science* **237:**49-55.

Furshpan, E. J., and Potter, D. D., 1959, Transmission at the giant motor synapses of the crayfish, *J. Physiol.* **145:**289-325.

Garfield, R. E., Sims, S., Kannan, M. S., and Daniel, E. E., 1977, Gap junctions: their presence and necessity in myometrium during parturition, *Science* **198:**958-960.

Gimlich, R. L., Kumar, N. M., and Gilula, N. B., 1988, Sequence and developmental expression of mRNA coding for a gap junction protein in *Xenopus*, *J. Cell Biol.* **107:**1065-1073.

Green, C. R., and Bergquist, P. R., 1978, Cell membrane specialisations in the Porifera, in: *Sponge Biology* (C. Levi and N. Boury-Esnault, eds.), Colloques Internat. du CNRS, No 291, Paris.

Green, C. R., and Severs, N. J., 1984, Gap junction connexon configuration in rapidly frozen myocardium and isolated intercalated disks, *J. Cell Biol.* **99:**453-463.

Green, C. R., Harfst, E., Gourdie, R. G., and Severs, N. J., 1988, Analysis of the rat liver gap junction protein: clarification of anomalies in its molecular size, *Proc. R. Soc. Lond.* **B233:**165-174.

Hertzberg, E. L., and Skibbens, R. V., 1984, A protein homologous to the 27,000 Dalton liver gap junction protein is present in a wide variety of species and tissues, *Cell* **39:**61-69.

Hoffmeister, S. A. H., and Schaller, H. C., 1987, Head activator and head inhibitor are signals for nerve cell differentiation in Hydra, *Devel. Biol.* **122:**72-77.

Keane, R. W., Mehta, P. P., Rose, B., Honig, L. S., Loewenstein, W. R., and Rutishauser, U., 1988, Neuronal differentiation, nCAM-mediated adhesion and gap junctional communication in neuroectoderm. A study in vivo, *J. Cell Biol.* **106:**1307-1319.

Kistler, J., Kirkland, B., and Bullivant, S., 1985, Identification of a 70,000-D protein in lens membrane junctional domain, *J. Cell Biol.* **101:**28-35.

Kistler, J., Christie, D., and Bullivant, S., 1988, Homologies between gap junction proteins in lens, heart and liver, *Nature* **331:**721-723.

Kumar, N. M., and Gilula, N. B., 1986, Cloning and characterization of human and rat liver cDNAs coding for a gap junction protein, *J. Cell Biol.* **103:**767-776.

Lane, N. J., and Skaer, H. Le B., 1980, Intercellular junctions in insect tissues, *Adv. Insect Physiol.* **15:**35-213.

Larsen, W. J., 1983, Biological implications of gap junction structure, distribution and composition: a review, *Tissue Cell* **15:**645-671.

Lawrence, T. S., Beers, W. H., and Gilula, N. B., 1978, Transmission of hormonal stimulation by cell-to-cell communication, *Nature* **272:**501-506.

Lee, S., Gilula, N. B., and Warner, A. E., 1987, Gap junctional communication and compaction during preimplantation stages of mouse development, *Cell* **51:**851-860.

Lethias, C., Garrone, R., and Mazzanorana, M., 1983, Fine structure of sponge cell membranes: comparative study with freeze-fracture and conventional thin section methods, *Tiss. Cell* **15:**523-535.

Llinás, R. R., 1985, Electronic transmission in the mammalian central nervous system, in: *Gap Junctions* (M. V. L. Bennett and D. C. Spray, eds.), Cold Srping Harbor, New York.

Lo, C. W., 1988, Communication compartments in insect and mammalian development, in: *Modern Cell Biology*, vol 7 (B. H. Satir, ed.), Alan R. Liss, Inc., New York.

Loewenstein, W. R., 1966, Permeability of membrane junctions, *Ann. N.Y. Acad. Sci.* **137**:441-472.

Loewenstein, W. R., 1967, On the genesis of cellular communication, *Devel. Biol.* **15**:503-520.

MacWilliams, H. K., 1983a, Hydra transplantation phenomena and the mechanism of Hydra head regeneration. I. Properties of head inhibition, *Devel. Biol.* **96**:217-238.

MacWilliams, H. K., 1983b, Hydra transplantation phenomena and the mechanism of Hydra head regeneration. II. Properties of head activation, *Devel. Biol.* **96**:239-272.

Maldonado, P. E., Rose, B., and Loewenstein, W. R., 1988, Growth factors modulate junctional cell-to-cell communication, *J. Memb. Biol.* **106**:203-210.

Manjunath, C. K., Goings, G. E., and Page, E., 1984, Cytoplasmic surface and intramembrane components of rat heart junctional proteins, *Am. J. Physiol.* **246**:H865-H875.

Marcum, B. A., and Campbell, R. D., 1978, Development of Hydra lacking nerve and interstitial cells, *J. Cell Sci.* **29**:17-33.

Mehta, P. P., Bertram, J. S., and Loewenstein, W. R., 1986, Growth inhibiton of transformed cells correlates with their junctional communication with normal cells, *Cell* **44**:187-196.

Mehta, P. P., Bertram, J. S., and Loewenstein, W. R., 1989, The actions of retinoids on cellular growth correlate with their actions on gap junctional communication, *J. Cell Biol.* **108**:1053-1065.

Meinhardt, H., and Gierer, A., 1974, Applications of a theory of biological pattern formation based on lateral inhibition, *J. Cell Sci.* **15**:321-346.

Milks, L. C., Kumar, N. M., Houghten, R., Unwin, N., and Gilula, N. B., 1988, Topology of the 32-kd liver gap junction protein determined by site-directed antibody localizations, *The EMBO J.* **7**:2967-2975.

Miller. T. M., and Goodenough, D. A., 1985, Gap junction structures after experimental alteration of junctional channel conductance, *J. Cell Biol.* **101**:1741-1748.

Nagy, J. I., Yamamoto, T., Shiosaka, S., Dewar, K. M., Whittaker, M. E., and Hertzberg, E. L., 1988, Immunohistochemical localization of gap junction protein in rat CNS: a preliminary account, in: *Modern Cell Biology*, vol 7 (B. H. Satir, ed.), Alan R. Liss, Inc., New York.

Neyton, J., and Trautmann, A., 1985, Single channel currents of an intercellular junction, *Nature* **317**:331-335.

Nicholson, B. J., Dermietzel, R., Teplow, D., Traub, O., Willecke, K., and Revel, J.-P., 1987, Two homologous protein components of hepatic gap junctions, *Nature* **239**:732-734.

Noma, A., and Tsuboi, N., 1987, Dependence of junctional conductance on proton, calcium and magnesium ions in cardiac paired cells of guinea pig, *J. Physiol. (Lond.)* **382**:193-212.

Paul, D. L., 1986, Molecular cloning of cDNA for rat liver gap junction protein, *J. Cell Biol.* **103**:123-134.

Pavans de Ceccatty, M., 1974, Coordination in sponges. The foundations of integration, *Amer. Zool.* **14**:895-903.

Peracchia, C., 1988, The calmodulin hypothesis for gap junction regulation six years later, in: *Modern Cell Biology*, vol 7 (B. H. Satir, ed.), Alan R. Liss, Inc., New York.

Piccolino, M., Neyton, J., and Gerschenfeld, H. M., 1984, Decrease of gap junction permeability induced by dopamine and cyclic adenosine 3':5'-monophosphate in horizontal cells of turtle retina, *J. Neurosci.* **4**:2477-2488.

Pitts, J. D., 1971, Molecular exchange and growth control in tissue culture, in: *Growth Control in Cell Cultures* (G. E. W. Wolstenholme and J. Knight, eds.), CIBA Foundation Symposium, Churchill, London.

Rose, B., and Loewenstein, W. R., 1976, Permeability of a cell junction and the local cytoplasmic concentration of free ionized calcium concentration: a study with aequorin, *J. Memb. Biol.* **28**:87-119.

Rubin, D. I., and Bode, H. R., 1982, Both the epithelial cells and the nerve cells are involved in the head inhibition properties in *Hydra attenuata*, *Devel. Biol.* **89**:332-338.

Ryerse, J., and Nagel, B. A., 1984, Gap junction distribution in the Drosophila wing disc mutants vg, lgd, and 1c43, *Dev. Biol.* **105**:396-403.

Saez, J. C., Spray, D. C., Nairn, A. C., Hertzberg, E., Greengard, P., and Bennett, M. V. L., 1986, cAMP increases junctional conductance and stimulates phosphorylation of the 27-kDa principal gap junction polypeptide, *Proc. Natl. Acad. Sci. USA* **83**:2473-2477.

Schaller, H. C., 1973, Isolation and characterization of a low molecular weight substance activating head and bud formation in Hydra, *J. Embryol. Exp. Morphol.* **29**:27-38.

Schaller, H. C., and Bodenmuller, H., 1981, Isolation and amino acid sequence of a morphogenetic peptide from Hydra, *Proc. Natl. Acad. Sci. USA* **78:**7000-7004.

Severs, N. J., 1985, Intercellular junctions and the cardiac intercalated disk, *Adv. Myocardiol.* **5:**223-242.

Sims, S. M., Daniel, E. E., and Garfield, R. E., 1982, Improved electrical coupling is associated with increased numbers of gap junctions in uterine smooth muscle at parturition, *J. Gen. Physiol.* **80:**353-375.

Spray, D. C., and Bennett, M. V. L., 1985, Physiology and pharmacology of gap junctions, *Ann. Rev. Physiol.* **47:**281-303.

Spray, D. C., Harris, A. L., and Bennett, M. V. L., 1981, Equilibrium properties of a voltage dependent junctional conductance, *J. Gen. Physiol.* **77:**75-94.

Spray, D. C., Stern, J. H., Harris, A. L., and Bennett, M. V. L., 1982, Gap junctional conductance: comparison of sensitivities to H and Ca ions, *Proc. Natl. Acad. Sci. USA* **79:**441-445.

Stoker, M. G. P., 1964, Regulation of growth and orientation in hamster cells transformed by polyoma virus, *Virology* **24:**165-174.

Subak-Sharpe, H., Burk, R. R., and Pitts, J. D., 1966, Metabolic cooperation by cell to cell transfer between genetically different mammalian cells in tissue culture, *Heredity* **21:**342-323.

Subak-Sharpe, H., Burk, R. R., and Pitts, J. D., 1969, Metabolic cooperation between biochemically marked mammalian cells in culture, *J. Cell Sci.* **4:**353-367.

Traub, O., Druge, P. M., and Willecke, K., 1983, Degradation and resynthesis of gap junction protein in plasma membranes of regenerating liver after partial hepatectomy or cholestasis, *Proc. Natl. Acad. Sci. USA* **80:**255-259.

Traub, O., Look, J., Dermietzel, R., Brummer, F., Hulser, D., and Willecke, K., 1989, Comparative characterization of the 21-kD and 26-kD gap junction proteins in murine liver and cultured hepatocytes, *J. Cell Biol.* **108:**1039-1051.

Turin, L., and Warner, A. E., 1977, Carbon dioxide reversibly abolishes ionic communication between cells of early amphibian embryos, *Nature* **270:**56-57.

Van den Biggelaar, J. A. M., and Serras, F., 1988, Determinative decisions and dye-coupling changes in the molluscan embryo, in: *Modern Cell Biology*, vol 7 (B. H. Satir, ed.), Alan R. Liss, Inc., New York.

Warner, A. E., 1988, The gap junction, *J. Cell Biol.* **89:**1-7.

Warner, A. E., Guthrie, S. E., and Gilula, N. B., 1984, Antibodies to gap junctional protein selectively disrupt junctional communication in the early amphibian embryo, *Nature* **311:**127-131.

Warner, A. E., and Lawrence, P. A., 1982, Permeability of gap junctions at the segmental border in insect epidermis, *Cell* **28:**243-252.

Webster, G., and Wolpert, L., 1966, Studies on pattern regulation in Hydra, *J. Embryol. Exp. Morphol.* **16:**91-04.

Wood, R. L., 1977, The cell junctions of Hydra as viewed by freeze-fracture replication, *J. Ultrastruct. Res.* **58:**299-315.

Wood, R. L., and Hageman, G. S., 1982, The fine structure of cellular junctions in a marine bryozoan: gap junctions, *J. Ultrastruct. Res.* **79:**174-188.

Wood, R. L., and Kuda, A. M., 1980, Formation of junctions in regenerating Hydra: Gap junctions, *J. Ultrastruct. Res.* **73:**350-360.

Young, J. D.-E., Cohn, Z. A., and Gilula, N. B., 1987, Functional assembly of gap junction conductance in lipid bilayers: demonstration that the major 27kD protein forms the junctional channel, *Cell* **48:**733-743.

Zimmer, D. B., Green, C. R., Evans, W. H., and Gilula, N. B., 1987, Topological analysis of the major protein in isolated intact rat liver gap junctions and gap junction-derived single membrane structures, *J. Biol. Chem.* **262:**7751-7763.

Chapter 2

Intercellular Junctions in Ctenophore Integument

MARI-LUZ HERNANDEZ-NICAISE, GHISLAIN NICAISE and THOMAS S. REESE

"In essence we would argue here that a critical step in metazoan evolution was the development of 'tight' external epithelia enclosing inner spaces well isolated from the external medium" (Mackie, 1984).

1. Introduction

The ctenophoran integument consists basically of a single-layered epidermis which covers the entire body, including appendages such as the tentacular apparatus, the lobes and auricles, and lines the stomodeal cavity (generally referred to as the gastric cavity). This integument is always devoid of any cuticle or hard secretion, but is permanently covered by a film of mucus. It rests on a gelatinous mesoglea, which is an unusual connective tissue devoid of collagen and elastin fibers (Franc et al., 1976), and harboring mesenchymal cells and numerous true muscle cells. The mesoglea may be considered as the internal milieu of the ctenophore; it is fed and oxygenated by a system of gastrovascular channels, which may be very elaborate in large species. Franc (1972) demonstrated that specialized ciliated cells, located in the canal walls and grouped in ciliated rosettes, regulated the relative ionic and osmotic composition of the mesoglea.

The basic framework of the integument consists of epithelial cells specialized into gland cells, a great variety of ciliated cells, supporting cells, colloblasts (exclusive of the

MARI-LUZ HERNANDEZ-NICAISE[1], GHISLAIN NICAISE[1], and THOMAS S. REESE[2] • [1]Cytologie Expérimentale, Université de Nice, Parc Valrose, 06034, Nice, France, and [2]Laboratory of Neurobiology, N.I.N.D.S., N.I.H., Bethesda, Maryland 20892, U.S.A.

Evolution of the First Nervous Systems
Edited by P.A.V. Anderson
Plenum Press, New York

phylum), chromatophores, etc. In most species the epidermis comprises a basal layer of parietal, true muscle cells, which are separated from the mesoglea by the basal lamina of the integument.

The epidermis harbors most of the sensory and nervous elements of the organism. Various types of sensory neuron are interspersed among the epithelial cells, except at the aboral pole where they are grouped into a complex sensory area. The sensory neurons are synaptically connected with the nerve net(s) located between the epithelial and muscle cells. In contrast with the situation found in Cnidaria, the nerves and neurons are tightly wrapped by extensions of the epithelial cells which can, therefore, be considered as an epithelial glia (Hernandez-Nicaise, 1973a). Some nerves have been found in the mesoglea, where they make synaptic contacts with the mesogleal muscle fibers and with mesenchyme cells (Hernandez-Nicaise, 1973b).

At the ultrastructural level, the nervous system of ctenophores has been described as a synaptic nerve net. The presynaptic terminals of neuro-neuronal and neuromuscular synpases are organized identically in a manner typical of the phylum (Hernandez-Nicaise, 1973c). There are no ganglia, although the sensory aboral organ may be viewed as a kind of ganglion-like concentration of nervous and sensory structures.

The nerve net was thought to be of a "through-conducting" type (see review in Bullock and Horridge, 1965), mainly on the basis of behavioral observations. If any point of the epidermis (not the pharynx lining) is stimulated (mechanically or electrically), the comb rows stop beating simultaneously, as if the stimulus were propagated uniformly along every direction (Horridge, 1966; Tamm, 1982). Bipolarized (or symmetrical) synapses have been found in the ctenophore nervous system (Hernandez-Nicaise, 1968, 1973c), and were proposed as the possible structural correlate of this through conduction. This point of view was strengthened by the demonstration by Anderson (1985) that the morphologically symmetrical giant synapses of the nerve net of the scyphomedusa *Cyanea* are functionally "two way" synapses.

Meanwhile, other studies (see review in Anderson, 1980) have shown that the epithelia of many invertebrates propagate action potentials and cooperate with the nervous system in controlling behavior. In some cases, the structural correlates of this ability to propagage electrical activity are communicating junctions (or gap junctions). Such electrophysiological studies have not been performed on ctenophores.

Considering the intimate spatial relationship between neurons and epithelial cells in ctenophores, it is now critical to answer the questions: 1) Does the integument regulate the circulation and/or composition of the extracellular fluid bathing the nerve net by way of specialized junctions? 2) Has the integument differentiated an alternative manner appropriate for signals (yet to be defined) through gap-like junctions?

With these questions in mind, we have been looking for likely candidates for occluding and communicating junctions in ctenophore epithelia, and present here morphological evidence for these types of junctions: apical zonular junctions of a unique design, and gap junctions. Both possess similar junctional intramembrane particles.

2. Material and Methods

The data presented here have been obtained from tissues of three species, from three orders of the phylum: *Pleurobrachia rhodopis* (Cydippa) and *Beroe ovata* (Beroida) collected at the Station Marine, Villefranche-sur-Mer, France and *Mnemiopsis leydii* (Lobata), an oceanic neritic species, collected at the Marine Biological Laboratory, Woods Hole, Massachusetts.

Tissues prepared for conventional transmission electron microscopy (TEM) were fixed with 3 to 5% glutaraldehyde in artificial, cacodylate-buffered, sea water (see Hernandez-Nicaise, 1973c and Hernandez-Nicaise et al., 1984 for details), followed by 2% osmium tetroxyde in the same buffer. Some specimens of *Beroe* and *Pleurobrachia* were block-stained with 2% uranyle acetate in maleate buffer. Tissues and whole larva of *Beroe* were infiltrated with colloidal lanthanum which was added to the fixatives, and *Mnemiopsis* tissues were block-stained with 1% tannic acid in 0.1 M sodium cacodylate following osmication.

For freeze-fracture studies, pieces of body wall of adult *Beroe* and *Mnemiopsis*, and whole, newly-hatched larvae of *Beroe* were fixed as above in isosmotic cacodylate-buffered glutaraldehyde for variable times. The samples were rinsed subsequently in buffered saline and infiltrated with 30% glycerol in the same buffer for 3 hours. Tissue blocks were then placed in specimen holders, rapidly (manually) frozen at -210°C in nitrogen slush and transferred to a "Reichert Jung CF 250" unit. They were shadowed with carbon/platinum at an angle of 45°. The replicas were cleaned with NaClO and mounted on copper grids.

Complementary data were obtained from unfixed *Mnemiopsis* auricles which were directly frozen using the liquid helium-cooled copper block/slam freezing method (Heuser et al., 1979). Replicas were obtained from freeze-fractured specimens and rotary shadowed. Other specimens were freeze-substituted in 10% osmium in acetone, stained en bloc sequentially with hafnium chloride and uranyl acetate, embedded in Epon-araldite and sectioned.

3. Results

3.1. Macular Gap-like Junctions

On TEM preparations, close gap-like junctions link a variety of epithelial cell types of *Beroe*, *Pleurobrachia* and *Mnemiopsis*, and are particularly abundant in the floor of the aboral organ, in the labial epidermis and between the cells of the locomotory tracts (polster cells and cells of the ciliated grooves) (Satterlie and Case, 1978; Hernandez-Nicaise, 1974b, 1989), where they alternate with microtubule-associated desmosomes. They have also been reported to occur between muscle cells of *Beroe* (Hernandez-Nicaise and Amsellem, 1980) and *Mnemiopsis* (Hernandez-Nicaise et al.,

1984) and between photocytes and other gastrodermal cells in the meridional canals of *Mnemiopsis* (Anctil, 1985).

In thin sections from tissues block-stained with uranyl acetate, each junction appears as an area of close contact, from 0.3 to 0.8 μm long, where the intercellular space is reduced to a uniform width of 3 nm (Fig. 1). A similar aspect is observed in thin sections of cryo-substituted fast-frozen tissues of *Mnemiopsis*. In tissues permeated with lanthanum or tannic acid (Fig. 1A), unstained structures bridge the intercellular gap. Symmetrical cisternae flank both sides of these junctions in *Mnemiopsis* gastrodermal epithelia (Anctil, 1985) and pharynx integument, and between *Beroe* and *Mnemiopsis* muscles. In replicas from glutaraldehyde-fixed and glycerol-protected tissues, we have found macular junctions which may be correlated with the close junctions seen in TEM sections. Typically, each junction appears as an elevated oval patch bearing clusters of particles on the exoplasmic fracture face (E face), and as a depressed area bearing small pits and circled by a row of particles in the protoplasmic fracture face (P face). When the fracture plan jumps from a P face to an E face across a junction, these designs are displayed on each face. Complementary fractures of these gap-like junctions have been obtained in muscles of *Mnemiopsis* (Hernandez-Nicaise et al., 1984). The junctional particles are heterogenous in size, ranging from 9 to 13 nm in diameter, and display an irregular, loose lattice. The particles are prone to fuse in anastomosed short strands.

Similar oval junctional areas have been found in fast-frozen, unfixed tissues of *Mnemiopsis* (Fig. 1B). Some of them cleave with the P-fracture face, while in other cases they are retained with the E face. With this technique, the outer row of particles

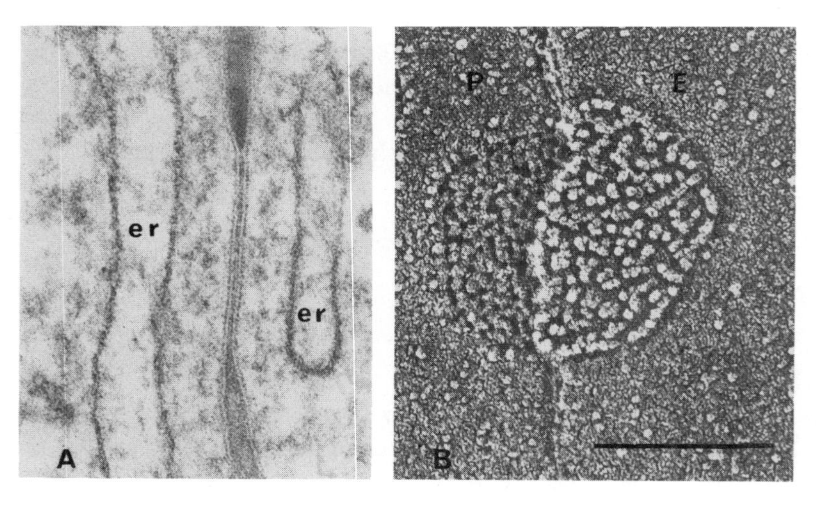

Figure 1. Gap junctions in *Mnemiopsis leydii* epithelium. **A.** Gap junction in pharynx epithelium fixed with glutaraldehyde, osmicated and inflitrated with tannic acid. Irregularly-spaced bridges cross the junctional gap. er: endoplasmic reticulum. **B.** Gap junction in the auricle epithelium, after fast-freezing and rotary shadowing. The particles have cleaved with the E face (E) and complementary discrete pits appear on the P face (P). Negative print: the particles appear white. Bar: 0.2 μm.

retained around the pits is composed of distinctly smaller particles, 8 nm in diameter, and the complementary fracture face displays a groove or a row of pits around the patch of junctional particles. In the unfixed material, the junctional particles exhibit an obvious central pore, 2-3 nm wide, surrounded by a rosette of subunits. On replicas from unfixed tissue, the non-junctional particles appear more evenly distributed in the membrane and are present on both faces. Their density per surface unit is highly variable, ranging from 480 to 880 per μm^2 for the E face, and from 560 to 1840 per μm^2 for the P face. This variability probably results from differences in the quality of preservation of the initial structure, as a good freezing is achieved only at the surface of the specimens. The majority of the non-junctional particles of the P face appear smaller than the junctional particles, with a mean diameter of 8 nm (range of 7.5 to 9 nm); however, a small proportion of bigger intramembrane components, 11 to 12 nm in diameter, appears randomly distributed all over the membrane.

3.2. Apical Belt Junctions

Our previous studies on ctenophore ultrastructure had shown the existence of apical junctions between the different components of the integument, particularly between the ciliated locomotory cells and between the cells of the floor of the sensory organ of *Beroe* and *Pleurobrachia* (Hernandez-Nicaise, 1974). These cells appear to be linked by a series of focal contacts. On sections from tissues stained en bloc with uranyl-acetate, each contact appears as a close apposition of the two plasma membranes, which remain separated by a narrow gap of 2-3 nm, the total width of the heptalaminar structure being 15-20 nm. The junctional membranes do not fuse nor are they linked by septa. In most cases, the junctional membranes are lined by a dense fuzzy coat, 20 nm thick, in which numerous microfilaments are embedded. At that stage it had been proposed that these junctions were mini-gap junctions (Hernandez-Nicaise, 1974). The same features have been reported in the epithelial lining of the meridional canals (of endodermal origin) of *Mnemiopsis*, and have been termed septate desmosome-like junctions (Anctil, 1985).

In lanthanum-infiltrated preparations of *Beroe*, the intercellular spaces of the epidermis are filled by lanthanum, "down" to the basal lamina which stops the tracer molecule unless the mesoglea has been partially digested by hyaluronidase. No septa are visible between the cells, and in tangential sections, lanthanum does not outline any particular pattern. The same results are obtained if whole, intact larvae are treated with lanthanum.

In freeze-fracture replicas from adult and larval stages of *Beroe ovata* and tissues of adult *Mnemiopsis*, the apical junctions have been found to be zonulae. We have observed them in every location examined. Apical belt junctions link the cells of the floor of the aboral organ, of the lips, of the epidermis between the comb rows, of the pharynx epithelium, of the endodermic canals, and the ciliated cells of a comb plate.

The junctions are constituted by rows of closely-packed, discrete particles on the E face, and complementary pits on the P face. The array of junctional particles displays a unique pattern which looks similar in *Beroe* and *Mnemiopsis*: the belt is delineated

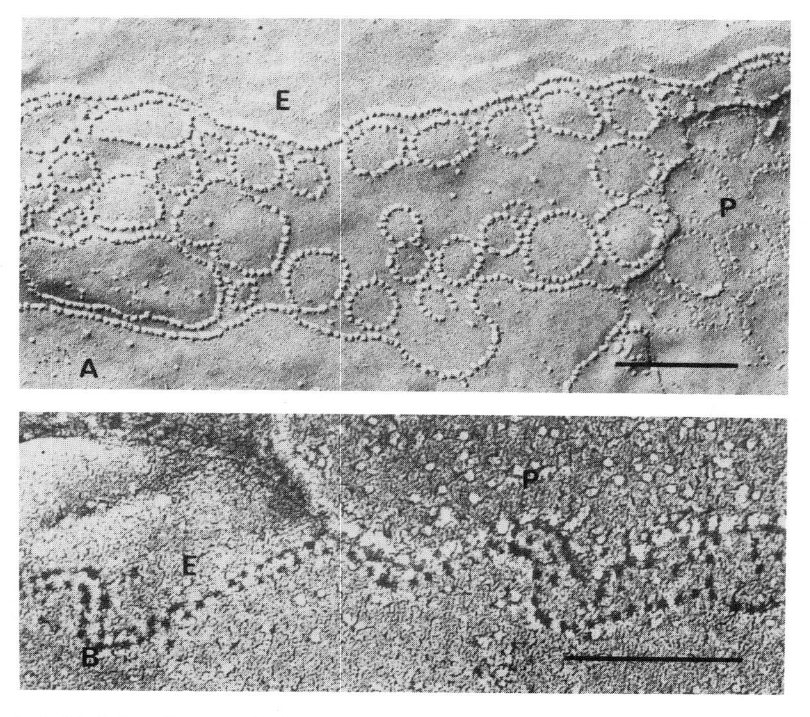

Figure 2. Apical belt junctions. **A.** Belt junction at the apex of innerlabial epithelium of *Beroe ovata*. Replica from conventionally fixed, cryoprotected and freeze-fractured tissue. The pattern of rows and loops of particles on the E face (E) is continuous with the pattern of pits on the P face (P) of the membrane of the adjacent cell. Bar: 0.2 μm. **B.** Two strand belt junction at the apex of auricle epithelium of *Mnemiopsis*, in a rotary-shadowed replica of fast-frozen, cryo-fractured tissue. The particles have cleaved with the P face (P), and the pits are most conspicuous on the E face (E). Negative print. Bar: 0.2 μm.

by two wavy strands of particles, roughly parallel to the surface, an apical or outer ring and a basal or inner ring (Fig. 2). These two rings enclose annular loops of particles, which are more or less tightly packed, depending on their number and the total width of the junction. The membrane enclosed by each of the intercalated loops is generally poor in particles in either face. Additional particles may completely fill the space between the outer and inner rings and the loops. When the fracture plane jumps from a P face to an E face, the intercellular space appears reduced along the chains of particles and the design of the loops and strands is continuous from one cell membrane to the other, the strands of pits of the P face displaying the same distribution as the corresponding particles. At low magnification, the whole belt junction appears as a slightly elevated, wavy ridge on the P face, and a shallow depressed area on the E face. The total width of the junctional strand varies from 25 nm to 1 μm. The simplest arrangement found is a belt composed of two closely apposed rings of particles. It has been observed in a majority of epithelial cells in the newly hatched larva of *Beroe*, but occurs also in adult tissues, especially in *Mnemiopsis*.

The junctional particles have a diameter ranging from 9 to 11 nm, and a pit may be seen at their center. These particles may occasionally appear fused, building a short rod-like strand. We have not found a noticeable difference in diameter and "height" between the particles and rods of the limiting rings, and the particles of the inner loops. The center-to-center spacing of the particles of the limiting strands ranges from 11 to 16 nm. The junctional particles are larger in diameter and apparent "height" than the non-junctional particles which characterize the P-fracture face.

After fast freezing with no prior chemical treatment, the epidermal cells of the auricles of *Mnemiopsis* appear linked by apical zonular junctions, the design of which is identical to the peculiar pattern described above for fixed tissues (Fig. 2B). However the chains of particles cleave with the P-fracture face and the corresponding discrete pits appear on the E-fracture face, with a few exceptions, where the distribution of particles is as in fixed tissues. As in fixed tissues, fusion of several junctional particles may occur; the frequency and length of the resulting rods may be correlated with inadequate freezing as only the 10 upper μm are correctly frozen with this method. The junctional particles diameter ranges from 9 to 12 nm and most of them bear a conspicuous central pore 2-3 nm wide.

In thin sections of cryosubstituted tissues of *Mnemiopsis*, single focal points of contact, with no intervening gap, have been observed between the apices of the flat epidermal cells of the auricule. We propose that they correlate with belts composed of two strands of particles observed in replicas.

3.3. Annular Junctions

In replicas from both fixed and unfixed tissues, we have found a variety of annular junctional patches made of concentric loops of particles, with the same pattern and size as loops of the belt junctions. These macular junctions exhibit a great variety of size and pattern, and seem restricted to the apical and upper half of the lateral membrane of the cells. In replicas from fixed tissues, the junctional particles cleave with the E face, while in replicas from fast-frozen tissues the particles appear preferentially located on the P face, although some have been observed on the E face.

4. Discussion

4.1. Gap Junctions

The freeze-fracture features of gap junctions are now well established and considered specific enough to be used as reliable criteria to identify gap junctions (see reviews in Perrachia, 1980; Larsen, 1983; Spray and Bennett, 1985).

Replicas of both fixed and unfixed freeze-fractured tissues of ctenophores give evidence for macular arrays of intramembrane particles that are coextensive in the apposed membrane leaflets of two adjacent cells, and present features identical to the E-type gap junctions, which occur in a majority of invertebrates tissues (Flower, 1977) and in one genera of Cnidaria, *Hydra* (Filshie and Flower, 1977; Wood, 1977; Wood and Kuda, 1980). Our observations on unfixed, fast-frozen epithelia of *Mnemiopsis* suggest, however, that the distinction between E and P types may not be taken as an absolute criterion of classification, as junctional particles seem to cleave with both faces in unfixed tissue.

It is noteworthy that gap junctions appear particularly abundant in three locations: ciliated cells of the floor of the sensory organ, ciliated cells of the inner labial integument, and ciliated cells bearing the comb plate cilia. The two first locations are sensory areas containing specialized sensory neurons. A synchronization of the balancer cells, which support the statolith and act as pace-makers for the comb rows, may be achieved by such junctions. Similarly, synchronization of cells of the lip, which respond to contact with prey, may be of importance in triggering feeding behavior. As for the ciliated cells of the comb plates, the question has been adressed by Tamm (1982). The coordination of the ciliary activity is not based on electrotonic coupling, but on mechanical coupling. However, the comb cells receive synapses from different types of neurons (Hernandez-Nicaise, 1973a, 1974) and an inhibitory control of their activity by the nerve net is well documented (Tamm, 1982). The numerous gap junctions that we have identified may be necessary to synchronize the cells, if a small number of them in each polster receive synapses.

4.2. Apical Belt Junctions

Apical zonular junctions link the cells of most vertebrate and invertebrate epithelia (except sponges). A zonula occludens, or tight junction, connects the cell apices and seals, more or less tightly, the intercellular spaces in vertebrate epithelia (see reviews in Pinto da Silva and Kachar, 1982; Schneeberger and Lynch, 1984; Gumbiner, 1987; Stevenson et al., 1988) and in some specialized tissues of a few invertebrate groups, namely insects and arachnids (Lane, 1981) and tunicates (Georges, 1979; Green and Bergquist, 1982; Lane et al., 1986). In most epithelia of invertebrates, another type of zonular apical junction, the septate desmosome, connects epithelial cells and is thought to restrict to various degrees, the circulation of ions and small molecules through the intercellular route (see reviews in Green and Bergquist, 1982; Green, 1984).

We give evidence in this paper that Ctenophora epithelia possess a zonular apical junction. Its unique morphology on freeze-fracture replicas of fixed or unfixed epithelia does not allow it to be labelled as tight junction or as septate desmosome.

Are Ctenophora Apical Belt Junctions a Variety of Septate Desmosomes?

The particulate nature of the junctional strands and their pattern are reminiscent of the convoluted strands of pleated septate junctions of the annelids and molluscs and, to a lesser extent, of the closed network of evenly sized particles found in septate junctions of echinoderms. However, the intercellular gap of ctenophoran apical junctions is never crossed by septa and, on the contrary, appears conspicuously narrower than the neighboring non-junctional intercellular space. Moreover, the junctional particles differ from their counterparts in lower invertebrate desmosomes by their uniformity in size and shape, and discrete complementary pits are never found in other invertebrates, even in unfixed, fast-frozen tissues (Kachar et al., 1986).

Are Ctenophora Apical Belt Junctions Tight-like Junctions?

Functionally speaking we know that these junctions are permeated by lanthanum so that they are likely to be permeable to ions. Franc (1985) has demonstrated that the epithelial cells take up significant amounts of tritiated proline (a precursor of collagen) from sea water, so a variety of metabolites are likely to be transported through the cells membrane rather than follow the paracellular route. To date, no data are available on the effectiveness of the ctenophoran epidermis as a barrier and/or as an actively transporting cell layer.

Ctenophora apical junctions in either fixed or unfixed tissues differ radically from the current model of vertebrate tight junction with its anastomosed straight cylindrical ridges. Vertebrate tight-junction ridges may resolve into discrete particles, if the tissue is unfixed (Staehelin, 1973; van Deurs et al., 1979), but it has been shown that the ridges are typical features of tight junctions in unfixed fast-frozen tissue (Hirokawa, 1982; Kachar and Reese, 1982). Ctenophore junctional particles retain their morphology in fixed and unfixed specimens and probably differ from the ridges of higher phyla, although we have to keep in mind that freeze-fracture methods do not provide any biochemical information on intramembrane particles. However, two types of tight junction, described in tunicates (Georges, 1979; Green and Bergquist, 1982; Lane et al., 1986) and mammals (Simionescu et al., 1976), bear some resemblance with ctenophore junctions.

Tunicates possess apical junctions that appear, on freeze-fracture replicas, to consist of a reticular network of either ridges or discrete particles, 8-12 nm in diameter. As in ctenophores and unlike tight junction ridges of mammals, the particles cleave on the E face in fixed tissues and on the P face in unfixed tissue and are irregular in size and shape. The complementary face of this junction displays shallow grooves in both cases. In some species, the junctions are permeable to lanthanum (Georges, 1979; Green and Bergquist, 1982).

In the rat, the "special endothelium junction, type I," described by Simionescu et al. (1976) in mesenteric arteries, consists of a complex of occluding and communicating junctions. The junction consists of parallel, closely-spaced rows of occluding particles alternating with or running along strands of communicating particles. The occluding particles are larger and "higher," and may fuse into short ridges. The geometry of these junctions is strongly reminiscent of the figures observed in ctenophores.

Are the Apical Belt Junctions of Ctenophora Zonular Gap Junctions?

There is no example of zonular communicating junctions in the animal kingdom, but communicating junctions, with a pattern similar to that of the apical junction of ctenophores, occur in several excitable tissues. These unusual, communicating junctions have been termed "fenestrated gap junctions" in frog spinal motoneurons (Taugner et al., 1978), circular "nexuses" in frog myocardium, where they are the predominant type of junction (Kensler et al., 1977; Mazet, 1977), and "anastomosing strands," in the septum of the lateral axon of the earthworm (Kensler et al., 1979). As a consequence, all these junctions correlate with punctate membrane appositions in transverse sections.

It is clear that the preceding questions cannot be answered at this stage and that immunocytochemical studies using antibodies raised against known junctional proteins, together with electrophysiological experiments, are needed. Whatever the nature of the proteins (or the lipids ?) involved in the apical junctions described here, it is logical to suppose that they fulfill a role similar to that attributed to septate desmosomes and tight junctions in the other phyla. Even if it is leaky to lanthanum and probably other ions and molecules, this zonular junction restricts the intercellular spaces of the covering and inner epithelia in a ctenophore. In addition, it probably also serves an adhesive role, in conjunction with desmosomes, in these fragile organisms. We should view the junctions of extant Ctenophora, as a successful evolutionary experiment of the diploblastic genome. Some of its potentialities were also "retained" by the ancestors of the triploblastic phyla, and followed parallel pathways.

References

Anctil, M., 1985, Ultrastructure of the luminescent system of the ctenophore *Mnemiopsis leidyi*, *Cell Tissue Res.* **242**:333-340.

Anderson, P. A. V., 1980, Epithelial conduction: its properties and functions, *Prog. Neurobiol.* **15**:161-203.

Anderson, P. A. V., 1985, Physiology of a bidirectional, excitatory, chemical synapse, *J. Neurophysiol.* **53**:821-835.

Bullock, T. H., and Horridge, G. A., 1965, *Structure and Function in the Nervous Systems of Invertebrates. 8; Coelenterata and Ctenophora*, W. H. Freeman, San Francisco and London.

Filshie, B. K., and Flower, N. E., 1977, Junctional structures in *Hydra*, *J. Cell Sci.* **23**:151-172.

Flower, N. E., 1977, Invertebrate gap junctions, *J. Cell Sci.* **25**:163-171.

Franc, J. M., 1972, Activités des rosettes ciliées et leurs supports chez les Cténaires, *Z. Zellforsch.* **130**:527-544.

Franc, J. M., 1985, *La mésoglée des Cténaires; approches ultrastructurale, biochimique et métabolique*, Doctoral diss. no. 8534, Univ. Claude Bernard-Lyon I, France.

Franc, S., Franc, J. M., and Garrone, R., 1976, Fine structure and cellular origine of collagenous matrices in primitive animals: Porifera, Cnidaria and Ctenophora, in: *Burkitt Lymphoma, Hemostasis and Intercellular Matrix. Frontiers of Matrix Biology*, vol. 3, pp. 143-156 (A. M. Robert and L. Robert, eds.), Karger, Basel.

Georges, D., 1979, Gap and tight junctions in Tunicates. Study in conventional and freeze-fracture techniques, *Tissue Cell* **11**:781-792.

Green, C. R., 1984, Intercellular junctions, in: *Biology of the Integument, Vol. I. Invertebrates*, pp. 5-16 (J. Bereiter-Hahn, A. G. Matoltsy and K. S. Richards, eds.), Springer-Verlag, Berlin, Heidelberg.

Green, C. R., and P. R. Bergquist, 1982, Phylogenetic relationships within the invertebrata in relation to the structure of septate junctions and the development of "occluding" junctional types, *J. Cell Sci.* **53**: 279-305.

Gumbiner, B., 1987, Structure, biochemistry, and assembly of epithelial tight junctions, *Am. J. Physiol.*, **253C**:749-758.

Hernandez-Nicaise, M. L., 1968, Distribution et ultrastructure des synapses symétriques dans le système nerveux des Cténaires, *C. R. Acad. Sci.* **267**:1731-1734.

Hernandez-Nicaise, M. L., 1973a, Le système nerveux des Cténaires. I. Structure et ultrastructure des réseaux épithéliaux, *Z. Zellforsch.* **137**:223-250.

Hernandez-Nicaise, M. L., 1973b, Le système nerveux des Cténaires. II. Les éléments nerveux intra-mésogléens chez les Béroidés et les Cydippidés, *Z. Zellforsch.* **143**:117-133.

Hernandez-Nicaise, M. L., 1973c, The nervous system of Ctenophora. III. Ultrastructure of synapses, *J. Neurocytol.* **2**:249-243.

Hernandez-Nicaise, M. L., 1974, *Système nerveux et intégration chez les Cténaires. Etude ultrastructurale et comportementale,* Doctoral diss. no. 278, Univ. Claude Bernard-Lyon I, France.

Hernandez-Nicaise, M. L., 1989, Ctenophora, in: *Microscopic Anatomy of Invertebrates* (G. Harrisson ed.), A. R. Liss, New York.

Hernandez-Nicaise, M. L., and Amsellem, J., 1980, Ultrastructure of the giant smooth muscle fiber of the Ctenophore *Beroe ovata*, *J. Ultrastruct. Res.* **72**:151-158.

Hernandez-Nicaise, M. L., Nicaise, G., and Malaval, L., 1984, Giant smooth muscle fibers of the Ctenophore *Mnemiopsis leidyi*: ultrastructural study of in situ and isolated cells, *Biol. Bull.* **167**:210-228.

Heuser, J. E., Reese, T. S., Dennis, M. J., Jan, V., Jan, L., and Evans, L., 1979, Synaptic vesicle exocytosis captured by quick freezing correlated with quantal transmiter release, *J. Cell Biol.* **81**:275-300.

Hirokawa, N., 1982, The intramembrane structure of tight junctions: an experimental analysis of the single-fibril and two-fibrils models using the quick-freeze method, *J. Ultrastruct. Res.* **80**:288-301.

Horridge, G. A., 1966, Pathways of co-ordination in ctenophores. In: *The Cnidaria and their Evolution* (W.J. Rees, ed.), *Symp. Zool. Soc. Lond.* **16**:247-266.

Kachar, B., and Reese, T. S., 1982, Evidence for the lipidic nature of tight junction strands, *Nature* **296**:464-466.

Kachar, B., Christakis, N. A., Reese, T. S., and Lane, N. J., 1986, The intramembrane structure of septate junctions based on direct freezing, *J. Cell Sci.* **80**:13-28.

Kensler, R. W., Brink, P. R., and Dewey, M. M., 1977, The nexus of frog ventricle, *J. Cell. Biol.* **73**:768-781.

Kensler, R. W., Brink, P. R. and Dewey, M. M., 1979, The septum of the lateral axon of the earthworm: a thin section and freeze-fracture study, *J. Neurocytol.* **8**: 565-590.

Lane, N. J., 1981, Tight junctions in arthropod tissues, *Int. Rev. Cytol.* **73**:243-318.

Lane, N. J., Dallai, R., Burighel, P., and Martinucci, G. B., 1986, Tight and gap junctions in the intestinal tract of tunicates (Urochordata) : a freeze-fracture study, *J. Cell Sci.* **84**:1-17.

Larsen, W. J., 1983, Biological implications of gap junction structure, distribution and composition: a review, *Tissue Cell* **15**:645-671.

Mackie, G. O., 1984, Introduction to the diploblastic level, in: *Biology of the Integument, Vol. I. Invertebrates*, pp. 43-46 (J. Bereiter-Hahn, A. G. Matoltsy and K. S. Richards, eds.), Springer-Verlag, Berlin, Heidelberg.

Mazet, F., 1977, Freeze-fracture studies of gap junction in the developing and adult amphibian cardiac muscle, *Dev. Biol.* **60**:139-152.

Perrachia, C., 1980, Structural correlates of gap junction permeation, *Int. Rev. Cytol.* **66**:81-146.

Pinto da Silva, P., and Kachar, B., 1982, On tight-junction structure, *Cell* **28**:441-450.

Satterlie, R. A., and Case, J. F., 1978, Gap junctions suggest epithelial conduction within the comb plates of the Ctenophore *Pleurobrachia bachei*, *Cell Tissue Res.* **193**:87-91.

Schneeberger, E. E., and Lynch, R. D., 1984, Tight junctions. Their structure, composition and function, *Circul. Res.* **55**:723-733.

Simionescu, M., Simionescu, N., and Palade, G. E., 1976, Segmental differentiations of cell junctions in the vascular epithelium. Arteries and veins, *J. Cell Biol.* **68**:705-723.

Spray, D. C., and Bennett, M. V. L., 1985, *Gap Junctions.* Cold Spring Harbor Laboratory Press, New York.

Staehelin, L. A., 1973, Further observations on the fine structure of freeze-cleaved tight junctions, *J. Cell Sci.* **13**:763-786.

Stevenson B. R., Anderson, J. M., and Bullivant, S., 1988, The epithelial tight junction: structure, function and preliminary biochemical characterization, *Mol. cell. Biochem.* **83**:129-145.

Tamm, S. L., 1982, Ctenophora, in: *Electrical Conduction and Behaviour in "Simple" Invertebrates*, pp. 266-358 (G. A. B. Shelton, ed.), Clarendon Press, Oxford.

Taugner, R., Sonnhof, U., Richter, D. W., and Schiiler, A., 1978, Mixed (chemical and electrical) synapses on frog spinal motorneurones, *Cell tiss. Res.* **193**:41-59.

Van Deurs, B., and Luft, J. H., 1979, Effects of glutaraldehyde fixation on the structure of tight junctions. A quantitative freeze-fracture analysis, *J. Ultrastruct. Res.* **68**:160-172.

Wood, R. L., 1977, The cell junctions of hydra as viewed by freeze-fracture replication, *J. Ultrastruct. Res.* **58**:299-315.

Wood, R. L., and Kuda, A. M., 1980, Formation of junctions in regenerating Hydra: gap junctions, *J. Ultrastruct. Res.* **73**:350-360.

Chapter 3

Chemical and Electrical Synaptic Transmission in the Cnidaria

ANDREW N. SPENCER

1. Introduction

It was probably in an ancestral cnidarian that the earliest evolutionary experiments in neuro-neuronal and neuro-effector communication were played out. By studying present day cnidarians, we hope we are examining those synaptic mechanisms which were selected for, perhaps as far back as the Pre-Cambrian era, and which have been conserved with only minimal modification. Of course, we cannot be certain that physiological evolution proceeded at the same rate as morphological changes, nevertheless, the close resemblance of extant forms to fossilized imprints of cnidarians from this era (for example the Ediacara fauna of Australia) hint of slow rates of evolution.

What will become apparent in this chapter is that many of the basic synaptic mechanisms and properties that we associate with more "advanced" nervous systems, such as Excitatory and Inhibitory Post-Synaptic Potentials (EPSPs and IPSPs) and Minature End Plate Potentials (MEPPs), facilitation, temporal and spatial summation, and Ca^{++}-dependent release of transmitter, can be demonstrated in the Cnidaria. With some danger of oversimplifying, one could say that it was in this phylum that most of the important properties of synapses evolved, and that since that time, most evolutionary change in higher nervous systems (major protostome phyla and the chordates) has been with respect to the complexity of connections.

ANDREW N. SPENCER • Department of Zoology, University of Alberta, Edmonton, Alberta, TG6 2E9, Canada.

Evolution of the First Nervous Systems
Edited by P.A.V. Anderson
Plenum Press, New York

The three cnidarian classes, Anthozoa, Scyphozoa and Hydrozoa, show major differences in the organization of their nervous systems and in the functioning of their synapses and, therefore, generalizations that imply that there is a single "cnidarian" solution to an evolutionary problem are not valid. For example, gap junctions have not been identified in either the Anthozoa or Scyphozoa, whereas nearly all hydrozoan tissues have high densities of these junctions (Mackie et al., 1984). Thus, we are forced to conclude that electrical synapses and, hence, electrical coupling are absent in these two classes.

Alternation of generation between polypoid and medusoid individuals is a common feature of scyphozoans and hydrozoans, while anthozoans exist as polyps throughout their life history. It is important that we consider whether a particular nervous system exists in an attached polyp or a planktonic medusa, since the selection pressures on these stages are very different and, thus, we can expect the organization and functioning of their nervous systems to be different. Finally, although the physiological data presented here are relevant for any discussion of the ancestry of the Cnidaria or of the phylogeny of the three classes, we do not yet have enough comparative data or data from the various developmental stages in any one species to draw any firm conclusions.

2. Chemical Synaptic Transmission

2.1. Ultrastructure of Chemical Synapses

In this chapter there is not sufficient space to give a complete description of the ultrastructural diversity of chemical synapses in the Cnidaria. For a concise review see Westfall (1987). In most respects cnidarian synapses are conventional. They all show straight, parallel, electron-dense membranes of variable length with a synaptic cleft of 12-20 nm. The shapes and sizes of the vesicles are particularily variable with electron-lucent, electron-dense and dense-cored vesicles being seen. In the anthomedusae, such as *Sarsia*, *Polyorchis* and *Stomotoca*, electron-dense and dense-cored vesicles from 60-150 nm in diameter are found at neuro-neuronal synapses, while at neuromuscular synapses, vesicles are more irregular (often appearing as horseshoe-shaped sections) and do not stain heavily (Jha and Mackie, 1967; Mackie and Singla, 1975; Spencer, 1979). In *Hydra*, there are a similar range of synapse types (Westfall et al., 1970, 1971; Kinnamon and Westfall, 1982). Anthozoan neuro-neuronal synapses also tend to have densely-staining vesicles, while neuromuscular synapses have tiers of clear vesicles (Westfall, 1970; Peteya, 1973). Because of considerable interest in the scyphozoan medusa *Cyanea* as a physiological model, its synapses have been well described. Particular attention has been given to the *en passant* synapses that the giant fiber nerve net neurons (GFNN) make with each other. These synapses, originally described as symmetrical synapses by Horridge and Mackay (1962), have vesicles on both sides of the synapse. Anderson and Grünert (1988), using three-dimensional reconstruction of these synapses, have shown that a typical terminal contains 30 to 40 relatively spherical, translucent vesicles as well as irregular bulbous cisternae. A single thin section normally shows only 2 or 3 vesicles. Such a paucity of vesicles at synapses appears to be a common feature in cnidarians. As will be described later, these synapses in *Cyanea* are able to conduct in either direction and are now called bidirectional synapses (Anderson, 1985). Neuromuscular synapses from GFNN neurons to the striated swimming muscle are also electron lucent, but synapses onto smooth radial muscle tend to be dense cored (Anderson and Schwab, 1981). Symmetrical neuro-neuronal synapses are not restricted to the Scyphozoa. They have been reported in the hydromedusa *Sarsia* (Jha and Mackie, 1967) and the hydropolyp *Hydra* (Westfall et al., 1970).

2.2. Physiology of Scyphozoan Synapses

Conduction in the subumbrellar nerve net of scyphomedusae and the physiology of their neuromuscular systems has held a particular fascination for physiologists for more than 100 years. In fact the "entrapped wave preparation" using *Cassiopea* (Mayer, 1906) was the only practical nerve preparation, apart from the frog sciatic nerve, available at that time for physiological studies of the basic mechanisms responsible for the action potential. In 1954, Horridge recorded nerve impulses in the GFNN preceding contractions of the subumbrellar muscle of *Aurelia*, unambiguously showing that conduction was neuronal in this nerve-net of mostly bipolar neurons. Later, it was shown that these spikes originated in rhopalial pacemakers (Yamashita, 1957). By 1980, using intracellular recording, it was clearly established that action potentials in the GFNN of both cubomedusae and scyphomedusae were conventional, even though their waveform was complex (Satterlie and Spencer, 1979; Schwab and Anderson, 1980; Anderson and Schwab, 1983). As regards the mechanism of conduction between neurons of the GFNN, little was known. Using our present knowledge of synaptic physiology we can interpret experiments by Bullock (1943), examining the effect of excess Mg^{++} on conduction and contraction of subumbrellar strip preparations from *Aurelia* and *Cyanea*, as indicating that both the neuromuscular and neuro-neuronal synapses require Ca^{++} for transmitter release. When Mg^{++} was added, facilitation of neuromuscular synapses was prevented followed by conduction block in the nerve net which could be overcome with repetitive stimulation.

In *Cyanea* it is possible to obtain isolated portions of the GFNN, by exposing a section of the subumbrellar perirhopalial tissue to a strong oxidizing agent, such as sodium hypochlorite, for a few seconds (Anderson and Schwab, 1984). This treatment removes the overlying epithelial cells, and neurons can be maintained in culture for a few days using a *Cyanea* saline supplemented with glucose and fetal calf serum. The arrangement of neurons resulting from this treatment is similar to that seen *in vivo*, though the density of neurons was sometimes much reduced. GFNN neurons make either *en passant* synapses, where processes cross one another, or parallel appositions. Synapses are not restricted to processes, but also occur on cell bodies. Using patch electrodes in the whole-cell, current-clamp configuration, Anderson (1985) was able to

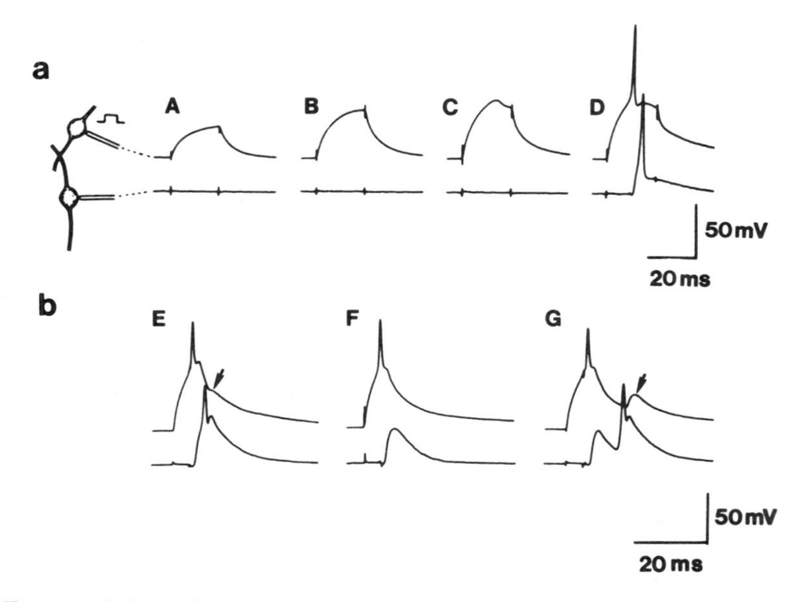

Figure 1. Transynaptic intracellular recording from a pair of GFNN neurons from the perirhopalial tissue of the scyphomedusa *Cyanea capillata*. **(a)** On the left is a diagram of the recording situation, with one electrode (upper trace) used to inject depolarizing current and record membrane potential in the presynaptic cell. A second electrode (lower trace) recorded the voltage response in the postsynaptic cell. Subthreshold depolarizations **(A-C)** of the presynaptic cell did not produce any response in the postsynaptic cell, while a spike in the presynaptic cell **(D)** elicited an EPSP and spike in the postsynaptic cell. **(b)** Recording situation as before showing that although the postsynaptic cell produced a spike at first **(E)**, the EPSP amplitude decreased until a spike was no longer produced in the postsynaptic cell **(F)**. When the postsynaptic cell fired **(E)** it produced a notch (arrow) with a constant latency of about 1 ms on the falling phase of the presynaptic spike. This is presumed to be the "echo" potential resulting from release of transmitter by the "postsynaptic" cell onto the "presynaptic" cell. An EPSP alone did not cause release of transmitter **(F)**. **G** shows a recording in which the first EPSP failed to elicit a spike but a second, delayed EPSP, which summed on the first, produced a spike. This spike produced a larger "echo" potential than was produced in E, due to the increased input resistance as the K^+ conductance decreases during the repolarizing phase (from Anderson, 1985).

make transynaptic recordings from pairs of neurons that made an obvious contact with one another. When increasing depolarizing current pulses were injected into one of the cells and the membrane voltage of both cells was monitored, the following response was always seen (Fig. 1a). Subthreshold depolarizations never produced depolarizations in the second cell. However, if the presynaptic cell fired an action potential (AP), then a rapid, large EPSP with a superimposed AP was recorded from the postsynaptic cell. The synaptic delay between the peak of the presynaptic AP and the onset of the EPSP was constant (mean = 0.98 ms). In almost every transynaptic recording, it was possible to reverse the direction of synaptic transmission such that the presynaptic neuron became postsynaptic. The mean delay times of synaptic transmission in either direction were essentially the same. Repetitive stimulation of the presynaptic cell caused a progressive decline in the amplitude of the EPSP, until they failed to elicit an AP. The EPSP seen when the AP first disappeared had an amplitude of some 70 mV, took 2.4 ms to peak and declined in 25 ms (Fig. 1b). The decrease in EPSP amplitude that was seen with repetitive stimulation (10 Hz) was assumed to be due to either a reduction in the amount of transmitter released or desensitization of postsynaptic receptors, since there was no change in the input resistance of the postsynaptic neuron. As can be seen from figure 1b an EPSP was often produced in the "presynaptic" neuron by release of transmitter by the "postsynaptic" neuron. This "echo potential" appeared as a small inflection on the repolarizing phase of the presynaptic spike, about 1 ms after the postsynaptic spike. It did not appear if the "postsynaptic" cell failed to fire. The amplitude of the echo potential varied, depending on when it appeared on the repolarizing phase (Fig. 1b). It is assumed that as the AP repolarizes the input resistance increases with the reduction in K^+ conductance and, therefore, the effect of the synaptic current is increased.

This study by Anderson clearly established that scyphozoan nerve nets depend on chemical synapses for propagation of action potentials throughout the network rather than electrical synapses. It also demonstrated, that many aspects of their physiology are similar to synapses in other phyla. For example he was able to show that: a) the synaptic delay was in the same range as delays measured at other invertebrate and vertebrate chemical synapses (Takeuchi and Takeuchi, 1962; Martin and Pilar, 1964), b) although he was unable to reverse the synaptic potential, he did show that EPSP amplitude was dependent on the membrane potential of the postsynaptic cell, and c) by using lidocaine to remove the AP, a steep transfer relationship between EPSP amplitude and presynaptic membrane potential could be seen (a 14 mV increase in membrane potential produced a 10-fold increase in EPSP amplitude). As might be expected there were some unusual features which are most probably a reflection of the particular requirements for chemical synapses in a nerve net. Despite symmetrical or at least reciprocal synapses having been reported from ultrastructural studies in many phyla, such as annelids (Hama, 1959), molluscs (McCarragher and Chase, 1983); crustaceans (Hama, 1961) and chordates (Kohno, 1970), this study by Anderson was the first physiological demonstration of a truly bidirectional chemical synapse. Although at first glance it would appear that the ability of a synapse to conduct in either direction is an obvious adaptation for a nerve net (where the direction of propagation of a wave is unpredictable), it has been pointed out by Anderson and Spencer (1989) that bidirectionality is not required if the nerve net is sufficiently dense (see Josephson et al., 1961). These authors thought it likely that bidirectionality may be a developmental consequence of neurons being programmed to form a synapse wherever they contact another neuron, under circumstances where neither partner has precedence. The high threshold for release of transmitter (0 mV or above) is probably an adaptation to bidirectionality, as a low threshold would mean that there would be a continuous reverberatory depolarization of the pair of terminals, until the neurons recovered from their refractoriness and a second spike ensued (Anderson, 1985).

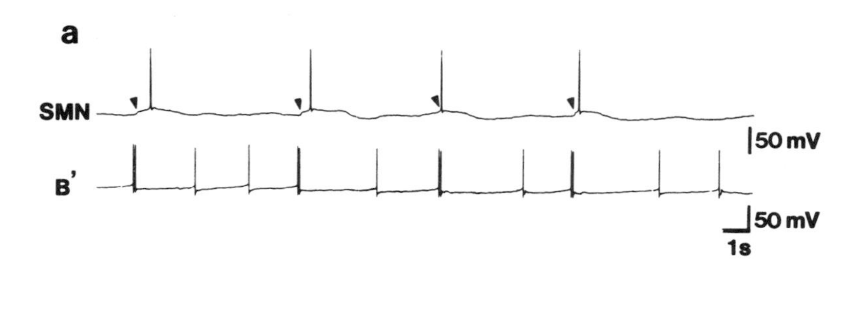

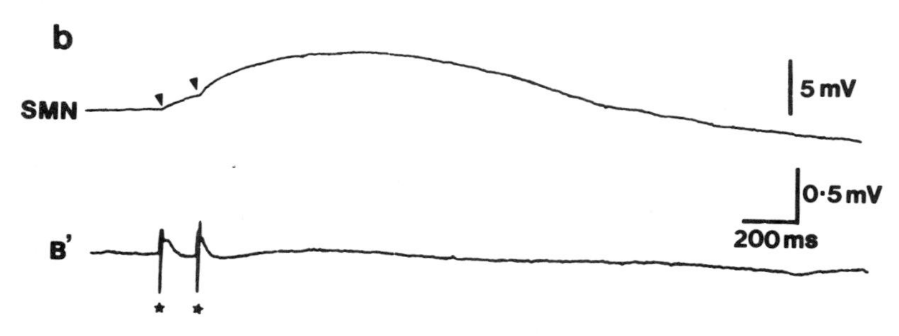

Figure 2. Facilitation and summation of excitatory chemical synapses between "B" system neurons and swimming motor neurons, recorded from the nerve rings of the hydromedusa *Polyorchis penicillatus*. **(a)** Intracellular recordings from the swimming motor neuron network (upper trace) showing action potentials elicited by spontaneous spikes in the "B" system (lower trace). Single spikes in the "B" system produced EPSPs that were barely detectable, while the second of a pair of "B" spikes produced a facilitated EPSP (arrows) that exceeded threshold and resulted in SMN spikes. **(b)** Intracellular recording from an SMN (upper trace) while recording extracellularly from the outer nerve ring (lower trace). The first spontaneous "B" system spike (asterisks) produced a small amplitude EPSP, while the second spike elicited a facilitating and summing EPSP (from Spencer and Arkett, 1984).

By attaching a patch recording pipette to a single soma in the *Cyanea* GFNN system, on one side of a synapse, it is possible to voltage clamp both the presynaptic and postsynaptic terminals simultaneously (Anderson and Spencer, 1989). Preliminary experiments using this technique showed that a voltage protocol which produces a large depolarization from a holding potential of -70 mV, elicited large tail currents, followed by a large, transient inward current that is believed to be the "postsynaptic" current resulting from the neuron detecting release of its own transmitter.

The assumption is made that transmitter release in cnidarians is vesicular and, indeed, Anderson (1988) established that transmitter release at GFNN synapses is quantal; yet the paucity of synaptic vesicles at these synapses brings the vesicular hypothesis of transmitter release into question. Anderson has attempted to explain this anomoly. The average amplitude of an EPSP at spiking threshold is 60 mV which when corrected for the non-linearities that occur as the EPSP approaches its reversal potential (Hubbard et al., 1969), becomes 960 mV. Since the unitary postsynaptic events are 600-800 μV, at least 1200 vesicles must be released, assuming each quantum represents one vesicle. As there are only 30-40 true vesicles present in most synapses, Anderson suggests that the possibility that there is non-vesicular release must be seriously considered. This phenomenon could be widespread in the Cnidaria, as the lack of a large releasable pool, as judged by ultrastructural criteria, is a common feature. One might easily be drawn to the conclusion that non-vesicular release is a feature of "primitive" synapses, but it should be remembered that non-vesicular release has been suggested for both molluscs and vertebrates (Tauc, 1982; Dunant, 1986).

Despite considerable effort (Anderson, personal communication), the identity of the transmitter substance used at GFNN synapses is unknown. Major transmitter candidates, including ACh, amines and neuropeptides, have not shown any electrophysiological effect on GFNN neurons.

2.3. Physiology of Hydrozoan Chemical Synapses

Two species of hydromedusae have been used extensively for cellular studies of synaptic function in the Hydrozoa; these are the anthomedusa *Polyorchis penicillatus* and the trachymedusa *Aglantha digitale*. Unfortunately, similar experiments have not been possible using hydroid polyps and thus, for the present, we have to be satisfied that these medusae are good representatives of the hydrozoan condition. In both these medusae, neuro-neuronal and neuromuscular synapses have been examined.

There is strong evidence for chemical synapses between identified networks of neurons in *Polyorchis* (Spencer and Arkett, 1984; Arkett and Spencer, 1986a,b). The first example is found between a network of neurons, the burster or "B" system (located in the outer nerve ring and the ectoderm of the tentacles), and a network of neurons in the inner nerve ring, the swimming motor neurons or SMNs (Spencer, 1981; Spencer and Arkett, 1984).

Dual intracellular recordings of spontaneous activity in "B" and "SMN" neurons showed that the two networks were not electrically coupled, and that "B" system action potentials produced EPSPs that could cause spiking in SMNs (Fig. 2a). These synaptic potentials, up to 10 mV in amplitude, followed the peak of "B" spikes with a delay of 5-8 ms (Fig. 2b). Since the network of "B" neurons is electrically coupled, spikes occurred synchronously throughout and, therefore, EPSPs also appeared synchronously in the SMN network, resulting in both spatial and temporal summation. In this way, the electrical loading that a postsynaptic neuron experiences is reduced and the synaptic current in any one neuron has a greater effect on membrane potential (Getting, 1974). Often, single "B" spikes did not produce discernable EPSPs (Fig. 2a), while a burst gave a summing staircase of synaptic potentials (Fig. 2b) that was often sufficient to bring the motor neuron network to threshold. Synaptic potentials were abolished by excess Mg^{++} in the bathing solution. Blockade by Mg^{++} was also used by Roberts and Mackie (1980), to establish that synapses between the ring giant axon and the radial motor giant axons in the trachymedusa *Aglantha* are chemical. Those authors measured a synaptic delay of 1.6 - 1.8 ms, and showed that EPSPs in motor giant axons could sum and facilitate. If these two synapses in *Polyorchis* and *Aglantha* are typical of excitatory, inter-network synapses of hydrozoans, then it is apparent that

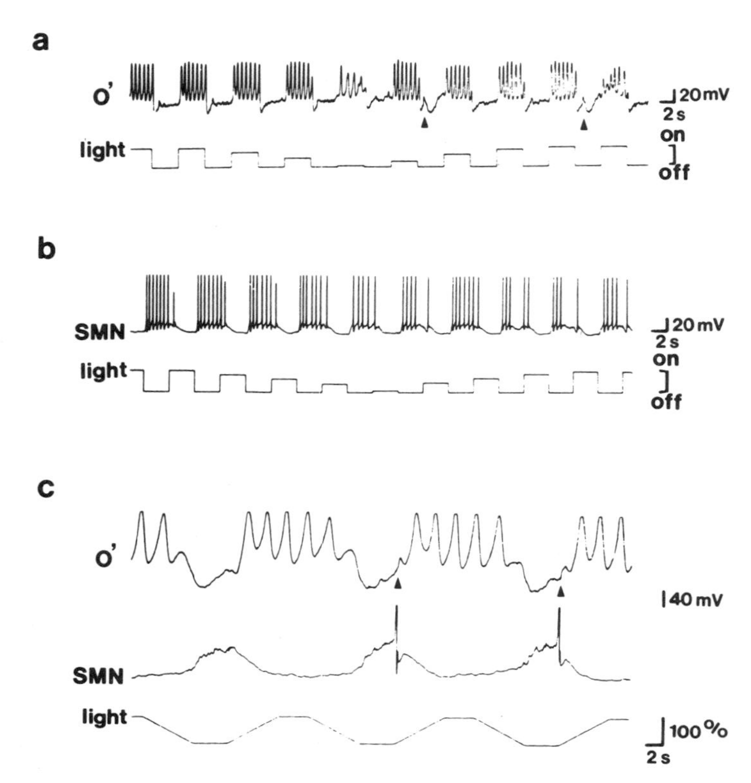

Figure 3. Synaptic interactions between the electrically-coupled network of oscillator ("O") neurons and swimming motor neuron (SMN) network, during photic stimulation in the hydromedusa *Polyorchis penicillatus*. **(a)** Intracellular recording from the "O" system (upper trace), while subjecting the nerve ring to 100% "instantaneous" shadows of varying intensities, monitored by measuring the current to an LED (lower trace). Onset of a shadow, regardless of the magnitude of the intensity change, produced a rapid hyperpolarization and cessation of the oscillations. The response to light "on" consisted of a depolarization and recommencement of the oscillations. The occasional transient depolarizations (arrows) were EPSPs from the SMN system. **(b)** Similar protocol as in **(a)** except that the intracellular recording is from the SMN system (upper trace). With each shadow, the SMN network depolarized and produced a burst of spikes. Each increase in light intensity caused a hyperpolarization and termination of the burst. **(c)** Simultaneous intracellular recordings from the "O" (upper trace) and SMN (middle trace) systems, while altering the light intensity (bottom trace). As the light intensity was reduced, the "O" system hyperpolarized coincidently with a slow depolarization and high frequency EPSPs in the SMN network. Two of these episodes resulted in SMN spikes which produced EPSPs (arrows) in the "O" system. These depolarizations of the SMN network were interpreted as disinhibiton via presynaptic "O" neurons (from Arkett and Spencer, 1986b).

chemical transmission at polarized synapses is quite conventional. These studies have also provided confirmation, at a cellular level, that facilitation can occur at neuro-neuronal synapses in the Cnidaria (cf. Pantin, 1935).

In the outer nerve ring of *Polyorchis*, there is circumstantial evidence for inhibitory chemical synapses between neuronal networks. Within the outer nerve ring there is a

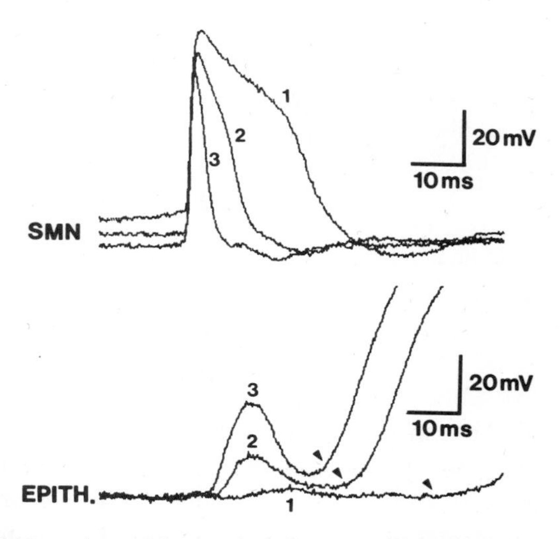

Figure 4. Chemical synaptic transmission at a neuromuscular junction in the hydromedusa *Polyorchis penicillatus*, showing the relationship between presynaptic spike duration, EPSP amplitude and the delay to the muscle action potential. Simultaneous, superimposed intracellular recordings of spontaneous swimming activity from a swimming motor neuron (upper trace) and an overlying myoepithelial cell (lower trace). When a spike (AP 1) originated at a site close to the recording electrode it appeared from a relatively depolarized membrane potential and had a long duration, due to steady-state inactivation of the fast, transient, K^+ current (I_k fast). Conversely, spikes which originated at increasingly more distant sites (APs 2 and 3) arose from more negative resting potentials and had shorter durations, due to the increased I_k fast, which caused more rapid repolarization. The corresponding EPSPs and initial portions of the muscle action potentials recorded in a myoepithelial cell are shown in the lower trace. Long duration presynaptic spikes produced small amplitude EPSPs and a long delay to the initiation of the muscle AP (arrows), and the reverse was true for short duration spikes (from Spencer et al., 1989).

condensed network of small neurons, the oscillator or "O" system, that also extends into the ocellar cup (Spencer and Arkett, 1984). These non-spiking neurons responded to changes in light intensity by immediate, graded changes in the membrane potential (Fig. 3a,c). Decreases in light intensity caused hyperpolarization and cessation of oscillatory events, while increases in intensity brought about depolarization and a resumption of the oscillations (Arkett and Spencer, 1986b). For these reasons, "O" neurons are believed to be primary photoreceptors. Swimming motor neurons (Fig. 3b) and "B" system neurons also responded to changes in light intensity, but the polarity of the membrane potential responses was opposite to that of "O" neurons, and followed the "O" system response with a slight delay (Fig. 3c). The spiking frequencies of both SMN and "B" neurons reflected these changes in membrane potential. Arkett and Spencer interpreted these results as evidence that in the light, the "O" neurons tonically release an inhibitory transmitter onto both these postsynaptic networks, while a decrease in light intensity hyperpolarizes "O" neurons, reducing the amount of transmitter released thus allowing the SMN and "B" neurons to be released from inhibition, to depolarize and fire. Such disinhibition is quite common in photoreceptor systems of invertebrates and vertebrates (Fain et al., 1983; Laughlin, 1981).

One of the most completely described cnidarian chemical synapses is that between the swimming motor neurons and overlying epithelial cells in *Polyorchis* (Spencer, 1982). This can be considered a neuromuscular synapse as the postsynaptic epithelial cells are electrically coupled to the myoepithelial cells that form the sheets of swimming muscle. Simultaneous intracellular recordings from motor neurons and overlying epithelial cells, in half-bell preparations, showed the basic properties of these synapses during spontaneous swimming (Fig. 4). Three lines of evidence supported chemical transmission: synaptic delay was fairly constant with a mean of 3.2 ms and a minimum delay of 0.9 ms, excess Mg^{++} caused a reduction in junctional potential amplitude, until transmission failed, and the membrane potential of postsynaptic epithelial cells could not be altered by injecting current into motor neurons (Spencer, 1982). The waveform of the response of epithelial cells to SMN spikes was quite complex, consisting of an initial excitatory junctional potential (EJP) which occurred at a constant delay, followed by the muscle action potential which was usually in two components. The first of these components was the muscle action potential generated in the velum, propagating back into the synaptic region, and the second was a similar AP from the subumbrellar muscle sheet. Muscle action potentials were not initiated in the immediate region of the synapses over the inner nerve ring, but at some distance out onto the muscle sheets of the velum and subumbrella. For this reason, action potentials did not arise from the peaks of the junctional potentials. The most surprising finding was that the amplitude of junctional potentials and the delay to the appearance of the muscle action potential, depended on the duration of the presynaptic action potentials (Spencer, 1982). The extraordinary variability of the duration of SMN action potentials had been previously reported by Spencer (1981), who showed that as an action potential propagates through the SMN network, away from its initiation site, its duration decreases from about 30 to 5 ms, due to a loss of the plateau phase. Simultaneous intracellular recordings from SMNs and epithelial cells showed that short duration presynaptic APs gave large amplitude junctional potentials (up to some 40 mV) and short delays, while the longest duration APs produced small amplitude EJPs (less than 5 mV) and long delays (Fig. 4). This modulation of the delay of the muscle action potential by the distance the motor spike has travelled is of extreme functional significance. It is a mechanism that automatically compensates for the conduction delay of the motor spike as it propagates through the nerve ring, resulting in synchronous contraction of the swimming muscle sheets.

With the recent development of a technique to culture swimming motor neurons obtained from the nerve rings of *Polyorchis* (Przysiezniak and Spencer, 1989), it became possible to examine this phenomenon in more detail (Spencer et al., 1989). Isolated swimming motor neurons can be cultured on a substratum of extracted mesoglea and identified by their characteristic morphology. Using patch pipettes, it is possible to record from these neurons in the whole-cell configuration either under current-clamp or voltage-clamp conditions. The first objective was to determine the basis for the decrease in the duration of action potentials. It soon became apparent that the major factor determining whether the plateau was expressed or not was the resting membrane potential immediately prior to spike initiation; this had been suspected by Anderson (1979). Voltage clamp experiments showed that a rapidly-activating, transient K^+ current ($I_{K\text{-fast}}$), reminiscent of an A-current, which is partly responsible for AP repolarization, was almost completely inactivated at depolarized resting potentials (10% of maximal current at -30 mV), thus allowing expression of a

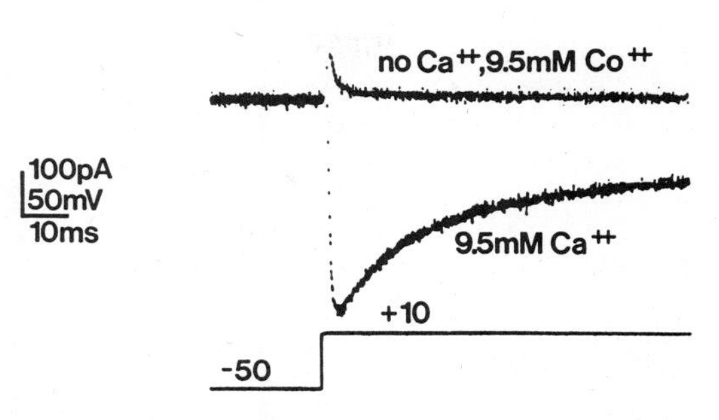

Figure 5. Voltage-activated Ca^{++} current (I_{Ca}) recorded from isolated, cultured swimming motor neurons of *Polyorchis penicillatus*, using the whole-cell, voltage-clamp technique, with Gigaohm seals. The middle trace shows the I_{Ca} elicited by a 200 ms voltage step from - 50 to +10 mV, which was blocked (upper trace) by perfusing with an extracellular solution containing 9.5 mM Co^{++}. The bottom trace shows the voltage protocol. Note that there are two components to the Ca^{++} current, one with a short time constant (approx. 20 ms), which inactivated quite rapidly, and a second, slowly inactivating current with a longer time constant (approx. 65 ms). The Na^{+}-free electrode solution contained (in mM): CsCl, 109; TEA-Cl, 100; Dextrose, 536; $MgCl_2$, 2; $CaCl_2$, 1; EGTA, 11; HEPES, 10; CsOH, 31 (from Spencer et al., 1989).

plateau which is carried by Ca^{++} and Na^{+}. At relatively more polarized resting potentials this plateau was obliterated by the outward current corrected by $I_{K\text{-fast}}$ (65% maximal current at -50 mV). It should be pointed out that as motor spikes propagate *in vivo* they are always invading regions of the network that are relatively more hyperpolarized than the region they just left (Spencer, 1981).

The next step was to determine what voltage-dependent Ca^{++} currents (that might mediate transmitter release) are present in the motor neuron, and to establish whether they were sensitive to changes in action potential duration. Figure 5 shows that there are probably two such Ca^{++} currents present. The larger current (up to 500 pA) activated rapidly (within 4 ms) and inactivated with a time constant of approximately 20 ms. A much smaller current which inactivated far more slowly (65 ms) was also apparent. The Ca^{++} current showed peak activation at +10 mV and half inactivation at -27 mV. The rapid activation and inactivation of the major inward Ca^{++} current is unusual for a Ca^{++} current at a presynaptic terminal (Augustine, et al., 1987). As explained below, the dynamics of I_{Ca}, especially its rapid inactivation, may be an adaptation for the modulation of transmitter release caused by changes in action potential duration. It is important to note that because SMNs in culture lack long processes, they are electrically compact and, therefore, it is reasonable to assume that the whole-cell Ca^{++} currents recorded are a true representation of terminal Ca^{++} currents. In addition, ultrastructural data suggest that the SMN/epithelial cell synapses are located throughout the neurons and are not restricted to terminals (Spencer, 1979). In order to examine the dynamics of the voltage-gated Ca^{++} influx when action potentials of varying duration invade a presynaptic release site, Spencer et al. (1989) stimulated motor neurons *in vitro* using voltage commands shaped like APs. Figure 6 shows that a long duration AP elicited a small amplitude, slow Ca^{++} current that, *in vivo,* would have resulted in a small amplitude EJP. In contrast, a short duration AP elicited a larger, more transient Ca^{++} current that, *in vivo*, would have produced a larger EJP. Voltage-clamp stimuli of intermediate durations elicited Ca^{++} currents of intermediate amplitudes and times to peak (Fig. 6).

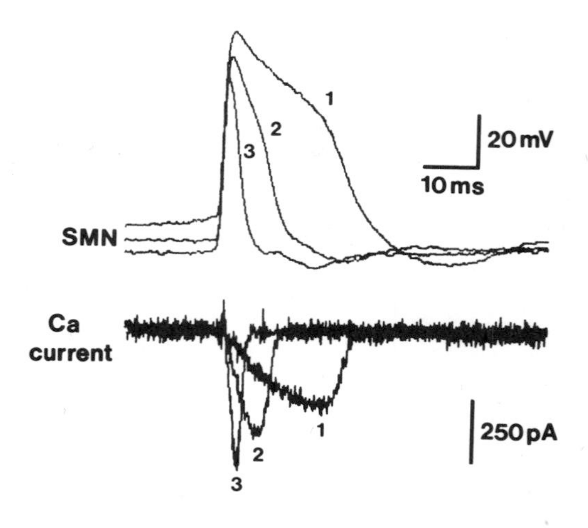

Figure 6. Relationship between presynaptic spike duration and the dynamics of the presynaptic Ca^{++} current in cultured swimming motor neurons of *Polyorchis penicillatus*. Three prerecorded SMN spikes of varying duration were used as the voltage commands to elicit presynaptic Ca^{++} currents in a voltage-clamped neuron. The long duration spike command (1) elicited a slowly-developing, small amplitude Ca^{++} current (I_{Ca}), while progressively shorter spike commands (2 and 3) produced more transient but larger amplitude currents. Contaminating Ca^{++}-independent currents were subtracted using leakage- and capacitance-corrected responses obtained from the same cell in Ca^{++}-free, Co^{++}-substituted saline. Electrode solution as for figure 5 (from Spencer et al., 1989).

The control of Ca^{++} influx by action potential duration can be explained in the following way. The number of channels available for activation is determined by the holding potential preceding the spike and, therefore, short duration spikes which arise from a more polarized baseline are able to activate more channels. All available channels are activated by the rising phase of the spike, but no current flows until a net inward driving force develops during spike repolarization. With short duration APs, the driving force on Ca^{++} develops more rapidly than channels are closing and inactivating, resulting in a phasic Ca^{++} influx, analogous to a "tail" current. On the other hand, the slower rate of repolarization of long duration APs means that the driving force increases more slowly, resulting in a slower, smaller amplitude Ca^{++} current. Although the integrals of the currents show that more Ca^{++} enters during long spikes than short spikes (see Fig. 6), it is the short spikes which release more transmitter (as judged by the amplitude of EJPs). Why, then, is a phasic influx of Ca^{++} so effective in mediating transmitter release?

Fast transmitter release is thought to depend on the Ca^{++} concentration at cytosolic binding sites, which regulate the rate of vesicle fusion to the plasma membrane (Zucker and Lando, 1986). The Ca^{++} concentration at these sites, which are just beneath the membrane, is determined by the difference between the rate of Ca^{++} influx and the rate of its removal by diffusion, sequestration and active pumping (Augustine et al., 1987). It seems likely that, in *Polyorchis* motor neurons, removal of Ca^{++} from just below the membrane is sufficiently effective to counter the increased $[Ca^{++}]$ produced by a slow influx and yet allows Ca^{++} to build up during a rapid influx. These results are in contrast to what has become the dogma for Ca^{++}-mediated release in other systems, namely that long duration action potentials will potentiate release due to the larger total influx of Ca^{++} in the terminals (Kusasno et al., 1967; Fuchs et al., 1982; Lin and Faber, 1988). For example, in the sensory neurons involved in short term sensitization of the gill withdrawal reflex of *Aplysia*, presynaptic spike duration is modulated by the degree of development of the serotonin-sensitive S current, which is a K^+ current responsible for repolarization. In the presence of serotonin the resulting long-duration action potential produced an increased Ca^{++} transient (Boyle, et al., 1984) and an increase in the amplitude of the resulting PSP in follower neurons (Hochner et al., 1986). Presumably in this system, the mechanisms for removal of Ca^{++} as it enters through voltage-gated Ca^{++} channels are slower than for *Polyorchis* motor neurons. This voltage-clamp study of *Polyorchis* motor neurons shows that cnidarian presynaptic neurons have Ca^{++} currents that, although unusual with respect to their kinetics, can account for the modulation of transmitter release. Ca^{++}-dependent neurotransmission was also noted by Kerfoot et al. (1985) in *Aglantha*. Thus, there is no evidence yet that the molecular mechanisms of transmitter release in cnidarians are anything but conventional.

Another well-studied neuromuscular synapse is that between motor giant axons and myoepithelium in *Aglantha* (Kerfoot et al., 1985). As might be expected, this synapse shows many similarities to the neuromuscular synapse in *Polyorchis*. The mean synaptic delay of 0.7 ms in *Aglantha* was less than that in *Polyorchis* (3.2 ms) and is probably related to the fact that this synapse is used for escape swimming in *Aglantha*. If myoepithelial cells that are close to synapses are penetrated with microelectrodes, spontaneous MEPPs (up to 2 mV) are often recorded, giving support to the supposition that transmitter release is quantal, if not vesicular (see section on *Cyanea* above).

Besides these two hydromedusae, electrophysiological evidence is available for chemical transmission at neuromuscular synapses in the leptomedusa *Aequorea* (Satterlie, 1985) and various siphonophores (Mackie, 1978).

Little is known about the chemical nature of transmitters used at both neuro-neuronal and neuromuscular synapses in hydromedusae. Using HPLC and GC/mass spectrometry, dopamine and an unidentified catecholamine related to norepinephrine has been found in the nerve rings of *Polyorchis* (Chung et al., 1989). Some preliminary experiments (Chung and Spencer, in preparation) show that swimming motor neurons have fairly non-selective receptors for catecholamines that, when activated at the resting potential of the cell, produce a long duration outward current (Fig. 7) and, consequently, inhibition. The ionic dependence of this current has not been fully investigated, but it is assumed that there is a conductance increase to K^+.

Besides the suggestion that catecholamines are involved in neurotransmission, there is evidence, though somewhat contradictory, that a family of peptides with carboxy terminal Arg-Phe-amide have similar functions. Grimmelikhuijzen and collaborators have shown by immunohistochemistry that there are distinct subpopulations of neurons containing these peptides in a wide range of cnidarians (Grimmelikhuijzen et al., 1988a), including hydromedusae such as *Polyorchis* (Grimmelikhuijzen and Spencer, 1984). Several of these peptides have been isolated (Grimmelikhuijzen et al., in this book), including one (Pol RFamide) from *Polyorchis* (Grimmelikhuijzen et al., 1988b). Although it had been shown that RFamides tend to

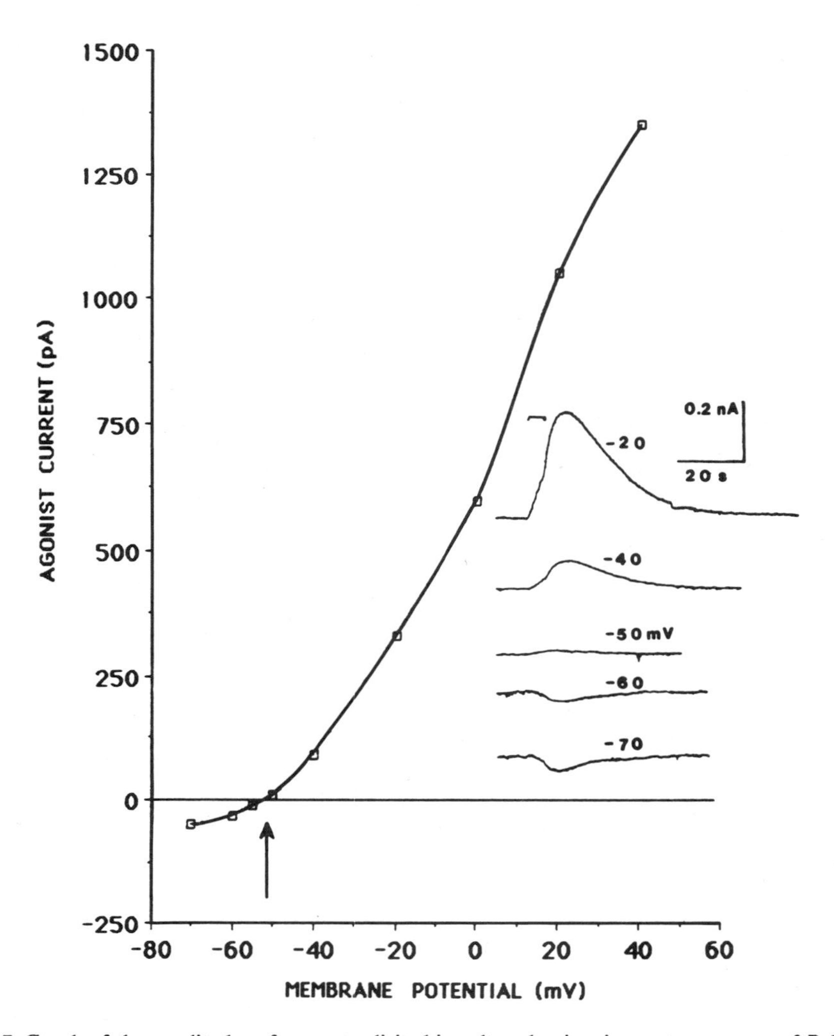

Figure 7. Graph of the amplitudes of currents elicited in cultured swimming motor neurons of *Polyorchis penicillatus*, by application of dopamine at different holding potentials. Neurons were voltage clamped in the whole-cell configuration using patch electrodes with seal resistances in excess of 1 Gigaohm. Dopamine at 10^{-4} M in artificial seawater (pH 7.4) was pressure injected (20 PSI, 5 s duration) through a glass pipette (methanol "bubble number" of 6.4) positioned approx. 15 μm from the soma. The current traces produced by application of dopamine at different holding potentials, close to the reversal potential (arrow) of approximately -50 mV, are shown in the inset. The non-linearity of the response is probably due to some contamination by a non-inactivating, or slowly inactivating voltage-dependent current (Chung and Spencer, in preparation).

produce long duration excitation of swimming motor neurons *in vivo* (Spencer, 1988), these neurons, when isolated in culture, produced outward currents when Pol RFamide is pressure injected onto them (Chung and Spencer, unpublished). This may mean that the excitatory effect of RFamides *in vivo* is produced by removal of a tonic inhibitory influence from a system presynaptic to the SMNs.

3. Electrical Synaptic Transmission

The ultrastructure of cnidarian electrical synapses, that is gap junctions, is discussed in the chapter by C. Green. This section will deal with the electrical properties of gap junctions between neurons. The properties of electrically-coupled epithelial cells have been described in studies by Josephson and Schwab (1979) on the exumbrellar epithelium of the hydromedusa *Euphysa*, by Chain et al. (1981) on the myoepithelium of the siphonophore *Chelophyes*, and by Kerfoot et al. (1985) on the subumbrellar myoepithelium of *Aglantha*.

As described in the introduction, there is no direct evidence for electrical coupling in the Anthozoa and Scyphozoa. However, in the Hydrozoa, each nerve net is formed from electrically-coupled member neurons (Spencer and Arkett, 1984). A good example of an electrically-coupled nerve net is the swimming motor neuron network in *Polyorchis*. The neurons of this compressed nerve net, which is located in the inner nerve ring and radial nerves, are the motor neurons that innervate the striated swimming muscle via chemical synapses (see section above on chemical synapses). Simultaneous intracellular recordings from pairs of motor neurons showed that neurons spiked almost synchronously, even when electrodes were positioned at opposite sides of the nerve ring (Spencer, 1981). When depolarizing current pulses were injected into one neuron, the network could be driven to fire at a much higher frequency than the spontaneous frequency (Fig. 8). This is suggestive of very strong electrical coupling,

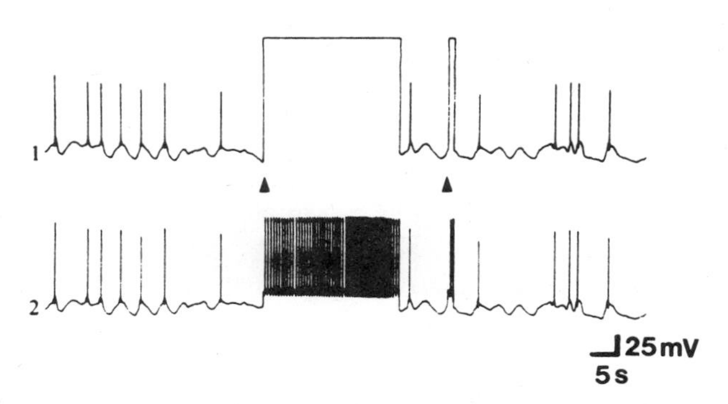

Figure 8. Electrical coupling in the swimming motor neuron network of *Polyorchis penicillatus*. Spontaneous activity recorded intracellularly at two sites at a 4 mm separation. Note that spiking was synchronous and that identical, slow membrane potential oscillations were recorded at both sites. One long and one short depolarizing current pulse (40 nA) was injected at the arrows into electrode 1. These pulses drove the network to fire at a high frequency. Because of the large current needed to overcome the considerable electrical load resulting from extensive coupling, the bridge circuit of electrode 1 could not be balanced (from Spencer, 1981).

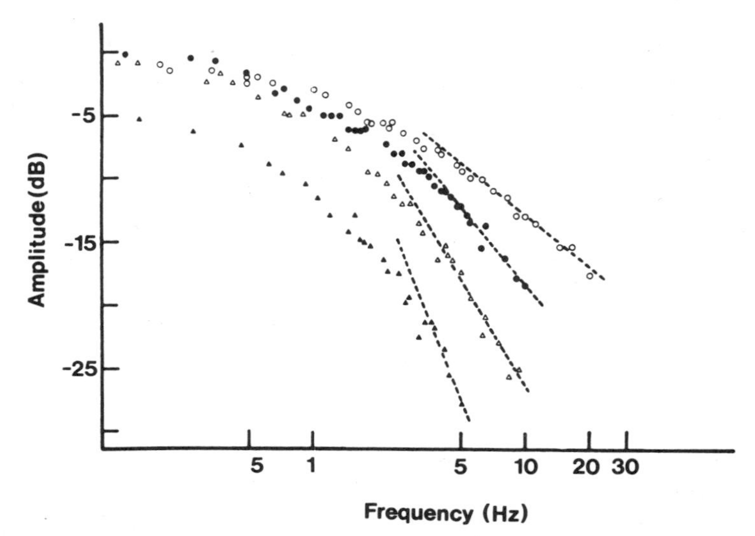

Figure 9. Bode plot showing the filtering properties of the swimming motor neuron network of *Polyorchis penicillatus*. Pairs of intracellular microelectrodes were placed in the network at the following separations: 0.92 mm for open circles, 3.0 mm for closed circles, 4.8 mm for open triangles and 8.7 mm for closed triangles. A 12 nA sinusoidal current of varying frequency was injected through one electrode and the amplitude of the voltage response was measured at the second electrode. 0 dB is the d.c. voltage recorded at the current injecting electrode for a current of 12 nA. Straight line approximations (shown as dotted lines) of the high frequency asymptotes of these curves at the four distances are -13.5 dB, -20 dB, -30 dB and -43 dB, respectively (from Spencer, 1981).

which was demonstrated by injecting hyperpolarizing current pulses through one electrode and recording the voltage response at increasing distances along the network with a second electrode. Assuming that current flow in the network is equivalent to flow in a single core conductor, and that the network behaves as a linear, one-dimensional cable (Jack et al., 1975), an average space constant of 7.1 mm was calculated for medium-sized medusae (Spencer, 1981). This strong coupling has important functional significance, since it allows spikes to be conducted at the high velocities needed for synchronization. Electical coupling appears to be a common adaptation for synchronizing activity in groups of neurons (Bennett, 1977). Specific examples would be pyramidal neurons of the mammalian hippocampus (MacVicar and Dudek, 1981), and pre-motor neurons in the buccal ganglia of molluscs (Kaneko et al., 1978). Conduction velocity in the SMN network is variable, from about 112 to 200 cm s^{-1}, depending on the membrane potential ahead of the advancing spike. If the membrane potential is relatively hyperpolarized, then the conduction velocity is decreased due to the increased charging time required for a spiking cell to depolarize neurons to threshold ahead of the advancing spike front. The average membrane potential of the entire nerve ring depends on the amount of excitatory input from sensory systems, such as the ocelli, located throughout the nerve ring. Thus, in an "unaroused" state, with the membrane potential far from threshold conduction velocities are low. A second property of this electrically-coupled network is that it acts as a low-pass filter, progressively attenuating passive signals with frequencies greater than approximately 1 Hz (Fig. 9). Low-pass filtering has also been noted in other electrically-coupled neuronal groups (Getting, 1974; MacVicar and Dudek,1981) The consequence of this filtering is that any postsynaptic potentials resulting from input from local sensory or interneurons should only have local influences because the PSPs would be shunted through the network. It was, therefore, surprising to record simultaneous EPSPs at widely separated sites in the SMN network (Spencer, 1981). The only neuroanatomical arrangement that could explain this paradox was the existence of presynaptic networks with multiple synapses onto the SMN network, which released transmitter synchronously. Such systems ("B" and "O") were discovered in the outer nerve ring, and, not surprisingly, also consist of electrically-coupled member neurons which explains the synchrony of PSPs (Spencer and Arkett, 1984). Thus the ring structure of the SMN network and its low-pass filtering properties would allow it to distinguish between local sensory input and general input occurring throughout 360° via the concentric presynaptic systems. However, it is not apparent how it might use local input since output of the SMN network to swimming muscle is not restricted to small radii.

4. Is Electrical Or Chemical Transmission More Primitive?

Inherent in any discussion of what the primitive mode of intercellular communication might have been in the first nervous system are the questions - from what ancestral cell type did neurons originate and did neurons evolve more than once? An intuitively attractive and popular argument has been put forward by several authors, notably Horridge (1968) and Mackie (1970), that propagative electrical signalling, using low resistance pathways between cells, must have appeared very early in the evolution of the Metazoa. It is easy to imagine that in an ancestral metazoan, the covering epidermal epithelium and perhaps internal epithelial layers were used as conducting pathways. The supporting evidence for this view comes from the numerous examples of epithelial conducting systems that have been described in both adults and particularily embryos, in a wide range of phyla (see reviews by Mackie, 1970; Spencer, 1974 and Anderson, 1980). Lacking any endocrine or nervous systems for coordinating metabolic activities, one could imagine that in this archetype, communication between cells forming a homogenous tissue was achieved by an interchange of chemical signals. Information could have been in the form of shared intracellular messengers or enzymes that were transferred by discrete, specialized pathways, such as gap junctions that bridged the intercellular gap. These pathways would also have had the property of allowing ionic flow and, hence, electrical current. All that was then required for epithelial conduction was the appearence of voltage-gated channels, probably for Ca^{++} ions (Hille, 1984), that enabled electrical signals to be propagated. On the other hand, chemical communication might have taken the form of messengers being secreted into extracellular space, diffusing locally, and then being taken up by or binding to receptors on neighboring cells. If so, the intercellular pathways that were required to channel messengers between cells would not be required and epithelial conduction by propagation of action potentials could not occur, since current would be shunted through low resistance extracellular pathways (Loewenstein, 1975).

In primitive metazoans the selection pressure for external epithelial layers to take on functions other than their primary role as a protective layer must have been considerable. For example, without the benefit of sensory innervation of the skin, epithelial cells must have developed the ability to transduce mechanical and chemical signals. Later, to aid in the rapid transmission of sensory information to effectors, definitive pathways developed, which in their final form, involved the evolution of neurons with long, polarized processes. Such a scenario, which was derived from various histological arrangements found in cnidarians and ctenophores, has been depicted by Horridge (1968). There is no reason to suppose that this could not have happened several times during the evolution of early metazoans and, thus, neurons might have multiple origins. The mode of intercellular communication in these early neuronal networks, whether electrical or chemical, would depend on what type of communication had pre-existed in the ancestral epithelium.

5. Conclusions

Only in the last few years have the full possibilities of examining the electrophysiological properties of cnidarian synapses been realized. This has been a consequence of the development of techniques for isolating identified neurons with known synaptic connections (Anderson and Schwab, 1984; Przysiezniak and Spencer, 1989). Not only do these preparations give us insights into the possible sequence of evolutionary change in synaptic functions in the early Metazoa, but will probably be useful for examining basic synaptic properties, since there are very few preparations in other phyla where there is accessibility to pre- and postsynaptic cells, both *in vivo* and *in vitro*. The electrophysiological data obtained so far indicate that, in most respects both chemical and electrical synapses in cnidarians are "conventional" when compared with those in other phyla. Nevertheless, there are several examples (such as strong electrical coupling and modulation of synaptic delay by a presynaptic Ca^{++} current in *Polyorchis* SMNs; bidirectional synapses in *Cyanea*) where there has been specialization to meet the particular needs of these radially symmetrical animals.

References

Anderson, P. A. V., 1979, Ionic basis of action potentials and bursting activity in the hydromedusan jellyfish *Polyorchis penicillatus*, *J. exp. Biol.* **78**:299.

Anderson, P. A. V., 1980, Epithelial conduction: its properties and functions, *Prog. Neurobiol.* **15**:161.

Anderson, P. A. V., 1985, Physiology of a bidirectional, excitatory chemical synapse, *J. Neurophysiol.* **53**:821.

Anderson, P. A. V., 1988, Evidence for quantal transmitter release at a cnidarian synapse, *Neurosci. Abstr.* **14**:1092.

Anderson, P. A. V., and Grünert, U., 1988, Three-dimensional structure of bidirectional, excitatory chemical synapses in the jellyfish *Cyanea capillata*, *Synapse* **2**:606.

Anderson, P. A. V., and Schwab, W. E., 1981, The organization and structure of nerve and muscles in the jellyfish *Cyanea capillata* (Coelenterata:Scyphozoa), *J. Morphol.* **170**:383.

Anderson, P. A. V., and Schwab, W. E., 1983, Action potential in neurons of the motor nerve net of *Cyanea* (Coelenterata), *J. Neurophysiol.* **50**:671.

Anderson, P. A. V., and Schwab, W. E., 1984, An epithelial-free preparation of the motor nerve-net of *Cyanea* (Coelenterata), *Biol. Bull.* **166**:396.

Anderson, P. A. V., and Spencer, A. N., 1989, The importance of cnidarian synapses for neurobiology. *J. Neurobiol.* **20**:435-457.

Arkett, S. A., and Spencer, A. N., 1986a, Neuronal mechanisms of a hydromedusan shadow reflex. I. Identified reflex components and sequence of events, *J. Comp. Physiol. A* **159**:201.

Arkett, S. A., and Spencer, A. N., 1986b, Neuronal mechanisms of a hydromedusan shadow reflex. II. Graded response of reflex components, possible mechanisms of photic integration, and functional significance, *J. Comp. Physiol. A* **159**:215.

Augustine, G. J., Charlton, M. P., and Smith, S. J., 1987, Calcium action in synaptic transmitter release, *Ann. Rev. Neurosci.* **10**:633.

Bennett, M. V. L., 1977, Electrical transmission: a functional analysis and comparison to chemical transmission, in: *Structure and Function of Synapses* (G. D. Pappas and D. P. Purpura, ed.), Raven Press, New York.

Boyle, M. B., Klein, M., Smith, S. J., and Kandel, E. R., 1984, Serotonin increases intracellular Ca^{++} transients in voltage clamped sensory neurons of *Aplysia californica*, *Proc. Natl. Acad. Sci.* **81**:7642.

Bullock, T. H., 1943, Neuromuscular facilitation in Scyphomedusae, *J. Cell. Comp. Physiol.* **22**:251.

Chain, B. M., Bone, Q., and Anderson, P. A. V., 1981, Electrophysiology of a myoid epithelium in *Chelophyes* (Coelenterata: Siphonophora), *J. Comp. Physiol.* **143**:329.

Chung, J. M., Spencer, A. N., and Gahm, K. H., 1989, Dopamine in tissues of the hydrozoan jellyfish *Polyorchis penicillatus* as revealed by HPLC and GC/MS, *J. Comp. Physiol. B* **159**:173-181.

Dunant, Y., 1986, On the mechanisms of acetylcholine release, *Prog. Neurobiol.* **26**:55.

Fain, G. L., Ishida, A. T., and Callery, S., 1983, Mechanisms of synaptic transmission in the retina, *Vis. Res.* **23**:1239.

Fuchs, P. A., Henderson, L. P., and Nicholls, J. G., 1982, Chemical transmission between individual Retzius and sensory neurones of the leech in culture, *J. Physiol. (Lond.)* **323**:195.

Getting, P. A., 1974, Modification of neuron properties by electronic synapses. I. Input resistance, time constant, and integration, *J. Neurophysiol.* **37**:846.

Grimmelikhuijzen, C. J. P., and Spencer, A. N., 1984, FMRFamide immunoreactivity in the nervous system of the medusa *Polyorchis penicillatus*, *J. Comp. Neurol.* **230**:361.

Grimmelikhuijzen, C. J. P., Graff, D., and Spencer, A. N., 1988a, Structure, location and possible actions of Arg-Phe-amide peptides in coelenterates, in: *Neurohormone in Invertebrates*, Soc. Exp. Biol. Sem. Ser. 33, pp. 199-217 (M. C. Thorndyke and G. J. Goldsworthy, ed.), Cambridge University Press.

Grimmelikhuijzen, C. J. P., Hahn, M., Rinehart, K. L., and Spencer, A. N., 1988b, Isolation of <Glu-Leu-Leu-Gly-Gly-Arg-Phe-NH_2 (Pol RFamide), a novel neuropeptide from hydromedusae. *Brain Res.* **475**:198.

Hama, K., 1959, Some observations on the fine structure of the giant nerve fibers of the earthworm *Eisenia foetida*, *J. Biophys. Biochem. Cytol.* **6**:61.

Hama, K., 1961, Some observations on the fine structure of the giant fibers of the crayfish (*Cambarus virilus* and *Camabarus clarkii*) with reference to the submicroscopic structure of the synapse, *Anat. Rec.* **141**:275.

Hille, B., 1984, *Ionic Channels of Excitable Membranes*, Sinauer Assoc., Sunderland, Mass.

Hochner, B., Klein, M., Schacher, S., and Kandel, E. R., 1986, Action-potential duration and the modulation of transmitter release from the sensory neurons of *Aplysia* in presynaptic facilitation and behavioral sensitization, *Proc. Natl. Acad. Sci.* **83**:8410.
Horridge, G. A., 1954, The nerves and muscles of medusae. I. Conduction in the nervous system of *Aurellia aurita* Lamarck, *J. exp. Biol.* **33**:366.
Horridge, G. A., 1968, *Interneurons*, W. H. Freeman, London and San Francisco.
Horridge, G. A., and Mackay, B., 1962, Naked axons and symmetrical synapses in coelenterates, *Quart. J. Micr. Sci.* **103**:531.
Hubbard, J. I., Llinas, R., and Quastel, D. M. J., 1969, *Electrophysiological Analysis of Synaptic Transmission*, Williams and Wilkins, Baltimore.
Jack, J. J. B., Noble, D., and Tsien, R. W., 1975, *Electric Current Flow in Excitable Cells*, Clarendon Press, Oxford.
Jha, R. K., and Mackie, G. O., 1967, The recognition, distribution and ultrastructure of hydrozoan nerve elements, *J. Morphol.* **123**:43.
Josephson, R. K., Reiss, R. F., and Worthy, R. M., 1961, A simulation study of a diffuse conducting system based on coelenterate nerve nets, *J. Theor. Biol.* **1**:460.
Josephson, R. K., and Schwab, W. E., 1979, Electrical properties of an excitable epithelium, *J. gen. Physiol.* **74**:213.
Kaneko, C. R. S., Merickel, M., and Kater, S., 1978, Centrally programmed feeding in *Helisoma*: Identification and characteristics of an electrically coupled pre-motor neuron network, *Brain Res.* **146**:1.
Kerfoot, P. A. H., Mackie, G. O., Meech, R. W., Roberts, A., and Singla, C. L., 1985, Neuromuscular transmission in the jellyfish *Aglantha digitale*, *J. exp. Biol.* **116**:1.
Kinnamon, J. C. and Westfall, J. A., 1982, Type of neurons and synaptic connections at hypostome-tentacle junctions in *Hydra*, *J. Morphol.* **173**:119.
Kohno, K., 1970, Symmetrical axo-axonic synapses in the axon cap of the goldfish Mauthner cell, *Brain Res.* **23**:255. Kusano, K., Livengood, D. R., and Werman, R., 1967, Tetraethylammonium ions: effect of presynaptic injection on synaptic transmission, *Science* **155**:1257.
Laughlin, S. B., 1981, Neural principles in the visual system, in: *Vision in Invertebrates. Handbook of Sensory Physiology*, Volume VII/6B (H. Autrum, ed.), Springer, Berlin, Heidelberg, New York.
Lin, J. W., and Faber, D. S., 1988, Synaptic transmission mediated by single club endings on the goldfish Mauthner cell. II. Plasticity of excitatory postsynaptic potentials, *J. Neurosci.* **8**:1313.
Loewenstein, W. R., 1975, Permeable junctions. *Symp. quant. Biol.* **40**:49.
Mackie, G. O., 1970, Neuroid conduction and the evolution of conducting tissue, *Quart. Rev. Biol.* **45**:319.
Mackie, G. O., 1978, Coordination in physonectid siphonophores, *Mar. Behav. Physiol.*, **5**:325.
Mackie, G. O., and Singla C. L., 1975, Neurobiology of *Stomotoca*. I. Action systems, *J. Neurobiol.* **6**:339.
Mackie, G. O., Anderson, P. A. V., and Singla, C. L., 1984, Apparent absence of gap junctions in two classes of Cnidaria, *Biol. Bull.* **167**:120.
MacVicar, B., and Dudek, F. E., 1981, Electronic coupling between pyramidal cells: a direct demonstration in rat hippocampal slices, *Science* **213**:782.
Martin, A. R., and Pilar, R., 1964, Quantal components of the synaptic potential in the ciliary ganglion of the chick, *J. Physiol. (Lond.)* **175**:1.
Mayer, A. G., 1906, Rhythmical pulsation in scyphomedusae, *Carnegie Inst. Wash. Publ.*, **47**:1.
McCarragher, G., and Chase, R., 1983, Morphological evidence for bidirectional chemical synapses, *Neurosci. Abst.* **9**:1025.
Pantin, C. F. A., 1935, The nerve-net of the Actinozoa. IV. Facilitation, *J. exp. Biol.* **12**:119.
Passano, L. M., 1965, Pacemakers and activity patterns in medusae: homage to Romanes, *Am. Zool.* **5**:465.
Peteya, D. J., 1973, A light and electron microscope study of the nervous system of *Ceriantheopsis americanus* (Cnidaria, Ceriantharia), *Z. Zellforsh. Mikroskop. Anat.* **141**:301.
Przysiezniak, J. and Spencer, A. N., 1989, Primary culture of identified neurones from a cnidarian, *J. exp. Biol.* **142**:97.
Roberts, A., and Mackie, G. O., 1980, The giant axon escape system of a hydrozoan medusa, *Aglantha digitale*, *J. exp. Biol.* **84**:303.

Satterlie, R. A., 1985, Central generation of swimming activity in the hydrozoan jellyfish *Aequorea aequorea*, *J. Neurobiol.* **16:**41.

Satterlie, R. A., and Spencer, A. N., 1979, Swimming control in a cubomedusan jellyfish, *Nature* **281:**141.

Schwab, W. E. and Anderson, P. A. V., 1980, Intracellular recordings of spontaneous and evoked electrical events in the motorneurons of the jellyfish *Cyanea capillata*, *Am. Zool.* **20:**941.

Spencer, A. N., 1974, Non-nervous conduction in invertebrates and embryos, *Amer. Zool.* **14:**917.

Spencer, A. N., 1979, Neurobiology of *Polyorchis*. II. Structure of effector systems, *J. Neurobiol.* **10:**95.

Spencer, A. N., 1981, The parameters and properties of a group of electrically coupled neurones in the central nervous system of a hydrozoan jellyfish, *J. exp. Biol.* **93:**33.

Spencer, A. N, 1982, The physiology of a coelenterate neuromuscular synapse, *J. Comp. Physiol.* **148:**353.

Spencer, A. N., 1988, Effects of Arg-Phe-amide peptides on identified motor neurons in the hydromedusa *Polyorchis penicillatus*, *Can. J. Zool.* **66:**639.

Spencer, A. N., and Arkett, S. A., 1984, Radial symmetry and the organization of central neurones in a hydrozoan jellyfish, *J. exp. Biol.* **110:**69.

Spencer, A.N., Przysiezniak, J., Acosta-Urquida, j. and Basarsky, T.A., 1989, Presynaptic spike broadening reduces junctional potential amplitude, *Nature* **340**:636-638.

Tauc, L., 1982, Non-vesicular release of neurotransmitter, *Phys. Rev.* **62:**857.

Takeuchi, A., and Takeuchi, N., 1970, Electrical changes in pre- and postsynaptic axons of the giant synapse of *Loligo*, *J. gen. Physiol.* **45:**1181.

Westfall, J. A., 1970, Synapses in a sea-anemone, *Metridium* (Anthozoa), *Electron Microsc. Proc. Int. Congr. 7th, Société Française de Microscopie Electronique, Paris* **3:**717.

Westfall, J. A., 1987, Ultrastructure of invertebrate synapses, in: *Nervous Systems in Invertebrates* (M. A. Ali, ed.), Plenum Press, New York and London.

Westfall, J. A., Yamataka, S., and Enos, P. D., 1970, Ultrastructure of synapses in *Hydra*, *J. Cell Biol.* **47:**226.

Westfall, J. A., Yamataka, S., and Enos, P. D., 1971, Ultrastructural evidence of polarized synapses in the nerve-net of *Hydra*, *J. Cell Biol.* **51:**318.

Yamashita, T., 1957, Das aktionspotential der Sinneskörper (Randkörper) der Meduse *Aurelia aurita*, *Z. Biol.* **109:**116.

Zucker, R. S., and Lando, L., 1986, Mechanism of transmitter release: voltage hypothesis and calcium hypothesis, *Science* **231:**574.

Chapter 4

Control of Morphogenesis by Nervous System-derived Factors

S. A. H. HOFFMEISTER and S. DÜBEL

1. Introduction

Intercellular communication is an essential acquisition of metazoans which allows a coordinated existence of individual cells in multicellular organisms. The invention of a nervous system during evolution made intercellular communication very rapid and effective. In addition to functioning to transmit changes in membrane potential, the role of the nervous system gains increasing interest in embryogenesis, and in regenerative processes. *Hydra* provides an ideal system to investigate the roots and the evolutionary development of these diverse tasks of the nervous system. Being evolutionary very old, and belonging to the first organisms to develop a nervous system, *Hydra* is an exciting animal for the study of the early functions of the nervous system.

In *Hydra*, the nervous system is directly involved in the development and maintenance of the ordered spatial and temporal patterns of cellular differentiation that result in morphogenesis, because morphogenetically-active substances are produced and released by nerve cells (Schaller et al., 1979). For pattern formation, basically two types of signals are required: positive signals which induce specific, local, differentiation events, and negative signals which prevent the system overshooting and locally restrict the inductive stimulus. To achieve this, the diffusion properties of substances responsible for induction should be limited, thereby ensuring a local response, whereas those substances serving as inhibitors should be easily diffusible and, therefore, be able to mediate communication between cells over greater distances

S. A. H. HOFFMEISTER and S. DÜBEL • Zentrum für Molekulare Biologie, Im Neuenheimer Feld 282, 6900 Heidelberg, Federal Republic of Germany.

Evolution of the First Nervous Systems
Edited by P.A.V. Anderson
Plenum Press, New York

(Gierer and Meinhardt, 1972; Kemmner, 1984). We have found that at least two inducing substances in *Hydra*, the head activator and the foot activator, are neuropeptides. The advantages and versatility of peptides for these functions will be illustrated in the following account, by describing the role of the head activator and foot activator in the development and cellular differentiation of *Hydra*. The special molecular properties of the head activator, namely its binding to large molecular weight carriers and self-inactivation by dimerization, predestine it for its function as a locally-acting signal for growth and differentiation. The two inducing substances are antagonized by at least two inhibiting substances, the head inhibitor and the foot inhibitor, both of which are nonpeptidergic, small, hydrophilic molecules. These molecular properties make them easily diffusible and enable them to act over comparatively long distances. All four substances are produced and secreted by nerve cells.

Hydra contains only a few cell types, with short differentiation pathways, and is very amenable to experimental manipulations, such as tissue grafting, regeneration and reaggregation of cells. Like all coelenterates, *Hydra* consists of only two cell layers, the ectoderm and the endoderm, which are separated by a collagenous extracellular structure, the mesogloea. Both layers are made up of epithelio-muscular cells. Interstitial cells and their derivatives are located in the intercellular space between the epithelial cells. It has to be emphasized that the stem cells are concentrated in the middle part of the animal, in the gastric column. The structures at the ends, the head with hypostome and tentacles, and the foot with peduncle and basal disc, contain predominately committed cells on their way to, or in, terminal differentiation. Therefore, upon cutting *Hydra* horizontally into two parts, gastric cells will differentiate to the specific cells of the missing structure, the head or the foot. In the case of epithelial cells, this differentiation process occurs within a single cell-cycle, which lasts 24-72 hours (David and Campbell, 1972). Foot-specific epithelial cells, for example, which are characterized by the production of a peroxidase activity (Hoffmeister and Schaller, 1985), develop from epithelial cells 22-24 hours after initiation of foot regeneration by cutting. Similarly, head-specific ectodermal epithelial cells, characterized by a head-specific monoclonal antibody (Javois et al., 1986), appear at about the same time after initiation of head regeneration by cutting, each marker thereby being restricted to the appropriate regenerating tissue (Dübel et al., 1987). Ectodermal and endodermal epithelial cells do not interconvert and must, therefore, derive from different gastric stem cells (Smid and Tardent, 1982).

In *Hydra*, the stem cell population of interstitial cells is especially interesting because these cells are able to differentiate to nerve cells, nematocytes, endodermal gland and mucus cells, and, in the sexual cycle, also to oocytes and sperm cells (David and Gierer, 1974; David and Murphy, 1977; Bode and David, 1978; Bode et al., 1987; Bosch and David, 1987). Nerve cells differentiate from stem cells within a single cell-cycle, one stem cell giving rise to two mature nerve cells. In contrast, differentiation to nematocytes requires 2-4 cell cycles, leading to 4, 8, 16, or 32 clonally-derived nematocytes. The differentiation of both interstitial and epithelial stem cells is

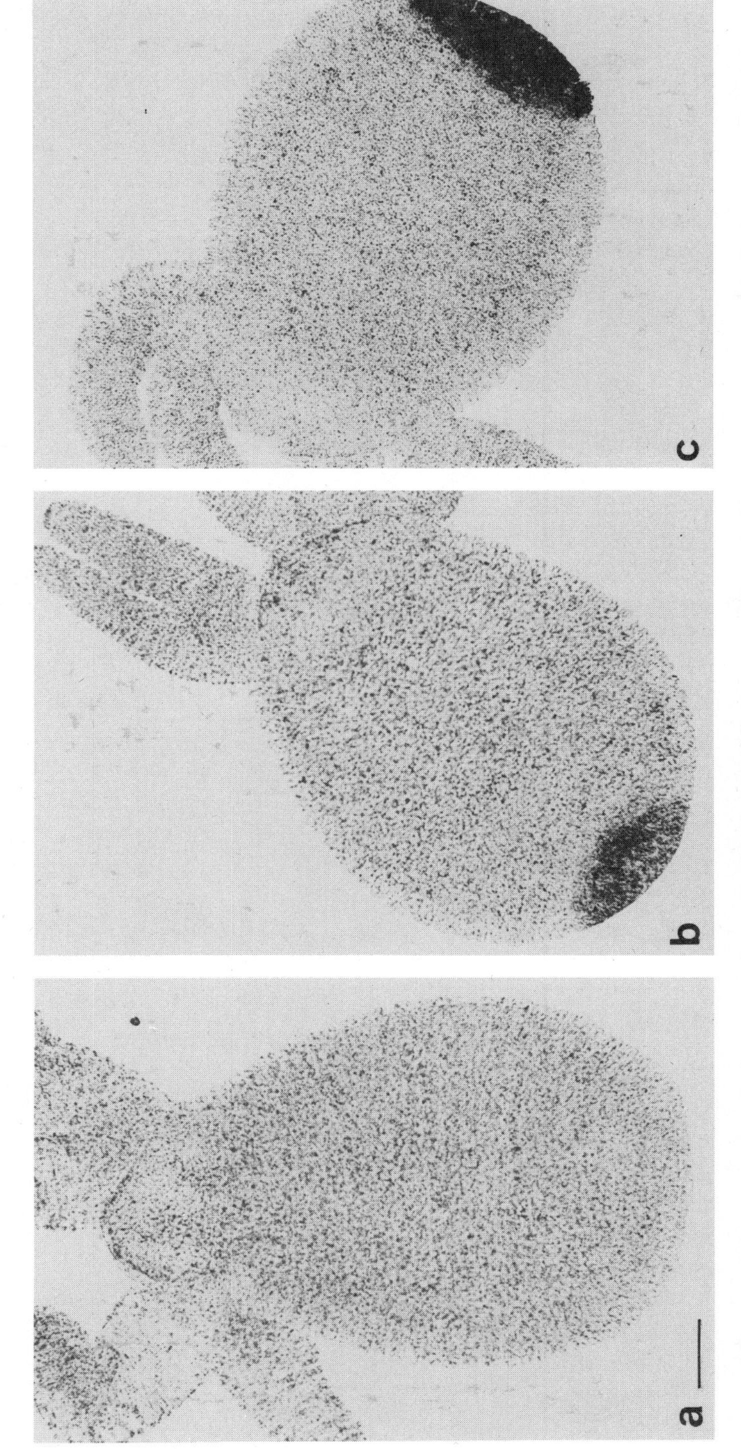

Figure 1. Reappearance of the peroxidase activity during foot regeneration in *Hydra attenuata*. Whole mount preparations were stained with diaminobenzidine in the presence of H_2O_2 **(a)** 20 hours, **(b)** 26 hours, and **(c)** 36 hours after the foot was cut off. Bar corresponds to 500 μm.

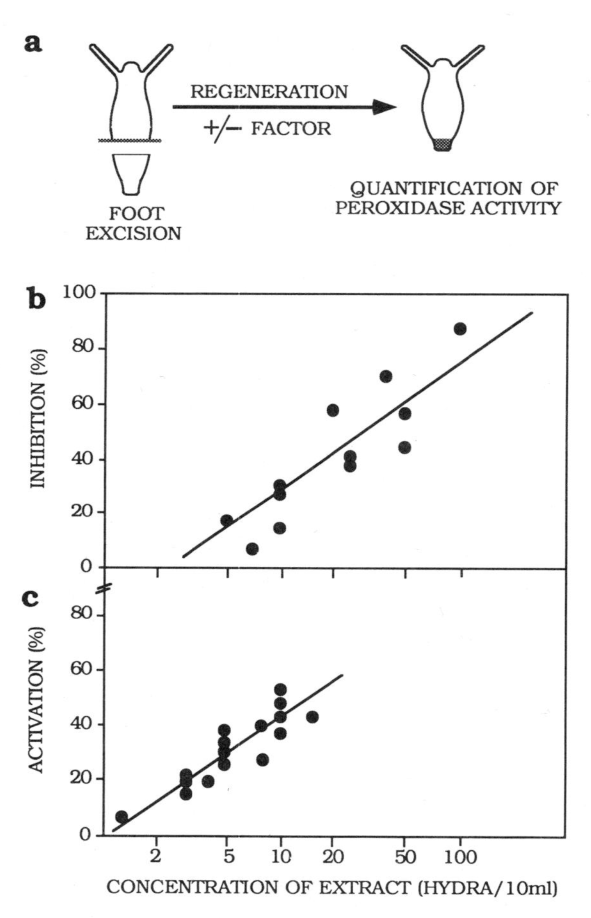

Figure 2. Biological assays for the foot factors. **(a)** Experimental procedure. b) and c) effect of increasing concentrations of foot inhibitor **(b)** and foot activator **(c)** on foot regeneration. The decrease or increase of peroxidase activity, relative to the untreated controls, was determined 24 hours **(b)** or 22 hours **(c)** after foot excision. Each point is the mean of 60 regenerating animals, assayed on the same day.

dependent on their position within the animal. Under steady-state conditions, a constant ratio between stem cells and differentiating cells is strictly maintained (Bode and Flick, 1976; Bode et al., 1976).

2. Assay Systems for Head and Foot Factors

Quantitative assays are an essential prerequisite for the purification of biologically active substances. The effect of substances which induce or inhibit head or foot formation in *Hydra* can be measured as acceleration or inhibition of head or foot regeneration (Schaller et al., 1979). An example for the time course of foot regeneration in animals of *Hydra attenuata* is shown in figure 1. The time-dependent increase in the peroxidase activity-containing foot mucus cells, in foot regenerating animals is shown here by staining of whole-mount preparations with diaminobenzidine in the presence of H_2O_2. The amount of peroxidase-like activity of foot mucus cells can be quantified photometrically by the use of a soluble substrate for peroxidases (Hoffmeister and Schaller, 1985). Determination of the enzyme activity, therefore, provides a rapid and reliable test for quantifying the differentiation of foot-specific cells. For assays quantifying the amount of head and/or foot factors in a given extract, heads or feet, respectively, are removed from intact animals, and the parts regenerating the respective extremity are incubated in medium containing the extracts to be assayed. The amount of the respective factors is measured as relative activation or inhibition over the controls. Figure 2 shows dose response curves for the foot activator and the foot inhibitor. In these assays, regenerated, foot-specific cells are measured 22 hours and 24 hours, respectively, after cutting off the feet, by quantifying the peroxidase-activity. In the presence of foot inhibitor, the reappearance of the peroxidase-activity and, thus, of foot-specific epithelial cells (foot mucus cells) is retarded, while it is accelerated in the presence of foot activator. Similarly, head inhibitor retards head regeneration, whereas, in the presence of head activator, the rate of head regeneration is accelerated. The respective effects are dose-dependent and specific (Schaller et al., 1979). This means, for example, that the head activator accelerates head regeneration, but not foot regeneration. The head factors not only influence head regeneration, but also affect bud induction and outgrowth.

3. Biochemical Characterization of the Factors

Using the assays described above, it has been possible to separate the four substances from each other. The two activators have been purified extensively, the two inhibitors to a lesser extent (Table 1). Both activators turn out to be small peptides, with molecular weights of around 1000. The sequence of the quite hydrophobic head activator peptide was determined to be pGlu-Pro-Pro-Gly-Gly-Ser-Lys-Val-Ile-Leu-Phe

Table 1. Properties of head and foot factors from *Hydra*

Morphogen	Molecular mass (D)	Purification (x-fold)	Active concentration
Head activator	1124	10^9	10^{-13} M
Head inhibitor	<500	10^5	$<10^{-9}$ M
Foot activator	~1000	10^8	$<10^{-12}$ M
Foot inhibitor	<500	10^4	$<10^{-8}$ M

(Schaller and Bodenmüller, 1981). Both activators can be separated using reverse phase columns from which foot activator elutes under neutral pH conditions in 25% methanol. In the same system, elution of head activator requires 60% methanol. This difference shows that foot activator is a more polar molecule than head activator. We are currently determining the amino acid sequence of this peptide. The two inhibitors are both non-peptidergic, hydrophilic molecules with molecular weights of less than 500. They are more difficult to purify because, besides the fact that they are slightly positively charged, we have not yet found any peculiar biochemical properties which could help to distinguish them from other small molecules.

The inhibitor fractions are sufficiently pure for biological experiments, but since all four factors occur in the animals in very low concentrations, we still have not obtained enough pure material for a chemical analysis of the inhibitors. As shown in Table 1, the activators are active at picomolar and the inhibitors at nanomolar concentrations. Therefore, any one animal needs and contains very small amounts of the respective substances. Head activator, for instance, occurs at a concentration of less than one femtomole per *Hydra attenuata*. Since at least one nanomole is required for a chemical analysis and with a yield of 10% in the ideal case, at least 10^7 *Hydra* have to be extracted. Sequence analysis of the head activator peptide was made possible by the fact that head activator also occurs in other coelenterates; 200 kg of the sea anemone *Anthopleura elegantissima* were processed to obtain 20 nanomoles (20 μg) of pure head activator (Schaller and Bodenmüller, 1981).

4. Action of Activators and Inhibitors in *Hydra*

In *Hydra*, all four substances occur as gradients, with maximal concentration of the head factors in the head region, and of the foot factors in the basal disc (Schaller et al., 1979). The fact that all four factors copurify with nerve cells was taken as evidence that, in normal *Hydra*, they all are products of nerve cells (Schaller and Gierer, 1973; Berking, 1977; Grimmelikhuijzen, 1979; Schmidt and Schaller, 1980). This has been confirmed by copurification of head activator with neurosecretory granules (Schaller and Gierer, 1973). Furthermore, putative head activator precursors have been localized immunocytochemically in developing nerve cells (Schawaller et al., 1988).

To become active on their target cells, all four factors have to be released from the nerve cells. The site where a certain structure will be induced is dependent on the amounts of released inhibitors, which, in turn, regulate the release of the activators, and their own release. For the head factors, we have shown that very low concentrations of head inhibitor inhibit the release of head activator, whereas at least 20-fold higher concentrations are required to inhibit its own release (Schaller, 1976c; Kemmner and Schaller, 1984). This control of release by the inhibitors prevents an overshoot of the inducing effects of the activators. Thus, under steady-state conditions, the high concentration of head inhibitor in the head prevents the release of large amounts of head activator from the highly concentrated head activator sources located in the head. The formation of another head close to the already existing one is, therefore, not possible. Further down the body axis, neurons containing both head inhibitor or head activator are less abundant. Because cells in the gastric region of *Hydra* proliferate continuously, the animal becomes larger and the concentration of head inhibitor is reduced by the lengthwise growth. At a certain distance from the head, a distance which is determined by the total amount of head inhibitor and the steepness of the gradient of inhibitor-containing neurons, the density of head inhibitor sources becomes so low that head activator can be released from some head activator-containing nerve cells in this area.

A direct consequence of this locally restricted release of head activator is the induction of a bud. The onset of budding is indicated by an accumulation of interstitial cells (Moore and Campbell, 1973), and this is followed by an increase in the density of nerve cells in the budding area (Bode et al., 1973). Due to the rearrangement and proliferation of gastric epithelial cells (Graf and Gierer, 1980), a bud protrusion is formed and this then develops tentacle buds and a hypostome. A constriction then occurs between the parent body and the bud, and foot mucus cells begin to appear in the junction between the parent animal and the bud (Hoffmeister and Schaller, 1985). Finally, the constriction narrows and the bud detaches.

The fact that head inhibitor effects size regulation and budding in *Hydra* is illustrated by the morphology and budding behavior of two mutants, mini and maxi. Maxi mutants have a high head inhibitor content, with the result that the distance between head and bud becomes very large, resulting in a maxi phenotype. In mutants with low head inhibitor content, the head/bud distance is smaller, leading to mini animals (Schaller et al., 1977). This implies that the larger head inhibitor concentration in maxi seems to inhibit a greater area. Moreover, it demonstrates that the role of head inhibitor in size regulation is due to its ability to diffuse over long distances.

The inhibitors also contribute to the maintenance of axial polarity, and this becomes particularly obvious during head and foot regeneration. If a *Hydra* is cut horizontally into two pieces of equal size, the upper half will regenerate a foot and the lower half will regenerate a head. Head factors are maximally concentrated in the head, foot factors in the foot. In the upper half, therefore, the density of head inhibitor sources is very high; the density of foot inhibitor sources very low. The first result of cutting is release of both inhibitors. The release of head inhibitor at the cut surface is not very significant, because the resulting decrease in the overall concentration of head inhibitor in the tissue of the upper half is quickly overcome by new release of head inhibitor from the highly abundant sources of the head. This leads to a large increase in the concentration of free (released) head inhibitor in the tissue. High concentrations of head inhibitor inhibit its own release and, at even lower concentrations, inhibit the release of head activator (Schaller, 1976c; Kemmner and Schaller, 1984). Consequently, the release of head inhibitor and head activator are blocked at the cut surface of the upper half. In contrast, the foot inhibitor concentration decreases markedly in the tissue at the cut surface (Schmidt and Schaller, 1980; Kemmner and Schaller, 1984) because, in the upper half of the animal, foot inhibitor sources are not very dense and the loss of foot inhibitor at the cut surface cannot be replenished. Because it is hydrophilic, the released foot inhibitor quickly diffuses out of the tissue, thus permitting the release of foot activator. Since foot activator is released in a high molecular weight form (unpublished results), it will not diffuse as quickly as foot inhibitor, but rather persists for a longer time at the cut surface. Thus, it will be able to induce foot-specific differentiation processes, which will lead to the regeneration of a new foot. In the headless, lower half of the cut animal, foot inhibitor is present in very high amounts, since the maximal density of foot inhibitor containing nerve cells occurs in the foot. The concentration of foot inhibitor at the cut surface does not decrease markedly because of replenishment from the sources in the foot, and consequently, the release of foot activator is not allowed. However, because head inhibitor is rapidly depleted by its diffusion from the cut surface, the release of head activator is permitted, allowing the induction of head-specific differentiation processes. In this way, the distribution of the various activating and inhibiting factors, their respective chemical properties, and their mode of interaction guarantee the maintenance of the polar organization of the animal.

5. Activators are Co-released with Carrier Molecules

The difference between the molecular weight of head activator (1124 Daltons) and head inhibitor (about 500 Daltons) does not explain their strikingly different diffusion properties. The explanation for the different diffusion behavior resides in the fact that head inhibitor is released in its low molecular weight form, whereas head activator is released bound to a large molecular weight carrier (Schaller et al., 1986). This high

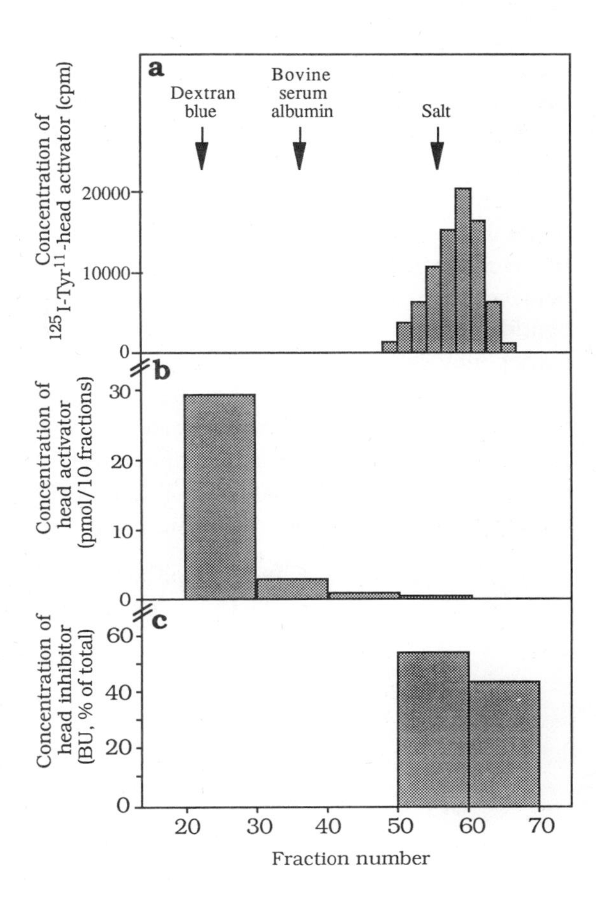

Figure 3. Gel filtration of medium collected from head-regenerating animals on a Sephacryl S-300 equilibrated with 25 mM ammonium bicarbonate pH 7.5, at 4°C (total volume 40 ml, 0.94 ml/fraction). **(a)** For the determination of the position where the free head activator peptide elutes from the column, [^{125}I] Tyr^{11} head activator was applied to the column. It elutes from the S-300 column in the low molecular mass region. In **(b)** and **(c)**, medium collected from head regenerating *Hydra* was applied to the column. **(b)** For the determination of the head activator concentration, fractions were lyophilized in batches of 10, extracted with methanol and purified over Sep Pak C_{18}, cartidges. The head activator peptide content was determined in the radioimmunoassay. **(c)** For the determination of the head inhibitor concentration, the S-300 fractions were tested directly in the biological assay.

molecular weight form of head activator, is detectable when head activator is extracted from *Hydra* tissue, using aqueous solvents, or if medium is collected from animals that are regenerating a head, and elutes from S-300 columns with an apparent molecular weight of several hundred kD (Fig. 3). Head activator binds to the carrier non-covalently, as demonstrated by the fact that following extraction with organic solvents such as methanol, or by treatment with 2 M NaCl, head activator peptide can be regained in its low molecular weight form (Schaller et al., 1986). The carrier-bound head activator is active in biological assays at a concentration of 10^{-13} M, indicating that binding to its carrier does not inhibit its biological effect. On the contrary, binding to the carrier seems to stabilize the biologically active form of the peptide by preventing the formation of biologically inactive peptide dimers (Bodenmüller et al., 1986). Furthermore, the head activator peptide is protected against enzymatic degradation when it is bound to its carrier (Roberge et al., 1984).

The potency of the head activator/carrier complex means either that its receptor binds to the complex, or that the the affinity of the head activator peptide for its receptor is much higher than its affinity to its carrier, with the result that the peptide detaches from its carrier upon encountering the receptor. These properties of the peptide provide an additional guarantee that its action is local; when bound to the carrier molecules, the diffusion range of the peptide is greatly reduced, but following its release from the carrier molecule, the peptide is quickly inactivated by dimerization or enzymatic degradation. In contrast, head inhibitor, which has a global action and a shorter half-life than head activator (Kemmner and Schaller, 1984), is released in a low molecular weight form (Schaller et al., 1986). This explains why head inhibitor diffuses rapidly over great distances and, therefore, exerts its effects over longer distances in

Table 2. Effect of foot activator on the mitotic index of single big interstitial cells and those in nests of 2*

Concentration of FA	Mitotic index interstitial cells
[BU]	[%]
0	4.2 ±0.6
0.75	4.7 ±0.3
1.00	5.5 ±0.3
2.00	5.9 ±0.3

*measured after an incubation time of 1.75 hr. The numbers given in the table are mean values ± standand deviation. BU: biological units.

the animal. Foot inhibitor is also released in a low molecular weight form, as evidenced by assaying S-300 column fractions for their biological activity. Foot activator, however, is released in a high molecular weight form in the same manner as head activator. Therefore, the binding to high molecular weight carrier molecules seems to be a general principle in *Hydra*. Binding to carrier molecules ensures local action, by preventing long range diffusion, and a longer half-life of activators over inhibitors, by providing protection from enzymatic degradation. The carrier molecules, thereby, provide the basis for a local action of inducing substances.

6. Action at the Cellular Level

Activators have two functions at the cellular level: they act as mitogens, and as terminal differentiation-inducing substances (Schaller, 1976a,b; Holstein et al., 1986; Hoffmeister and Schaller, 1987; Hoffmeister, 1989).

In *Hydra*, there are three major stem cell populations: the ectodermal epithelial stem cells, endodermal epithelial stem cells, and interstitial stem cells. Ectodermal epithelial stem cells can give rise to the terminally differentiated battery cells of the tentacles. During regeneration and budding they can also differentiate into the ectodermal epithelial cells of the hypostome. Under steady-state conditions this pathway is absent since the hypostome contains its own ectodermal epithelial stem cell population (Dübel et al., 1987). In addition to this head-specific differentiation, the ectodermal epithelial stem cells can also terminally differentiate to foot-specific cells, the foot mucus cells. Interstitial stem cells are able to differentiate into nerve cells or nematocytes, and can also give rise to gland cells (in the endoderm) and endodermal hypostomal mucus cells. They can also differentiate into gametes in response to environmental stresses, such as cold (4-10°C).

Head activator and foot activator act as mitogens on all proliferating cell types, such as epithelial cells and interstitial cells, and on those interstitial cells committed to the nerve cell or nematocyte pathway (Schaller, 1976a; Hoffmeister, 1989). Both activators exert their mitogenic effect on cells in the G_2 phase of the cell cycle, as evidenced by a very fast increase of the mitotic index 1-3 h after the addition of the activators. This mode of action distinguishes them from most of the known growth factors which act during the transition between G_0 to G_1. Table 2 gives an example of the effect of foot activator on interstitial cells, showing that this effect is dose-dependent. The same is true for head activator (Hoffmeister and Schaller, 1987). Head inhibitor acts antagonistically by blocking mitosis of proliferating cells, such as epithelial cells, interstitial cells, and those interstitial cells which are on their way to differentiate into nerve cells or nematocytes (Berking, 1974).

In addition to their mitogenic effect, the activators also stimulate terminal differentiation processes. Ectodermal epithelial stem cells differentiate into head-specific epithelial cells in the presence of head activator, but in the presence of foot activator, they differentiate into foot mucus cells. The differentiation of nerve cells is also influenced by both factors. Interstitial stem cells have to become determined for nerve cell development. Head activator, but not foot activator (Hoffmeister, 1989), acts as a signal for nerve cell development, and the interstitial cells are sensitive to this stimulus during the early S-phase (Schaller, 1976b) and this effect of head activator is antagonized by high concentrations of head inhibitor (Berking, 1979). In the subsequent G_2-phase, the nerve cell precursors require a second stimulus to go through a final mitosis and terminally differentiate into two mature nerve cells per nerve cell precursor (Hoffmeister and Schaller, 1987). This second signal can be the presence of either head activator or foot activator, the former leading to the differentiation of nerve cells which are head-specific, the latter to the maturation of nerve cells which are foot-specific. Once again, head inhibitor acts antagonistically, keeping these cells arrested in G_2. Thus, both head activator and foot activator are used as signals for the final differentiation. Head inhibitor, and maybe also foot inhibitor, block the final differentiation.

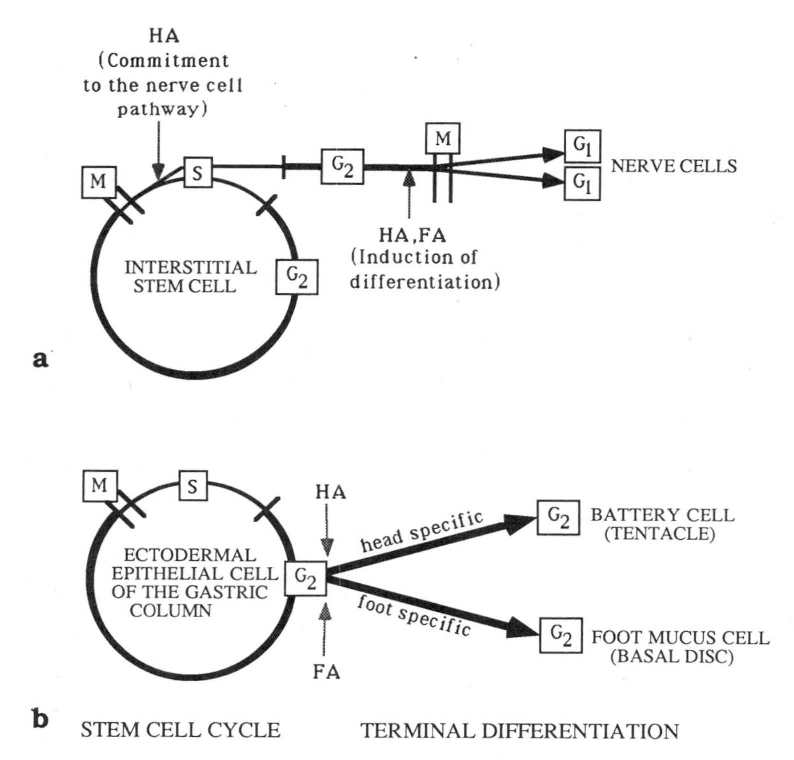

Figure 4. Action of head activator and foot activator on **(a)** nerve cell development and **(b)** ectodermal, epithelial stem cells.

These findings support the following model: in gastric tissues adjacent to the head region, where head activator concentration is highest, many interstitial stem cells are determined to become nerve cells and complete terminal differentiation as head-specific nerve cells. Further from the head, fewer stem cells are determined to become nerve cells, and most are prevented from terminal differentiation into nerve cells by head inhibitor. As a consequence of the gradual displacement of tissue from the subhypostomal to the basal region, these cells reach the foot where the concentration of foot activator is highest, and under the influence of the foot activator they differentiate into nerve cells of the foot (Fig. 4a). Since head activator and head inhibitor are produced by nerve cells of the head, and foot activator and foot inhibitor by nerve cells located in the foot, this regulation of nerve cell differentiation represents an autocrine control loop which ensures that head-specific differentiation is maintained in the head region, whereas foot-specific differentiation is restricted to the foot.

The polar distribution of the head and foot factor-containing nerve cells establishes a concentration gradient of head factors, which decreases in the head to foot direction, and an opposite gradient of foot factors, which decreases in the foot to head direction. The position of the epithelial stem cells within the body axis of the animals, relative to the two concentration gradients, determines whether they remain stem cells or undergo either head- or foot-specific differentiation. During tissue movement (Campbell, 1967) those epithelial stem cells which enter the head region will encounter higher concentrations of head activator and become head-specific epithelial cells; whereas those which enter the foot region will undergo foot-specific differentiation because of the presence there of the high foot activator (Fig. 4b).

Furthermore, the mode of action of head and foot activator on nerve cell development explains the existence of a dominance of the head system over the foot system that has been proposed by several authors (Hicklin et al., 1973; Hicklin and Wolpert, 1973; Bode and Bode, 1980). Only head activator is able to stimulate the first decisive step, the determination of nerve cells. Later, at the second control point during nerve cell development, foot activator can also act as a differentiation-inducing factor, by specifying the future nerve cell as head- or foot-specific.

ACKNOWLEDGEMENTS. This work was carried out in the lab of C. Schaller. The authors wish to thank C. Schaller for critical reading of the manuscript and I. Baro for typing. We are supported by the Deutsche Forschungsgemeinschaft (SFB 317), by the Bundesministerium für Forschung und Technologie (BCT 365/1), and by the Fonds der Deutschen Chemischen Industrie.

References

Berking, S., 1974, *Nachweis eines morphogenetisch aktiven Hemmstoffs in* Hydra attenuata *und Untersuchung seiner Eigenschaften und Wirkungen*, Doctoral thesis, Ekerhard-Karls Universität, Tübingen.

Berking, S., 1977, Bud formation in hydra: inhibition by an endogenous morphogen, *Wilhelm Roux's Arch.* **181**:215-225. Berking, S., 1979, Control of nerve cell formation from multipotent stem cells in hydra, *J. Cell Sci.* **40**:193-205.

Bode, H., Berking, S., David, C. N., Gierer, A., Schaller, H., and Trenkner, E., 1973, Quantitative analysis of cell types during growth and morphogenesis in hydra, *Wilhelm Roux's Arch.* **171**:269-285.

Bode, H. R., and Flick, K. M., 1976, Distribution and dynamics of nematocyte populations in *Hydra attenuata*, *J. Cell Sci.* **21**:15-34.

Bode, H. R., Flick, K. M., and Smith G. S., 1976, Regulation of interstitial cell differentiation in *Hydra attenuata*. I. Homeostatic control of interstitial cell population size, *J. Cell Sci.* **20**:29-46.

Bode, H. R., and David, C. N., 1978, Regulation of a multipotent stem cell, the interstitial cell of hydra, *Progr. Biophys. Mol. Biol.* **33**:198-206.

Bode, P. M., and Bode, H. R., 1980, Formation of patterns in regeneration tissue pieces of *Hydra attenuata*. I. Head-body proportion regulation, *Dev. Biol.* **78**:484-496.

Bode, H. R., Heimfeld, S., Chow, M. A., and Huang, L. W., 1987, Gland cells arise by differentiation from interstitial cells in *Hydra attenuata*, *Dev. Biol.* **122**:577-585.

Bodenmüller, H., Schilling, E., Zachmann, B., and Schaller, H. C., 1986, The neuropeptide head activator loses its biological activity by dimerisation, *EMBO J.* **5**:1825-1829.

Bosch, T., and David, C. N., 1987, Stem cells of *Hydra magnipapillata* can differentiate somatic cells and germ line cells, *Dev. Biol.* **121**:182-191.

Campbell, R. D., 1967, Tissue dynamics of steady state growth in *Hydra littoralis*. III. Behaviour of specific cell types during tissue movements, *J. Exp. Zool.* **164**:379-391.

David, C. N., and Campbell, R. D., 1972, Cell cycle kinetics and development of *Hydra attenuata*. I. Epithelial cells. *J. Cell Sci.* **11**:557-568.

David, C. N., and Gierer, A., 1974, Cell cycle kinetics and development of *Hydra attenuata*. III. Nerve and nematocyte differentiation, *J. Cell Sci.* **16**:359-375.

David, C. N., and Murphy, S., 1977, Characterisation of interstitial stem cells in hydra by cloning, *Dev. Biol.* **58**:373-383.

Dübel, S., Hoffmeister, S. A. H., and Schaller. H. C., 1987, Differentiation pathways of ectodermal epithelial cell in hydra, *Differentiation* **35**:181-189.

Dübel, S., 1989, Differentiation in the head of hydra, *Differentiation*, in press.

Gierer, A., and Meinhardt, H., 1972, A theory of biological pattern formation, *Kybernetik* **12**:30-39.

Graf, L., and Gierer, A., 1980, Size, shape and orientation of cells in budding hydra and regulation of regeneration in cell aggregates, *Wilhelm Roux's Arch.* **188**:141-151.

Grimmelikhuijzen, C. J. P., 1979, Properties of the foot activator from hydra, *Cell Differ.* **8**:267-273.

Hicklin, J., and Wolpert, L., 1973, Positional information and pattern regulation in hydra: formation of the foot end. *J. Embryol. exp. Morph.* **30**:727-740.

Hicklin, J., Hornbruch, A., Wolpert, L., and Clarke, M., 1973, Positional information and pattern regulation in hydra: the formation of boundary regions following axial grafts, *J. Embryol. exp. Morph.* **30**:701-725.

Hoffmeister, S. A. H., 1989, Action of foot activator on growth and differentiation of cells in hydra, *Dev. Biol.* **133**:254-261.

Hoffmeister, S. A. H., and Schaller, H. C., 1985, A new biochemical marker for foot-specific cell differentiation in hydra, *Wilhelm Roux's Arch.* **194**:433-461.

Hoffmeister, S. A. H., and Schaller, H. C., 1987, Head activator and head inhibitor are signal for nerve cell differentiation in hydra, *Dev. Biol.* **122**:72-77.

Holstein, T., Schaller, H. C., and David, C. N., 1986, Nerve cell differentiation in hydra requires two signals, *Dev. Biol.* **115**:9-17.

Javois, L., Wood, R. D., and Bode, H. R., 1986, Patterning of the head in hydra as visualised by a monoclonal antibody, *Dev. Biol.* **117**:607-618.

Kemmner, W., 1984, A model of head regeneration in hydra, *Differentiation* **26**:83-90.

Kemmner, W., and Schaller, H. C., 1984, Actions of head activator and head inhibitor during head regeneration in hydra, *Differentiation* **26**:91-96.

Moore, L. B., and Campbell, R. D., 1973, Bud initiation in a non-budding strain of hydra: role of interstitial cells, *J. Exp. Zool.* **184**:397-407.

Roberge, M., Escher, E., Schaller, H. C., and Bodenmüller, H., 1984, The hydra head activator in human blood circulation. Degradation of the synthetic peptide by plasma angiotensin-converting enzyme, *FEBS Lett.* **173:**307-313.

Schaller, H. C., and Gierer, A., 1973, Distribution of the head activating substance in hydra and its localisation in membranous particles in nerve cells, *J. Embryol. exp. Morph.* **29:**39-52.

Schaller, H. C., 1976a, Action of the head activator as a growth hormone in hydra, *Cell Diff.* **5:**1-11.

Schaller, H. C., 1976b, Action of the head activator on the determination of interstitial cells in hydra, *Cell Diff.* **5:**13-20.

Schaller, H. C., 1976c, Head regeneration in Hydra is initiated by the release of head activator and inhibitor, *Wilhelm Roux Archiv.* **180**:287-295.

Schaller, H. C., Schmidt, T., Flick, K., and Grimmelikhuijzen, C. J. P., 1977, Analysis of morphogentic mutants of hydra. II. The non-budding mutant, *Wilhelm Roux's Arch.* **183:**207-214.

Schaller, H. C., Schmidt, T., Flick, K., and Grimmelikhuijzen, C. J. P., 1977, Analysis of morphogenetic mutants of hydra. III Maxi and mini, *Wilhelm Roux's Arch.* **183:**215-222.

Schaller, H. C., Schmidt, T., and Grimmelikhuijzen, C. J. P., 1979, Separation and specificity of action of four morphogens from hydra, *Wilhelm Roux's Arch.* **186:**139-149.

Schaller, H. C., and Bodenmüller, H., 1981, Isolation and amino acid sequence of a morphogenetic peptide from hydra, *Proc. Natl. Acad. Sci. USA* **78:**7000-7004.

Schaller, H.C., Roberge, M., Zachmann, B., Hoffmeister, S., Schilling, E., and Bodenmüller, H., 1986, The head activator is released from regenerating hydra bound to a carrier molecule, *EMBO J.* **5:**1821-1824.

Schawaller, M., Schenk, K., Hoffmeister, S. A. H., Schaller, H., and Schaller, H. C., 1988, Production and characterisation of monoclonal antibodies recognizing head activator in precursor form and immunocytochemical localisation of head activator precursor and head activator peptide in the neural cell line NH15-CA2 and in hydra, *Differentiation* **38:**149-160.

Schmidt, T., and Schaller, H. C., 1980, Properties of the foot inhibitor from hydra, *Wilhelm Roux's Arch.* **188:**133-139.

Smid, I., and Tardent, P.. 1982, The influences of ecto- and endoderm in determining the axial polarity of *Hydra attenuata* Pall. (*Cnidaria, Hydrozoa*), *Wilhelm Roux's Arch.* **191:**64-67.

Chapter 5

Differentiation of a Nerve Cell-Battery Cell Complex in *Hydra*

ENGELBERT HOBMAYER and CHARLES N. DAVID

1. Introduction

Complex cell-cell interactions appeared early in the evolution of metazoans. One of the most interesting examples of such complexity is the battery cell in tentacles of cnidarians. This cell consists of a modified ectodermal epithelial cell which has nematocytes and sensory nerve cells embedded in it. To investigate the formation of this complex, we use the simple fresh water cnidarian *Hydra*. In this organism, epithelial cells of the gastric region are continuously displaced into tentacles (Campbell, 1967; Dübel et al., 1987), where they interact with sensory nerve cells and nematocytes to form battery cells.

Using the monoclonal antibody NV1 as a marker for tentacle-specific nerve cells (Hobmayer et al., in preparation) we have investigated formation of tentacle tissue on a cellular level. Formation of a NV1-battery cell complex occurs during head formation and is stimulated by treatment with the neuropeptide head activator (HA) (Schaller and Bodenmüller, 1981), which has been shown to stimulate tentacle (Schaller, 1973) and nerve cell formation (Holstein et al., 1986) in *Hydra*. Differentiation of NV1 immunoreactive (NV1+) nerve cells, however, does not appear to be stimulated directly by HA, but rather by cell-cell interactions with battery cell precursors during tentacle formation.

ENGELBERT HOBMAYER and CHARLES N. DAVID • Zoologisches Institut der Universität München, Luisenstrasse 14, 8000 München 2, Federal Republic of Germany.

Evolution of the First Nervous Systems
Edited by P.A.V. Anderson
Plenum Press, New York

2. Morphology of Battery Cells

Battery cells in the tentacles of *Hydra* constitute an association of different cell types (Hufnagel et al., 1985). As shown schematically in figure 1E, 15-20 nematocytes and one epidermal sensory nerve cell are embedded in an ectodermal epitheliomuscular cell, in a typical arrangement: one stenotele or one or two isorhizas lie in the center of a ring of desmonemes. The body of the sensory nerve cell is located to the side of the central nematocyte.

Using a monoclonal antibody, NV1, we were able to identify these tentacle-specific nerve cells in *H. magnipapillata* (Hobmayer et al., in preparation). With the exception of a few ganglion cells in the lower peduncle, no NV1+ cells occur in the rest of the body column. In *H. oligactis*, the same type of nerve cell is recognized by the monoclonal antibody JD1 (Dunne et al., 1985).

Based on *in situ* observations, using indirect immunofluorescence, on either NV1-stained whole mounts or maceration preparations NV1+ cells can be classified as bipolar and multipolar epidermal sensory nerve cells (Fig. 1A; Yu et al., 1986). They have an apical cilium which extends to the surface of the surrounding epithelial cell. Two or more processes extend laterally from the basal part of the cell body (Fig. 1C). They run along the base of the battery cell adjacent to the mesoglea and innervate several neighboring battery cells; short sidebranches make contact with the battery cell's nematocytes (Fig. 1A,B).

3. Development of NV1+ Nerve Cells During Head Formation

In both budding and head regeneration, the first NV1+ cells appear at the time of evagination of short tentacle tips (Fig. 2). Earlier stages of head formation, when the prospective head is only discernible as a rounded protrusion, contain no NV1+ cells and no battery cells. During outgrowth of tentacles, the density of newly formed NV1+ cells remains constant along the entire length of the tentacles. Thus, in general, differentiation of NV1+ cells shows a strong correlation with the formation of battery cells.

This dependence of NV1+ differentiation on battery cell formation is also clearly demonstrated in a regeneration deficient mutant, reg-16 (Sugiyama and Fujisawa, 1977). Animals of strain reg-16 are blocked at an early stage of head regeration, and do not form tentacles. To investigate whether such animals form NV1+ cells during head regeration, it was necessary to introduce interstitial cells of *H. magnipapillata* wild-type strain into reg-16, because reg-16 nerve cells do not express the NV1 antigen. Such reg-16/105 chimeras are defective in head regeration, like the reg-16 parent (Wanek et al., 1986), but can differentiate NV1+ nerve cells from wild-type strain 105

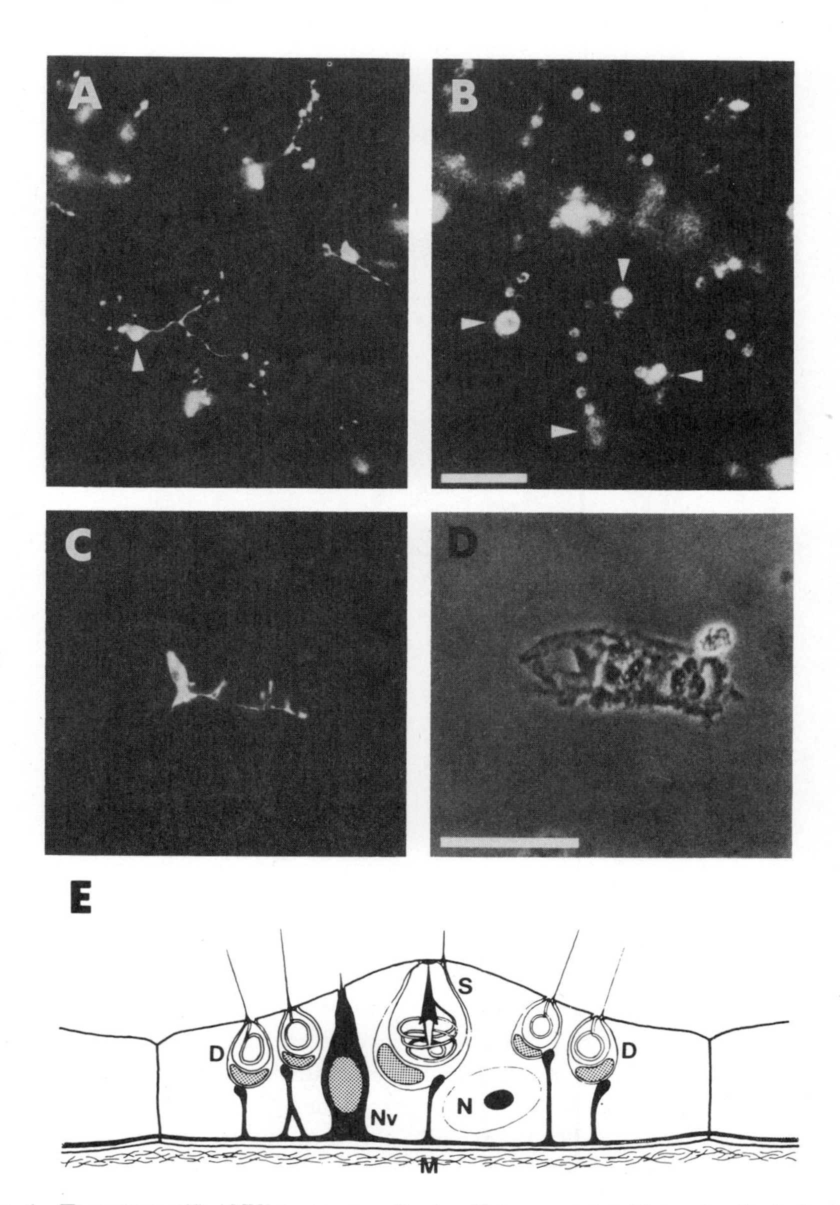

Figure 1. Tentacle-specific NV1+ nerve cells in *Hydra magnipapillata* visualized by indirect immunofluorescence. (A). NV1+ nerve cells in tentacles *in situ*. (B). Double staining with the nematocyte-specific monoclonal antibody H22 shows innervation of nematocytes of several battery cells by one NV1+ sensory cell (arrows indicate NV1+ cell body (A) and battery cell's stenoteles (B)). (C). Single NV1+ nerve cell in maceration preparation. D. Surrounding battery cell in phase-contrast. E: Schematic representation showing the location of a NV1+ nerve cell within the battery cell. Nv, NV1+ nerve cell; N, battery cell nucleus; S, stenotele; D, desmonemes; M, mesoglea. Bars: 25 μm.

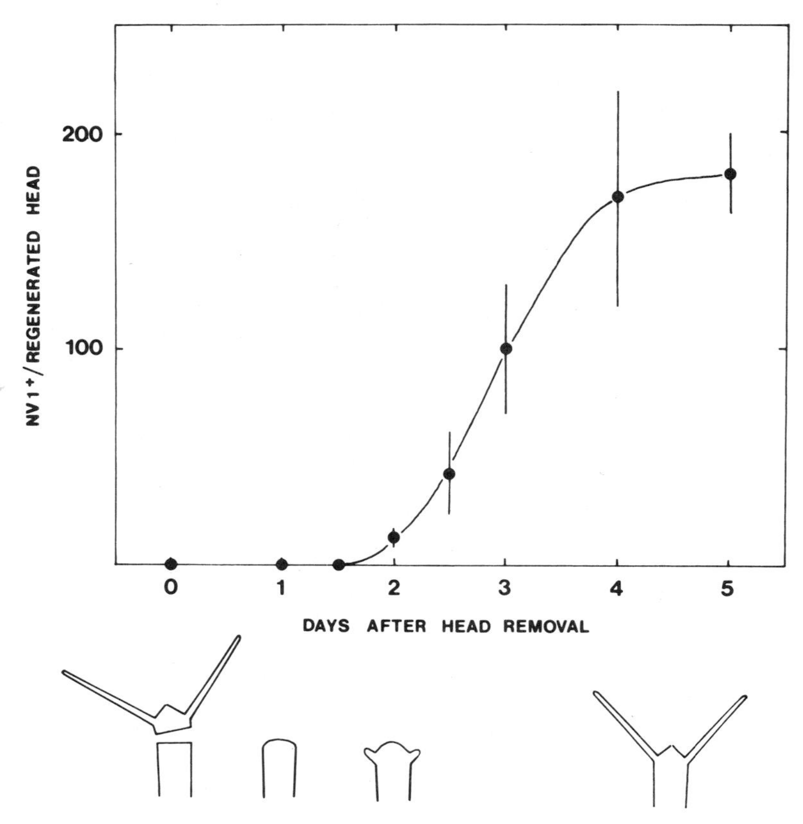

Figure 2. Reappearance of NV1+ nerve cells during head regeneration. Typical stages of head formation at the times indicated are given as schematic drawings.

interstitial cells. When chimeric animals were decapitated below the tentacle ring and allowed to regenerate, three types of regenerates were observed (Table 1): regenerates with completely normal heads (about 50%), incomplete regenerates having less than four tentacles per head (about 5%), and regenerates showing no regeneration of tentacle structures (about 45%). In the latter case, head formation was terminated by a rounded cap at the site of head removal.

In regenerates with normal heads, formation of NV1+ nerve cells was comparable to regeneration of the wild-type strain (Table 1). Tentacles contained normal numbers of NV1+ cells and the kinetics of appearance of these NV1+ cells was comparable to wild-type 105 (see Fig. 2). No NV1+ cells appeared in the regenerating tips of animals in which tentacle formation was inhibited (Table 1). NV1+ nerve cells formed, however, in partially inhibited animals with reduced numbers of tentacles (Fig. 3). There, NV1+ cells appeared only in tentacle tissue. Thus, formation of NV1+ nerve cells is tightly coupled with formation of tentacle morphology.

Table 1. Head Regeneration in Regeneration Deficient reg-16/105 Chimeras

regeneration of head structures	number of head regenerates	development of NV1+ nerve cells
complete	86	wild-type like reformation of NV1+ nerve cell pattern
incomplete	7	appearance of NV1+ nerve cells in tentacle structures
inhibited	66	no appearance of NV1+ nerve cells

Chimeras were decapitated below the tentacles, allowed to regenerate, and analyzed 7 days after head removal. Sample size: 159 head regenerates.

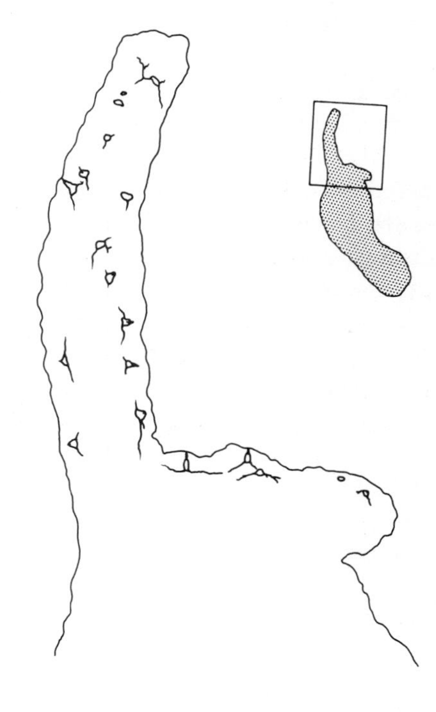

Figure 3. Camera lucida drawing showing NV1+ nerve cells in a partially inhibited reg-16/105 chimera 7 days after head removal. Inset indicates orientation of drawing.

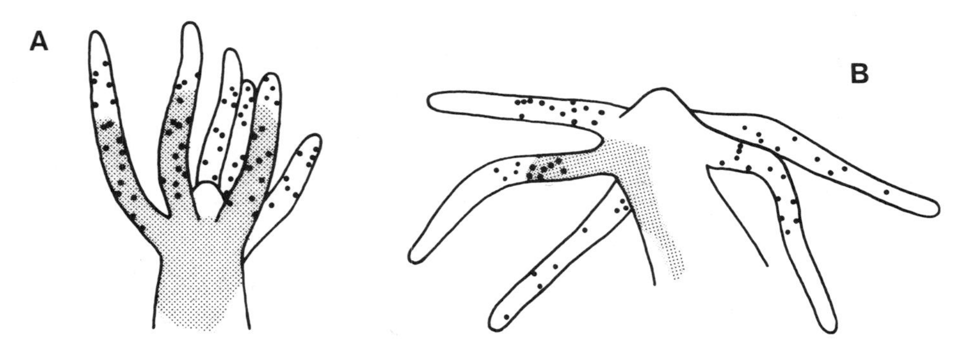

Figure 4. Camera lucida drawings showing a tentacle regenerating head (A) and an intact (B) NV1-free head, 4 days after transplantation. Black spots represent NV1+ cell bodies; stippled areas indicate the position of ink marked cells.

4. Requirements for Formation of a NV1-Battery Cell Complex

4.1. Formation of NV1+ Nerve Cells Requires Interstitial Cell Differentiation

Cell proliferation occurs continuously in the body column of *Hydra*. The new tissue is displaced into buds and into the head (tentacles) and foot, at either end of the body column (Campbell, 1967). During this displacement process, nerve cells from the body columnn become part of the head. Additional head-specific nerve cells also differentiate from interstitial cells at this time (Yaross et al., 1986). Thus, nerve cells in the head and tentacles are derived from two sources. These two sources can be distinguished by analyzing nerve cell formation in interstitial cell-free animals. Nerve cells, which appear in newly formed heads of such animals arise from nerve cells pre-existing in the body column; nerve cells which fail to form under these conditions must arise in normal animals by differentiation from interstitial cells.

To differentiate the source of NV1+ cells in head tissue interstitial cell-free polyps (Diehl and Burnett, 1964) were allowed to regenerate heads. After six days of regeneration no NV1+ cells appeared in the regenerated tentacles; other types of nerve cells could be recognized in these tentacles using a different monoclonal antibody (NV4; Hobmayer et al., in preparation). Thus, tentacle-specific NV1+ nerve cells arise only by differentiation from interstitial precursor cells.

4.2. Formation of NV1+ Nerve Cells Requires Differentiation of New Battery Cells

Since differentiation of NV1+ nerve cells is closely correlated to differentiation of tentacle structures, it appeared possible that NV1 formation only occurs during differentiation of new battery cells. To investigate this, we grafted NV1-free heads onto the body columns of normal animals and followed the appearance of newly differentiated NV1+ nerve cells in the NV1-free tentacles. To permit tracking of epithelial cell movement from the body column into tentacles, ectodermal epithelial cells in the body column were labelled with India Ink at the site of transplantation (Campbell, 1973). Some experimental animals were left intact; in others, the tentacles were excised to follow formation of new tentacle cells.

The appearance of newly differentiated NV1+ cells was the same in both intact and tentacle regenerating transplants, and the amount of tentacle tissue containing

Table 2. Stimulation of NV1+ and Tentacle Epithelial Cell Differentiation in HA Treated Tentacle Regenerates

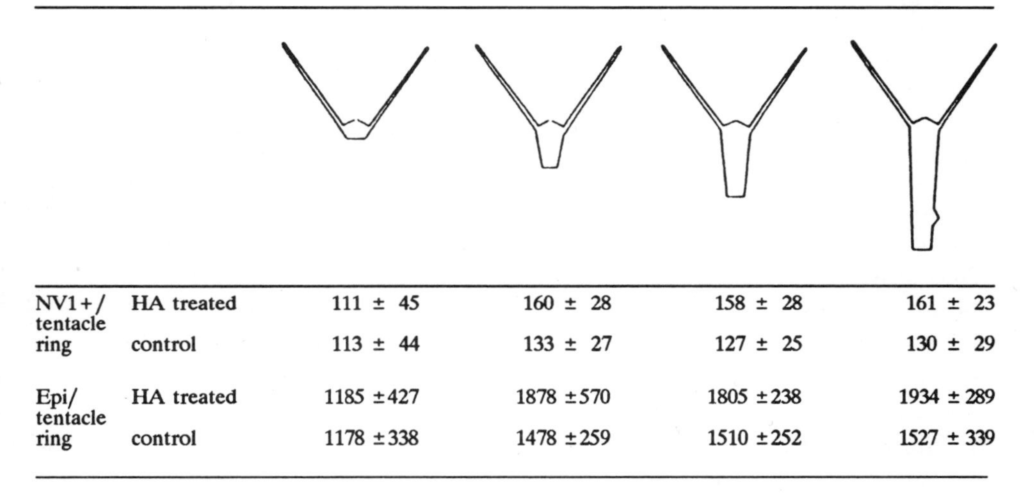

NV1+/ tentacle ring	HA treated	111 ± 45	160 ± 28	158 ± 28	161 ± 23
	control	113 ± 44	133 ± 27	127 ± 25	130 ± 29
Epi/ tentacle ring	HA treated	1185 ±427	1878 ±570	1805 ±238	1934 ± 289
	control	1178 ±338	1478 ±259	1510 ±252	1527 ± 339

Hydra were incubated in 1 pM HA for 18 hr. Pieces of different size (heads, distal 1/4, distal 1/2, and whole animals) were then cut, as shown above. Tentacles were excised from all pieces and the pieces were incubated in hydra medium for 2 days to permit tentacle regeneration. The regenerated tentacles were scored in whole mounts for NV1+ cells, by antibody staining, and for epithelial cells by staining their nuclei with the DNA-specific fluorochrome DAPI. The numbers in the Table refer to the total number of cells in all tentacles of a regenerate (the number of tentacles per regenerate varied from 4-7). The means of HA-treated and control animals differ (95% confidence limit, Students t-test) for distal 1/4, distal 1/2 and whole animal pieces.

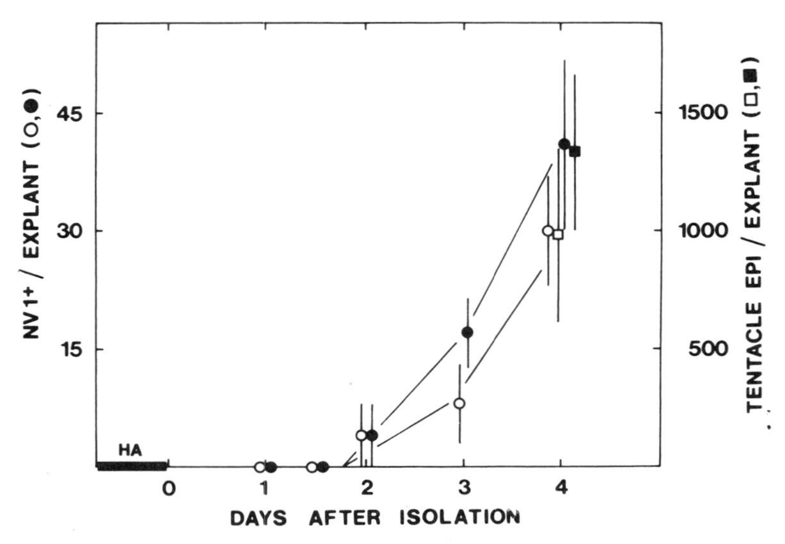

Figure 5. Kinetics of NV1+ and tentacle epithelial cell differentiation in distal gastric region explants of *Hydra* treated with 1 pM HA for 18 hr (closed symbols). Open symbols represent untreated control animals.

NV1+ cells was roughly the same. NV1+ cells filled almost the entire length of regenerated tentacles (Fig. 4A), while in intact animals there was a well defined distal boundary of NV1+ cells and an essentially empty area at the ends of the tentacles which corresponded to "old" tentacle tissue present at the time of grafting (Fig. 4B). Thus, NV1 precursors did not differentiate in association with already differentiated battery cells. Rather, it appears that the NV1-battery cell complex can only be formed by interaction between an interstitial cell precursor and a battery cell precursor.

5. Stimulation of NV1+ and Battery Cell Differentiation in Head Activator-treated Polyps

In order to characterize signals which control battery cell formation and to localize the site of their action in *Hydra* we have analyzed the effect of HA (Schaller, 1973) on differentiation of tentacle-specific NV1+ nerve cells and battery cells. Whole animals were incubated in 1 pM HA for 18 hr. Then tentacles were excised and after two days of regeneration in *Hydra* medium, the number of newly differentiated NV1+ nerve cells and tentacle epithelial cells was scored. In some animals, various amounts of proximal body column tissue were also removed. The results in Table 2 show that HA-treated animals contained about 25% more NV1+ nerve cells and tentacle epithelial cells than untreated control animals. Thus, HA stimulates formation of battery cells.

Truncated 1/4 and 1/2 animals differentiated the same number of NV1+ cells as intact animals and also showed the same HA effect (Table 2). In contrast, isolated head pieces regenerated reduced numbers of battery cells and showed no stimulation of NV1+ differentiation in HA treated animals (Table 2). Thus, head tissue itself is insensitive to HA.

Table 2 shows that pieces of *Hydra* which contained the distal gastric region responded to HA treatment with increased tentacle differentiation; head pieces which lacked this tissue did not respond. This suggests that tissue in the distal gastric region is the site of formation of the NV1-battery cell complex. To test this directly, we treated whole animals with 1 pM HA for 18 hr and then isolated the distal gastric region. Each isolated piece regenerated a small polyp with a head and tentacles. The first tentacle-specific NV1+ nerve cells appeared two days after isolation, coincident with the outgrowth of tentacle tips in both treated and untreated explants (Fig. 5). The number of NV1+ cells and the number of tentacle epithelial cells was about 30% higher in HA-treated animals than in control animals on day four.

In contrast, isolates from the proximal body column showed no stimulation of NV1+ differentiation by HA. From this we conclude that battery cell formation does not occur in this region. The distal gastric region seems to be the only site of battery cell formation in normal animals. In this region NV1 precursors and epithelial cell precursors interact to form a complex, which then differentiates to a battery cell during movement into the base of the tentacles.

References

Campbell, R. D., 1967, Tissue dynamics of steady state growth in *Hydra littoralis*, II. Patterns of tissue movement, *J. Morph.* **121**:19.

Campbell, R. D., 1973, Vital marking of single cells in developing tissues: India Ink injection to trace tissue movements in hydra, *J. Cell Sci.* **13**:651.

Diehl, F. A., and Burnett, A. L., 1964, The role of interstitial cells in the maintenance of hydra, I. Specific destruction of interstitial cells in normal, asexual, non-budding animals, *J. Exp. Zool.* **155**:253.

Dübel, S., Hoffmeister, S. A. H., and Schaller, C. H., 1987, Differentiation pathways of ectodermal epithelial cells in hydra, *Differentiation* **35**:181.

Dunne, J. F., Javois, L. C., Huang, L. W., and Bode, H. R., 1985, A subset of cells in the nerve net of *Hydra oligactis* defined by a monoclonal antibody: Its arrangement and development, *Dev. Biol.* **109**:41.

Holstein, T., Schaller, C. H., and David, C. N., 1986, Nerve cell differentiation in hydra requires two signals, *Dev. Biol.* **115**:9.

Hufnagel, L. A., Kass-Simon, G., and Lyon, M. K., 1985, Functional organization of battery cell complexes in tentacles of *Hydra attenuata*, *J. Morph.* **184**:323.

Schaller, H. C., 1973, Isolation and characterization of a low-molecular-weight substance activating head and bud formation in hydra, *J. Embryol. exp. Morph.* **29**:27.

Schaller, H. C., and Bodenmüller, H., 1981, Isolation and amino acid sequence of a morphogenetic peptide from hydra, *Proc. Natl. Acad. Sci. USA* **78**:7000.

Sugiyama, T., and Fujisawa, T., 1977, Genetic analysis of developmental mechanisms in hydra, I. Sexual reproduction of *Hydra magnipapillata* and isolation of mutants, *Development, Growth and Differentiation* **19**:187.

Wanek, N., Nishimiya, C., Achermann, J., and Sugiyama, T., 1986, Genetic analysis of developmental mechanisms in hydra, XIII. Identification of the cell lineages responsible for the reduced regenerative capacity in a mutant strain, reg-16, *Dev. Biol.* **115**:459.

Yaross, M. S., Westerfield, J., Javois, L. C., and Bode, H. R., 1986, Nerve cells in hydra: monoclonal antibodies identify two lineages with distinct mechanisms for their incorporation into head tissue, *Dev. Biol.* **114**:225.

Yu, S.-M., Westfall, J. A., and Dunne, J. F., 1986, Use of a monoclonal antibody to classify neurons isolated from the head region of hydra, *J. Morph.* **188**:79.

Chapter 6

Chemical Signaling Systems in Lower Organisms: A Prelude to the Evolution of Chemical Communication in the Nervous System

WILLIAM E. S. CARR

1. Introduction

The use of chemoreceptors to monitor chemicals appearing in the external environment has many similarities to the use of neuronal receptors to monitor chemicals appearing in a synaptic cleft. Both the chemoreceptors and neuronal receptors obtain information about specific signal molecules appearing in an aquatic milieu external to the receptor cell itself. Both receptor types include membrane-bound binding sites, whose occupancy is coupled to a transduction process which may affect the electrical properties of the membrane. Moreover, many examples are known wherein the same substance stimulates both external chemoreceptors and internal neuronal receptors (e.g., Carr et al., 1989). Earlier reviews describing analogies between external and internal receptors are provided by Kittredge et al. (1974), Lenhoff and Heagy (1977), Carr et al. (1987; 1989) and Janssens (1987).

Haldane (1954) was the first to propose that receptor systems for neurotransmitters and hormones may have evolved from the external chemosensory systems of "simple" unicellular organisms. Haldane's notion regarding the evolution of chemical signaling stemmed from observations that in certain protozoans,

WILLIAM E. S. CARR • Whitney Laboratory and Department of Zoology, University of Florida, 9505 Ocean Shore Blvd., St. Augustine, Florida 32086, USA.

Evolution of the First Nervous Systems
Edited by P.A.V. Anderson
Plenum Press, New York

conjugation occurs only between dissimilar mating types whose identity is signaled to neighboring cells by specific diffusible chemicals. He reasoned that chemical communication between these unicells might involve mechanisms similar to those whereby neurotransmitters and hormones are employed for chemical communication between neurons and other cells in higher organisms (*ibid.*).

In this review, we will do as Haldane first suggested 35 years ago, and assess similarities between the chemical communication systems of certain unicellular organisms and those of neurons and other internal cell types. Specifically, the molecular components of the chemical signaling systems of a slime mold (*Dictyostelium discoideum*), a yeast (*Saccharomyces cerevisiae*), and of nerve cells will be examined and compared. The following interrelated properties and gene products will be considered: (1) signal molecules and their synthetic pathways; (2) transmembrane signaling systems including receptors and transduction components; (3) mechanisms for inactivating signal molecules; and (4) evidence that components of an external chemical sensing system can be internalized.

Recent reviews that compare the components of transmitter receptor systems in invertebrates and vertebrates are provided by Venter et al. (1988) and Walker and Holden-Dye (1989).

2. Slime Molds and Yeast: Signal Molecules and Their Synthesis

During its unicellular stage, amoeboid cells of the slime mold *D. discoideum* feed on bacteria and multiply by mitotic divisions. When the food supply becomes limited, thousands of amoeboid cells aggregate into a multicellular mass that will later differentiate into a spore-producing body (see Bonner, 1983). In *D. discoideum*, aggregation of the unicellular amoebae in preparation for the multicellular stage is induced by a chemotactic response to 3',5'-cyclic adenosine monophosphate (cAMP) that is released externally in a pulsatile manner by signaling cells (*ibid.*). Hence, cAMP, the ubiquitous second messenger used intracellularly in many neurotransmitter systems of higher organisms, serves as an extracellular primary (first) messenger in the chemical signaling system of the slime mold. As in higher organisms, slime mold amoebae employ ATP and the membrane-bound enzyme adenylate cyclase to synthesize cAMP (e.g., Janssens and Van Haastert, 1987).

The yeast species, *S. cerevisiae*, has two haploid cell types (mating types) that conjugate to form diploid cells. Conjugation occurs only between dissimilar mating types that are designated as α cells and a̲ cells (e.g., Sprague et al., 1983). Both mating types release diffusible peptide mating factors that induce cells of the opposite mating type to undergo changes in preparation for conjugation. The mating factor released by α cells is composed of a tridecapeptide and a closely related dodecapeptide that are together termed α-factor (Fig. 1A) (Sprague et al., 1983). Both peptides are synthesized

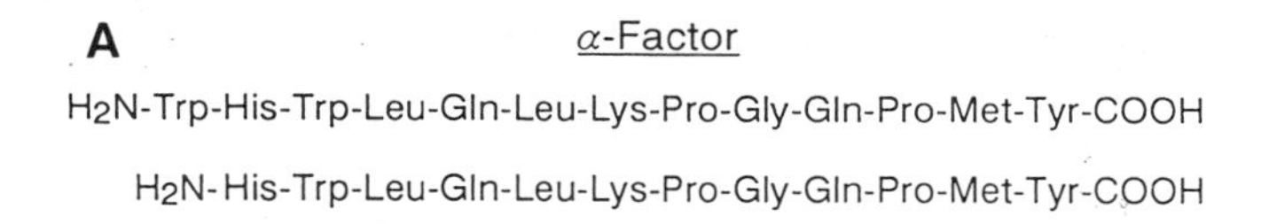

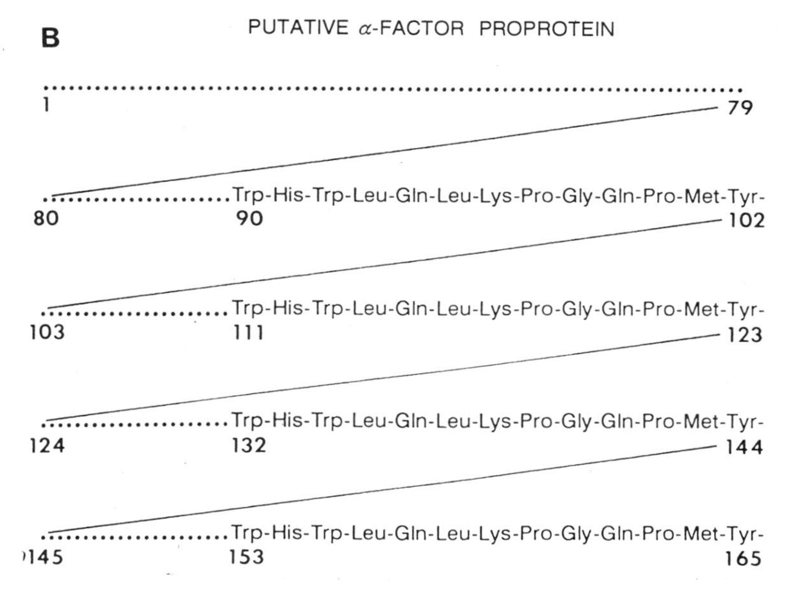

Figure 1. Mating factor of yeast, *S. cerevisiae*. **A.** Structure of tridecapeptide and dodecapeptide comprising α-factor. **B.** Four copies of α-factor contained within a putative proprotein composed of 165 amino acids. Deduced amino acid sequence of proprotein obtained after cloning and sequencing the α-factor gene (from Kurjan and Herskowitz, 1982).

in a proprotein that is later cleaved to provide four copies of the active α-factor (Fig. 1B) (Kurjan and Herskowitz, 1982). Hence, yeast produce these bioactive peptides by the post-translational processing of an inactive proprotein in the same manner as other active peptides, such as oxytocin, vasopressin, and FMRFamide, are produced in mammals (e.g., Gainer et al., 1985) or molluscs (Taussig and Scheller, 1986). Interestingly, α-factor also has extensive amino acid sequence identity and cross reactivity with gonadotropin-releasing hormone, a major reproductive hormone in mammals (Loumaye et al., 1982).

It is clear from the above, that the capacity of biological systems to produce cAMP from ATP, and specific peptides from proproteins, and then to employ these molecules as integral components of chemical signaling systems, did not have to await the advent of multicellularity and the eventual evolution of a primitive nervous system. The capacity to synthesize and employ these molecules is evident in slime molds and yeast.

See the following for broader reviews of chemoresponses in slime molds (McRobbie 1986; Devreotes and Zigmond 1988), yeast (Sprague et al., 1983; Cross et al., 1988), and other microbes (Van Houten and Preston, 1987).

3. Transmembrane Signaling Systems

The nervous system of a multicellular organism may employ many different neurotransmitters and hormones to regulate the activities of its neurons and effectors. Most of these transmitters and hormones affect target cells, without the signal molecule itself entering the cell (e.g., Berridge, 1985). This is accomplished by the binding of the signal molecule to a receptor protein on the outer cell membrane and the subsequent initiation of a series of transduction events that transform the chemical signal into a cellular response. One well-known transmembrane signaling system involves the integrated activity of several proteins (Fig. 2). In this system, the binding of the signal molecule to the receptor protein affects structural changes in another membrane protein, called a guanine nucleotide-binding protein (G-protein). The activated G-protein then effects structural changes in an effector enzyme, such as adenylate cyclase, which, in turn, catalyzes the production of a second messenger substance. The second messenger may then directly evoke a cellular response (e.g., Nakamura and Gold,

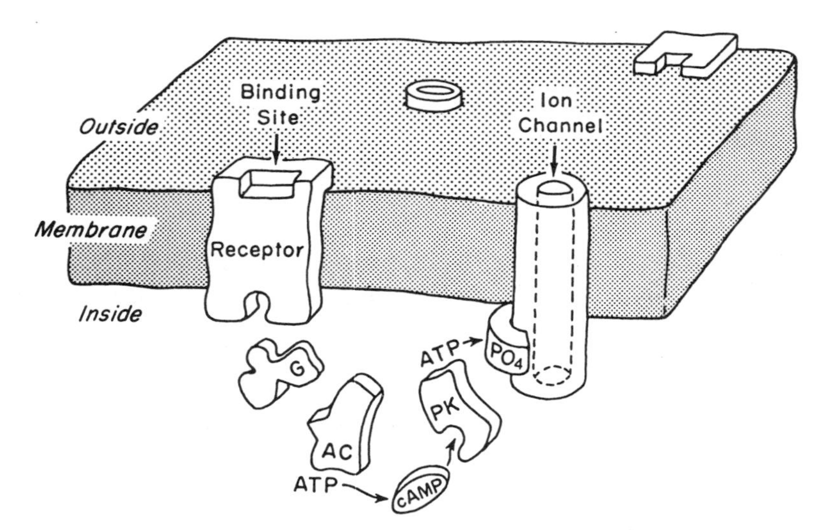

Figure 2. Generalized scheme of transmembrane signaling system that utilizes a G-protein (G) to couple the membrane-bound receptor to intracellular effectors. Other depicted components include the effector enzymes adenylate cyclase (AC) and protein kinase (PK), the second messenger cAMP, and an ion channel that is phosphorylated (PO_4) by PK (from Carr et al., 1989).

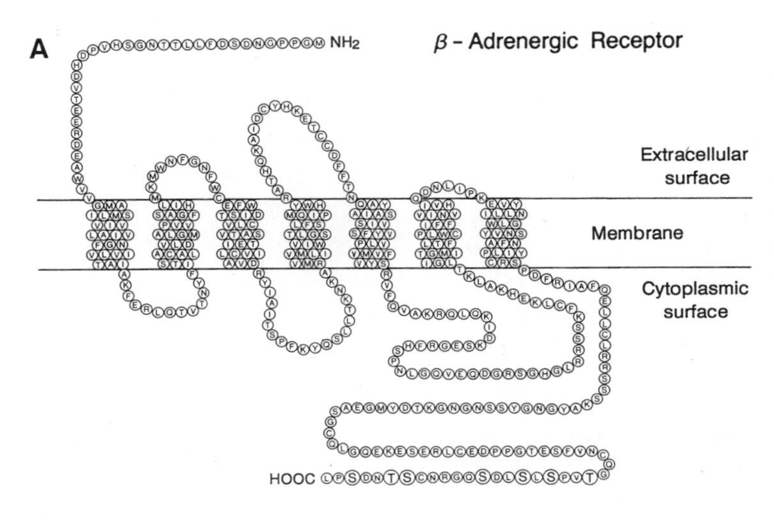

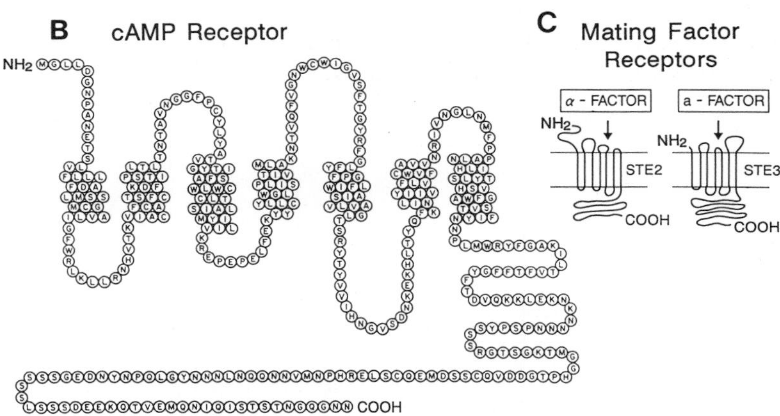

Figure 3. Schematic structures proposed for ß-adrenergic receptor (A), cAMP receptor of slime mold (B), and peptide mating factor receptors of yeast (C). All receptors are integral membrane proteins with 7 transmembrane segments, extracellular N-terminus, intracellular C-terminus, and serine (S) and threonine (T) residues near C-terminus. Further details in text (ßAR structure from Sibley et al., 1987; cAMP receptor structure from Klein et al., 1988; mating factor receptors from Herskowitz and Marsh, 1987).

1987), or it may activate an additional effector enzyme, such as a protein kinase, which can phosphorylate another effector element such as an ion channel. The latter event may then affect ion movements and lead to an overt cellular response (e.g., depolarization, hyperpolarization, secretion, contraction or relaxation). As described below, utilization of a G-protein to couple receptor activation to effector enzymes, second messengers, and ion channels is known to occur with the ß-adrenergic receptor and other members of a distinct family of proteins which includes the cAMP receptor of slime molds and the peptide receptors of yeast.

3.1. Structure of Receptors Coupled to G-proteins

The ß-adrenergic receptor (ßAR), activated by epinephrine (adrenalin) and other defined pharmacological agents, represents the most thoroughly studied member of the G-protein coupled family of receptors for chemical transmitters (for reviews, see Levitzki, 1986, 1988). The ßAR is an integral membrane protein that is embedded in, and spans, the entire width of the phospholipid bilayer of the plasma membrane. As depicted in figure 3A, the ßAR exhibits four distinct features that relate to its structure and juxtapostion (topography) vis-à-vis the membrane: (1) the protein has 7 transmembrane-spanning segments of hydrophobic amino acids; (2) the amino-terminal end of the protein is extracellular; (3) the carboxyl-terminus is intracellular; and (4) several serine or threonine residues with free hydroxyl groups occur near the carboxyl-terminus. These hydroxylated amino acids are thought to provide sites for phosphorylation by a receptor kinase whose activity affects ligand affinity and receptor desensitization (Sibley et al., 1987). In addition to the ßAR, the following additional membrane receptors are also coupled to G-proteins and exhibit the same basic topographical and structural features described above for the ßAR: muscarinic acetylcholine (Gocayne et al., 1987), substance K (Masu et al., 1987), and the vertebrate light receptor, rhopodsin (e.g., Hall, 1987). These topographical and

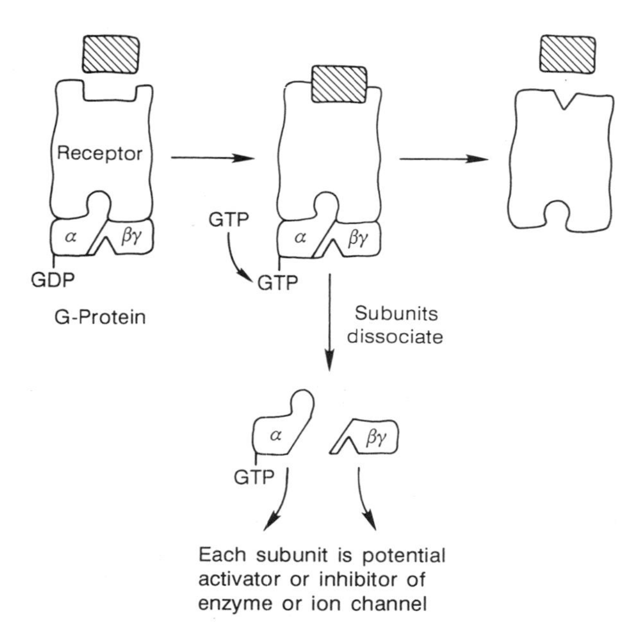

Figure 4. Schematic model of G-protein coupling a receptor to intracellular effectors. Receptor activation by ligand results in exchange of GDP for GTP at nucleotide binding site on α-subunit of the heterotrimeric G-protein. Activated G-protein dissociates into α- and ßγ-subunits. Either or both of dissociated subunits may participate in next step of tranduction process (see text).

structural similarities, plus a high degree of amino acid sequence identity, have led to the proposal that these proteins comprise a closely related receptor family that may have evolved from a common ancestral gene (*ibid.*). As shown below, recent evidence indicates that the slime mold cAMP receptor and the yeast peptide receptors belong to this same protein family.

The cAMP receptor of *D. discoideum* and the peptide receptors of *S. cerevisiae* exhibit the same four topographical and structural features described above for the ßAR family (see Fig. 3B, C). These features include the 7 transmembrane-spanning segments of hydrophobic amino acids, the extracellular N-terminus, the intracellular C-terminus, and the serine and threonine residues near the C-terminus (for cAMP receptor, see Klein et al., 1988; for yeast peptide receptors, see Burkholder and Hartwell, 1985; Hagen et al., 1986; Herskowitz and Marsh, 1987). In addition to the above similarities, the cAMP receptor also shows significant amino acid sequence identity with members of the ßAR family, particularly with bovine rhodopsin (22 to 32% sequence identity; Klein et al., 1988). Curiously, despite the marked topographical similarities, only slight sequence identity occurs between the two yeast peptide receptors themselves or between these receptors and other members of the ßAR family (Herskowitz and Marsh, 1987; Cross et al., 1988).

3.2. Structure and Role of G-proteins in Signal Transduction

G-proteins associated with the ßAR family of membrane receptors are heterotrimeric proteins composed of α-, ß- and γ-subunits. The α-subunit reversibly binds the guanine nucleotides, GDP and GTP (see recent reviews by Dolphin, 1987; Gilman, 1987; Neer and Clapham, 1988). Receptor activation by ligand binding results in the G-protein exchanging its bound GDP for GTP and dissociating into a hydrophilic α-subunit and a hydrophobic ßγ-subunit (Fig. 4). Following dissociation, either the α- or the ßγ-subunit may function to activate or inhibit particular effector proteins which initiate the next step in the transduction cascade (see Bourne, 1989). Employment of either the α- or the ßγ-subunits has contributed to the evolution of two distinct strategies of signal tranduction by G-proteins. The strategy employing the α-subunit is most commonly seen, and includes activation or inhibition of adenylate cyclase in many tissues, and stimulation of cGMP diesterase in the retina (*ibid.*); employment of the ßγ-subunit is observed with certain cardiac K^+-channels (Kim et al., 1989) and with phospholipase A_2 stimulation in the retina (Jelsema and Axelrod, 1987).

As with other members of the ßAR family, G-proteins composed of α-, ß- and γ-subunits also serve to couple the activation of chemotactic receptors of slime molds and yeast to intracellular effectors (Klein et al., 1988; Firtel et al., 1989; Whiteway et al., 1989). Interestingly, when considered together, the signal transduction processes of slime molds and yeast reveal apparent examples wherein both the α- and the ßγ-subunits of G-proteins are engaged in receptor-effector coupling. In *D. discoideum*, there is evidence that it is the dissociated α-subunit that couples activation of the cAMP receptor with the effector enzymes adenylate cyclase, and phospholipase C, and with cGMP-mediated chemotaxis and gene expression (Kumagai et al., 1989; Firtel et al., 1989). Conversely, in *S. cerevisiae*, the dissociated ßγ-subunit apparently couples receptor activation to the unidentified intracellular effectors that prepare the haploid cells for conjugation (Whiteway et al., 1989).

In both *D. discoideum* and *S. cerevisiae*, the α- and the ß-subunits of the G-proteins noted above show significant amino acid sequence identity with α- and ß-subunits of mammalian G-proteins (for slime mold, see Klein et al., 1988; for yeast, see Dietzel and Kurjan, 1987; Miyajima et al., 1987; Whiteway et al., 1989).

It is clear that hypotheses about the origin of chemical communication in the nervous system must acknowledge that, even today, unicellular eukaryotes produce and employ functional transmembrane signaling systems closely akin to the ßAR family of G-protein coupled systems. Indeed, the basic molecular components of a "ßAR-like" system may have originated and evolved in concert with other adaptations that increased the fitness of a unicellular species by helping to integrate its intracellular activities with extracellular events.

4. Inactivation of Signal Molecules

Following release into the synaptic cleft, molecules of the neurotransmitter acetylcholine (ACh) remain only very briefly (< 1 msec) in the receptor environment. The residence time of these signal molecules is delimited by the hydrolytic ectoenzyme acetylcholinesterase (AChE) that occurs on the outer membranes of synaptic cells (Hall, 1973; Weinberg et al.,1981). Hydrolysis of ACh by AChE is an inactivation process wherein inactive products (choline and acetate) are produced from the signal molecule. This inactivation process is essential to chemical signaling because it curtails the receptor desensitization occurring when the stimulus has a prolonged residence time. Hence, inactivation of "old" signal molecules facilitates renewed receptor activation by subsequent pulses of the signal.

Receptor desensitization is not a unique phenomenon to chemical communication between neuronal cells. The evolution of efficient chemical signaling systems for communication between unicells also required mechanisms to prevent desensitization to long-lived stimuli. Indeed, prolonged exposure of slime mold amoebae to cAMP, or yeast cells to intact peptide mating factor, results in desensitization to subsequent pulses of signal molecules (Janssens and Van Haastert, 1987; Moore, 1984). Moreover, both slime molds and yeast employ hydrolytic ectoenzymes to inactivate the signal molecules and limit their residence time in the receptor environment. In *D. discoideum*, both a membrane-bound and a soluble phosphodiesterase function to cleave extracellular cAMP molecules into inactive products (e.g., Gerisch et al., 1974; Nanjundiah and Malchow, 1976). Likewise, *S. cerevisiae* expresses an external peptidase that cleaves α-factor into inactive products (Ciejek and Thorner, 1979; Finkelstein and Strausberg, 1979). The critical role of the hydrolytic enzymes in the signaling systems of these unicells is evident from the fact that enzyme-deficient strains, or cells with the enzymes inhibited, become desensitized and fail to aggregate or conjugate (Janssens and Van Haastert, 1987; Darmon et al., 1978; Chan and Otte, 1982; Moore, 1984). Clearly, the capacity to express and employ hydrolytic enzymes to delimit the residence time of signal molecules occurs in unicellular eukaryotes. Indeed, it is likely that the enzymes for inactivating specific signal molecules have co-evolved with the enzymes for synthesizing these molecules, and with the components of the transmembrane signaling systems that inform cells of their presence (see also Carr et al., 1990).

5. Internalization of Chemical Sensing Machinery

The life cycle of the slime mold, *D. discoideum*, has both unicellular and multicellular stages (e.g., Bonner, 1983). During this life cycle the signaling system for cAMP does more than just mediate the aggregation and assemblage of the free-living amoebae into the migratory slug. This same signaling system also persists in the multicellular stage. Within a forming and maturing slug composed of as many as 100,000 cells, cAMP is secreted in an oscillatory manner to function as one of several diffusible transmitter/hormones affecting cell movements, cell-cell interactions, and the differential gene expression which culminates in the differentiation of a fruiting body with stalk and spore cell types (Kay, 1983; Schaap and Wang, 1986; Schaap, 1986; Gerisch, 1987; Williams, 1988; Firtel et al., 1989). Hence, the ontogeny of the slime mold demonstrates that a successful external chemotactic system can indeed be internalized to function as an "archetypical transmitter system". Internalization of this system encompasses multiple gene products including: adenylate cyclase to synthesize cAMP, the complete G-protein coupled receptor system, phosphodiesterase to inactivate cAMP, and probably also the effector enzymes guanylate cyclase and phospholipase C, to synthesize the second messengers cGMP and inositol trisphosphate (*ibid.*).

6. Conclusions

When Haldane (1954) suggested that we look to unicellular organisms for clues about chemical signaling in the nervous system, he had no real knowledge of the molecular structure of receptor and transduction components. He simply reasoned that chemical communication between unicells and nerve cells might involve similar mechanisms. His reasoning was indeed prophetic!

As described in this review, modern techniques have revealed that some of today's "current generation" of eukaryotic unicells, as represented by the slime mold, *D. discoideum*, and the yeast, *S. cerevisiae*, still employ chemical signaling systems with components extremely similar to those existing in synapses and elsewhere in higher organisms (also see Janssens, 1987; Klein et al., 1988; Carr et al., 1989). As summarized below, these similarities include several levels of complexity that range from overall pathway organization to primary gene structure; the extensiveness of these similarities make it difficult to assume that they simply represent multiple examples of convergent evolution (Fig. 5). It seems more plausible that at least certain neurotransmitter systems do, indeed, employ assemblages of associated gene products that have descended and co-evolved together from the successful signaling system of a microbial progenitor.

The similarities between the external chemical signaling systems of slime molds and yeast and the internal transmitter systems of higher organisms begin with the utilization of signal molecules, such as cAMP and bioactive peptides, and also include the mechanisms and pathways whereby these signals are synthesized (Section 2). With the peptides, the similar synthetic pathways include both the pre- and post-translational

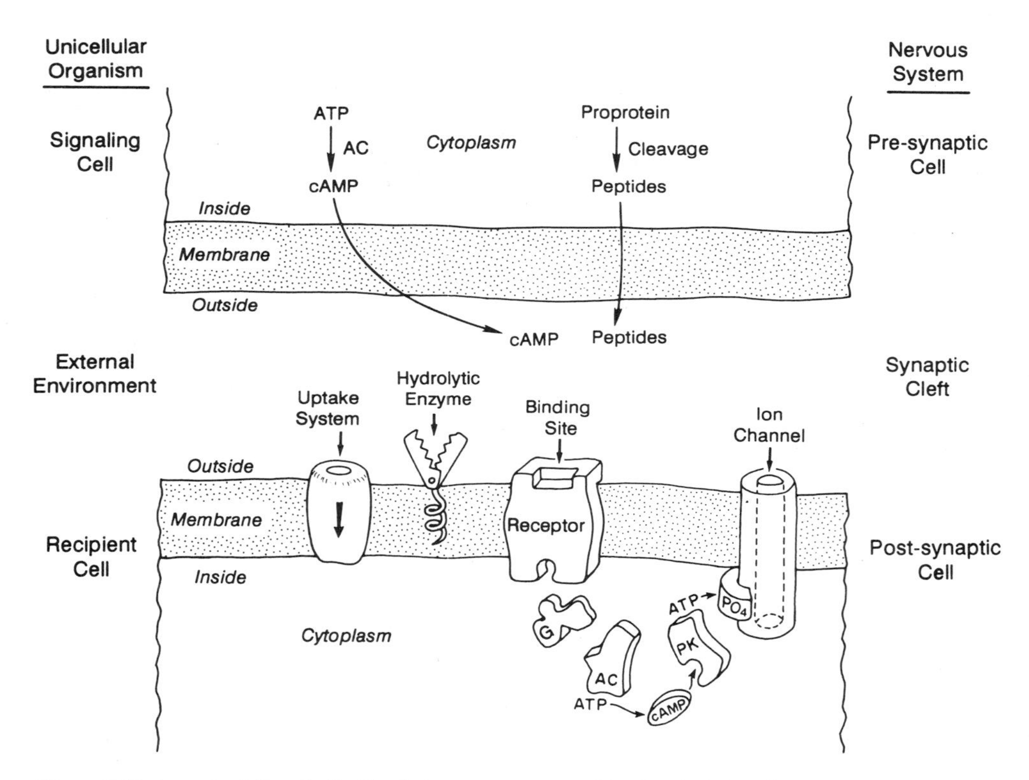

Figure 5. Cartoon depicting similarities between systems used in chemical signaling between unicells, and between pre- and post-synaptic members of a chemical synapse. Similarities may include signal molecules and synthetic pathways, G-protein coupled receptors and transduction components, effector enzymes and second messengers, and hydrolytic enzymes and uptake systems to limit receptor desensitization. Components of transmembrane signaling system as in figure 2.

components. Regarding receptors for the signal molecules, the microbial genomes contain information to express and employ membrane receptors with marked topographical and structural similarities to the ßAR family of G-protein-coupled receptors (Section 3.1). Further, for the cAMP receptor of the slime mold, the similarities to the ßAR family include a significant degree of primary amino acid sequence identity. Progressing to transduction components, it is evident that the genomes of both the slime mold and yeast have the information to express and employ G-proteins with α- and ß-subunits showing significant primary sequence identity to the analogous subunits of higher organisms (Section 3.2). Also, the capacity to employ either the dissociated α- or $\beta\gamma$-subunits to couple receptor activation to intracellular effectors appears in both unicells and higher organisms. Moreover, with *D. discoideum*, the intracellular effectors (i.e., adenylate cyclase, guanylate cyclase, and phospholipase C) include some of the very same G-protein coupled effectors and second messengers employed in higher organisms (see also Newell et al., 1987). Regarding other aspects of receptor function, it is evident that the signaling systems of both unicells and nerve cells become desensitized when signal molecules remain in the receptor environment for prolonged periods (Section 4). In both cases the residence time of signals is regulated with hydrolytic enzymes. Finally, it seems relevant that the life cycle of a slime mold illustrates that a G-protein coupled transmitter/hormonal system occurring in a multicellular stage can, indeed, trace its immediate ancestry to a preceding unicellular "progenitor" (Section 5).

As a correlate to the internalization of the G-protein coupled signaling system in the slime mold, it may be relevant that a somewhat similar event occurs during vertebrate development. Surface membranes of unfertilized amphibian or mammalian oocytes express receptors for the transmitters ACh, epinephrine and GABA (Kusano et al., 1982; Eusebi et al., 1984; Dolci et al., 1985); these external receptors are presumed to have some role in fertilization or development. Moreover, the ACh receptors on frog oocytes exhibit properties of muscarinic receptors, including sensitivities to inositol trisphosphate and Ca^{++}, which suggests that the same G-protein coupled transduction pathway occurs in the oocyte as in the cells of the differentiated nervous system (Gillo et al., 1987). Hence, the unicellular stage of the frog, like that of the slime mold, possesses functional transmembrane signaling systems that are initially positioned to monitor chemical signals in the external environment. In the life cycle of both the microbe and the vertebrate, the persistence of the signaling system in the multicellular stage positions it to next mediate chemical communication between neighboring cells. Hence these disparate organisms each express developmental events consistent with the hypothesis that the molecular components of a G-protein coupled signaling system evolved from a unicellular progenitor.

ACKNOWLEDGEMENTS. This work was supported by NSF Grant Nos. BNS-8607513 and -8908340. I am grateful to my research collaborators Drs. Richard A. Gleeson and Henry G. Trapido-Rosenthal for discussions of many of the ideas appearing in this review. Illustrations were prepared by Ms. Marsha Lynn Milstead.

References

Berridge, M. J., 1985, The molecular basis of communication within the cell, *Sci. Amer.* **253:**142-152.

Bonner, J. T., 1983, Chemical signals in social amoebae, *Sci. Amer.* **248:**114-120.

Bourne, H. R., 1989, G-protein subunits: who carries what message? *Nature* **337:**504-505.

Burkholder, A. C., and Hartwell, L. H., 1985, The yeast α-factor: structural properties deduced from the sequence of the STE2 gene, *Nucleic Acids Res.* **13:**8463-8475.

Carr, W. E. S., Ache, B. W., and Gleeson, R. A., 1987, Chemoreceptors of crustaceans: similarities to receptors for neuroactive substances in internal tissues, *Environ. Health Perspect.* **71:**31-46.

Carr, W. E. S., Gleeson, R. A., and Trapido-Rosenthal, H. G., 1989, Chemosensory systems in lower organisms: correlations with internal receptor systems for neurotransmitters and hormones, *Adv. Environ. Comp. Physiol.* **5:**25-52.

Carr, W. E. S., Trapido-Rosenthal, H. G., and Gleeson, R. A., 1990, The role of degradative enzymes in chemosensory processes, *Chem. Senses* **15:**, in press.

Chan, R. K., and Otte, C. A., 1982, Physiological characterization of *Saccharomyces cerevisiae* mutants supersensitive to G1 arrest by **a** factor and α factor pheromones, *Mol. Cell. Biol.* **2:**21-29.

Ciejek, E., and Thorner, J., 1979, Recovery of *S. cerevisiae* **a** cells from G1 arrest by α factor pheromone requires endopeptidase, *Cell* **18:**623-635.

Cross, F., Hartwell, L. H., Jackson, C., and Konopka, J. B., 1988, Conjugation in *Saccharomyces cerevisiae*, *Ann. Rev. Cell Biol.* **4:**429-457.

Darmon, M., Barra, J., and Brachet, P., 1978, The role of phosphodiesterase in aggregation of *Dictyostelium discoideum*, *J. Cell Sci.* **31:**233-243.

Devreotes, P. N., and Zigmond, S. H., 1988. Chemotaxis in eukaryotic cells: a focus on leukocytes and *Dictyostelium*, *Ann. Rev. Cell Biol.* **4:**649-686.

Dietzel, C., and Kurjan, J., 1987, The yeast SCG1 gene: a Gα-like protein implicated in the a- and α-factor response pathway, *Cell* **50:**1001-1010.

Dolci, S., Eusebi, F., and Siracusa, G., 1985, γ-Amino butyric-n-acid sensitivity of mouse and human oocytes, *Dev. Biol.* **109:**242-246.

Dolphin, A. C., 1987, Nucleotide binding proteins in signal transduction and disease, *Trends Neurosci.* **10:**53-57.

Eusebi, F., Pasetto, N., and Siracusa, G., 1984, Acetylcholine receptors in human oocytes, *J. Physiol. (Lond.)* **346:**321-330.

Finkelstein, D. B., and Strausberg, S., 1979, Metabolism of α-factor by **a** mating type cells of *Saccharomyces cerevisiae*, *J. Biol. Chem.* **254:**796-803.

Firtel, R. A., van Haastert, P. J. M., Kimmel, A. R., and Devreotes, P. N., 1989, G protein linked signal transduction in development: *Dictoystelium* as an experimental system, *Cell* **58:**235-239.

Gainer, H., Russell, J. T., and Loh, Y. P., 1985, The enzymology and intracellular organization of peptide precursor processing: the secretory vesicle hypothesis, *Prog. Neuroendocrinol.* **40:**171-184.

Gerisch, G., 1987, Cyclic AMP and other signals controlling cell development and differentiation in *Dictyostelium*, *Ann. Rev. Biochem.* **56:**853-879.

Gerisch, G., Malchow, D., and Hess, B., 1974, Cell communication and cyclic-AMP regulation during aggregation of the slime mold, *Dictyostelium discoideum*, in: *Biochemistry of Sensory Functions*, pp. 279-298 (L. Jaenicke, ed.), Springer-Verlag, New York.

Gillo, B., Lass, Y., Nadler, E., and Oron, Y., 1987, The involvement of inositol 1,4,5-trisphosphate and calcium in the two-component response to acetylcholine in *Xenopus* oocytes, *J. Physiol. (Lond.)* **392:**349-361.

Gilman, A. G., 1987, G proteins: transducers of receptor-generated signals, *Ann. Rev. Biochem.* **56:**615-649.

Gocayne, J., Robinson, D. A., FitzGerald, M. G., Chung, F.-Z., Kerlavage, A. R., Lentes, K.-U., Lai, J., Wang, C.-D., Fraser, C. M., and Venter, J. G., 1987, Primary structure of rat cardiac ß-adrenergic and muscarinic cholinergic receptors obtained by automated DNA sequence analysis: further evidence for a multigene family, *Proc. Natl. Acad. Sci. USA* **84:**8296-8300.

Hagen, D. C., McCaffrey, G., and Sprague, G. F., Jr., 1986, Evidence the yeast STE3 gene encodes a receptor for the peptide pheromone a factor: gene sequence and implications for the structure of the presumed receptor, *Proc. Natl. Acad. Sci. USA* **83**:1418-1422.

Haldane, J. B. S., 1954, La signalisation animale, *Année Biol.* **58**:89-98.

Hall, Z. W., 1973, Multiple forms of acetylcholinesterase and their distribution in endplate and non-endplate regions of rat diaphragm muscle, *J. Neurobiol.* **4**:343-361.

Hall, Z. W., 1987, Three of a kind: the ß-adrenergic receptor, the muscarinic acetylcholine receptor, and rhodopsin, *Trends Neurosci.* **10**:99-101.

Herskowitz, I., and Marsh, L., 1987, Conservation of a receptor/signal transduction system, *Cell* **50**:995-996.

Janssens, P. M. W., 1987, Did vertebrate signal transduction mechanisms originate in eukaryotic microbes? *Trends Biochem. Sci.* **12**:456-459.

Janssens, P. M. W., and Van Haastert, P. J. M., 1987, Molecular basis of transmembrane signal transduction in *Dictyostelium discoideum*, *Microbiol. Rev.* **51**:396-418.

Jelsema, C. L., and Axelrod, J., 1987, Stimulation of phospholipase A_2 activity in bovine rod outer segments by the ß subunits of transducin and its inhibition by the α subunit, *Proc. Natl. Acad. Sci. USA* **84**:3623-3627.

Kay, R., 1983, Cyclic AMP and development in the slime mould, *Nature* **301**:659.

Kim, D., Lewis, D. L., Graziadei, L., Neer, E. J., Bar-Sagi, D., and Clapham, D. E., 1989, G-protein ß - subunits activate the cardiac muscarinic K^+-channel via phospholipase A_2, *Nature* **337**:557-560.

Kittredge, J. S., Takahashi, F. T., Lindsey, J., and Lasker, R., 1974, Chemical signals in the sea: marine allelochemics and evolution, *Fishery Bull.* **72**:1-11.

Klein, P. S., Sun, T. J., Saxe, C. L. III, Kimmel, A. R., Johnson, R. L., and Devreotes, P. N., 1988, A chemoattractant receptor controls development in *Dictyostelium discoideum*, *Science* **241**:1467-1472.

Kumagai, A., Pupillo, M., Gundersen, R., Miake-Lye, R., Devreotes, P. N., and Firtel, R. A., 1989, Regulation and function of Gα protein subunits in *Dictyostelium*, *Cell* **57**:265-275.

Kurjan, J., and Herskowitz, I., 1982, Structure of a yeast pheromone gene (MF): a putative α-factor precursor contains four tandem copies of mature α-factor, *Cell* **30**:933-943.

Kusano, K., Miledi, R., and Stinnakre, J., 1982, Cholinergic and calecholaminergic receptors in the *Xenopus* oocyte membrane, *J. Physiol. (Lond.)* **328**:143-170.

Lenhoff, H. M., and Heagy, W., 1977, Aquatic invertebrates: model systems for the study of receptor activation and evolution of receptor proteins, *Ann. Rev. Pharmacol. Toxicol.* **17**:243-258.

Levitzki, A., 1986, ß-Adrenergic receptors and their mode of coupling to adenylate cyclase, *Physiol. Rev.* **66**:819-854.

Levitzki, A., 1988, From epinephrine to cyclic AMP, *Science* **241**:800-806.

Loumaye, E., Thorner, J., and Catt, K. J., 1982, Yeast mating pheromone activates mammalian gonadotrophs: evolutionary conservation of a reproductive hormone? *Science* **218**:1323-1325.

Masu, Y., Nakayama, K., Tamaki, H., Harada, Y., Kuno, M., and Nakanishi, S., 1987, cDNA cloning of bovine substance-K receptor through oocyte expression system, *Nature* **329**:836-838.

McRobbie, S. J., 1986, Chemotaxis and cell mobility in the cellular slime molds, *CRC Crit. Rev. Microbiol.* **13**:335-375.

Miyajima, I., Nakafuku, M., Nakayama, N., Brenner, C., Miyajima, A., Kaibuchi, K., Arai, K.-I., Kaziro, Y., and Matsumoto, K., 1987, GPA1, a haploid-specific essential gene, encodes a yeast homolog of mammalian G protein which may be involved in mating factor signal transduction, *Cell* **50**:1011-1019.

Moore, S. A., 1984, Yeast cells recover from mating pheromone α factor-induced division arrest by desensitization in the absence of α factor destruction, *J. Biol. Chem.* **259**:1004-1010.

Nakamura, T., and Gold, G. H., 1987, A cyclic nucleotide-gated conductance in olfactory receptor cilia, *Nature* **325**:442-444.

Nanjundiah, V., and Malchow, D., 1976, A theoretical study of the effects of cyclic AMP phosphodiesterases during aggregation in *Dictyostelium*, *J. Cell Sci.* **22**:49-58.

Neer, E. J., and Clapham, D. E., 1988, Roles of G protein subunits in transmembrane signalling, *Nature* **333**:129-134.

Newell, P. C., Europe-Finner, G. N., and Small, N. V., 1987, Signal transduction during amoebal chemotaxis of *Dictyostelium discoideum*, *Microbiol. Sci.* **4**:5-11.

Schaap, P., 1986, Regulation of size and pattern in the cellular slime molds, *Differentiation* **33**:1-16.

Schaap, P., and Wang, M., 1986, Interactions between adenosine and oscillatory cAMP signaling regulate size and pattern in *Dictyostelium*, *Cell* **45:**137-144.

Sibley, D. R., Benovic, J. L., Caron, M. G., and Lefkowitz, R. J., 1987, Regulation of transmembrane signaling by receptor phosphorylation, *Cell* **48:**913-922.

Sprague, G. F. Jr, Blair, L. C., and Thorner, J. 1983, Cell interactions and regulation of cell type in the yeast *Saccharomyces cerevisiae*, *Ann. Rev. Microbiol.* **37:**623-660.

Taussig, R., and Scheller, R. H., 1986, The *Aplysia* FMRFamide gene encodes sequences related to mammalian brain peptides, *DNA* **5:**453-462.

Van Houten, J., and Preston, R. R., 1987, Chemoreception in single-celled organisms, in: *Neurobiology of Taste and Smell*, pp. 11-38 (T. E. Finger and W.L. Silver, eds.), John Wiley & Sons, New York.

Venter, J. C., DiPorzio, U., Robinson, D. A., Shreeve, S. M., Lai J., Kerlavage, A. R., Fracek, S. P., Lentes, K.-U., and Fraser, C.M., 1988, Evolution of neurotransmitter receptor systems, *Prog. Neurobiol.* **30:**105-169.

Walker, R. J., and Holden-Dye, L., 1989, Commentary on the evolution of transmitters, receptors and ion channels in invertebrates, *Comp. Biochem. Physiol.* **93A:**25-39.

Weinberg, C. B., Sanes, J. R., and Hall, Z. W., 1981, Formation of neuromuscular junctions in adult rats: accumulation of acetylcholine receptors, acetylcholinesterase, and components of synaptic basal lamina, *Dev. Biol.* **84:**255-266.

Whiteway, M., Hougan, L., Dignard, D., Thomas, D. Y., Bell, L., Saari, G. C., Grant, F. J., O'Hara, P., and MacKay, V.L., 1989, The STE4 and STE18 genes of yeast encode potential ß and γ subunits of the mating factor receptor-coupled G protein, *Cell* **56:**467-477.

Williams, J. G., 1988, The role of diffusible molecules in regulating the cellular differentiation of *Dictyostelium discoideum*, *Development* **103:**1-16.

Chapter 7

Neurons and Their Peptide Transmitters in Coelenterates

C. J. P. GRIMMELIKHUIJZEN, D. GRAFF, O. KOIZUMI, J. A. WESTFALL and I. D. McFARLANE

1. Introduction

Coelenterates have the simplest nervous system in the animal kingdom, and it was probably within this group of animals that nervous systems first evolved. Present day coelenterates are diverse and comprise two phyla. The classes Hydrozoa (for example *Hydra*), Cubozoa ("box jellyfishes"), Scyphozoa ("true jellyfishes") and Anthozoa (for example sea anemones and corals) constitute the phylum Cnidaria. A companion phylum is the Ctenophora ("combjellies") or Acnidaria. Most Hydrozoa, Cubozoa and Scyphozoa have a life cycle including a polyp (sessile) and medusa (mobile) form. In Anthozoa, the medusa is lacking, whereas in Ctenophora no polyps occur. Coelenterates can either live individually or in colonies. Many Hydrozoa and Anthozoa form colonies of polyps (e.g. corals), but also mixed colonies of polyps and medusae exist. The physonectid siphonophores, for example, are swimming hydrozoans consisting of a long stem to which numerous medusae and various forms of polyps are attached.

The basic plan of the coelenterate nervous system is that of a nerve net. Although often diffuse, this nerve net is sometimes condensed to form linear or circular tracts,

C. J. P. GRIMMELIKHUIJZEN[1], D. GRAFF[1], O. KOIZUMI[2], J. A. WESTFALL[2], and I. D. McFARLANE[3] • [1]Centre for Molecular Neurobiology, University of Hamburg, Martinistr. 52, 2000 Hamburg 20, Federal Republic of Germany; [2]Department of Anatomy and Physiology, College of Veterinary Medicine, Kansas State University, Manhattan, Kansas 66506, USA; and [3]Department of Applied Biology, University of Hull, Hull HU6 7RX, United Kingdom.

or even ganglion-like structures. In the stem of physonectid siphonophores, for example, there is a longitudinal "giant axon," consisting of fused neurons, which mediates fast escape reactions (Mackie, 1973, 1990). Examples of circular condensations are the outer and inner nerve rings at the bell margin of hydrozoan medusae. These rings, which consist of electrically-coupled neurons, are capable of integrating a variety of sensory inputs and of transmitting these signals rapidly throughout the margin (Spencer and Arkett, 1984; Spencer, 1989).

The morphology of coelenterate neurons has been investigated by both light and electron microscopy. The visualization of neurons by light microscopy, however, has always been difficult, as no good method exists which stains coelenterate neurons well. The most commonly used stain so far has been methylene blue, which was already applied more than 80 years ago by Hadži (1909). Using this method, Hadži found that two types of neurons exist in *Hydra*: (1) "sensory cells"; these are long, slender neurons which are orientated perpendicular to the mesoglea (a layer of collagen-like material located between the ectoderm and endoderm) and which reach out to the surface of one of the epithelial cell layers. Near this surface the "sensory cell" forms a cilium which is in contact with the outer medium, or projects into the lumen of the gastric cavity, (2) "ganglion cells"; these are neurons with a more roundish perikaryon, located at the basal part of either ecto- or endoderm. Westfall has extended these early findings using electron microscopy (Westfall, 1973a; Westfall and Kinnamon, 1978). For *Hydra* neurons, she found that there is no essential difference between "sensory" and "ganglion cells". Both types of neurons contain a cilium (are "sensory"), store dense-cored vesicles at non-synaptic regions (are "neurosecretory"), and form synapses with both epitheliomuscular cells and neighboring neurons (are both "motor- and inter-neurons"). Westfall proposes that these primitive, multifunctional nerve cells are the ancestors of the more specialized neurons that we find in higher animals today. However, whether neurons in the other coelenterates also have such multifunctional features has yet to be established.

Chemical synapses are not restricted to the nervous system of *Hydra*. Dense-cored vesicles (70-150 nm), often associated with pre- and postsynaptic membrane specializations, have been seen in species belonging to all coelenterate classes (Horridge and Mackay, 1962; Jha and Mackie, 1967; Westfall, 1973a,b; Hernandez-Nicaise, 1973; for a review see Westfall 1987). Physiological studies demonstrating blockage by depletion of Ca^{++}, or by addition of excess Mg^{++}, also indicate that classical, exocytotic release of transmitter substances occurs (McFarlane, 1973; Satterlie, 1979; Spencer, 1982; Anderson and Schwab, 1982). The presence of chemical synapses was further confirmed by simultaneous intracellular recordings at both pre- and postsynaptic neurons, which displayed the expected EPSPs with constant latency from the presynaptic spike (Spencer, 1982; Spencer and Arkett, 1984; Anderson, 1985).

In addition to chemical synapses, electrical synapses also exist in coelenterates. This was demonstrated by the presence of electrical and dye coupling between neurons, and of structures which had the morphological characteristics of gap junctions (Anderson and Mackie, 1977; Spencer, 1978, 1979; Spencer and Satterlie, 1980; Westfall

Figure 1. Whole mount staining of hydroid polyps with RFamide antiserum 146 II (1:1000). **(a)**. Staining of the nervous system in the hypostome of *Hydra attenuata*. Numerous "sensory cells" form a cluster around the mouth opening (x150). **(b)**. The nervous system of a gastrozooid of *Hydractinia echinata*. A very dense plexus of immunoreactive processes occurs throughout the hydranth and numerous "sensory cells" are present around the mouth (the proboscis is broken somewhat by pressure of the coverslip) (x150) (adapted from Grimmelikhuijzen, 1985).

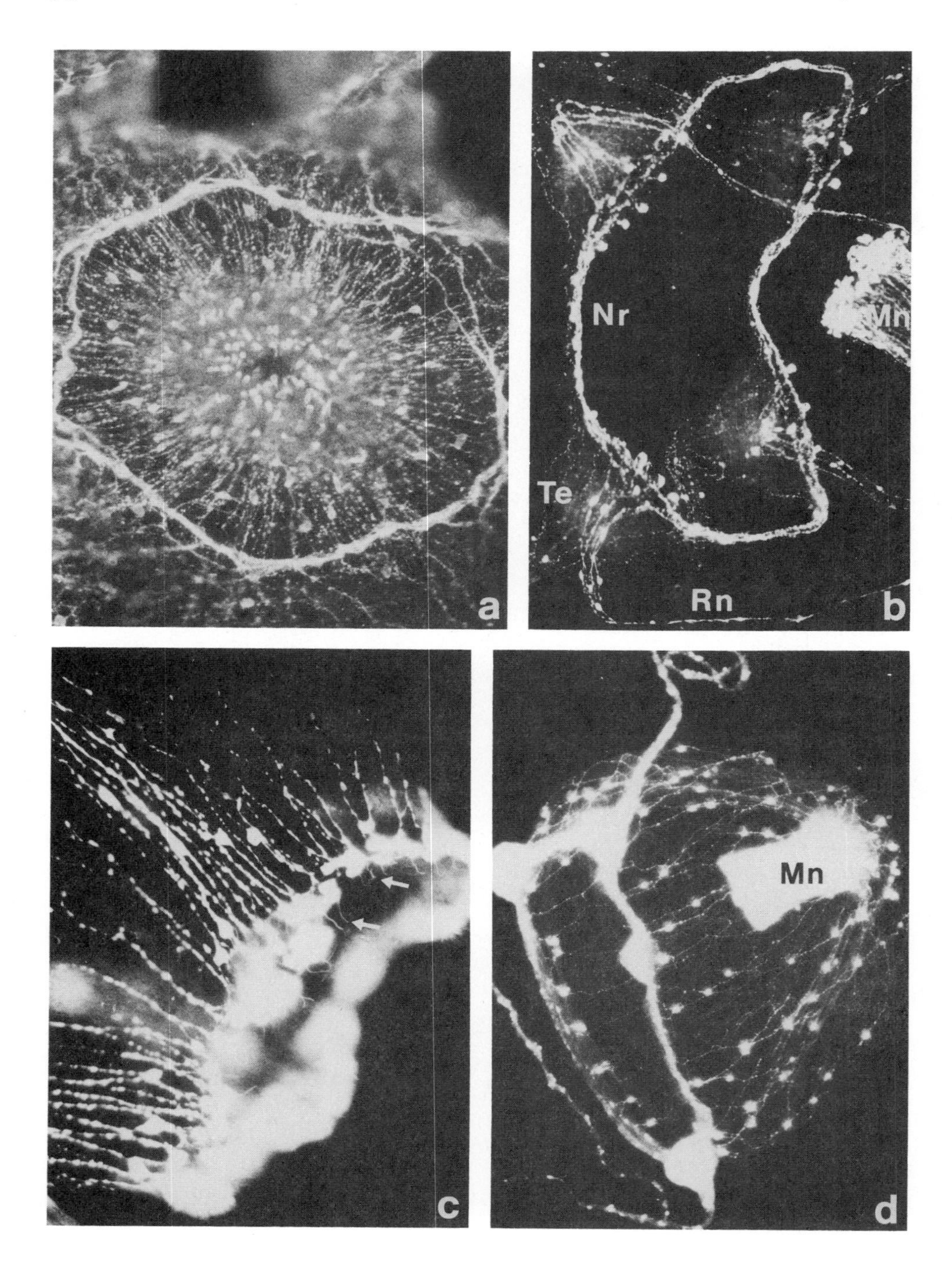
a
Nr
Mn
Te
Rn
b
c
Mn
d

et al., 1980; Spencer and Arkett, 1984). In some cases, such as in the nerve net and "giant axon" of the stem of physonectid siphonophores, the neurons have apparently fused with each other to form true syncytia (Mackie, 1973; Grimmelikhuijzen et al., 1986; Mackie et al., 1988).

2. Immunocytochemical Staining of Neurons in Coelenterates

As we mentioned above, there is no doubt that most neurons in coelenterates use transmitters or locally-acting hormones for signal transmission. Until recently, however, the nature of such transmitter substances has remained unknown. Using the Falck-Hillarp method, or using specific staining with antisera, Grimmelikhuijzen (1986) was unable to find catecholamines or serotonin in neurons of *Hydra*. In addition, it has not been possible to demonstrate acetycholine in extracts of *Hydra*, using a very sensitive bioassay (this collaborative work with Prof. K. Wächtler, Hannover, F.R.G., was mentioned in Grimmelikhuijzen, 1986). More recently, Koizumi was unable to demonstrate choline acetyltransferase in *Hydra* neurons, using immunocytochemical staining, or acetylcholinesterase using an ordinary, histochemical staining method. In addition, negative results were obtained for tyrosine hydroxylase and serotonin using specific staining with antisera (Koizumi, unpublished). These findings show that acetylcholine, catecholamines and serotonin are not generally occurring in coelenterate neurons and further suggest that these substances are not the evolutionary "oldest" transmitters.

Using immunocytochemistry and radioimmunoassays, we have found that substances occur in coelenterate neurons that are related to vertebrate or invertebrate neuropeptides (see Grimmelikhuijzen, 1984 for a review). In particular, peptides resembling the molluscan neuropeptide Phe-Met-Arg-Phe-NH_2 (FMRFamide, Price and Greenberg, 1977) are ubiquitous in coelenterates (Grimmelikhuijzen et al., 1982; Grimmelikhuijzen, 1983; Grimmelikhuijzen and Spencer, 1984; Mackie et al., 1985; Weber, 1989). Of the various antisera raised against different fragments of FMRFamide, those raised against RFamide were the most potent in staining

Figure 2. Whole mount staining of various hydrozoans with RFamide antiserum 146 II. Mn, manubrium; Nr, outer nerve ring; Rn, radial nerve; Te, tentacle. **(a)** The head region of *Hydra oligactis*, showing an obvious nerve ring located between hypostome and tentacle bases, and a cluster of "sensory cells" (somewhat out of focus) around the mouth opening (x200). **(b)** A 2-day-old medusa of *Podocoryne carnea* (Anthomedusa). The outer nerve ring, the four radial nerves, and the nerve nets in manubrium and tentacles are all stained (x75). **(c)** The manubrium of a medusa of *Podocoryne carnea*. Cilia (arrows) belonging to "sensory cells" located at the margin, can clearly be seen (x300). **(d)** A 2-day-old medusa of *Eirene sp.* (Leptomedusa). Note the immunoreactive subumbrellar nerve net (x80) (adapted from Grimmelikhuijzen et al., 1988a, 1989).

coelenterate nervous tissue (Grimmelikhuijzen and Spencer, 1984; Grimmelikhuijzen, 1985; Grimmelikhuijzen and Graff, 1985; Grimmelikhuijzen et al. 1986, 1987, 1988a, 1989). This suggested that the coelenterate peptides have the sequence RFamide in common with FMRFamide, but that the amino terminus is different (see below).

Staining with RFamide antisera has proven to be an invaluable tool for determining the organization of the nervous system of coelenterates at the light microscope level. With this technique, coelenterate neurons can be visualized in fine detail, which has previously not been possible. Of course, it has to be realized that probably only a portion of all coelenterate neurons may be stained by the RFamide antisera, but this same drawback might hold for any other staining method. Many coelenterates, such as hydrozoan polyps and medusae, are transparent and can be stained as whole mounts by the RFamide antisera. With this technique we have obtained many new details of the hydrozoan nervous system (Grimmelikhuijzen, 1985; Grimmelikhuijzen et al., 1986). Staining with methylene blue or other dyes during the last 100 years has often led to the assumption that hydroid polyps have "an elementary network of loosely interconnected neurons" with "no morphological definable center of coordination" (Tardent and Weber, 1976), despite the fact that in the original papers a concentration of neurons in hypostome and foot was often recognized (Schneider, 1890; Hadzi 1909; McConnell, 1932; Spangenberg and Ham, 1960; Burnett and Diehl, 1964). Whole mount staining with RFamide antisera shows that *Hydra attenuata* has a strong aggregation of neuronal perikarya and processes in the hypostome (Fig. 1a). In a related species, *Hydra oligactis*, there is an obvious nerve ring lying at the border of hypostome and tentacles (Fig. 2a). Polyps of *Hydractinia echinata* have a very dense neuronal plexus in the body column and numerous neurons around the mouth opening (Fig. 1b). Conventional histological and ultrastructural techniques have not previously demonstrated these centralizations to their full extent (cf. Davis et al., 1968; Kinnamon and Westfall, 1981; Matsuno and Kageyama, 1984).

The immunoreactive neurons around the mouth opening of *Hydra* and *Hydractinia* are typical "sensory cells", with slender perikarya and cilia projecting to the outer surface. The processes of these "sensory cells" innervate the radially orientated processes of the epitheliomuscular cells in the hypostome (cf. also Westfall and Kinnamon, 1984), and it is easy to imagine how a local reflex pathway could elicit feeding movements after appropriate signals (e.g., amino acids from the prey) have reached the cilium. This local reflex pathway is even simpler than the well known monosynaptic reflex arc, as only one, multifunctional neuron is involved.

Whole mount staining of hydrozoan medusae with RFamide antisera reveals an immunoreactive outer nerve ring, four radial nerves, and an immunoreactive nerve net in tentacles and manubria of Anthomedusae (Fig. 2b). In Leptomedusae, the subumbrellar nerve net, which replaces the radial nerves (Hertwig and Hertwig, 1878), becomes very obvious when stained with an RFamide antiserum (Fig. 2d). Again, many of the stained neurons are "sensory cells". In particular, the "sensory cells" at the lip of the manubrium have well developed cilia, which can easily be seen under the light microscope (Fig. 2c).

3. Ultrastructural Localization of RFamide-like Peptides

Using immuno-electron microscopy with RFamide antisera and gold-labeled secondary antibodies, RFamide-like material was found in the granular cores of neuronal dense-cored vesicles of *Hydra* (Fig. 3). The gold-labeled vesicles were located close to the Golgi complex, in neuronal processes paralleling the myonemes of epitheliomuscular cells, and in nerve endings terminating on the myonemes (Koizumi et al., 1989). Some vesicles in the nerve terminals lacked granules and formed closely packed tubular cisternae, indicating a site of exocytosis (Fig. 3; Koizumi et al., 1989). Unfortunately, gold-labeled vesicles have not been found at synapses, but this might be due to an insufficient number of sections being examined. The distribution of immunoreactive dense-cored vesicles is consistent with the following picture: RFamide-like material is synthesized on the rough endoplasmic reticulum (RER), subsequently packed into dense-cored vesicles in the Golgi complex, and transported along the neurites to the terminals. After adequate stimuli, the RFamide-like peptide is released by exocytosis. As in higher animals, once released, the peptide might exert its action as a hormone- or transmitter-like substance.

4. Isolation of Neuropeptides from Coelenterates

Staining with RFamide antisera is an excellent technique with which to visualize a major portion of the nervous system in coelenterates. In addition, it also gives a clear indication of the transmitter- or hormone-like substances being used. In order

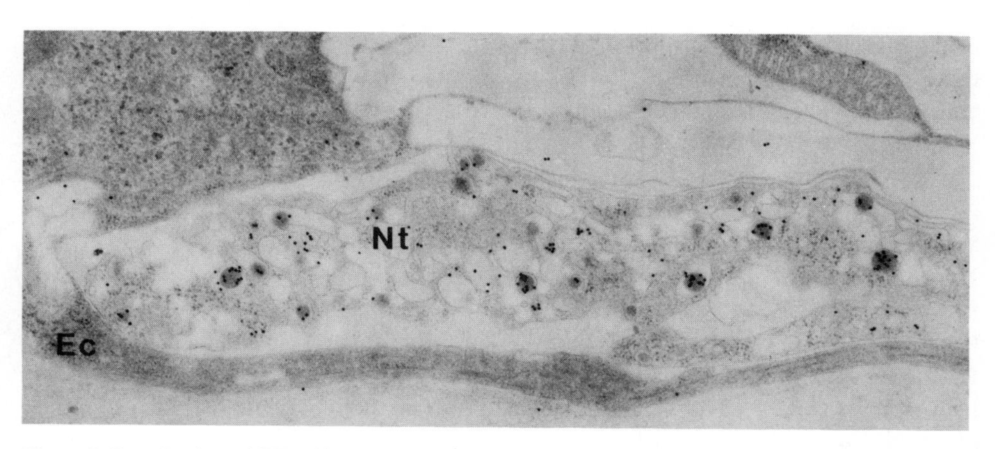

Figure 3. Localization of RFamide-like material in neuronal dense-cored vesicles of *Hydra littoralis*. Nt, nerve terminal; Ec, epitheliomuscular cell. Staining was carried out with RFamide antiserum 146 II. The micrograph reveals 15 nm-gold labeling of dense-cored in vesicles of a nerve terminal ending on the basal myoneme of a epitheliomuscular cell (x24,000) (from Koizumi et al., 1989).

to isolate these substances, an RFamide radioimmunoassay has been developed (Grimmelikhuijzen and Graff, 1986). This assay recognizes free RFamide, and the aminoterminal-elongated peptides containing the RFamide moiety. With much lower efficiency, the radioimmunoassay also recognizes other peptides, provided they contain a carboxyterminal amidation and an uncharged aromatic or aliphatic amino acid in the carboxyterminal position, preceded by Arg or Lys. Using the RFamide radioimmunoassay to monitor their purification, three peptides have been isolated from the sea anemone *Anthopleura elegantissima*. These are <Glu-Gly-Arg-Phe-NH_2 (Antho-RFamide), <Glu-Ser-Leu-Arg-Trp-NH_2 (Antho-RWamide I) and <Glu-Gly-Leu-Arg-Trp-NH_2 (Antho-RWamide II) (Grimmelikhuijzen and Graff, 1986; Graff and Grimmelikhuijzen, 1988a,b). Antho-RFamide has also been isolated from the pennatulid *Renilla köllikeri* and appears to occur generally in anthozoans (Grimmelikhuijzen and Groeger, 1987). Antho-RFamide, however, probably does not occur in hydrozoans (Grimmelikhuijzen, unpublished). Instead of this tetrapeptide, two Antho-RFamide-related heptapeptides have been isolated from the hydromedusa *Polyorchis penicillatus* (Grimmelikhuijzen et al., 1988b). These are <Glu-Leu-Leu-Gly-Gly-Arg-Phe-NH_2 (Pol-RFamide I) and <Glu-[Trp,Lys,Leu]-Gly-Arg-Phe-NH_2 (Pol-RFamide II). The structures of Antho-RFamide, Antho-RWamide I and II and Pol-RFamide I have been determined after digestion with pyroglutamate aminopeptidase to remove <Glu, followed by straightforward sequencing using the Edman degradation (Gray and Smith, 1970) or the DABITC method (Chang et al., 1978). Pol-RFamide II has been investigated with both classical methods and fast-atom-bombardment mass spectrometry (in collaboration with K. L. Rinehart, Urbana, U.S.A.). The molecular weight of Pol-RFamide II has been established as 915, but the definitive sequences of the amino acids in positions 2, 3 and 4 have yet to be determined.

All five sequenced coelenterate peptides have novel structures and have not been isolated earlier from vertebrates or higher invertebrates. Among themselves, however, these peptides have several features in common: they are small (4-7 amino acids), contain only basic or neutral amino acids, and their structures can be described as <Glu...Arg-X-NH_2, where X is an aromatic amino acid.

The new coelenterate peptides can be assigned to a superfamily of <Glu...Arg-X-NH_2 peptides which, in addition, can be sub-divided into several smaller families. Such a smaller family can occur within one species. This is the case for the Antho-RWamide family in *Anthopleura elegantissima*, and the Pol-RFamide family in *Polyorchis penicillatus* (see Table 1). Family relationships, however, can also exist between peptides of species belonging to different classes of coelenterates: Antho-RFamide, which occurs in anthozoans is related to the Pol-RFamides which occur in hydrozoans (Table I). This situation is similar to that in higher animals, where families of neuropeptides occur, both within one species and across the different classes or phyla (cf. Grimmelikhuijzen et al., 1987). The mechanism by which a peptide family emerges is believed to be by gene duplication, followed by mutation of the duplicated gene. At present it is hard to say which is the most primitive or ancestral peptide of

Table 1. Neuropeptide families in coelenterates

Species	Structure	Name
Anthopleura elegantissima[1]	<Glu -Ser- Leu-Arg-Trp-NH_2	Antho-RWamide I
Anthopleura elgantissima[2]	<Glu -Gly- Leu-Arg-Trp-NH_2	Antho-RWamide II
Renilla kölliker[3]	<Glu- Gly-Arg-Phe-NH_2	Antho-RFamide
Anthopleura elegantissima[4]	<Glu- Gly-Arg-Phe-NH_2	Antho-RFamide
Polyorchis penicillatus[5]	<Glu-Leu-Leu-Gly- Gly-Arg-Phe-NH_2	Pol-RFamide I
Polyorchis penicillatus[6]	<Glu-[Trp,Lys,Leu]- Gly-Arg-Phe-NH_2	Pol-RFamide II

[1] Graff and Grimmelikhuijzen, 1988a.
[2] Graff and Grimmelikhuijzen, 1988b.
[3] Grimmelikhuijzen and Groeger, 1987.
[4] Grimmelikhuijzen and Graff, 1986.
[5] Grimmelikhuijzen et al., 1988b.
[6] Grimmelikhuijzen et al., unpublished.

the <Glu...Arg-X-NH_2 family. An answer to this interesting question might be gained by the isolation and sequencing of <Glu...Arg-X-NH_2 peptides from species belonging to several classes or subclasses of coelenterates and by comparison of the genes encoding for these peptides. Such an investigation may eventually lead to the animal group which is, neurochemically speaking, the most primitive coelenterate. This animal group might be closely related to the animals in which the first nervous systems evolved.

The presence of the <Glu group (which originates from Gln by cyclization) at the amino terminus, and of the amide group at the carboxy terminus, is quite common in biologically active peptides of higher animals (cf. Grimmelikhuijzen et al., 1987). A consequence of these substitutions is that <Glu does not contain a positive charge, rendering the peptide resistent towards normal amino peptidases, while the amide group does not contain a negative charge, making the peptide more resistent to ordinary carboxy peptidases. From the structures of the five sequenced coelenterate peptides, therefore, one can already anticipate that they must have biological activity.

Antisera have been raised against the amino and carboxy termini of Antho-RFamide and against the carboxy termini of Antho-RWamide I, II and Pol-RFamide I. Staining of sea anemones and of *Polyorchis penicillatus* with these antisera showed numerous immunoreactive neurons, whereas non-neuronal cells were not stained (cf.

Fig. 4). Thus, all peptides are produced by neurons and therefore are neuropeptides. Antho-RFamide, Antho-RWamide I, II and Pol-RFamide I do not have free amino and carboxy groups. Therefore, these peptides cannot be fixed to the surrounding proteins and membranes with normal fixatives. What we see with carboxyterminal antisera (Fig. 4a), then, are aminoterminal elongated, fixable forms of the coelenterate neuropeptides. When we use aminoterminal directed antisera (Fig. 4b), we see carboxyterminal elongated forms. This means that the coelenterate neuropeptides are synthesized within a precursor protein, which is subsequently processed to give one or more copies of the peptide. This conclusion is confirmed by a western blot of sea anemone extract (Fig. 5). This shows a band at 35 kD reactive with an antiserum against the carboxy terminus, and a band at 20 kD reactive with both an antiserum against the carboxy and the amino terminus of Antho-RFamide (Graff, 1988). These data indicate that the Antho-RFamide precursor is ≥ 35 kD and that multiple copies of Antho-RFamide are incorporated within this protein. This is a situation quite similar to that in *Aplysia*, where a precursor of approximately 58 kD contains 28 copies of immature FMRFamide (Taussig and Scheller, 1986), and *Drosophila*, where a precursor of 39 kD accommodates 13 copies of FMRFamide-related peptides (Schneider and Taghert, 1988).

Many of the sea anemone neurons stained by Antho-RFamide antisera are typical "sensory cells" (Fig. 4b). The processes of these "sensory cells" are often associated with smooth muscle fibers, suggesting that these cells may be both "sensory" and "motor neurons". The neurons which are stained by the antisera against Antho-RWamide I and II appear to be of a different population. Again, these neurons are "sensory" and their processes are associated with smooth, endodermal muscles (Fig. 4a). These morphological data suggest that Antho-RFamide and Antho-RWamide I and II could act as transmitters or modulators at neuromuscular junctions.

In an accompanying paper, the actions of Antho-RFamide and of Antho-RWamide I and II have been investigated (McFarlane et al., 1990; cf. McFarlane et al., 1987). The results described there also strongly indicate that the three anthozoan peptides have a transmitter-like function.

Figure 4. Immunoreactive neurons in a cross section through the upper body wall of the sea anemone *Calliactis parasitica*. Me, mesoglea; Sm, sphincter muscle. **(a)** Staining with an antiserum (195 II, 1:1000) directed against the carboxyterminus (Arg-Trp-NH_2) of both Antho-RWamide I and II. Immunoreactive "sensory cells" (arrows) are located in the endoderm. Long, immunoreactive processes project through the mesoglea to a dense plexus of neurites associated with fibers of the sphincter muscle (x300). **(b)** Staining with an antiserum (code 193 I; dilution 1:200) directed against the amino terminus (<Glu-Gly-Arg-Phe-) of Antho-RFamide. Immunoreactive "sensory cells" (arrow) are located in the endoderm. Their processes are associated with endodermal muscle fibers which are orientated either in a circular or longitudinal direction. The muscle fibers of the sphincter muscle are not associated with immunoreactive neurites. A similar picture is obtained with antisera directed against the carboxyterminus (-Gln-Gly-Arg-Phe-NH_2) of Antho-RFamide (x300).

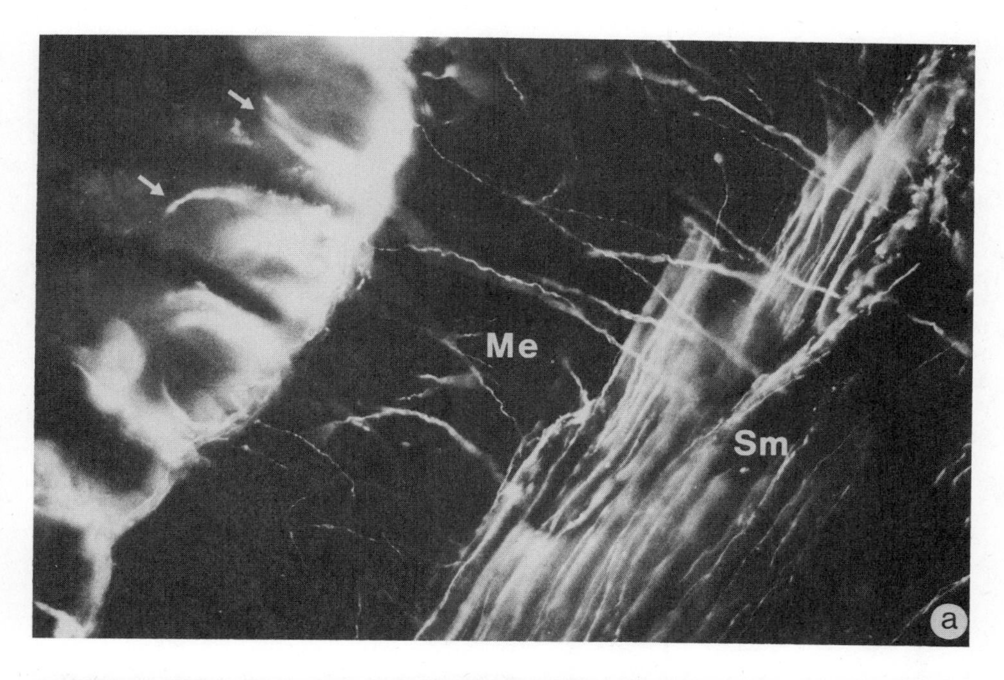
Me
Sm
a

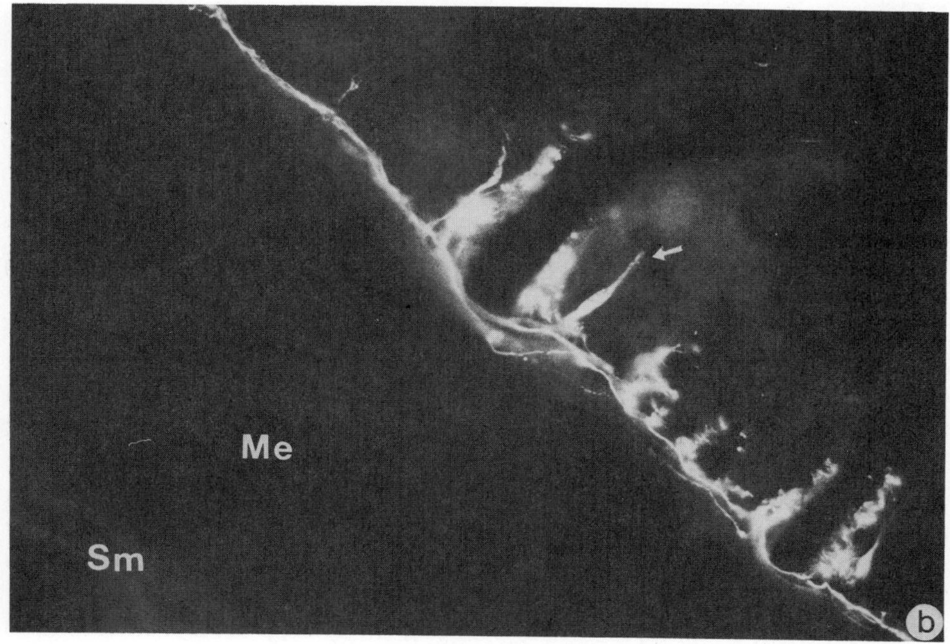
Me
Sm
b

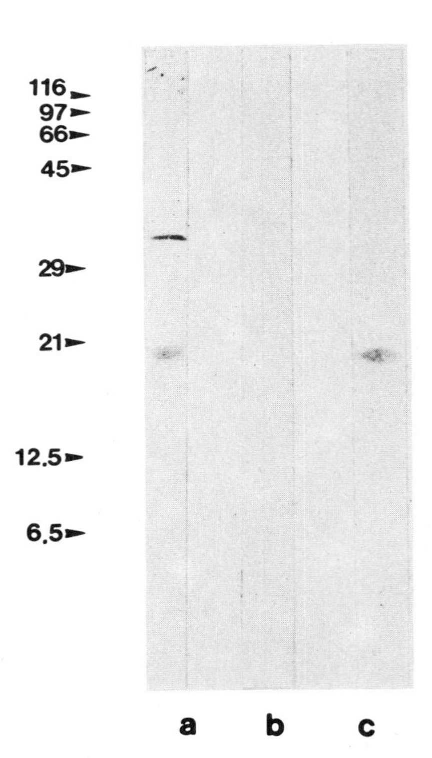

Figure 5. Western blot of an extract of the sea anemone *Calliactis parasitica* which was electrophoresed on 15% polyacrylamide. **(a)** Staining with antiserum 177 V (diluted 1:150) directed against the carboxyterminus (-Gln-Gly-Arg-Phe-NH_2) of Antho-RFamide. Two protein bands of approximately 35 kD and 20 kD are stained. **(b)** Preabsorption of antiserum 177 V with Sepharose-bound -Gln-Gly-Arg-Phe-NH_2, abolished all staining. **(c)** Staining with antiserum 193 I (diluted 1:150) directed against the aminoterminus (<Glu-Gly-Arg-Phe-) of Antho-RFamide. This antiserum stains a protein of approximately 20 kD, which might be the same protein stained in **a** (from Graff, 1988).

5. Discussion

Hydra possesses a class of neurons, the "secretory-sensory-motor-interneurons," which may be the ancestors of the more specialized neurons and paraneurons which arose later in evolution (Westfall, 1973a; Westfall and Kinnamon, 1978; Fujita and Kobayashi, 1979). Although the occurrence of multifunctional neurons in other coelenterates has not been carefully investigated by electron microscopy, our own light microscopic studies indicate that other hydrozoans (Grimmelikhuijzen et al., 1986) and anthozoans (Fig. 4) may also have such neurons.

During the last 50 years, several reports have described the occurrence of acetylcholine, serotonin and catecholamines in *Hydra* neurons (reviewed by Martin and Spencer, 1983), but none have been really convincing. Independently of each other, two of us (C.J.P. G. and O. K.) have investigated the presence of these "classical" transmitters in neurons of *Hydra*, and we have been unable to demonstrate these substances. Of course, we did not check the presence of the "classical" transmitters in many of the other coelenterate species (because negative results are not particularly exciting), and it is possible that these transmitters eventually occur elsewhere in coelenterates. What our data suggest, however, is that "classical" transmitters are not generally present in coelenterates and that they are, therefore, probably not the "oldest" transmitters.

RFamide-like peptides, on the other hand, occur generally in coelenterates (Grimmelikhuijzen, 1983; Grimmelikhuijzen et al., 1988a; Figs. 1-4). This suggests that these substances have emerged early in coelenterate evolution and makes them likely candidates for being the "first" transmitters. Preliminary biochemical work indicates that the RFamide-like peptides are synthesized within a large precursor molecule which, in subsequent steps, is processed to give multiple copies of biologically active peptides (Fig. 5). Ultrastructural studies suggest that the precursor is packed into dense-cored vesicles by the Golgi complex, and that these vesicles are transported along neurites to their sites of release (Fig. 3; Koizumi et al., 1989). All this is similar to processes that occur in peptidergic neurons of higher organisms (Brownstein et al., 1980; Taussig and Scheller, 1986; Schneider and Taghert, 1988). Also, the presence of the <Glu group at the amino and the amide group at the carboxy terminus (Table 1) is a common feature of peptide transmitters and hormones in higher animals (cf. Grimmelikhuijzen et al., 1987). The biosynthesis of biologically active peptides, then, appears to be an evolutionary old, and well conserved process.

ACKNOWLEDGEMENTS. We thank Susanne Raabe for typing the manuscript, Springer Verlag and Alan R. Liss, Inc. for their kind permission to reproduce figures 1 and 3, and the Bundesministerium für Forschung und Technologie and the Deutsche Forschungsgemeinschaft (Gr 762/7-2) for financial support.

References

Anderson, P. A. V., 1985, Physiology of a bidirectional, excitatory, chemical synapse, *J. Neurophysiol.* **53**:821-835.

Anderson, P. A. V., and Mackie, G. O., 1977, Electrically coupled, photosensitive neurons control swimming in a jellyfish, *Science* **197**:186-188.

Anderson, P. A. V., and Schwab, W. E., 1982, Recent advances and model systems in coelenterate neurobiology, *Prog. Neurobiol.* **19**:213-236.

Brownstein, M. J., Russel, J. T., and Gainer, H., 1980, Synthesis, transport, and release of posterior pituitary hormones, *Science* **207**:373-378.

Burnett, A. L., and Diehl, N. A., 1964, The nervous system of *Hydra*. I. Types, distribution and origin of nerve elements, *J. Exp. Zool.* **157**:217-226.

Chang, J. Y., Brauer, D., and Wittmann-Liebold, B., 1978, Microsequence analysis of peptides and proteins using 4-N,N-dimethylaminoazobenzene 4'-isothiocyanate/phenylisothiocyanate double coupling method, *FEBS Lett.* **93**:205-214.

Davis, L. E., Burnett, A. L., and Haynes, J. F., 1968, Histological and ultrastructural study of the muscular and nervous system in *Hydra*. II. Nervous system, *J. Exp. Zool.* **167**:295-332.

Fujita, T., and Kobayashi, S., 1979, Current views on the paraneurone concept, *Trends Neurosci.* **2**:27-30.

Graff, D., 1988, *Isolierung und Sequenzierung von Neuropeptiden aus Seeanemonen*, Ph.D. Thesis, University of Heidelberg.

Graff, D., and Grimmelikhuijzen, C. J. P., 1988a, Isolation of <Glu-Ser-Leu-Arg-Trp-NH_2, a novel neuropeptide from sea anemones, *Brain Res.* **442**:354-358.

Graff, D., and Grimmelikhuijzen, C. J. P., 1988b, Isolation of <Glu-Gly-Leu-Arg-Trp-NH_2 (Antho-RWamide II), a novel neuropeptide from sea anemones, *FEBS Lett.*, **239**:137-140.

Gray, W. R., and Smith, J. F., 1970, Rapid sequence analysis of small peptides, *Anal. Biochem.* **33:**36-42.

Grimmelikhuijzen, C. J. P., 1983, FMRFamide immunoreactivity is generally occurring in the nervous systems of coelenterates, *Histochem.* **78:**361-381.

Grimmelikhuijzen, C. J. P., 1984, Peptides in the nervous system of coelenterates, in: *Evolution and Tumor Pathology of the Neuroendocrine System*, pp. 39-58 (S. Falkmer, R. Håkanson, and F. Sundler, eds.), Elsevier, Amsterdam.

Grimmelikhuijzen, C. J. P., 1985, Antisera to the sequence Arg-Phe-amide visualize neuronal centralization in hydroid polyps, *Cell Tissue Res.* **241:**171-182.

Grimmelikhuijzen, C. J. P., 1986, FMRFamide-like peptides in the primitive nervous systems of coelenterates and complex nervous systems of higher animals, in: *Handbook of Comparative Opioid and Related Neuropeptide Mechanisms*, pp. 103-115 (G. Stephano, ed.), CRC Press, Boca Raton.

Grimmelikhuijzen, C. J. P., and Graff, D., 1985, Arg-Phe-amide-like peptides in the primitive nervous systems of coelenterates, *Peptides* **6** (Suppl. 3):477-483.

Grimmelikhuijzen, C. J. P., and Graff, D., 1986, Isolation of <Glu-Gly-Arg-Phe-NH_2 (Antho-RFamide), a neuropeptide from sea anemones, *Proc. Natl. Acad. Sci. USA* **83:**9817-9821.

Grimmelikhuijzen, C. J. P., and Groeger, A., 1987, Isolation of the neuropeptide pGlu-Gly-Arg-Phe-amide from the pennatulid *Renilla köllikeri*, *FEBS Lett.* **211:**105-108.

Grimmelikhuijzen, C. J. P. and Spencer, A. N., 1984, FMRFamide immunoreactivity in the nervous system of the medusa *Polyorchis penicillatus*, *J. Comp. Neurol.* **230:**361-371.

Grimmelikhuijzen, C. J. P., Dockray, G. J., and Schot, L. P. C., 1982, FMRFamide-like immunoreactivity in the nervous system of *Hydra*, *Histochem.* **73:**499-508.

Grimmelikhuijzen, C. J. P., Spencer, A. N., and Carré, D., 1986, Organization of the nervous system of physonectid siphonophores, *Cell Tissue Res.* **246:**463-479.

Grimmelikhuijzen, C. J. P., Graff, D., and McFarlane, I. D., 1987, Neuropeptides in invertebrates, in: *Nervous Systems in Invertebrates*, pp. 105-132 (M. A. Ali, ed.), Plenum Press, New York.

Grimmelikhuijzen, C. J. P., Graff, D., and Spencer, A. N., 1988a, Structure, location and possible actions of Arg-Phe-amide peptides in coelenterates, in: *Neurohormones in Invertebrates*, pp. 199-217 (M. C. Thorndyke and G. J. Goldsworthy, eds.), Cambridge University Press, Cambridge.

Grimmelikhuijzen, C. J. P., Hahn, M., Rinehart, K. L., and Spencer, A. N., 1988b, Isolation of <Glu-Leu-Leu-Gly-Gly-Arg-Phe-NH_2 (Pol-RFamide), a novel neuropeptide from hydromedusae, *Brain Res.* **475:**198-203.

Grimmelikhuijzen, C. J. P., Graff, D., and McFarlane, I. D., 1989, Neurones and neuropeptides in coelenterates, *Arch. Histol. Cytol.* **52** (Suppl.):0-0.

Hadži, J., 1909, Über das Nervensystem von *Hydra*, *Arb. Zool. Inst. Wien* **17:**225-268.

Hernandez-Nicaise, M. L., 1973, The nervous system of Ctenophora. III. Ultrastructure of synapses, *J. Neurocytol.* **2:**249-263.

Hertwig, O., and Hertwig, R., 1878, *Das Nervensystem und die Sinnesorgane der Medusen*, Vogel, Leipzig.

Horridge, G. A., and Mackay, B., 1962, Naked axons and symmetrical synapses in an elementary nervous system, *Nature* **193:**899-900.

Jha, R. K., and Mackie, G. O., 1967, The recognition, distribution and ultrastructure of hydrozoan nerve elements, *J. Morphol.* **123:**43-62.

Kinnamon, J. C., and Westfall, J. A., 1981, A three-dimensional serial reconstruction of neuronal distributions in the hypostome of a *Hydra*, *J. Morphol.* **168:**321-329.

Koizumi, O., Wilson, J. D., Grimmelikhuijzen, C. J. P., and Westfall, J. A., 1989, Ultrastructural localization of RFamide-like peptides in neuronal dense-cored vesicles in the peduncle of *Hydra*, *J. Exp. Zool.* **249:**17-22.

Mackie, G. O., 1973, Report on giant nerve fibres in *Nanomia*, *Publ. Seto Marine Labs.* **20:**745-756.

Mackie, G.O., 1990, Evolution of cnidarian giant axons, in: *Evolution of the First Nervous Systems*, (P.A.V. Anderson, ed.), Plenum Press, New York, in press.

Mackie, G. O., Singla, C. L., and Stell, W. K., 1985, Distribution of nerve elements showing FMRFamide-like immunoreactivity in Hydromedusae, *Acta. Zool. Stockh.* **66:**199-210.

Mackie, G. O., Singla, C. L., and Arkett, S. A., 1988, On the nervous system of *Vellela* (Hydrozoa: Chondrophora), *J. Morphol.* **198:**15-23.

Martin, S. M., and Spencer, A. N., 1983, Neurotransmitters in coelenterates, *Comp. Biochem. Physiol.* **74c:**1-14.

Matsuno, T., and Kageyama, T., 1984, The nervous system in the hypostome of *Pelmatohydra robusta*: The presence of a circumhypostomal nerve ring in the epidermis, *J. Morphol.* **182:**153-168.

McConnell, C. H., 1932, The development of the ectodermal nerve net in the buds of hydra, *Quart. J. Microsc. Sci.* **75:**495-509.

McFarlane, I. D., 1973, Spontaneous contractions and nerve-net activity in the sea anemone *Calliactis parasitica*, *Mar. Behaviour Physiol.* **2:**97-113.

McFarlane, I. D., Graff, D., and Grimmelikhuijzen, C. J. P., 1987, Excitatory actions of Antho-RFamide, an anthozoan neuropeptide, on muscles and conducting systems in the sea anemone *Calliactis parasitica*, *J. exp. Biol.* **133:**157-168.

McFarlane, I. D., Graff, D., and Grimmelikhuijzen, C. J. P., 1990, Evolution of conducting systems and neurotransmitters in the Anthozoa, in: *Evolution of the First Nervous Systems*, (P. A. V. Anderson, ed.), Plenum Press, New York, in press.

Price, D. A., and Greenberg, M., 1977, Structure of a molluscan cardioexcitatory neuropeptide, *Science* **197:**670-671.

Satterlie, R. A., 1979, Central control of swimming in the cubomedusan jellyfish *Charibdea rastonii*, *J. Comp. Physiol.* **133:**357-367.

Schneider, K. C., 1890, Histologie von *Hydra fusca* mit besonderer Berücksichtigung des Nervensystems der Hydropolypen, *Arch. mikr. Anat.* **35:**321-379.

Schneider, L. E., and Taghert, P. H., 1988, Isolation and characterization of a *Drosophila* gene that encodes multiple neuropeptides related to Phe-Met-Arg-Phe-NH_2 (FMRFamide), *Proc. Natl. Acad. Sci. USA* **85:**1993-1997.

Spangenberg, D. E., and Ham, R. G., 1960, The epidermal nerve net of *Hydra*, *J. Exp. Zool.* **143:**195-201.

Spencer, A. N., 1978, Neurobiology of *Polyorchis*. I. Function of effector systems, *J. Neurobiol.* **9:**143-157.

Spencer, A. N., 1979, Neurobiology of *Polyorchis*. II. Structure of effector systems, *J. Neurobiol.* **10:**95-117.

Spencer, A. N., 1982, The physiology of a coelenterate neuromuscular synapse, *J. Comp. Physiol.* **148:**353-363.

Spencer, A.N., 1990, Electrical and chemical synaptic transmission in the Cnidaria, in: *Evolution of the First Nervous Systems*, (P. A. V. Anderson, ed.), Plenum Press, New York, in press.

Spencer, A. N., and Arkett, S. A., 1984, Radial symmetry and the organization of central neurones in a hydrozoan jellyfish, *J. exp. Biol.* **110:**69-90.

Spencer, A. N., and Satterlie, R. A., 1980, Electrical and dye coupling in an identified group of neurons in a coelenterate, *J. Neurobiol.* **11:**13-19.

Tardent, P., and Weber, C., 1976, A qualitative and quantitative inventory of nervous cells in *Hydra attenuata* Pall., in: *Coelenterate Ecology and Behaviour*, pp. 501-512 (G. O. Mackie, ed.), Plenum Press, New York.

Taussig, R., and Scheller, R. H., 1986, The *Aplysia* FMRFamide gene encodes sequences related to mammalian brain peptides, *DNA* **5:**453-461.

Weber, C., 1989, Smooth muscle fibers of *Podocoryne carnea* (Hydrozoa) demonstrated by a specific monoclonal antibody and their association with neurons showing FMRFamide-like immunoreactivity, *Cell Tissue Res.* **255:**275-282.

Westfall, J. A., 1973a, Ultrastructural evidence for a granule-containing sensory-motor-interneuron in *Hydra littoralis*, *J. Ultrastruct. Res.* **42:**268-282.

Westfall, J. A., 1973b, Ultrastructural evidence for neuromuscular systems in coelenterates, *Am. Zool.* **13:**237-246.

Westfall, J. A., 1987, Ultrastructure of invertebrate synapses, in: *Nervous Systems in Invertebrates*, pp. 3-28 (M. A. Ali, ed.), Plenum Press, New York.

Westfall, J. A., and Kinnamon, J. C., 1978, A second sensory-motor-interneuron with neurosecretory granules in *Hydra*, *J. Neurocytol.* **7:**365-379.

Westfall, J. A., and Kinnamon, J. C., 1984, Perioral synaptic connections and their possible role in feeding behavior of *Hydra*, *Tissue Cell* **16:**355-365.

Westfall, J. A., Kinnamon, J. C., and Sims, D. E., 1980, Neuro-epitheliomuscular cell and neuro-neuronal gap junctions in *Hydra*, *J. Neurocytol.* **9:**725-732.

Chapter 8

Evolution of Conducting Systems and Neurotransmitters in the Anthozoa

I. D. McFARLANE, D. GRAFF, and
C. J. P. GRIMMELIKHUIJZEN

1. Introduction

We can justify studies of the sea anemone nervous system in two ways. First, as biologists we are interested in how any animal detects its surroundings and produces appropriate behavior. Secondly, more significant but difficult to prove, is that by studying the nervous system in a simple form, we will be able to discuss how the nervous system evolved. This article considers if studies of the sea anemone nervous system can contribute to this discussion.

We are investigating the location and function of sea anemone neuropeptides, but to demonstrate the relevance and potential of this work, and to raise questions about the evolution of nerve cells, we must first describe the organization of the sea anemone nervous system. We shall then ask if this differs significantly from that of other cnidarians and other invertebrates. Finally, we will describe our research on anthozoan neuropeptides and other putative neurotransmiters.

I. D. McFARLANE[1], D. GRAFF[2], and C. J. P. GRIMMELIKHUIJZEN[2] • [1]Department of Applied Biology, University of Hull, Hull HU6 7RX, United Kingdom. [2]Centre for Molecular Neurobiology, University of Hamburg, Martinistrasse 52, 2000 Hamburg 20, Federal Republic of Germany.

Evolution of the First Nervous Systems
Edited by P.A.V. Anderson
Plenum Press, New York

2. Organization of the Sea Anemone Nervous System

As far as function is concerned, no one has made intracellular recordings from anthozoan neurons and the little we have gleaned about neural function comes from extracellular techniques. As far as structure is concerned, silver (e.g., Batham, 1965) or methylene blue (e.g., Robson, 1965) stains neurons in the ectoderm and endoderm. These histological technques are, however, limited: they neither reveal connections across the mesogloea where neurophysiological evidence says connections must exist (Lawn, 1980), nor do they detect neurons in the body wall ectoderm, an area where electrophysiology demands the presence of a conducting system (McFarlane, 1969). Furthermore, these techniques show no functional divisions within the nerve net.

Electrophysiology, however, reveals three separate conducting systems (McFarlane, 1982). A single shock to the column evokes three pulses that can be recorded extracellularly at a tentacle (Fig. 1). The fastest pulse is in the through-conducting nerve net (TCNN), the slower pulses are in slow system 1 (SS1) and slow system 2 (SS2). As all are blocked by excess magnesium, they may be conventional nerve nets with chemical synapses. The conducting systems are separate: each has a distinct stimulus threshold and a distinct distribution. Expressed simply, the SS1 and SS2 are ectodermal and endodermal, respectively, whereas the TCNN occurs everywhere except the body wall ectoderm. Josephson (1974) asked if nerve nets or nerve cells are the functional units of the cnidarian nervous system, but we have no evidence for independent action of subcomponents of anthozoan conducting systems. In each system, activity spreads everywhere without restriction. Other conducting systems have been proposed (e.g., Jackson and McFarlane, 1976; Pickens, 1988), mostly to explain local reflexes. Neurons with a restricted spread of activity must exist, but we have few descriptions of local electrical responses (Shelton and Holley, 1984).

As for sense cells, anemones have only mechanoreceptors and chemoreceptors, and a general light sensitivity (Marks, 1976). Receptors may be simple, but Watson and Hessinger (1989) alert us to unexpected complexity by showing that chemical stimuli can tune cnidoblast mechanoreceptors to the vibration frequencies of their prey. Indeed, *Stomphia coccinea* will not feed while swimming (Ross and Sutton, 1964): this is reminiscent of receptor response modification in the sea slug *Tritonia diomedia* (Audesirk and Audesirk, 1980). Local sense cells may directly innervate nearby muscles (Parker, 1919), but we only know about receptors connected with the TCNN, SS1, and SS2. The TCNN responds to mechanical stimulation (Josephson, 1966), and the alarm reaction in *Anthopleura elegantissima* may be a TCNN chemosensory response (Howe and Sheikh, 1975). The SS1 in some species is touch-sensitive. SS1 chemoreceptors to food include betaine receptors on the column of *Tealia felina* (Boothby and McFarlane, 1986). In *Stomphia*, SS1 chemoreceptors on the tentacles respond to certain starfish, and column receptors detect predatory nudibranchs (Lawn, 1976c). SS2 chemoreceptors located around the mouth of *Calliactis parasitica* detect

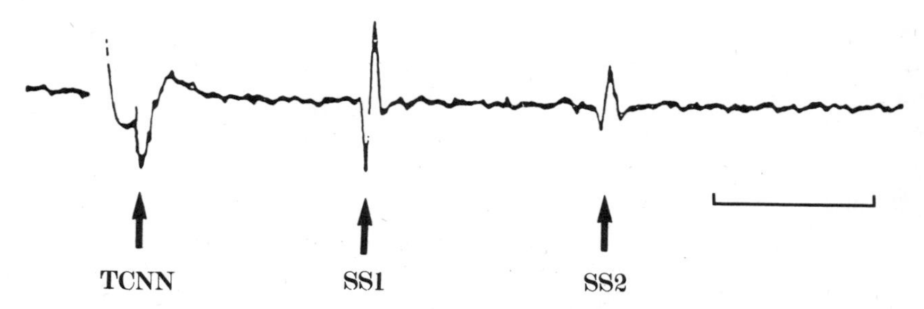

Figure 1. Sea anemones have at least three separate conducting systems. In response to a single shock to the column a suction electrode attached to a tentacle records three pulses, in the through-conducting nerve net (TCNN), slow system 1 (SS1), and slow system 2(SS2). This record is from *Calliactis parasitica*. Scale = 250 msec.

various amino acids (McFarlane, 1975). Tentacular SS2 receptors respond to shell when *Stomphia* climbs onto shells (Lawn and McFarlane, 1976), but climbing *Calliactis* detect shells with SS1 and SS2 chemoreceptors (McFarlane, 1976).

The TCNN is spontaneously active. The SS1 is not, but in detached *Calliactis parasitica* single shocks evoke delayed bursts of SS1 pulses (Jackson and McFarlane, 1976). Spontaneous SS2 pulses (McFarlane, 1973) might come from pacemakers, but are probably from unidentified sense cells. TCNN bursts come at intervals that vary according to behavioral phase (McFarlane, 1983). TCNN pacemakers are diffusely distributed (McFarlane, 1974b). In *Phyllactis concinnata* there may be different TCNN pacemakers, with different firing rates (Pickens, 1988).

Some behavioral outputs of the conducting systems are known. TCNN activity makes muscles contract, although in the primitive anemone *Protanthea simplex* the TCNN does not seem to innervate tentacle ectodermal muscles (McFarlane, 1984b). Individual muscle groups give either tonic or phasic contractions, or both (Ross, 1957). TCNN pulse frequency determines the type of contraction, and which muscle group contracts. Most muscles show spontaneous contractions, sometimes following TCNN bursts, but at other times independent of TCNN activity (McFarlane, 1974b). Presumably, local neural pacemakers innervate the muscles but do not connect with the TCNN. The TCNN also inhibits some muscles (Ewer, 1960; Lawn, 1976a,b).

SS1 activity can evoke pedal disc detachment (McFarlane, 1969), oral disc ectodermal muscle inhibition and subsequent oral disc expansion (McFarlane and Lawn, 1972), and column extension, presumably due to endodermal circular muscle contraction (Lawn, 1980). The SS1 has a weak excitatory coupling with ectodermal tentacle muscles in *Calliactis parasitica* (McFarlane, 1984a).

Established SS2 actions are inhibitory. SS2 pulses inhibit endodermal muscles and TCNN pacemakers in *Calliactis parasitica* (McFarlane, 1974a,b). Inhibition is dramatic in the primitive anemone *Protanthea simplex*, where SS2 activity makes the anemone "collapse" (McFarlane, 1985). The SS2 may also be excitatory: in *Calliactis*, SS2 activity is accompanied by contraction of transverse muscles in the perfect mesenteries (McFarlane, 1975).

Behavior can often be linked to activity in the three known conducting systems, but the connection between sensory input and behavioral output is sometimes obscure. For example, in *Stomphia coccinea* escape swimming a comparatively brief SS1 sensory response evokes repeated column flexions that may continue for several minutes without accompanying pulses in the TCNN, SS1, or SS2 (Lawn, 1976c).

3. Comparison with Other Cnidaria

Sea anemones may not be typical of all Anthozoa (Shelton, 1982), although there are three conducting systems in some hard corals (McFarlane, 1978) and pennatulids (Anderson and Case, 1975). Only one conducting system has been found in cerianthids (Arai, 1985; McFarlane, 1988).

Hydrozoa may be unique among the Cnidaria with their mixture of nerve nets and non-nervous (neuroid) conducting systems (Josephson, 1974; Anderson, 1980). Suggestions that the SS1 and SS2 may be neuroid systems (McFarlane, 1974a, 1978) have not been substantiated. On the contrary, our immunohistochemical studies convince us that earlier failures to detect neurons in places where physiological evidence shows a conducting system exists were due simply to the limitations of previous staining techniques. Mackie, Anderson, and Singla (1984) propose that only the Hydrozoa, and not Scyphozoa or Anthozoa, possess gap junctions, the essential components of cell-to-cell communication. This raises the problem of developmental regulation as gap junctions are important for patterning processes (Guthrie and Gilula, 1989). Mackie et al. (1984) suggest alternative mechanisms by which Scyphozoa and Anthozoa may control development.

Neural organization in Anthozoa more resembles that of the Scyphozoa than the Hydrozoa. Scyphozoa have two nerve nets in the subumbrellar ectoderm, the giant fiber nerve net (GFNN) and the diffuse fiber nerve net (DNN), and a third net in the endoderm (Passano, 1982). It is tempting, though of questionable value, to suggest that the GFNN, DNN, and endodermal net are equivalent to the TCNN, SS1, and SS2 respectively. In support of this, we note that scyphozoan pacemaker cells are connected to the GFNN, and pacemaker output can be DNN-modulated (Passano, 1973). Another interesting comparison, though again possibly coincidental, is that the DNN sometimes (but not always) excites waves of tentacle contraction (Romanes, 1878; Passano, 1982). This is reminiscent of the labile excitatory coupling of the SS1 to tentacle muscles in *Calliactis parasitica* (McFarlane, 1984a).

4. Comparison with Higher Invertebrates

By higher invertebrates we refer to molluscs and arthropods, the groups most intensively studied; other phyla may have unique solutions to problems of coordination.

Obviously, sea anemones lack a brain, but this is associated more with radial symmetry than with presumed simplicity of the nervous system. Do sea anemones have ganglia? Scyphomedusae and hydromedusae certainly do, though perhaps hydrozoan polyps do not (Josephson, 1974). If we define ganglia physiologically as groups of nerve cells that cooperate to analyze sensory information and coordinate discrete components of behavior, then anemones may have ganglia: the dense aggregations of nerve cells where the mesenteries insert onto the body wall are ideally situated to act as "linear ganglia".

If we consider ganglionic functions according to current concepts of the organization of invertebrate nervous systems, we see few significant differences between sea anemones and other invertebrates. For example, rhythmic and repetitive behaviors result from activity of central pattern generators or CPGs (Delcomyn, 1980); these may be single cells or neural networks showing emergent properties (Getting, 1983). TCNN bursts may originate in single pacemaker neurons: in *Calliactis parasitica* the bursts resemble the output of some molluscan oscillators (Berridge and Rapp, 1979) in showing parabolic bursting, with pulse frequency rising and falling during the course of the burst. The same membrane characteristics revealed in molluscan pacemakers may be responsible for TCNN burst production in sea anemones.

TCNN pacemakers can be classed as CPGs and, as in higher invertebrates (Pearson, Reye, and Robertson, 1983), sensory feedback from their driven behavioral outputs can modulate output patterns. In sea anemones, SS2 activity, possibly arising from endodermal sense cells that detect mechanical stresses between antagonistically-acting longitudinal and circular muscles of the body wall (McFarlane, 1974b), can slow TCNN bursts. This alters the behavioral output as TCNN frequency determines which muscles contract. The SS2 inhibits some TCNN pacemaker cells in *Phyllactis concinnata*, but not others (Pickens, 1988). In *Calliactis parasitica*, there is also a mechanism that varies the inter-burst interval: simultaneous SS1 and TCNN stimulation reduces the interval between TCNN bursts (McFarlane, 1983). The SS2- and SS1-associated modulations may be produced in very different ways: SS2 inhibition is quick-acting but decays within a few seconds (McFarlane, 1974b); whereas the SS1-associated action has a slow onset but may last more than an hour (McFarlane, 1983).

Apart from housing CPGs, invertebrate ganglia process primary sensory information and use it, together with commands coming from higher centers, to determine appropriate patterns of motor output. In arthropods, these functions are subserved by local interneurons (nerves whose axons are restricted to the ganglion), both spiking cells that analyze sensory information, and non-spiking cells that determine motoneuron firing patterns (Burrows, 1985). The terms "higher centers" and "local interneurons" are not applicable to sea anemones, and without intracellular recordings we cannot say if both propagated and decremental conduction are found in the nerve net.

Ganglionic output is in part coordinated by command fibers. A command fiber (Kupfermann and Weiss, 1978) is a neuron capable of driving many other neurons to produce a coordinated behavior: it is both sufficient and necessary for the behavior. Although made up of a great many neurons, sea anemone conducting systems act as command fibers, e.g., the SS1 in *Stomphia coccinea* is the escape swimming command fibre. The SS1, however, can command more than one type of behavior (McFarlane, 1973b).

What of differences at the synaptic level? The simplest form of learning, habituation, was shown in a behavioral study of *Anthopleura elegantissima* (Logan, 1975). Other examples of synaptic plasticity are less certain. There are no convincing demonstrations of sensitization or associative conditioning. Without intracellular techniques, the only way to identify higher forms of learning may be to find specific chemical pathways leading to phosphorylation or other modification of membrane channels; this would show pathways similar to those responsible for plastic changes in *Aplysia* synapses (Byrne, 1985).

5. Functions of Anthozoan Neuropeptides

Neurophysiological studies show us that the "simple" nervous systems of anthozoans are actually quite complex. There are interacting multiple conducting systems with both excitatory and inhibitory outputs (McFarlane, 1982), multiple pacemakers that act locally or connect with conducting systems, and a great variety of sensory inputs. Many muscles have multiple innervation, e.g., tentacle ectodermal longitudinal muscles in *Calliactis parasitica* are innervated by the TCNN, which elicits fast symmetrical contractions. They are also innervated by local neurons, as tentacles can show local contractions. They also receive labile, excitatory connections from the SS1 (McFarlane, 1984a), and perhaps inhibitory input from the same system (McFarlane, 1982). Again, *Calliactis* sphincter muscle has multiple innervation, or at least, has multiple transmitters acting on it, as we know that the sphincter gives both phasic and tonic contractions (Ross, 1957), and can be inhibited (Lawn, 1976a,b).

We know nothing about the mechanisms of transmitter release or uptake in the sea anemone, but if we prohibit any possibility of cross-talk in the nervous system, we can predict total numbers of neurotransmitters. If fast and slow contractions are to be evoked separately, and if each of the three conducting systems has excitatory and inhibitory actions, there must be seven different transmitters. There will be fewer required if conducting systems share the same transmitter (e.g., actions at well-separated sites will not suffer from cross-talk), or if two systems act the same way at the same site (e.g., both the SS1 and the TCNN can evoke contraction of body wall circular muscles). More transmitters will be needed if a conducting system does not utilize the same transmitter at all its synapses, or if we allow for local conducting systems and the existence of neuromodulators. The rest of this chapter will consider possible neurotransmitter/neuromodulator roles for anthozoan neuropeptides and "classical" transmitters.

The accompanying chapter (see Grimmelikhuijzen et al. in this book) describes the extraction and identification of three endogenous neuropeptides from the sea anemone *Anthopleura elegantissma*. The functions of these peptides, Antho-RFamide (pyroGlu-Gly-Arg-Phe-amide), Antho-RWamide I (pyroGlu-Ser-Leu-Arg-Trp-amide) and, Antho-RWamide II (pyroGlu-Gly-Leu-Arg-Trp-amide), have been investigated using isolated preparations from several species of sea anemone. Apart from describing the physiological actions of these neuropeptides, we aim to link our physiological knowledge of the conducting systems with a detailed immunohistochemical study of peptide distribution. We can then identify the neurons that make up the various nerve nets and we can map the sea anemone nervous system. Some of the Antho-RFamide results have already been published (McFarlane et al., 1987); the Antho-RWamide I and Antho-RWamide II results will appear shortly.

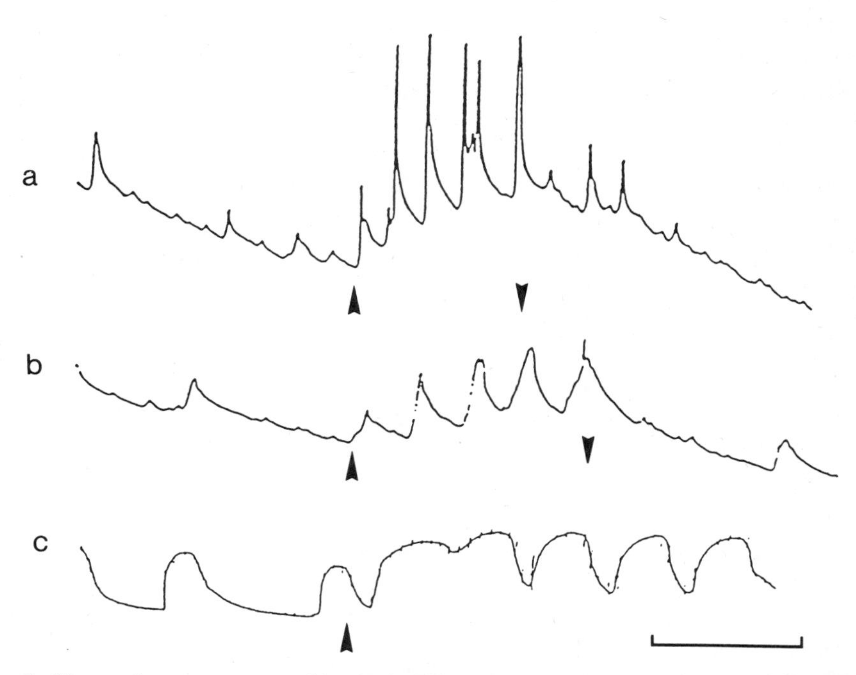

Figure 2. The anthozoan neuropeptide Antho-RFamide increases the frequency of spontaneous contractions in **(a)** *Calliactis parasitica* sphincter muscle rings, **(b)** *Calliactis* mid-column circular muscle rings, and **(c)** isolated tentacles of *Actinia equina*. Final concentrations were 10^{-6} M. Scale: (a,b) = 20 min, (c) = 5 min.

We have used several isolated preparations, including sphincter muscle and circular muscle rings from *Calliactis parasitica*. One difficulty encountered has been the high concentrations of neuropeptide sometimes required to evoke responses (up to 10^{-5} M). This may be a penetration problem, as trimming of the preparation reduces the threshold, in some instances, to only 10^{-8} M. The extent and influence of permeability barriers in cnidarian preparations has been discussed by Martin and Spencer (1983). Another problem is that anthozoans may show seasonal variation in sensitivity. *Renilla köllikeri* preparations were much more sensitive to FMRFamide during April - September than during the rest of the year (Anctil, 1987).

The first neuropeptide studied, Antho-RFamide, has similar effects on all preparations (McFarlane et al., 1987): it increases the frequency, and sometimes the amplitude of spontaneous contractions, and induces spontaneity in quiescent preparations (Fig. 2). This is true whether the preparation is of endodermal circular muscles (sphincter and body wall circulars), or of endodermal longitudinal muscles (body wall parietal muscles). Sometimes the evoked spontaneous contractions are superimposed upon a tonic contraction. A trimmed sphincter muscle preparation has a threshold of 10^{-7} M.

Antho-RFamide also excites spontaneous activity in ectodermal muscles. There are two groups of ectodermal muscles in most sea anemones, tentacle longitudinal muscles and oral disc radial muscles; they play important roles in feeding behavior. Antho-RFamide (threshold about 10^{-8} M) increases the frequency of spontaneous contractions (Fig. 2) in isolated tentacles of *Actinia equina*. Once again, it can induce spontaneous contractions in quiescent preparations. As with endodermal muscles, there is evidence of an early tonic contraction before the tentacle settles into a faster rhythm.

Antho-RFamide may be a neurotransmitter or neuromodulator, but we do not know how or where it acts. Similar activation of spontaneous contraction occurs with related invertebrate neuropeptides; for example, FMRFamide at 1.5×10^{-7} M causes rhythmicity in *Helix aspersa* pharyngeal retractor muscles (Lehmann and Greenberg, 1987). In other anthozoa, FMRFamide and RFamide (neither of these are the endogenous peptide) induce tonic contractions in preparations of *Renilla köllikeri* (Anctil, 1987) at a threshold concentration of around 10^{-6} M but at higher concentrations FMRFamide induces peristalsis-like rhythmicity in quiescent preparations. Here, there is evidence that FMRFamide acts directly on muscle cells. FMRFamide and related neuropeptides are common in invertebrates and actions on both muscles (e.g., Price and Greenberg, 1980) and neurons (e.g. Cottrell et al., 1984) have been noted. The great variety of effects of FMRFamide-like peptides and the range of molecular mechanisms by which invertebrate neuropeptides act (O'Shea and Schaffer, 1985) make it difficult to suggest where Antho-RFamide acts in sea anemones. Electrophysiological recordings showed that Antho-RFamide does not stimulate the TCNN. When applied to the ectoderm, it evokes SS1 activity; when applied to the endoderm, it elicits both SS1 and SS2 activity. We do not know, however, whether Antho-RFamide stimulates SS1 and SS2 sensory receptors or synapses in the nerve nets. The SS1 and SS2 can modulate TCNN pacemakers; one system excites (McFarlane, 1983), the other inhibits (McFarlane, 1974b), but it is not easy to relate these actions to the observed increase in spontaneity in preparations: SS2 pulses inhibit, not stimulate, spontaneous contractions (McFarlane, 1974a). Antho-RFamide action is consistent with excitation of pacemaker neurons, yet it appears not to affect TCNN pacemaker neurons. We conclude that Antho-RFamide either directly induces rhythmicity by acting on muscle membranes, or it stimulates local neural pacemakers; if the latter, then these pacemakers clearly respond differently from TCNN pacemakers.

What is the behavioral significance of Antho-RFamide? We assume that it is delivered by directed synapses, but more widespread actions via undirected synapses, or even release into the coelenteron (digestive enzymes will not always be present), cannot be ruled out. Certainly the physiological evidence suggests that Antho-RFamide-containing neurons must be widespread in the nervous system of *Calliactis parasitica*. In intact animals, as opposed to preparations, increases in spontaneous activity normally follow feeding and are associated with digestive and circulatory movements of coelenteric fluid (McFarlane, 1982). It may be worth noting that the SS1 and SS2, the conducting systems stimulated by Antho-RFamide, show sensory responses during feeding (McFarlane, 1975).

We might expect Antho-RWamide I and Antho-RWamide II to have similar actions, as they only differ by the substitution of glycine for serine. We cannot presently distinguish them immunohistochemically, so we do not know if they have different distributions. We presume that there are separate membrane receptors for the two peptides, but there may be cross-reaction in our pharmacological experiments, particularly at high concentrations. This was seen with two RFamide peptides in *Helix aspersa* (Lehman and Greenberg, 1987).

Antho-RWamide I and Antho-RWamide II both elicit slow tonic contractions of endodermal muscles in *Calliactis parasitica* (Figs. 3 and 4), at a threshold concentration of around 10^{-6} M and 10^{-7} M, respectively. The frequency of spontaneous contractions is not increased during Antho-RWamide I/II-induced contractures, and these peptides, unlike Antho-RFamide, do not induce spontaneity in quiescent preparations. These peptides do not evoke TCNN, SS1, or SS2 pulses. It is, therefore, unlikely that they act at interneural synapses in any of the known conducting systems. Antho-RWamide I and Antho-RWamide II may act directly on muscles: there are abundant Antho-RWamide I/II-containing neurons close to the sphincter muscle of *Calliactis* (Grimmelikhuijzen, in preparation). Therefore we propose that Antho-RWamide I and/or Antho-RWamide II may be neuromodulators or neurotransmitters for slow contraction of endodermal muscles in *Calliactis*. The sphincter muscle, but not the body wall circular muscles, also gives fast contractions. We do not yet know if either Antho-RWamide I or Antho-RWamide II excite fast contractions. The actions of Antho-RWamide I and II might be better compared in *Metridium senile*, where fast and slow contractions are easier to separate physiologically than in *Calliactis* (Ross, 1960b).

Isolated tentacles of *Actinia equina* are usually spontaneously active: both Antho-RWamide I and II inhibit these contractions (Figs. 3 and 4). Both peptides can also

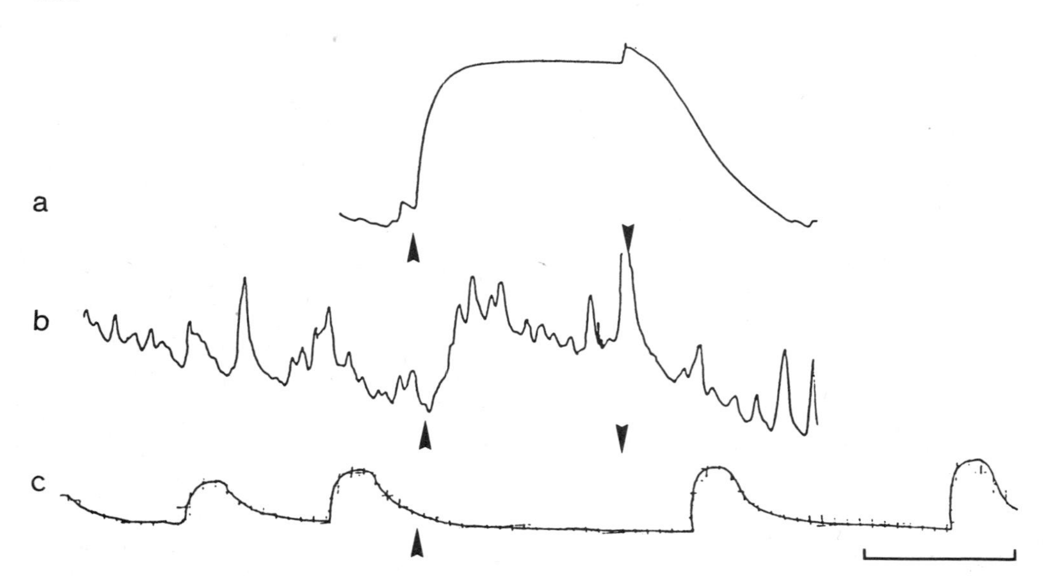

Figure 3. The anthozoan neuropeptide Antho-RWamide I stimulates slow contractions in **(a)** *Calliactis parasitica* sphincter muscle rings, and **(b)** *Calliactis* mid-column circular muscle rings, but it inhibits spontaneous contractions of **(c)** isolated tentacles of *Actinia equina*. Final concentrations were 10^{-6} M. Scale: **(a,b)** = 15 min, **(c)** = 5 min.

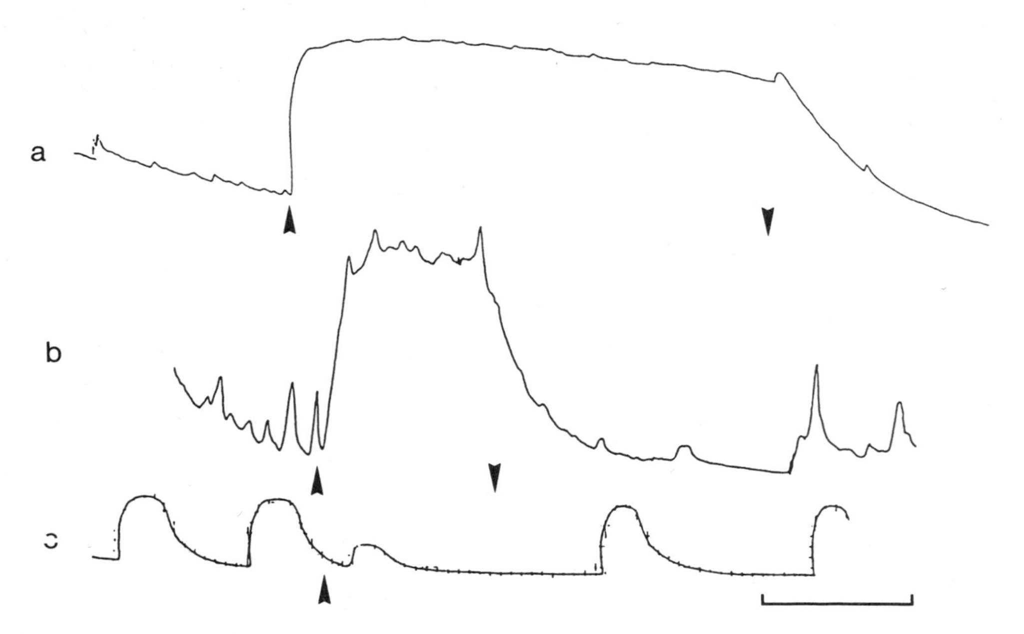

Figure 4. The anthozoan neuropeptide Antho-RWamide II stimulates slow contractions in **(a)** *Calliactis parasitica* sphincter muscle rings, and **(b)** *Calliactis* mid-column circular muscle rings, but it inhibits spontaneous contractions of **(c)** isolated tentacles of *Actinia equina*. Note the early contraction before the preparation became inhibited. Final concentration was 10^{-6} M. Scale: **(a,b)** = 15 min, **(c)** = 5 min.

produce a temporary inhibition of Antho-RFamide-enhanced rhythmicity in tentacles. The thresholds are similar (about 10^{-6} M) and the only observed difference is that in about 50% of trials Antho-RWamide II also evokes an early direct contraction (Fig. 4). There may be cross-reaction to the two peptides at the membrane receptor sites. We do not know where these peptides act; they might inhibit ectodermal longitudinal muscles directly, or inhibit pacemaker neurons, or excite inhibitory neurons. As in *Calliactis parasitica*, Antho-RWamide I and II evoke tonic contractions of endodermal sphincter and circular muscles in *Actinia*. Thus, we have demonstrated that in *Actinia* these neuropeptides have excitatory actions on endodermal neuromuscular systems, but inhibitory actions on ectodermal neuromuscular systems.

Antho-RWamide I and/or II may be endodermal slow muscle neurotransmitters. This presumably means that they are released by the TCNN, the conducting system that evokes endodermal muscle contractions. The TCNN, however, does not inhibit ectodermal muscles. The SS1, on the other hand, does inhibit ectodermal oral disc radial muscles in *Tealia felina* (McFarlane and Lawn, 1972), so we suspect that tentacle inhibition is mediated by the SS1. If so, then both the TCNN and the SS1 may contain Antho-RWamide I/II in some of their component neurons. This warns us to be careful when we attempt to relate physiological and immunohistochemical results. A further complication is that identified neuropeptides may have different functions in different species.

Although this is a "simple" nervous system, we must not assume that all sea anemones share a common physiology. Although the neuropeptides we study have been extracted for *Anthopleura elegantissima*, we have so far assumed that we can select the most suitable species for a given preparation and still make general statements about peptide action. It now looks as though we may be mistaken. In *Calliactis parasitica* tentacle preparations, although Antho-RWamide II inhibits spontaneous contractions as it does in *Actinia equina*, Antho-RWamide I causes direct tonic contraction. This suggests that muscles or nerves in *Calliactis* tentacles have membrane receptors that can distinguish between Antho-RWamide I and Antho-RWamide II. The difference between the species might be related to the observation that in *Calliactis* the SS1 has labile excitatory connections with tentacle muscles (McFarlane, 1982). We have never seen this in *Actinia*.

We have found physiological actions for the three known neuropeptides, but there are still physiological phenomena, e.g., inhibition of endodermal muscles, that lack a corresponding activator. Either we can expect further peptides to be isolated or we must turn to the consideration of possible non-peptide transmitters.

6. Physiology of Other Putative Transmitters

Neuropeptides do not have a monopoly as putative neurotransmitters in anthozoans. Ross (1960a,b) found that adrenaline at 10^{-5} M caused direct contractions in sphincter and circular muscle preparations of *Calliactis parasitica* and *Metridium senile*. Normal adrenaline-blocking agents were ineffective, so Ross was reluctant to suggest that adrenaline was a slow muscle transmitter. It is conceivable, however, that in sea anemones the mechanisms of receptor coupling and transmitter breakdown differ significantly from those in vertebrates. Here, then, we have three competing candidates for sphincter muscle excitatory neurotransmitters: Antho-RWamide I, Antho-RWamide II, and adrenaline. Assuming that fast and slow contractions are separately controlled, we can provide employment for only two. Of course, there may be coexistence of transmitters and modulation of contractions in ways not yet revealed in physiological studies. The co-existence of "classical" transmitters and neuropeptide transmitters is an established phenomenon (Adams and O'Shea, 1983): we need to compare responses induced by adrenaline and by the neuropeptides and we must see if adrenaline is found in synapses that contain Antho-RWamide I and/or Antho-RWamide II.

Van Marle (1977) did not find adrenaline in *Tealia felina* tentacles, but it does occur in the mesenteries of *Metridium senile* (Wood and Lentz, 1964). Tentacles contain noradrenaline (Van Marle, 1977), but Ross (1960a,b) found that noradrenaline did not affect sphincter or circular muscle preparations of *Calliactis parasitica*. The action of noradrenaline on tentacles is not known. Van Marle (1977) provides histochemical evidence that ATP is a transmitter in sea anemone tentacles, but again, the possibility of purinergic transmission in tentacles or in other muscle systems has yet to be tested physiologically.

Carlyle (1974) showed that glutamic acid reversibly inhibits electrically-evoked contraction of various circular muscle preparations in *Actinia equina*. We have confirmed this observation with sphincter muscle preparations of *Calliactis parasitica*

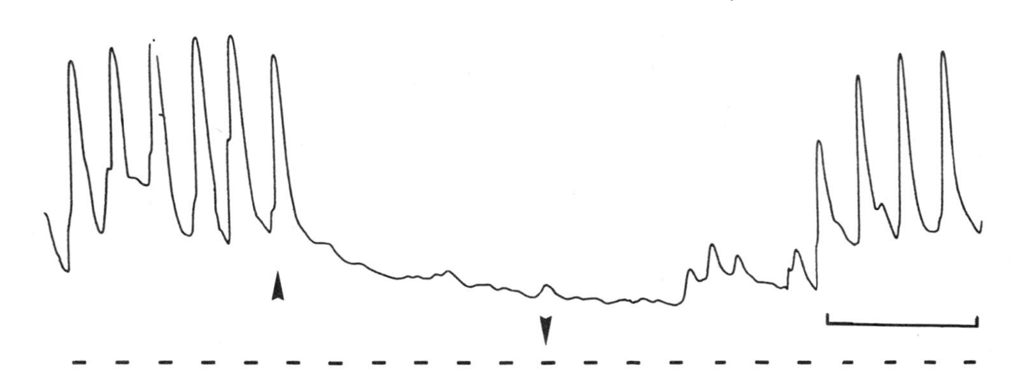

Figure 5. Glutamic acid (final concentration 5 x 10^{-3} M) inhibits electrically-evoked contractions of a sphincter muscle ring from the sea anemone *Tealia felina*. The stimuli (bars) were 30 shocks at 3 Hz, every 9 minutes. Scale = 30 min.

and *Tealia felina* (Fig. 5). Furthermore, we have shown that preparations inhibited with glutamic acid will still contract to Antho-RWamide II, so the glutamic acid action is not a non-specific depression of muscular responses. The inhibitory action of glutamic acid appears to be situated close to the muscle itself (Carlyle, 1974).

There is physiological evidence for the inhibitory actions of the TCNN. In *Calliactis parasitica* sphincter muscle (Lawn, 1976a,b) and body wall circular muscles (Ewer, 1960) electrical stimulation both excites and inhibits. The balance of excitation and inhibition may vary regionally (Ewer, 1960) or with the frequency of stimulation (Lawn, 1976b). In both studies the inhibitory and excitatory actions could not be separated, i.e., they were evoked at identical stimulus thresholds. This suggests that excitatory and inhibitory transmitters are released together, either from the same synaptic terminals, or from separate excitatory and inhibitory cells, both groups of neurons innervated by the TCNN. If we can identify Antho-RWamide I or II synapses on sphincter muscle, then these should be the first sites to seek localization of glutamic acid.

7. Conclusions

Electrophysiological studies of cnidaria have not convincingly answered the question "which is the stem group?". Although it is tempting to regard the hydrozoan neuroid systems as pre-dating nervous systems (Horridge, 1968; Mackie, 1970), there seem no firm reasons for so doing, indeed neuroid systems occur in widely-dispersed groups, including larval amphibia (Roberts, 1969).

What is clear from comparison with other cnidaria is that in organizational terms, the nervous systems of the Anthozoa more closely resemble those of the Scyphozoa than the Hydrozoa. In terms of basic physiological differences, we are on less firm ground. In Hydrozoa, there is anatomical (Westfall, 1973) and physiological (Spencer and Schwab, 1982) evidence for multifunctional neurons, single cells that perform sensory, interneuronal, and motor functions. We know of no such neurons in sea anemones, but we feel sure they await discovery. In Hydrozoa, there are chemical and electrical synapses, and the latter may be excitatory or inhibitory (Spencer and Schwab, 1982). Again, we have no direct evidence for electrical synapses in the Anthozoa, but we have no reason to doubt their existence.

An important function of the TCNN, SS1, and SS2 may be to coordinate local nerve cells, i.e., they act as command fibers. If we seek "primitiveness" in the cnidarian nervous system, perhaps we should look at local nerve cells in sea anemones, not the arguably more advanced TCNN, SS1, and SS2. Such nerve cells, possibly with decremental conduction, would be adequate for the coordination purposes of a small multicellular organism, where most of the body may be within reach of a single nerve cell. With increase in size would come a clearer distinction into local responses and symmetrical responses.

Compared with higher invertebrates, the simplicity of sensory receptors and the small number of conducting systems must place restrictions on the behavioral repertoire of sea anemones. These restrictions are partly overcome by the use of range fractionation in both sensory input and effector output (McFarlane, 1973a; Ross, 1957). This enables each conducting system to function in more than one behavioral response. Also the systems can interact to further increase the range of possible behaviors. This organizational difference reflects the simplicity of the anthozoan nervous system, but provides no sharp distinction between anemones and other invertebrates. Indeed, many features of the organization of the nervous system in higher invertebrates, for example sensory modulation of central pattern generators, are to be found in sea anemones.

With regard to the identities of neurotransmitters, we must aim to combine physiological and anatomical techniques to show the presence of physiologically-active substances in anthozoan synapses. We are close to this goal in the case of the neuropeptides Antho-RWamide I and Antho-RWamide II at neuromuscular junctions on *Calliactis parasitica* sphincter muscle. Glutamic acid and adrenaline await anatomical evidence of their presence at synapses, whilst ATP lacks physiological evidence of any action. Co-existence of "classical" and neuropeptide transmitters in sea anemones has yet to be established or disproven; until then there is little we can usefully contribute to any debate on possible evolutionary precedence of neuropeptide transmitters. Logically, we must expect that slow muscle contractions are the most primitive movements of sea anemones (Ross, 1957), and that fast contractions and inhibitory modulations are later sophistications. Evolution of phasic responses may have been accompanied by the evolution of separate fast and slow muscle neurotransmitters, possibly giving rise to two very closely-related neuropeptides.

We have shown that both Antho-RWamide I and II have opposite actions on ectodermal and endodermal muscle groups, stimulating one and inhibiting the other. This arrangement would be fitting for an archetypal anthozoan, where there may have been antagonistically-arranged ectodermal longitudinal and endodermal circular muscles; it would allow a single nerve net to coordinate both muscle layers, releasing the same transmitter at all its terminals. But overall coordination would require two independent nerve nets, one to excite ectodermal effectors and inhibit endodermal ones, the other to do the reverse. In most present-day sea anemones, however, the only ectodermal muscles lie in the tentacles and oral disc; endodermal circular muscles of the body wall are antagonized by endodermal parietal muscles. Thus, as we might expect, the actions of the three conducting systems are more complex than the arrangement in our archetype: the SS1 excites some endodermal muscles, and both inhibits and excites some ectodermal muscles. The SS2 inhibits endodermal muscles, and has no known effect on ectodermal muscles. The only action of the TCNN on ectodermal muscles is excitatory, but it both excites and inhibits endodermal muscles. Our archetype may gain credence after investigation of the action of anthozoan neuropeptides in the primitive sea anemone *Protanthea simplex*, a species that has ectodermal muscles in the body wall.

ACKNOWLEDGEMENTS. IDM received financial support from the Research Fund of the University of Hull. We are grateful for the help of Paul Currey and Lesley Sharp.

References

Adams, M., and O'Shea, M., 1983, Peptide cotransmitter at a neuromuscular juction, *Science* **221:**286-289.

Anctil, M., 1987, Bioactivity of FMRFamide and related peptides on a contractile system of the coelenterate *Renilla köllikeri*, *J. Comp. Physiol. B* **157:**31-38.

Anderson, P. A. V., 1980, Epithelial conduction: its properties and functions, *Prog. Neurobiol.* **15:**161-203.

Anderson, P. A. V., and Case, J. F., 1975, Electrical activity associated with luminescence and other colonial behavior in the pennatulid *Renilla köllikeri*, *Biol. Bull.* **149:**80-95.

Arai, M. N., 1985, Electrical activity associated with withdrawal and feeding of *Pachycerianthus fimbriatus* (Anthozoa, Ceriantharia), *Mar. Behav. Physiol.* **12:**47-56.

Audesirk, G., and Audesirk, T., 1980, Complex mechanoreceptors in *Tritonia diomedea*. I. Neuronal correlates of a change in behavioral responsiveness, *J. Comp. Physiol.* **141:**111-122.

Batham, E. J., 1965, The neural architecture of the sea anemone *Mimetridium cryptum*, *Am. Zool.* **5:**395-402.

Berridge, M. J., and Rapp, P. E., 1979, A comparative study of the function, mechanism and control of cellular oscillators, *J. exp. Biol.* **81:**217-279.

Boothby, K. M., and McFarlane, I. D., 1986, Chemoreception in sea anemones: Betaine stimulates the pre-feeding response in *Urticina eques* and *U. felina*, *J. exp. Biol.* **125:385-389.**

Burrows, M., 1985, Nonspiking and spiking local interneurones in the locust, in: *Model Neural Networks and Behavior* (A. I. Selverston, ed.), pp. 109-125, Plenum Press, New York.

Byrne, J. H., 1985, Neural and molecular mechanisms underlying information storage in *Aplysia*: implications for learning and memory, *Trends in Neurosci.* **8:**478-482.

Carlyle, R. F., 1974, The occurrence in and action of amino acids on isolated oral sphincter preparations of the sea anemone *Actinia equina*, *J. Physiol. (Lond.)* **236:**635-652.

Cottrell, G., A., Davies, N. N., and Green, K. A., 1984, Multiple actions of a molluscan cardioexcitatory neuropeptide and related peptides on identified *Helix* neurones, *J. Physiol. (Lond.)* **356:**315-333.

Delcomyn, F., 1980, Neural basis of rhythmic behavior in animals, *Science* **210:**492-498.

Ewer, D. W., 1960, Inhibition and rhythmic activity of the circular muscles of *Calliactis parasitica* (Couch), *J. exp. Biol.* **37:**812-831.

Getting, P. A., 1983, Interaction of network, synaptic, and cellular properties in pattern generation, *Symp. Soc. Exp. Biol.* **37:**89-128.

Guthrie, S. C., and Gilula, N. B., 1989, Gap junction communication and development, *Trends in Neurosci.* **12:**12-16.

Horridge, G. A., 1968, The origins of the nervous system, in: *The Structure and Function of Nervous Tissue*, Volume 1, pp. 1-31 (G. H. Bourne, ed.), Academic Press, New York.

Howe, N. R., and Sheikh, Y. M., 1975, Anthopleurine: a sea anemone alarm pheromone, *Science* **189:**386-388.

Jackson, A. J., and McFarlane, I. D., 1976, Delayed initiation of SS1 pulses in the sea anemone *Calliactis parasitica*: evidence for a fourth conducting system, *J. exp. Biol.* **65:**539-552.

Josephson, R. K., 1966, Neuromuscular transmission in a sea anemone, *J. exp. Biol.* **45:**305-319.

Josephson, R. K., 1974, Cnidarian neurobiology, in: *Coelenterate Biology: Reviews and New Perspectives*, pp. 245-280 (L. Muscatine and H. M. Lenhoff, eds.), Academic Press, New York.

Kupfermann, I., and Weiss, K. R., 1978, The command neurone concept, *Behav. Brain Sci.* **1:**3-39.

Lawn, I. D., 1976a, The marginal sphincter of the sea anemone *Calliactis parasitica*. I. Responses of intact animals and preparations, *J. Comp. Physiol.* **105:**287-300.

Lawn, I. D., 1976b, The marginal sphincter of the sea anemone *Calliactis parasitica*. II. Properties of the inhibitory response, *J. Comp. Physiol.* **105**:301-311.

Lawn, I. D., 1976c, Swimming in the sea anemone *Stomphia coccinea* triggered by a slow conduction system, *Nature* **262**:708-709.

Lawn, I., D., 1980, A transmesogloeal conduction system in the swimming sea anemone *Stomphia coccinea*, *J. exp. Biol.* **83**:45-52.

Lawn, I. D., and McFarlane, I. D., 1976, Control of shell settling in the swimming sea anemone *Stomphia coccinea*, *J. exp. Biol.* **64**:419-429.

Lehman, H. K., and Greenberg, M. J., 1987, The actions of FMRFamide-like peptides on visceral and somatic muscles of the snail *Helix aspersa*, *J. exp. Biol.* **131**:55-68.

Logan, C. A., 1975, Topographic changes in responding during habituation to water-stream stimulation in sea anemones (*Anthopleura elegantissima*), *J. Comp. Physiol. Psychol.* **89**:105-117.

McFarlane, I. D., 1969, Co-ordination of pedal-disk detachment in the sea anemone *Calliactis parasitica*, *J. exp. Biol.* **51**:387-396.

McFarlane, I. D., 1973a, Spontaneous electrical activity in the sea anemone *Calliactis parasitica*, *J. exp. Biol.* **58**:77-90.

McFarlane, I. D., 1973b, Multiple conduction systems and the behaviour of sea anemones, *Publ. Seto Mar. Biol. Lab.* **20**:513-523.

McFarlane, I. D., 1974a, Excitatory and inhibitory control of inherent contractions in the sea anemone *Calliactis parasitica*, *J. exp. Biol.* **60**:397-422.

McFarlane, I. D., 1974b, Control of the pacemaker system of the nerve net in the sea anemone *Calliactis parasitica*, *J. exp. Biol.* **61**:129-143.

McFarlane, I. D., 1975, Control of mouth opening and pharynx protrusion during feeding in the sea anemone *Calliactis parasitica*, *J. exp. Biol.* **63**:615-626.

McFarlane, I. D., 1976, Two slow conduction systems co-ordinate shell-climbing behaviour in the sea anemone *Calliactis parasitica*, *J. exp. Biol.* **64**:431-445.

McFarlane, I. D., 1978, Multiple conducting systems and the control of behaviour in the brain coral *Meandrina meandrites* (L.)., *Proc. R. Soc. Lond., B* **200**:193-216.

McFarlane, I. D., 1982, *Calliactis parasitica*, in: *Electrical Conduction and Behaviour in "Simple" Invertebrates*, pp. 243-265 (G. A. B. Shelton, ed.), Clarendon Press, Oxford.

McFarlane, I. D., 1983, Nerve net pacemakers and phases of behaviour in the sea anemone *Calliactis parasitica*, *J. exp. Biol.* **104**:231-246.

McFarlane, I. D., 1984a, Nerve nets and conducting systems in sea anemones: two pathways excite tentacle contractions in *Calliactis parasitica*, *J. exp. Biol.* **108**:137-149.

McFarlane, I. D., 1984b, Nerve nets and conducting systems in sea anemones: coordination of ipsilateral and contralateral contractions in *Protanthea simplex*, *Mar. Behav. Physiol.* **11**:219-228.

McFarlane, I. D., 1985, Collapse behaviour in the primitive sea anemone *Protanthea simplex*, *Mar. Behav. Physiol.* **11**:259-269.

McFarlane, I. D., 1988, Variability in the startle response of *Pachycerianthus multiplicatus* (Anthozoa: Ceriantharia), *Comp. Biochem. Physiol.* **89A**:365-370.

McFarlane, I. D., Graff, D., and Grimmelikhuijzen, C. J. P., 1987, Excitatory actions of Antho-RFamide, an anthozoan neuropeptide on muscles and conducting systems in the sea anemone *Calliactis parasitica*, *J. exp. Biol.* **133**:157-168.

McFarlane, I. D., and Lawn, I. D., 1972, Expansion and contraction of the oral disc in the sea anemone *Tealia felina*, *J. exp. Biol.* **57**:633-649.

Mackie, G. O., 1970, Neuroid conduction and the evolution of conducting tissues, *Quart. Rev. Biol.* **45**:319-332.

Mackie, G. O., Anderson, P. A. V., and Singla, C. L., 1984, Apparent absence of gap junctions in two clases of cnidaria, *Biol. Bull.* **167**:120-123.

Marks, P. S., 1976, Nervous control of light responses in the sea anemone, *Calamactis praelongus*, *J. exp. Biol.* **65**:85-96.

Martin, S. M., and Spencer, A. N., 1983, Neurotransmitters in coelenterates, *Comp. Biochem. Physiol.* **74C**:1-14.

O'Shea, M., and Shaffer, M., 1985, Neuropeptide function: The invertebrate contribution, *Ann. Rev. Neurosci.* **8:**171-198.

Parker, G. H., 1919, *The Elementary Nervous System,* Lippincott, Philadelphia.

Passano, L. M., 1973, Behavioral control systems in medusae: a comparison between hydro- and scyphomedusae, *Publ. Seto Mar. Biol. Lab.* **20:**615-645.

Passano, L. M., 1982, Scyphozoa and Cubozoa, in: *Electrical Conduction and Behaviour in "Simple" Invertebrates*, pp. 149-202 (G. A. B. Shelton, ed.), Clarendon Press, Oxford.

Pearson, K. G., Reye, D., N., and Robertson, R. M., 1983, Phase-dependent influences of wing stretch receptors on flight rhythm in the locust, *J. Neurophysiol.* **49:**1168-1181.

Pickens, P. E., 1988, Systems that control the burrowing behaviour of a sea anemone, *J. exp. Biol.* **135:**133-164.

Price, D. A., and Greenberg, M. J., 1980, Pharmacology of the molluscan cardioexcitatory neuropeptide FMRFamide, *Gen. Pharmacol.* **11:**237-241.

Roberts, A., 1969, Conducted impulses in the skin of young tadpoles, *Nature* **222:**1265-1266.

Robson, E. A., 1965, Some aspects of the structure of the nervous system in the anemone *Calliactis*, *Amer. Zool.* **5:**503-410.

Romanes, G. J., 1878, Further observations on the locomotor systems of medusae, *Phil. Trans. R. Soc. Lond.* **167:**659-752.

Ross, D. M., 1957, Quick and slow contractions in the isolated sphincter of the sea anemone, *Calliactis parasitica*, *J. exp. Biol.* **34:**11-28.

Ross, D. M., 1960a, The effects of ions and drugs on neuromuscular preparations of sea anemones. I. On preparations of the column of *Calliactis* and *Metridium*, *J. exp. Biol.* **37:**732-752.

Ross, D. M., 1960b, The effects of ions and drugs on neuromuscular preparations of sea anemones. II. On sphincter preparations of *Calliactis* and *Metridium*, *J. exp. Biol.* **37:**753-773.

Ross, D. M., and Sutton, L., 1964, Inhibition of the swimming response by food and of nematocyst discharge during swimming in the sea anemone *Stomphia coccinea*, *J. exp. Biol.* **41:**751-757.

Shelton, G. A. B., 1982, Anthozoa, in: *Electrical Conduction and Behaviour in "Simple" Invertebrates*, pp. 203-242 (G. A. B. Shelton, ed.), Clarendon Press, Oxford.

Shelton, G. A. B., and Holley, M. C., 1984, The role of a "local electrical conduction system" during feeding in the Devonshire cup coral *Caryophyllia smithii* Stokes and Broderip, *Proc. R. Soc. Lond. B* **200:**489-500.

Spencer, A. N., and Schwab, W. E., 1982, Hydrozoa, in: *Electrical Conduction and Behaviour in "Simple" Invertebrates*, pp. 73-148 (G. A. B. Shelton, ed.), Clarendon Press, Oxford.

Van Marle, J., 1977, Contribution to the knowledge of the nervous system in the tentacles of some coelenterates, *Bijdragen tot de Dierkunde* **46:**219-260.

Watson, G. M., and Hessinger, D. A., 1989, Cnidocyte mechanoreceptors are tuned to the movements of swimming prey by chemoreceptors, *Science* **243:**1589-1591.

Westfall, J. A., 1973, Ultrastructural evidence of a granule-containing sensory-motor interneuron in *Hydra littoralis*, *J. Ultrastruct. Res.* **42:**268-282.

Wood, J. G., and Lentz, T. L., 1964, Histochemical localization of amines in *Hydra* and in the sea anemone, *Nature* **201:**88-90.

Chapter 9

Catecholamines, Related Compounds and the Nervous System in the Tentacles of Some Anthozoans

J. VAN MARLE

1. Introduction

There are two separate nerve plexuses in the tentacles of Anthozoa. One is found in the endoderm close to the mesogloea; the other is located in the ectoderm, also close to the mesogloea. The morphology of their component neurons and the variety of transmitters they posses, have been studied most extensively (Van Marle, 1977; Van Marle et al., 1983) in the sea anemone *Tealia felina*, since in this animal, the mesoglea clearly separates the ectodermal nerve plexus and the muscle. All findings have, however, been confirmed in *Metridium senile, Anemonia sulcata* and *Cerianthus membranaceus* and, to a lesser degree, in *Hydra vulgaris*. As far as we could establish the two plexuses are not connected, for, in the tentacles at least, there are no morphological nor histochemical indications of any connections.

J. VAN MARLE • Department of Electron Microscopy, Faculty of Medicine, University of Amsterdam, Meibergdreef 15, 1105 AZ, Amsterdam, The Netherlands.

Evolution of the First Nervous Systems
Edited by P.A.V. Anderson
Plenum Press, New York

2. The Endodermal Plexus

The endodermal plexus consists of small bipolar or multipolar nerve cells which are dispersed throughout the plexus and connected by fine processes, or neurites. At the electron microscopice level, these neurites are morphologically similar to their counterparts in the ectoderm. They contain electron-lucent vesicles (150 nm) and/or electron-dense granules (110 nm). Unlike cells in the ectodermal plexus, however, where two morphologically distinct classes of neurites contain electron opaque granules, only one type of neurite in the endodermal plexus contains opaque granules. This is consistent with the observation that no Formol-Induced Fluorescence (FIF) occurs in the endodermal plexus (see below). Histochemistry has not provided any clues, as yet, as to the identity of neurotransmitters in this plexus, and all attempts to stain neurosecretory material have, so far, remained negative. Recent studies on the suject of neuropeptides, reveal the presence of one or more of these substances in the endodermal neurites (Grimmelikhuijzen, et al., 1987; Grimmelikhuijzen and Groeger, 1987; Grimmelikhuijzen and Graff, 1985, 1986; reviewed by Grimmelikhuijzen et al., in this book).

3. The Ectodermal Plexus

The ectodermal plexus is a much more elaborate structure than the endodermal plexus. However, owing to the small dimensions of the component neurites, and their predominantly multipolar perikarya, morphological and organizational information is difficult to obtain at the light microscope level. Electron microscopy of the contracted tentacles of sea anemones reveals that this plexus consists of numerous neurites, which are inevitably cut in cross section as a result of the tentacle being contracted. No concentrations of somata or neurites were ever observed in the tentacles of the anthozoan species examined, and no morphological evidence for the existence of the several conduction systems known to be present in these tissues (McFarlane, 1982 and in this book), has been found.

Three types of synaptic vesicle can be identified in neurites of the ectodermal plexus: electron-lucent vesicles (150 nm), dense-core granules (130 nm) and opaque granules (110 nm). Dense-core granules can be distinguished from opaque granules by the presence of a clear margin between the limiting membrane and the core: opaque granules have no such margin and are altogether electron dense. No neurites containing the small (50 nm) electron-lucent vesicles typical of cholinergic or glutamatergic systems were found, nor were the large, irregularly-shaped granules indicative of neurosecretory activity. Electron-lucent vesicles and either dense core or opaque granules were observed in the same neurite, but dense core and opaque granules were never found together. The distribution of vesicles and granules within a neurite was not random; sometimes cross sections through neurites contained only a few vesicles or granules, but in most cases a neurite contained either no granules at all or a relatively large number of granules, present in a so-called varicosity.

We concluded, therefore, that neurons in the ectodermal plexus possess two morphologically different types of neurite; one with dense-core granules, the other with opaque ones. Owing to the morphological complexity of the plexus, however, it was not possible to decide whether all processes of any one neuron contain the same type of granules or whether different neurites have different granules.

In spite of suggestions to the contrary (for reviews, see Westfall, 1987; Spencer, in this book), we have not been able to confirm the observation that real synapses (neuro-neuronal or neuro-muscular contacts delineated by pre- and postsynaptic membrane specializations) are present in the species investigated. In our hands, conventional morphological techniques, including the specific Ethanolic Phospho-Tungstic acid method, did not yield any indication of the presence of real synapses. This discrepancy between our observations and those reported elsewhere may be explained, to some extent, by our use of a goniometer stage, which allows a better investigation of parallel membranes than is possible with a fixed stage. The typical synaptic structure, namely a thickened, postsynaptic membrane running parallel to the terminal membrane for some distance and separated from it by an interval of about 30 nm, was never observed in either the anthozoans examined, or in *Hydra*. However, occasionally, close contacts were observed within the plexus, and in these instances the two opposed cell membranes ran parallel to one another over some distance, but the spacing was closer than usual for the normal inercellular gap (15 nm instead of 20 nm). Such close contacts were observed between neurites, and between neurites and the stalks of ectodermal myoepithelial cells. It appeared that close contacts were not restricted to one type of neurites, but they were also observed between neurites containing dense-core as well as opaque granules. Considering the structure of the plexus, the type of transmitters apparently involved and the physiological properties of the plexus, true synapses are hardly to be expected in a system which so closely resembles some parts of the vertebrate autonomic system (Smolen, 1988). It should be noted that elaborate cell-cell contacts do exist between the epithelial cells at the periphery of the tentacle, and between the muscle cells close to or in the mesogloea. A freeze-fracture study of the morphology of these structures is in progress.

Radial nerve fibers run from the periphery of the tentacle straight to the ectodermal plexus. Electron microscopy reveals that these are dispersed solitary sensory cells, whose cilia project into the surrounding sea water. The sensory cells connect to the plexus by way of a varicose neurite which contains dense-core granules. Nerve fibers which contain only opaque granules and/or electron-lucent vesicles extend from the plexus to the muscles. These neurites then form varicosities and terminate between the muscles. They do not form any kind of specialized contact, not even close contacts resembling those of the non-adrenergic, non-cholinergic fibers of the vertebrate autonomic innervation (Smolen, 1988).

4. Evidence for Cholinergic Mechanisms

The presence of a neurotransmitter can be revealed histochemically or cytochemically. For example, indicators of the presence of cholinergic mechanisms are the presence of acetylcholinesterase, the enzyme responsible for the degradation of released acetylcholine (ACh), and choline acetyltransferase, the enzyme responsible for its synthesis in the synaptic terminal.

We have been unable to demonstrate the presence of the choline acetyltransferase in the species examined, using either histochemical or immunocytochemical techniques (unpublished result), and have also been unable to demonstrate the presence of this enzyme in a biochemical assay (unpublished result). When acetylthiocholine iodide was provided as a substrate to reveal the presence of acetylcholinesterase, a brown precipitate was observed in several of the ectodermal cells. The cell bodies of the epithelial cells were slightly positive, as were some cells within the plexus; some cells between the muscles and especially the interstitial cells in the mesogloea stained heavily. However, no precipitate was formed in the presence of iso-ompa (10^{-4} M), a selective cholinesterase inhibitor, indicating that the brown precipitate was due to non-specific cholinesterases only, and that no acetylcholinesterase is present in the tentacles (Van Marle, 1977). Furthermore, the absence of the enzyme in the tentacles was verified biochemically (unpublished result).

Considering the morphological structure of the plexus and the absence of small clear vesicles, true synapses, and the two marker enzymes of the cholinergic transmission, it is highly unlikely that ACh is a transmitter in the Anthozoan species investigated, nor in *Hydra*. The pharmacological and physiological effects of ACh and some of its analogues on several hydrozoan species (Scemes and Mendes, 1986) may be open to various interpretations (but see chapter by Scemes in this book).

5. Evidence for GABA-ergic and Glutaminergic Mechanisms

Although both GABA and glutamate are reported to affect certain hydrozoans (Mendes and de Freitas, 1984; Van Marle, 1977), immunocytochemistry provides no evidence for the presence of either GABA or glutamate (unpublished result), nor could we establish, histochemically, that the enzymes typical of glutamate or GABA transmission are present (Van Marle, 1977; unpublished result). Another method for investigating the presence of glutamatergic or GABA-ergic transmission is to demonstrate a specific uptake of ^{3}H-glutamate or ^{3}H-GABA, under high affinity uptake conditions. After incubation in the presence of tritiated transmitters, it appeared that several parts of the ectoderm of the sea anemones were labelled, but specific and selective uptake of GABA or glutamate was not observed (unpublished result). The elements of the ectoderm which accumulated the transmitters were developing nematocytes and spirocytes and interstitial cells. Moreover, non-specific labelling of the mesogloea was observed. Here too, the absence of true synapses with small clear vesicles and the failure to demonstrate the localization of either the transmitters or their marker enzymes, make it unlikely that GABA and glutamate function as neurotransmitters in the Anthozoa. Again, the physiological effects described in the literature are open to more than one interpretation (Mendes and de Freitas, 1984; Van Marle, 1977).

6. Evidence for Catecholamines

The Formol-Induced Fluorescence (FIF) method is an easy, sensitive and specific method for demonstrating catechol- and indolamines and related compounds in tissue (Axelsson et al., 1972; Björklund et al., 1972; Carlberg, 1983; Elofsson et al., 1977). Frozen, dried sections of the tentacles of sea anemones and *Hydra* were exposed to formaldehyde vapor at 50°C, for 10 to 60 min. This resulted in an intense fluorescence of the ectodermal plexus and of the radial fibers running from the periphery of the tentacle to the plexus. Given that the FIF method is so simple, it is remarkable that catecholamines were reported absent in *Hydra* (Grimmelikhuijzen, 1985). In our hands, all *Hydra* specimens investigated exhibited a clearly observable fluorescence. However, the location of this fluorescence seems not to coincide with the known distribution of the nerves in the *Hydra* tentacles (David and Grimmelikhuijzen, personal communication). A non-neural distribution of catachol- and indolamines and related compounds in Coelenterates is discussed by Anctil (this book).

The fluorescence produced by the FIF method is white or blue, depending on the microscope and optics used. These colors indicate the presence of a catechol compound or a peptide with NH_2-terminal catechol compound. Yellow fluorescence indicative of indolamines was never observed, in either the sea anemones examined or in *Hydra* (Dahl et al., 1963; Van Marle, 1977; Van Marle et al., 1983). Since the white/bluish fluorescence develops within 15 minutes, we can assume that the catechol compound present in the plexus and the radial fibers is not adrenalin (Axelsson et al., 1972; Björklund et al., 1972), since it takes at least 60 minutes in 80°C aldehyde vapor to convert adrenalin to a fluorophore.

The presence of a catechol compound in the tentacles was confirmed using microspectrofluorimetry. The characteristics of both the excitation and emission spectra would, at first glance appear to indicate that noradrenalin is the only catecholamine occurring in both the plexus and the fibers of sea anemones and *Hydra* (Van Marle, 1977; Van Marle et al., 1983). However, this identification is incorrect. The excitation spectrum in the short ultraviolet region (320 nm) deviates from that expected for pure noradrenalin. Furthermore. the characteristics of the excitation spectrum obtained after acid exposure, differ markedly from all known catechol compounds (Elofsson et al., 1977). It would, therefore, be more appropriate to assume that an as yet unknown catechol compound is present. Its excitation spectrum suggests that this compound is ß-hydroxylated (Elofsson et al., 1977).

The FIF method cannot be used to demonstrate the presence of the catecholamine octopamine, since this substance cannot be converted to a fluorophore by aldehyde ring closure. However, this compound does not appear to be present in these tissues, since analysis of the metabolites of tissues and whole animals incubated in ^{3}H-tyramine did not reveal the presence of octopamine (Van Marle, 1977). If ^{3}H-tyrosine were added as a precursor, ^{3}H-noradrenalin was found, while the amounts of ^{3}H-dopa and ^{3}H-dopamine produced were too small to suggest a physiological significance (Van Marle, 1977). Therefore, it seems that the metabolic pathway of noradrenalin synthesis runs via dopa and dopamine, and not via tyramine and octopamine.

These observations were not, however, confirmed by later studies, which indicate that significant amounts of dopa and/or dopa related compounds are present in the plexus, and that the enzyme dopa-decarboxylase is absent. These later findings make the presence of catecholamines unlikely (Carlberg, 1983, 1988; Martin and Spencer, 1983), but not impossible, since an alternative metabolic pathway has been described (Carlberg and Rosengren, 1985).

7. Evidence for 5-hydroxytryptamine

Although 5-hydroxytryptamine (5-HT) has been reported in the nervous systems of both anthozoans and hydrozoans, the yellow fluorescence typical of 5-HT never developed when the FIF method was applied to the species investigated. Analysis of the FIF spectrum confirmed the absence of an indolamine in all species investigated, and the analysis of the metabolites of administered ^{3}H-tryptophane provided no evidence for the presence of an indolamine in the nervous system.

8. Cellular Localization of Transmitters

Whereas light microscopical identification of catechol compounds is relatively simple, electron microscopic identification has proved to be problematic. There are several methods by which catechol compounds can be visualized at the electron microscopic level. In our hands, these, and especially that of Tranzer (Tranzer and Richards, 1976), provide excellent results in vertebrate tissues, but in cnidarians examined, the results are invariably negative. Therefore, we have been unable to directly localize the catechol compounds to either the dense core or opaque granules. Such localization may be possible with immunological techniques. However, on the basis of the very distinct distribution of the FIF and the location of the neurites known to contain either dense-core or opaque granules, an indirect localization can be deduced. Dense-core granules are found in neurites of the plexus and in the radial fibers, but not in those neurites that lie between the muscles, and, as described above (Section 6), the FIF occurs in the ectodermal plexus and in the radial fibers that run between the periphery of the tentacle and the plexus, but it was never observed between the muscles, in cells lying in the mesogloea or in the endodermal plexus. The fact that the nerves that lie between the muscles are completely devoid of catechol compounds is most easily observed in *Tealia* since, in this species, the muscles are morphologically distinct from the plexus, being situated in long tunnels that are oriented lengthwise in the mesogloea of the tentacle. Thus, the distribution of the FIF and the dense core granules coincide.

From the above, we may infer that the neurites with dense-core granules contain a catechol compound. If this indirect identification is valid, it is in agreement with the situation found in the vertebrate autonomous system, where the dense-core granules contain biogene amines (Smolen, 1988). The identity of the transmitter located in the opaque granules in terminals that apparently innervate the muscles, remains to be established. Although a number of substances such as choline compounds, glutamate, GABA, 5-HT, purines (Martin and Spencer, 1983; Mendes and De Freitas, 1984; Scemes and Mendes, 1986) are known to be physiologically active on the muscles of Coelenterates, they are not likely candidates, either because they are normally associated with true synaptic transmission and are usually found only in small (50 nm) vesicles, which are not present in these tissues, or histochemical evidence for their presence is lacking. The most likely candidates for the transmitter present in the opaque granules are neuropeptides. These compounds are active on anemone muscle (McFarlane et al., 1987) and several neuropeptides have been isolated and sequenced from cnidarian tissues (Grimmelikhuijzen, 1983a,b, 1985; Grimmelikhuijzen et al., 1980, 1981, 1982; Grimmelikhuijzen and Graff, 1985, 1986; Grimmelikhuijzen et al., 1987; Koizumi et al., 1989; McFarlane et al., 1987; see chapter by Grimmelikhuijzen et al. in this book). It will be interesting to determine the exact cellular location of these various peptides since, in vertebrates, small neuropeptides are reported to act both as transmitters in their own right and as co-transmitters, with various other transmitters, i.e. ACh, glutamate, GABA, noradrenalin and 5-HT. Moreover, co-localization of two neuropeptides (gastrin releasing peptide and calcitonin) in the same dense core granule has been reported (Stahlman et al., 1985, 1987).

Although there are morphological similarities between the catechol-containing neurites of coelenterates, and those of the vertebrate autonomic nervous system (Smolen, 1988), there are clearly functional differences. For instance, monoamine oxidase, the only catecholamine degrading enzyme that can be demonstrated histochemically could not be visualized in the tissues examined (Van Marle, 1977), but has, however, been reported, using the same histochemical technique, in *Metridium*. There, it is found close to the mesogloea in both the ectoderm and the endoderm (Lenicque et al., 1977). Its localization in the endoderm in particular seems improbable, however, since FIF has never been observed or described in the endodermal plexus (Dahl et al., 1963; Elofsson et al., 1977; Van Marle, 1977; Van

Marle et al., 1983), although the presence of monoamine oxidase is not necessarily a requirement for aminergic transmission (Martin and Spencer, 1983).

A further difference between the neurites described here and those in vertebrates is that the vertebrate catecholaminergic systems are known to accumulate released or administered catecholamines, in order to remove released catecholamines from the interneural space. However, when ^{3}H-noradrenalin was supplied to anemones under high affinity uptake conditions, it was not possible to detect any specific accumulation in the autoradiographs, even after prolonged incubation times (up to 48 hours) (unpublished result). There was no preferential labelling of either the perikarya or the neurites of the plexus throughout the experiments. The labelling we observed at the light microscope level resembles that reported following high affinity uptake ^{3}H-noradrenalin by *Renilla* (Anctil et al., 1984). However, high resolution autoradiography revealed that, although a labelled, dense-core neurite was occasionally encountered, selective uptake by neural elements did not take place (Van Marle, 1977; Van Marle et al., 1983). The straight lines observed in the autoradiographs were due to uptake of isotope in the stalks of the epithelial cells (unpublished results). Some cellular elements of the ectoderm were heavily labelled, but they will accumulate large amounts of almost any substance applied, e.g., glutamate, GABA, noradrenalin, tryptophane. Elements which did show accumulation were developing nematocytes and spirocytes and, especially, interstitial cells in the mesogloea (Van Marle, 1977; Van Marle et al., 1983).

9. Pharmacology

Another approach that can be used to study the properties of the catecholaminergic system involves the use of pharmacological tools. However, any results obtained with coelenterates should be interpreted with caution, since some of the catechol compounds present may differ from those present in the vertebrates, where the pharmacological responses have been characterized.

Reserpine and guanethidine inhibit the uptake and accumulation of catecholamines by neurites, but they act at different sites (Goodman and Gillman, 1980); reserpine inhibits the uptake at the level of the neural membrane, whereas guanethidine inhibits the uptake at the level of the limiting membrane of the granules. However, neither prolonged incubation (up to one week), nor high doses of either reserpine (100 mg/ml) or guanethidine (1 mg/ml) had any effect on the intensity or distribution of FIF in the plexus or the radial fibers of sea anemones (Carlberg, 1983; Van Marle et al., 1983). It should be mentioned that the behavior of the animals, as well as their food consumption (i.e. 1 shrimp/week for the sea anemones), remained unaltered in the presence of either of the two drugs.

From studies on the vertebrate catecholaminergic system, we know that the uptake of so-called false transmitters such as 5OH- or 6OH- dopamine results in a rapid (within 30 minutes) disappearance of the FIF, followed by marked degeneration of the nerve fibers (Hökfelt et al., 1972). Although an uptake mechanism did not seem to be present in *Hydra* and the anthozoans studied, incubation of the animals in either 5OH- or 6OH-dopamine for 24 hours produced decreased FIF intensity, followed by a complete disappearance of the FIF (Van Marle et al., 1983). Another false transmitter, 5OH-dopa, was reported to be ineffective (Carlberg, 1983). In the case of 5OH- and 6OH-dopamine, incubation times shorter than 24 hours did not decrease fluorescence and a complete disappearance required an incubation period of at least 72 hours (Van Marle et al., 1983). After repeatedly rinsing of the animals in fresh seawater, the FIF reappeared within a week of its complete disappearance.

The effect of 5OH- or 6OH-dopamine was very noticeable behaviorally. Sea anemones contracted within 30 minutes of the addition of the false transmitter, and they remainined contracted until the FIF levels returned to normal, after rinsing. After the reappearance of the FIF, the animals looked normal, as did their food intake.

Although the disappearance of the FIF in vertebrate systems is followed (within 4 weeks) by a complete degeneration of the fibers involved (Hökfelt et al., 1972), no degeneration was observed in the cnidarians we investigated (Van Marle et al., 1983). The dense-core granules-containing neurites did not show any severe degeneration, nor did the number of dense-core granules or electron-lucent vesicles appear affected (there is, however, no clear relationship between the amount of granules in a neurite and the FIF intensity).

The fact that FIF disappears due to the accumulation of 5OH- and 6OH-dopamine is difficult to reconcile with the absence of an uptake system for released catecholamines. However, if we assume that there really is no uptake mechanism, then the long incubation periods and high concentrations of false transmitter required to produce a decrease of the FIF may be explained by the fact that the false transmitter is accumulated solely by diffusion, and not by active uptake. Furthermore, the absence of an uptake system might also explain why neural degeneration was never observed in the cnidarians examined, in contrast to the situation in vertebrates. Diffusion alone might never lead to the high concentrations necessary to induce such a degeneration. While the false transmitters had an obvious effect on the FIF, they had no effect on the dense-core granules we believe contain a β-hydroxylated catechol compound (see above). The formation or maintenance of the granules and their peptide or other content seems to continue, albeit without incorporation of a catechol compound, the synthesis of which is disturbed by a false transmitter.

Pharmacological effects have been reported in *Renilla*, on the basis of physiological experiments (Anctil et al., 1982). Although the false transmitter 6OH-dopamine inhibited the luminescence of *Renilla*, no effects on contractions were observed. In contrast, sea anemones react within 30 minutes, with a complete contracture, and a complete abolition of the FIF was observed after 24 h. Additionally, reserpine abolishes luminescence in *Renilla* but appears to have no effect whatsoever on sea anemones (Carlberg, 1983; Van Marle, 1983).

10. Conclusions

When one considers the morphological, physiological and pharmacological properties of the Anthozoan nervous system (*Hydra* does not seem to differ significantly in these respects), it may be argued that this nervous system is able to produce a long-lasting effect on all elements of the ectoderm, without the capacity for a discrete, localized response one of the type one might expect from the discrete chemical synapses that are apparently lacking. As regards transmitters, reliable evidence has been presented for the presence of only catechol compounds (Carlberg, 1983, 1988; Carlberg and Rosengren, 1985; Dahl et al., 1963; Elofsson et al., 1977; Martin and Spencer, 1983; Van Marle, 1977; Van Marle et al., 1983) and neuropeptides (Grimmelikhuijzen, 1983a,b, 1985; Grimmelikhuijzen et al., 1980; Grimmelikhuijzen et al., 1981; Grimmelikhuijzen et al., 1982; Grimmelikhuijzen and Graff, 1985; Koizumi et al., 1989; McFarlane et al., 1987). These catechol compounds must be acting within the ectodermal plexus (Anctil et al., 1984; Carlberg, 1983; Dahl et al., 1963; Elofsson et al., 1988; Van Marle, 1977; Van Marle et al., 1983), since they are not present in the neurons that lie between the muscles, and they do not affect the muscles whereas the neuropeptides do indeed innervate or otherwise influence the muscles (Anctil et al., 1984; McFarlane et al., 1987). Both catecholaminergic and peptidergic transmitters are known to act for a considerable time and, as a result of diffusion from the site of release, they will act over a considerable distance. The effects of the released transmitters will be prolonged, since no true synaptic transmission is present and, (diffusion excepted) no system for the elimination of the released catechol-related compounds seems to be present. A co-transmitter might be released simultaneously with the catechol compound and act as an inhibitor, in much the same way as occurs in the vertebrate lung (Barnes, 1987). Here VIP (vasointestinal peptide) simultaneously released with ACh counteracts the effects of ACh on smooth muscles. Specific degrading systems for the neuropeptides are unknown. Moreover, their chemical composition makes them hard to digest by peptidases (Grimmelikhuijzen, this book), which also suggests a prolonged action. To some extent our conclusion is supported by the physiological data from *Renilla* (Anctil et al., 1984). Here too, a propagation of activity within the plexus is thought to be due to a catechol compound, while the innervation of the muscles is considered nonadrenergic.

As far as we can understand, the Anthozoan ectodermal plexus resembles the autonomic nervous system of vertebrates (Smolen, 1988) - albeit without the capacity for a direct and localized response - rather than the dispersed neuroendocrine system (DNS) (Fujita and Kobayashi, 1979; Pearse, 1977) of the vertebrates. The latter also acts over a considerable distance. Its transmitters, catechol- and indolamines and various neuropeptides, are long acting, but the DNS cells are found either isolated or in small groups and they lack the typical structure of a neurite-bearing neuron. Moreover, in these cells, large amounts of neurosecretory material may be demonstrated at the light and electron microscope levels. It may be argued that the rapid responses of anthozoans originate in a rather autonomous reaction of the elaborately connected epithelial and muscle cells, while the plexus provides a "milieu intérieur" modulating their rather autonomous reaction. This may be an explanation for the two ectodermal conduction systems described in sea anemones (McFarlane, 1982), since, in the tentacles of the Anthozoa at any rate no morphological evidence has been found for the existence of different conduction systems.

ACKNOWLEDGEMENTS. Thanks are due to Dr. A. B. Marschall for supervising my use of the English language.

References

Anctil, M., Boulay, D., and Larivière, L., 1982, Monoaminergic mechanisms associated with control of luminescence and contractile activities in the Coeleterate, *Renilla köllikeri*, *J. Exp. Zool.* **223**:11-24.

Anctil, M., Germain, G., and Larivière L., 1984, Catecholamines in the Coelenterate *Renilla köllikeri*. Uptake and autoradiographic localization, *Cell Tissue Res.* **238**:69-80.

Axelsson, S., Björklund, A., and Lindvall, O., 1972, Fluorescence histochemistry of biogenic monoamines. A study of the capacity of various carbonyl compounds to form fluorophores with biogenic monoamines in gas phase reactions, *J. Histochem. Cytochem.* **20**:435-444.

Barnes, P. J., 1987, Regulatory peptides in the respiratory system, *Experientia* **43**:832-839.

Björklund, A., Elsinger, B., and Falck, B., 1972, Analysis of fluorescence excitation peak ratios for the cellular identification of noradrenalin, dopamine or their mixtures, *J. Histochem. Cytochem.* **20**:56-64.

Carlberg, M., 1983, Evidence of dopa in the nerves of sea anemones, *J. Neural Transmission* **57**:75-85.

Carlberg, M., 1988, Localization and identification of catechol compounds in the Ctenophore *Mnemiopsis leidyi*, *Comp. Biochem. Physiol.* **91C**:69-74.

Carlberg, M., Rosengren, E., 1985, Biochemical basis for adrenergic neurotransmission in Coelenterates, *J. Comp. Physiol.* **155B**:251-255.

Dahl, E., Falck, B., Von Mecklenburg, C., and Myrhberg, H., 1963, An adrenergic system in sea-anamones, *Q. J. microsc. Sci.* **104**:531-534.

Elofsson, R., Falck, B., Lindvall, O., and Myrhberg, H., 1977, Evidence for new catecholamines and related amino acids in some invertebrate sensory neurons, *Cell Tiss. Res.* **182**:525-536.

Fujita, T., and Kobayashi, S., 1979, Current views on the paraneuron concept, *Trends in Neurosciences* **2**:27-30.

Goodman, L. S., and Gillman, A., 1980, *The Pharmacological Basis of Therapeutics*, 6th ed., pp. 198-205.

Grimmelikhuijzen, C. J. P., 1983a, FMRFamide immunoreactivity is generally occurring in the nervous system of Coelenterates, *Histochem.* **78**:361-381.

Grimmelikhuijzen, C. J. P., 1983b, Coexistence of neuropeptides in Hydra, *Neuroscience* **9**:837-845.

Grimmelikhuijzen, C. J. P., 1985, FMRFamide-like peptides in the primitive nervous systems of Coelenterates and complex nervous systems of higher animals, in: *Handbook of Opioid and Related Neuropeptides* (G. Stephano, ed.), CRC Press, New York.

Grimmelikhuijzen, C. J. P., and Graff, D., 1985, Arg-Phe-amide-like peptides in the primitive nervous system of Coelenterates, *Peptides 6*, suppl. **3**:477-483.

Grimmelikhuijzen, C. J. P., and Graff, D., 1986, Isolation of <Glu-Gly-Arg-Phe-NH_2 (Antho-RFamide), a neuropeptide from sea anemones, *Proc. Natl. Acad. Sci. USA* **83**:9817-9821.

Grimmelikhuijzen, C. J. P., and Groeger, A., 1987, Isolation of the neuropeptide pGlu-Gly-Arg-Phe-amide from the pennatulid *Renilla köllikeri*, *FEBS Lett.* **211**:105-108.

Grimmelikhuijzen, C. J. P., Sundler, F., and Rehfeld, J. F., 1980, Gastrin/CCK-like immunoreactivity in the nervous system of Coelenterates, *Histochem.* **69**:61-68.

Grimmelikhuijzen, C. J. P., Dockray, G. J., and Yanaihara, N., 1981, Bombesin-like immunoreactivity in the nervous system of Hydra, *Histochem.* **73:**171-180.

Grimmelikhuijzen, C. J. P., Dockray, G. J., and Schot, L. P. C., 1982, FMRFamide-like immunoreactivity in the nervous system of Hydra, *Histochem.* **73:**499-508.

Grimmelikhuijzen, C. J. P., Graff, D., and McFarlane, I. D., 1987, Neuropeptides in invertebrates, in: *Nervous Systems in Invertebrates*, pp. 105-132 (M. A. Ali, ed.), Plenum Press, New York.

Hökfelt, T., Jonsson, G., and Sachs, Ch., 1972, Fine structure and fluorescence morphology of adrenergic nerves after 6 hydroxy dopamine in vivo and in vitro, *Z. Zellforsch.* **131:**529-543.

Koizumi, O., Wilson, J. D., Grimmelikhuijzen, C. J. P., Westfall, J. A., 1989, Ultrastructural localization of RF amide-like peptides in neuronal dense cored vesicles in the peduncle of Hydra, *J. Exp. Zool.* **249:**17-22.

Lenicque, P. M., Toneby, M. I., and Doumenc, D., 1977, Demonstration of biogenic amines and localization of monoamine oxidases in the sea anemone Metridium senile (Linné). *Comp. Biochem. Physiol.* **56C:**31-34.

Martin, S. M., and Spencer, A. N., 1983, Neurotransmitters in Coelenterates, *Comp. Biochem. Physiol.* **74C:**1-14.

McFarlane, I.D., 1982, *Calliactis parasitica*. In *Electrical Conduction and Behaviour in 'Simple' Invertebrates,* G.A.B. Shelton (Ed.), Clarendon Press, Oxford. 243-265.

McFarlane, I. D., Graff, D., and Grimmelikhuijzen, C. J. P., 1987, Excitatory actions of Antho-RF amide, an Anthozoan neuropeptide, on muscles and conducting systems in the sea anemone Calliactis parasitica, *J. exp. Biol.* **133:**157-168.

Mendes, E. G., and de Freitas, J. C., 1984, The responses of isolated preparations of *Bunodosoma caissarum* (Correa, 1964) (Cnidaria, Anthozoa) to drugs, *Comp. Biochem. Physiol.* **79C:**375-382.

Pearse, A. G. E., 1977, The diffuse neuroendocrine system and the APUD concept: related "endocrine" peptides in brain, intestine, pituitary and anuran cutaneous glands, *Med. Biol.* **55:**115.

Scemes, E., and Mendes, E. G., 1986, Cholinergic mechanism in *Lirope tetraphylla* (Cnidaria, Hydrozoa), *Comp. Biochem. Physiol.* **83C:**171-178.

Smolen, A. J., 1988, Morphology of synapses in the autonomous nervous system, *J. Electron Microscopy Technique* **10:**187-204.

Stahlman, M. T., Kasselberg, A. G., Orth, D. N., Gray, M. E., 1985, Ontogeny of neuroendocrine cells in the human fetal lung. II An immunohistochemical study, *Lab. Invest.* **52:**52-60.

Stahlman, M. T., Jones, M., Gray, M. E., Kasselberg, A. G., Vaughn, W. K., 1987, Ontogeny of neuroendocrine cells in the human fetal lung. III An electron microscopic immunohistochemical study, *Lab. Invest.* **56:**629-641.

Tranzer J. P., and Richards, J. G., 1976, Ultrastructural cytochemistry of biogenic amines in nervous tissue: methodologic improvements, *J. Histochem. Cytochem.* **24:**1178-1193.

Van Marle, J., 1977, Contribution to the knowledge of the nervous system in the tentacles of some Coelenterates, *Bydragen tot de Dierkunde* **46:**219-260.

Van Marle, J., Lind, A., and Van Weeren-Kramer, J., 1983, Properties of a catecholaminergic system in some Coelenterates. A histochemical and autoradiographic study, *Comp. Biochem. Physiol.* **76C:**193-197.

Westfall, J. A., 1987, Ultrastructure of invertebrate synapses, in: *Nervous Systems in Invertebrates* (M. A. Ali, ed.), Plenum Press, New York and London.

Chapter 10

The Antiquity of Monoaminergic Neurotransmitters: Evidence from Cnidaria

MICHEL ANCTIL

1. Introduction

When wrestling with the issue of the origin of nervous systems and their associated messenger molecules, conventional wisdom has historically dictated that one should look at coelenterates for experimental models (Robson, 1975; Anderson and Schwab, 1982). However, recent molecular approaches to constructing phylogenies suggest that Cnidaria evolved from a protist ancestor, along a line separate from that leading to other multicellular animals (Field et al., 1988). Regardless of the eventual "pedigree" of ancestral Cnidaria that may emerge from these analyses, it is still reasonable to view their nervous system as the most ancient, having evolved some 600 to 700 million years ago. The search for monoamine transmitters and their mechanisms of action in Cnidaria should be envisaged with that perspective in mind.

Following the introduction of the Falck-Hillarp histofluorescence technique (Falck et al., 1962; see Dahlstrom and Carlsson, 1986), biogenic monoamines became the focus of immense interest, and their widespread distribution in the mammalian nervous system was soon recognized. This histochemical technique was the first significant breakthrough that allowed the visualization of transmitter-specific neurons and the study of their distribution. The opportunity to track monoaminergic structures down the phylogenetic tree was quickly realized as Dahl et al. (1963) used the newly developed technique to demonstrate a catecholamine-like fluorescence in ectodermal neurons of sea anemone tentacles. Although the source of this histofluorescence was

MICHEL ANCTIL • Département de Sciences Biologiques and Centre de Recherche en Sciences Neurologiques, Université de Montréal, Montréal, Québec, H3C 3J7, Canada.

Evolution of the First Nervous Systems
Edited by P.A.V. Anderson
Plenum Press, New York

later attributed to other catechol-derived substances (Elofsson et al., 1977; Carlberg, 1983), that study prompted several investigations designed to identify monoaminergic mechanisms in coelenterates (for review, see Martin and Spencer, 1983).

It is not the purpose of this paper to provide a historical and critical analysis of the literature on monoamines in Cnidaria. In this respect, Martin and Spencer (1983) have stressed the ambiguities and inconclusiveness of past contributions. Instead, this chapter will review the biochemical, immunohistochemical, and pharmacological evidence for the existence of monoaminergic mechanisms in a single species, the sea pansy *Renilla köllikeri*. These studies will be set in the context of our current understanding of monoaminergic mechanisms and of the anthozoan nervous system. Furthermore, the relevence of the findings to the problem of the evolutionary emergence of neuroactive substances will be evaluated.

2. The Investigated Species: *Renilla köllikeri*

The sea pansy (Fig. 1) is a colonial anthozoan grouped with the sea pens in the Pennatulacea. The colony develops from the budding of a primary polyp whose progeny is represented by two types of secondary polyps, autozooids and siphonozooids (Wilson, 1883). These polyps are rooted in a flattened discoidal mass, the rachis. The peduncle, a muscular cylindrical mass used to anchor the colony in sandy bottoms, represents the foot of the primary polyp.

The nervous system of the sea pansy has the typical nerve net organization found in anthozoans and other cnidarians. It is set within the diploblastic tissue sheets that constitute the basic body plan of this phylum (Mackie, 1984). These sheets consist of

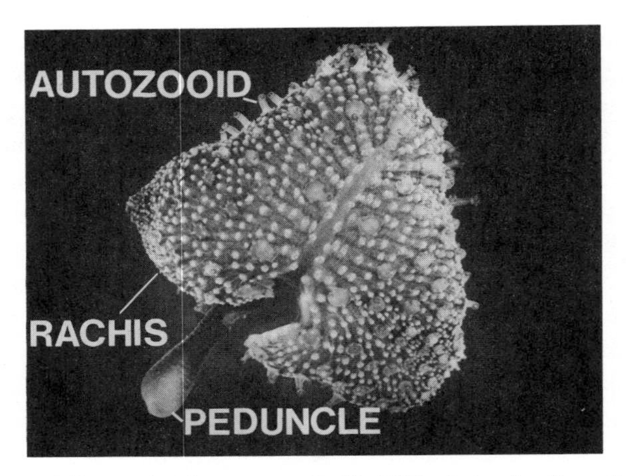

Figure 1. Dorsal view of a partially contracted sea pansy, *Renilla köllikeri*, showing the radial distribution of polyps over the flat colonial tissue (rachis). Some of the feeding polyps (autozooids) are profiled at the edge of the rachis.

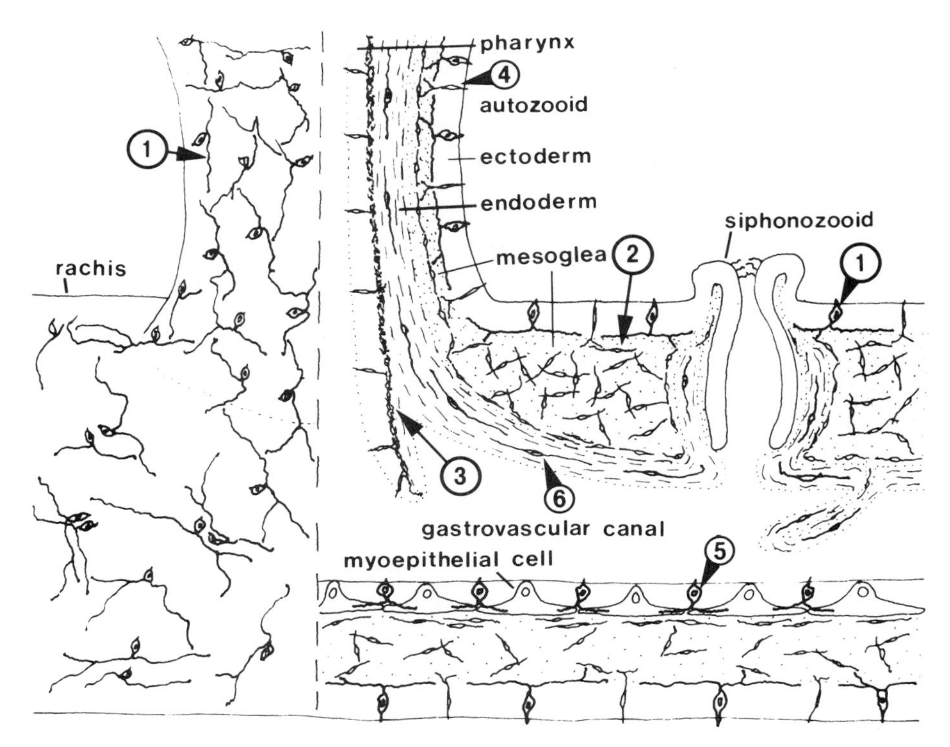

Figure 2. Schematic, two-dimensional representation of the tissue and neuronal organization of *R. köllikeri*. The surface of the rachis and an autozooid base is represented to the left of the dashed line, and the section through the tissue to the right. Note that endodermal layers are always myoepithelia, even though they are shown as such only in the ventral rachidial wall. 1, ectodermal nerve-net; 2, mesogleal nerve-net; 3, pharyngeal nerve-net; 4, ectodermal sensory cells; 5, endodermal ciliated neurons; 6, endodermal bipolar neurons. Drawing not to scale.

two epithelial layers, ectoderm and endoderm, sandwiching an extracellular matrix, the mesoglea (Fig. 2). The mesogleal nerve net, which extends throughout the mesoglea in the autozooids, rachis and peduncle, is the postulated substrate for the through-conducted impulses that coordinate luminescence, autozooid withdrawal and rachidial contraction (Anderson and Case, 1975). Neurites extend from this net toward muscular processes of the endodermal epithelia (Satterlie et al., 1980). There are also ectodermal ciliated neurons whose basiepithelial processes may form a nerve net from which neurites extend to the underlying mesogleal nerve net (Umbriaco et al., 1989). These are distributed throughout the rachis and in polyps. A nerve net composed of small, tightly intermeshed neurons lies at the base of the pharyngeal epithelium (Grimmelikhuijzen and Anctil, in preparation). Slender, ciliated sensory cells are present in the ectoderm and pharyngeal epithelium. In addition, large ciliated neurons embedded in endodermal epithelia of the rachis innervate the muscle feet of myoepithelial cells (Grimmelikhuijzen and Anctil, in preparation).

3. Evidence for Catecholamines

While dopamine and noradrenaline have been detected in many coelenterates by HPLC with electrochemical detection (HPLC-ED) (Venturini et al., 1984; Carlberg and Rosengren, 1985; Chung et al., 1989), only in *Renilla* was the presence of all three biogenic catecholamines, including adrenaline, demonstrated. This was confirmed by two different and sensitive analytical tools: radioenzymatic assays with thin-layer chromatography and HPLC-ED (Fig. 3) (De Waele et al., 1987). The use of these techniques also revealed considerable variations in the detectability of these substances among specimens sampled. Since adrenaline has not been convincingly identified in other invertebrates (Klemm, 1985), its presence in *Renilla* and in the ctenophore *Mnemiopsis* (Carlberg, 1988) appears to be a distinctive feature of coelenterates.

HPLC-ED chromatograms of some *Renilla* tissues also disclosed substantial peaks which co-elute with DOPAC, an oxidation product of dopamine, and with normetane-

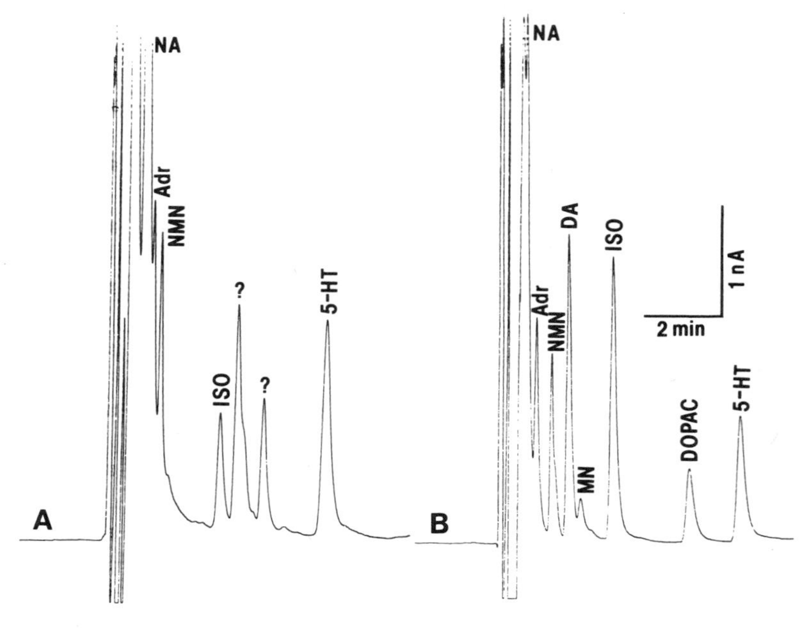

Figure 3. Electrochromatograms obtained by HPLC with a Bio-Sil ODS-5S reversed-phase column (Bio-Rad), a Model 600 HRLC workstation (Bio-Rad) and an EG&G-Princeton Model 400 electrochemical detector, operated at 750 mV. **(A)** Noradrenaline (NA), adrenaline (Adr), normetanephrine (NMN) and serotonin (5-HT) were recorded in this extract from the rachis of *R. köllikeri*. **(B)** Standards (1 ng in column) of NA, Adr, NMN, dopamine (DA), metanephrine (MN), 3,4-dihydroxyphenylacetic acid (DOPAC) and 5-HT. Isoproterenol (ISO, 3 ng in column) was used as an internal standard. Tissues were extracted and the standards were dissolved in 0.1 N perchloric acid. Mobile phase: 0.75 M Na^+ phosphate, 1 mM 1-octane sulfonic acid, 50 μM EDTA, 12% methanol and 3% acetonitrile, pH adjusted to 3.95 with phosphoric acid. Flow rate: 1 ml/min.

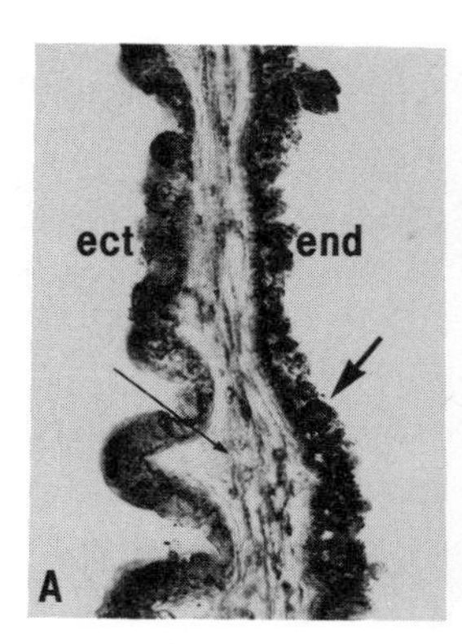

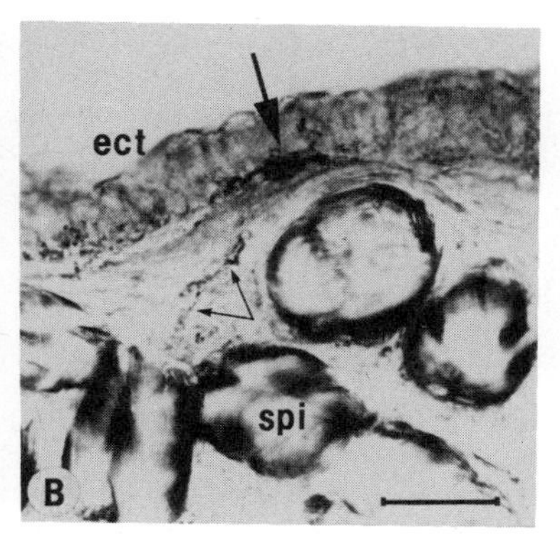

Figure 4. Noradrenaline immunoreactivity in 10 μm sections through tissues of *R. köllikeri*, revealed using the peroxidase-antiperoxidase (PAP) technique. **(A)** The wall of an autozooid column comprises the weakly reactive ectoderm (ect), the highly reactive endoderm (end), wherein individual myoepithelial cells are visible (thick arrow), and between these epithelia the mesoglea, in which numerous reactive neurites are present (thin arrow). **(B)** Dorsal rachis containing a reactive bipolar neuron at the base of the ectoderm (thick arrow) and reactive mesogleal neurites extending between calcareous spicules (spi). Bar: 50 μm **(A)** and 25 μm **(B)**. Tissues were fixed in paraformaldehyde and the sections were incubated in antiserum NA-P-1-18 raised in sheep (courtesy of Dr. A. A. J. Verhofstad), and diluted 1:600.

phrine, a methylation product of noradrenaline (Fig. 3) (Pani and Anctil, in preparation). These substances are indicators of the presence of monoamine oxidase and catechol-O-methyltransferase activities in vertebrates, and their presence in *Renilla* would suggest that such metabolic pathways for biogenic catecholamines are very ancient.

The Falck-Hillarp and derived histofluorescence localization techniques lack specificity with regard to individual catecholamines or their precursor DOPA, and have proved to be of little help in visualizing these amines in cnidarian cells (Elofsson et al., 1977; Carlberg, 1983). No discrete fluorescence has ever been generated in *Renilla* by these techniques, despite multiple attempts. Immunohistochemical studies with specific antibodies against dopamine (Geffard et al., 1984), noradrenaline and adrenaline (Verhofstad et al., 1983), provided the opportunity of a more reliable identification of catecholamine-containing cells. This approach demonstrated that mesogleal neurons and endodermal (myoepithelial) cells in *Renilla* were immunoreactive to all three antibodies (Fig. 4) (Umbriaco and Anctil, 1988, and in preparation). In addition, the ectodermal epithelium of polyp tentacles was strongly dopamine- and noradrenaline-immunoreactive, and some basiectodermal cells, which resemble bipolar neurons, were noradrenaline-immunoreactive. These observations (1) provide confirmation for the biochemical detection of catecholamines in *Renilla*, (2) indicate that the cellular distribution of the three catecholamine antigens partially overlaps, and more importantly, (3) suggest that these catecholamines are widely present in non-neuronal as well as neuronal cells. The presence of catecholamines in non-neuronal cells of *Renilla* is another distinctive feature for a cnidarian since only

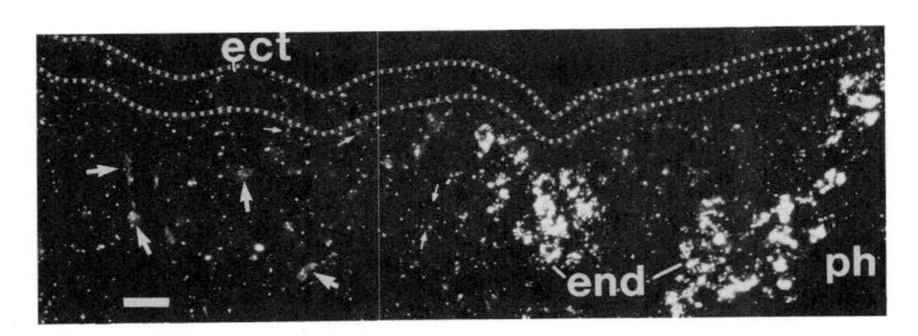

Figure 5. Light-microscopic audioradiograph (darkfield) of a section through the dorsal rachis of *R. kölliker*i which was incubated in 5 μM of tritiated adrenaline. Ectoderm is outlined by the broken line. Note the light labelling of amoebocytes (thick arrows) and presumptive neuronal tracts (thin arrows) in the mesoglea. Labelling is heavier in the endodermal layer (end), but absent in the pharynx (ph) of a siphonozooid. Bar: 20 μm (adapted from Anctil et al., 1984).

in the salivary gland cells of *Octopus*, among other invertebrates, have catecholamines been associated with non-neuronal cells (Matus, 1971; Juorio and Killick, 1973).

Kinetic and pharmacological analyses of catecholamine uptake provide useful information on an efficient mode of transmitter inactivation (re-uptake) exploited in vertebrate nervous systems (Paton, 1976). In *Renilla*, the existence of specific, temperature-sensitive, carrier-mediated transport mechanisms for noradrenaline and adrenaline was demonstrated, as was the presence of a low-affinity, high-capacity diffusion component (Anctil et al., 1984). A unique feature of the high-affinity uptake systems is their Ca^{++} dependence, which contrasts with the Na^{+}-dependent uptake systems of vertebrates (Paton, 1976) and other invertebrates (Osborne et al., 1975). This Ca^{++} substitution raises the possibility that monoamine transport mechanisms evolved from a Ca^{++}- to a Na^{+}-dependence in multicellular animals, and in this respect it may present some parallel with the evolution of Na^{+} channels of cnidarian neurons (see Anderson, this book).

Attempts to visualize cellular sites of catecholamine uptake by radioautography met with mixed results. Radioautographic analyses of *Renilla* tissues exposed to tritiated noradrenaline and adrenaline allowed the visualization of labelled mesogleal cells, but labelling of neuronal elements could not be satisfactorily resolved (Fig. 5) (Anctil et al., 1984). In addition, there was widespread labelling in the epithelial and muscle layers of the endoderm. It appears, therefore, that the overall distribution of radioautographic reactions largely mirrors that of the immunoreactivity (see above).

Adrenaline, but neither noradrenaline nor dopamine, induced bioluminescence in the sea pansy (Anctil et al., 1982), and this response was depressed by the beta-adrenergic blocker propranolol. Similarly, Ross (1960a,b) found that adrenaline was the only catecholamine capable of inducing contractions of slow column muscles in sea anemones. In both the sea pansy and anemones, some contractile activities were enhanced by noradrenaline and/or adrenaline, but only at relatively high concentrations (Anctil et al., 1982; Ross, 1960a,b). Should these amines prove to play a physiologically significant role in these activities, then the catecholamine-immunoreactive mesogleal neurons that lie in close proximity to endodermal musculature or photocytes, or the similarly immunoreactive myoepithelial cells, may provide pathways for their involvement. Dopamine and noradrenaline have been associated with other cnidarian activities, such as a depression of the feeding response of *Hydra* (Hanai and Kitajima, 1984) and stimulation of metamorphosis of larvae of the hydrozoan *Halocordyle* (Edwards et al., 1987). Electrophysiological recordings at the cellular level will be required to address the mechanisms associated with these adrenergic responses. In view of the well-known association of catecholamines with the regulation of carbohydrate metabolism in vertebrates, even in Protozoa (Ness and Morse, 1985a,b), the possibility of a similar role of these amines in Cnidaria should be explored.

4. Evidence for Serotonin

Serotonin (5-HT) has been detected in all major invertebrate phyla above the Cnidaria (Klemm, 1985), and 5-HT-immunoreactive neurons have been visualized in the lowest invertebrates possessing a "brain", the platyhelminths (Webb and Mizukawa, 1985; Reuter et al., 1986; Sukhdeo and Sukhdeo, 1988). One would expect, therefore, to find 5-HT in the nervous systems of Cnidaria. In fact, claims of the presence or absence of 5-HT in Cnidaria are many and often contradictory (Martin and Spencer, 1983). Although Carlberg and Rosengren (1985) surveyed many species from all three cnidarian classes by high performance liquid chromatography with electrochemical detection (HPLC-ED), 5-HT was detected only in one scyphozoan species, *Cyanea lamarcki*. Later, 5-HT immunoreactive ectodermal gland cells were visualized in the tentacles of this jellyfish, apparently accounting for the biochemical detection (Elofsson and Carlberg, 1989).

Variable amounts of 5-HT have been detected in the sea pansy by HPLC-ED (Fig. 3) (Pani and Anctil, in preparation). In addition to 5-HT, its metabolites, 5-hydroxyindoleacetic acid (5-HIAA) and N-acetylserotonin, were occasionally detected, suggesting the presence of oxidation and acetylation pathways, respectively, for 5-HT metabolism in *Renilla*. This finding, along with the detection of DOPAC (see above), indicates that cnidarians share with vertebrates the possession of an active monoamine oxidase, although specific enzyme activities remain to be investigated. In contrast, arthropods rely largely on acetylation (insects) or sulfation (crustaceans) for monoamine metabolism (Vaughan, 1988).

That the 5-HT detected in the sea pansy is primarily associated with nerve cells, was documented in an immunohistochemical study (Umbriaco et al., 1989). The ectoderm of both polyps and rachis contain 5-HT immunoreactive pseudounipolar neurons which are endowed with an apical (ciliary) knob and an axonal process. The bifurcated and beaded neurites from this process spread within the ectoderm-mesoglea interface, and in certain regions, appear to be organized into a nerve-net (Fig. 6). Some of these neurites cross into the mesogleal layer and are found in close proximity to 5-HT immunoreactive bipolar mesogleal neurons. The latter are less numerous than the 5-HT immunoreactive ectodermal neurons and their short neurites extend toward endodermal myoepithelial cells. In addition, in the oral complex of the autozooids, nematocytes, which, in *Hydra* are known to be derived, along with nervecells, from common multipotent stem cells (David, 1983), were also 5-HT immunoreactive. The distribution and cellular composition of the 5-HT immunoreactivity did not overlap significantly with that of catecholamines. The 5-HT immunoreactive ectodermal neurons of *Renilla* are morphologically similar to the RFamide immunoreactive sensory ectodermal neurons of *Hydra* (Grimmelikhuijzen et al., 1982) and neurites of both form nerve nets. The lack of detectable 5-HT immunoreactivity in *Hydra* (Grimmelikhuijzen, cited in Martin and Spencer, 1983) and conversely the failure to visualize RFamide immunoreactive ectodermal neurons of similar morphology in *Renilla* (Grimmelikhuijzen and Anctil, in preparation), suggest that these may be homologous neurons expressing different neuroactive substances in Hydrozoa and Anthozoa. In a representative of the platyhelminths, 5-HT and FMRFamide immunoreactivities were segregated in different neuronal populations (Reuter et al., 1986).

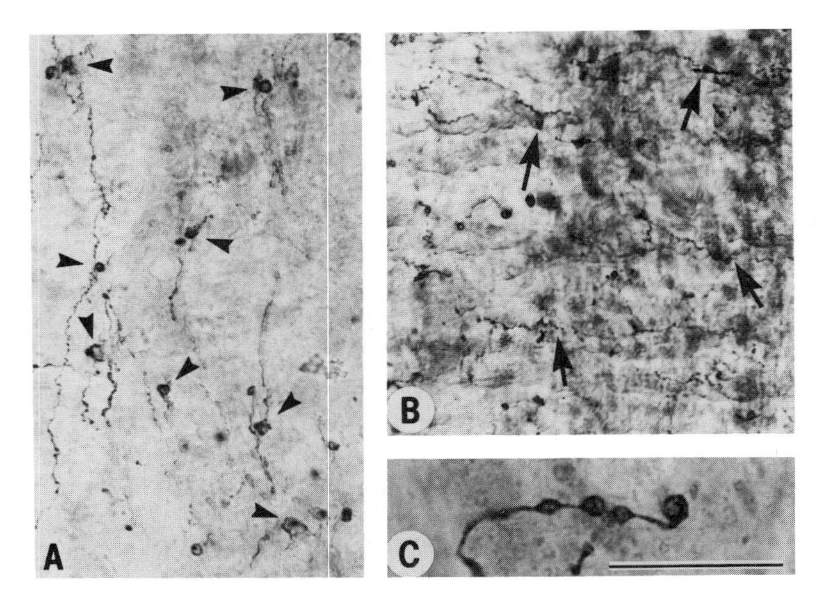

Figure 6. Ectodermal 5-HT-immunoreactive neurons in an autozooid of *R. kölliikeri* (wholemounts). **(A)** Upper region of the autozooid column (near oral complex), where processes of the pseudounipolar neurons (arrowheads) are mostly aligned longitudinally. **(B)** Lower region of the column where neurons (arrows) are aligned transversely. Visibility is reduced due to the presence of symbiotic algae in background. **(C)** Close-up of the terminal portion of a reactive neurite, showing the polymorphic varicosities. See Umbriaco et al. (1989) for methods. Bar: 80 μm (A), 130 μm (B) and 20 μm (C).

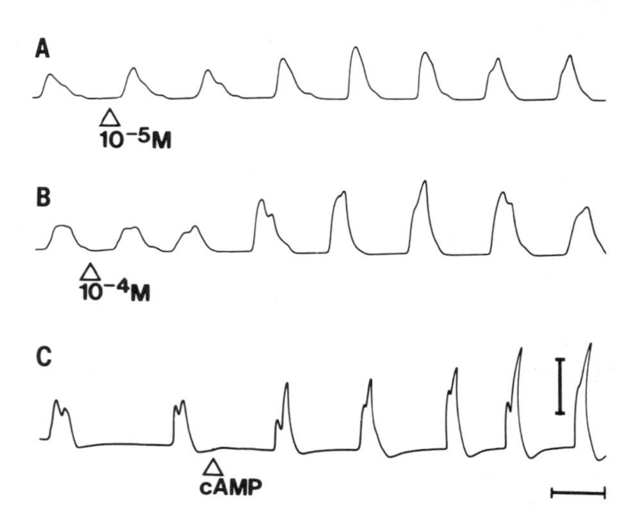

Figure 7. Effects of 5-HT (A,B) and 0.5 mM dibutyryl cAMP (C) on ongoing rhythmic contractions of two rachidial preparations of *R. köllikeri*. Arrowheads indicate time of addition of substance to the bath. Recording in B is from same preparation as in A, after recovery from initial response (to show concentration dependence). Vertical bar: 1.5 μN (A,B) or 3 μN (C); horizonal bar: 2 min (A,B) or 4.5 min (C) (adapted from Anctil, 1989).

A response of the sea pansy to water flow (e.g. enhancement of the amplitude of rhythmic contractions) was mimicked by application of 5-HT, in the micromolar range (Fig. 7) (Anctil, 1989). This response appeared to be pharmacologically specific for 5-HT, and drugs known to selectively disrupt serotonergic neurons (5,7-dihydroxy-tryptamine, p-chlorophenylalanine, reserpine) all had similar effects on the contractile activities of the sea pansy, namely a depression of rhythmic and electrically stimulated contractions (see Anctil, 1989 for a discussion of the limitations of this approach with respect to reserpine). It is reasonable, therefore, to suggest that the visualized 5-HT immunoreactive neurons are involved in the modulation of this rhythmic behavior. The nature of their involvement is unknown, but it is possible that they are mechanoreceptive neurons, since the ectodermal 5-HT immunoreactive neurons of *Renilla* possess an apical knob, typical of cnidarian sensory neurons (Umbriaco et al., 1989). The modulatory effect of 5-HT is probably exerted either on local neuronal circuits innervating muscle, or directly on muscles, but not on a through-conducting nerve net. Indeed, if a rachidial preparation is partitioned by a vaseline gap, the enhancing effect of 5-HT on peristaltic contractions fails to spread to the portion of the preparation unexposed to 5-HT (Anctil, 1989).

The physiological significance of the modulatory effect of 5-HT in *Renilla* was supported by demonstrating that cAMP analogues mimicked the 5-HT-induced enhancement of rhythmic contractions (Fig. 7), and that 5-HT elevated cAMP levels substantially in *Renilla* tissues (Anctil, 1989). Thus, cAMP appears to play a second messenger role in this activity. Although fluctuations of cAMP levels or adenylate cyclase activity have been associated with regeneration and the feeding response in Cnidaria (Lesh-Laurie, 1988, for review), this is the first evidence of a cAMP-mediated 5-HT response in a cnidarian. The association of this messenger with serotonergic mechanisms is common in invertebrates (Walker, 1984; Vaughan, 1988) as well as vertebrates, and its occurrence in the sea pansy is a testimony of its conserved character in phylogeny.

5. Functional Implications

The above evidence supports the existence of neuronal monoamines and monoaminergic mechanisms in a cnidarian species. Whether these substances act as neurotransmitters in the sea pansy, according to criteria established for vertebrate neurotransmitters, cannot be ascertained from our limited evidence. Table 1 shows that while some key criteria have been satisfied, data on the synthesis and neuronal release of monoamines are essential to consider these substances endogenously neuroactive. Moreover, without unequivocal evidence of synaptic release and evoked postsynaptic potentials, the neuronal monoamines of *Renilla* may be, at best, eligible for neuromodulatory or neurohormonal roles. As mentioned earlier, the physiological action of 5-HT reported in the sea pansy (Anctil, 1989) would suggest a modulatory influence on muscle effectors. Although axonal processes of the 5-HT immunoreactive

Table 1. The extent to which criteria for the identification of neurotransmitters are satisfied for monoamines in *Renilla köllikeri*

Criterion	CA	5-HT	Remarks
Presence of transmitters	yes[1]	yes[2]	Adrenaline not found in other invertebrates
Neuronal localization	yes[3]	yes[4]	Also non-neuronal
Synthetic machinery	?	?	--
Release upon neuronal stimulation	?	?	--
Inactivation mechanisms	yes[2,5]	yes[2]	Ca^{++}-dependent uptake; oxidation, methylation and acetylation products
Physiological action of test substance identical with that of native transmitter	?	yes?[6]	Difficult to assess in diffuse tissue systems of Anthozoa

[1]De Waele et al., 1987
[2]Pani and Anctil, in preparation
[3]Umbriaco and Anctil, 1988
[4]Umbriaco et al., 1989
[5]Anctil et al., 1984
[6]Anctil, 1989

neurons of *Renilla* are endowed with varicosities (Umbriaco et al., 1989), synaptic specializations at these sites have yet to be documented. In the mammalian CNS, the percentage of varicosities from 5-HT immunoreactive neurons with synaptic specializations is often rather low, as little as 15% of total (Soghomonian et al., 1988), suggesting that non-junctional release is the norm rather than the exception for these neurons. There is even a supra-ependymal meshwork of 5-HT immunoreactive neurons with fibers that criss-cross over the walls of the cerebral
ventricles (Takeuchi, 1988). This is strikingly reminiscent of the network of ectodermal 5-HT immunoreactive neurites in *Renilla*, and interestingly, varicosities in the supra-ependymal neurons also lack junctional specializations (Soghomonian et al., 1988). Whether the monoamine-containing neurons of *Renilla* interact with neurons and muscles in an analogous manner, through diffuse transmission of modulatory messages, remains to be explored.

Our understanding of the pharmacology of monoaminergic systems in Cnidaria is hampered by problems resulting from the diffuse, tissue-grade organization of these animals, particularly the problems of diffusion barriers to drugs and localization of their sites of action (Martin and Spencer, 1983). For instance, the depressive action of dopamine and noradrenaline on the *Hydra* feeding response, which presumably occurs through binding to surface receptors, is effective at concentrations as low as 1 nM (Hanai and Kitajima, 1984). In contrast, catecholamine concentrations necessary to induce significant flashing in *Renilla* (Anctil et al., 1982) or metamorphosis in *Halocordyle* (Edwards et al., 1987) were 10 μM or higher. Nonetheless, the few pharmacological studies, to date, indicate that the catecholaminergic receptors of cnidarians differ from those of vertebrates. For example, adrenaline, but not the closely related agonist noradrenaline, induced luminescence in *Renilla* (Anctil et al., 1982; Germain and Anctil, 1988). The depression of the *Hydra* feeding response induced by dopamine is blocked by propranolol, but not by known dopamine antagonists (Hanai and Kitajima, 1984). Recent radiobinding assays disclosed a propranolol- and methysergide-sensitive binding of tritiated dihydroalprenolol, a beta-adrenergic receptor ligand, to membrane preparations of *Renilla* (Teyssier and Anctil, in preparation). Kinetic analyses revealed the presence of a single type of binding sites, with a K_d in the 20-30 nanomolar range. The lack of selectivity between the beta blocker propranolol and the serotonergic antagonist methysergide for displacement of specific binding, suggests that these sites resemble mammalian 5-HT_{1b} receptors, which are identified by their high affinity for beta-adrenergic ligands (Pazos et al., 1985) and their coupling to adenylate cyclase (Hamon et al., 1988). Thus it is possible that *Renilla* possesses receptors sharing properties with known beta-adrenoceptors and serotonin receptors.

Alternatives to membrane receptor mechanisms may deserve consideration in cases where high concentrations (0.5-1.0 mM) of transmitter substances are required for activity; e.g., as muscle or photocyte responses of *Renilla* to noradrenaline or adrenaline (Anctil et al., 1982; Anctil, 1987; Germain and Anctil, 1988). These concentrations are similar to those required for cyclic nucleotide analogs to induce responses, suggesting that catecholamines may act as intracellular regulators in the sea pansy. One prerequisite for such a role is proof that these substances accumulate within the cells of interest. The visualization of immunoreactivity for all three catecholamines in endodermal cells, where muscle responses and luminescence occur (Umbriaco and Anctil, 1988), would indicate that this condition is met. It is interesting that an intracellular role for catecholamines in the regulation of expression of a metabolic enzyme has been proposed for the protozoan *Tetrahymena* (Ness and Morse, 1985a).

6. Evolutionary Implications

The evidence presented here, and that pertaining to platyhelminths (Reuter et al., 1986) argue in favor of the co-evolution of aminergic and peptidergic pathways in nervous systems (see Grimmelikhuijzen, Prosser, this book). In this respect, it is striking that the immunohistochemical distributions of monoamines (Umbriaco and Anctil, 1988; Umbriaco et al., 1989) and of the neuropeptide antho-RFamide (Grimmelikhuijzen and Anctil, in preparation) in *Renilla* do not overlap, except partly in the mesogleal nerve-net. Similarly, monoamines induce luminescence or modulate ongoing muscle activity (Anctil et al., 1982; Anctil, 1989), whereas antho-RFamide and related peptides induce muscle contractions but not luminescence in *Renilla* (Anctil, 1987; Anctil and Grimmelikhuijzen, 1989). Thus, the two pathways may have evolved independently, in terms of cellular localization and functional target tissues.

Table 1 highlights features of monoaminergic systems in *Renilla* that can be considered departures from those documented in other animals or in invertebrates: the presence of adrenaline, the association of monoamines with non-neuronal cells, and the existence of Ca^{++}-dependent catecholamine uptakes. Whether these departures reflect the independent evolution of emerging transmitter systems, along polyphyletic lines between Cnidaria and remaining animals, as suggested by recent molecular phylogenetic analyses (Field et al., 1988), is a matter of debate. Because these analyses suggest that Cnidaria and ciliate Protozoa have a common derivation from a protist ancestor, it is appropriate to ask whether monoaminergic mechanisms are shared by these two groups. Adrenaline, as well as noradrenaline, 5-HT (Janakidevi et al., 1966a, b; Brizzi and Blum, 1970), DOPA and dopamine (Ness and Morse, 1985b), has been reported in the protozoan *Tetrahymena*. Adrenaline and 5-HT are reported to stimulate adenylate cyclase activity in *Tetrahymena*, and the adrenaline effect was blocked by propranolol (Rosenzweig and Kindler, 1972). The presence of catecholamines in non-neuronal cells of *Renilla* would be a reasonable development from their presence in a protist ancestor. In this respect, the presence of these amines in neuronal as well as non-neuronal cells of the sea pansy suggests that the Cnidaria may have set the stage for the evolutionary transition from an autocrine, or paracrine role, to that of neuroactive transmitter for these substances. And finally,

Ca^{++}-dependent amine uptake mechanisms in the sea pansy may reflect on their probable protist ancestry since Ca^{++}-, but apparently not Na^{+}-dependent processes, are present in excitable membranes of Protozoa (Hille, 1984).

References

Anctil, M., 1987, Bioactivity of FMRFamide and related peptides on a contractile system of the coelenterate *Renilla köllikeri*, *J. Comp. Physiol.* **157**:31-38.

Anctil, M., 1989, Modulation of a rhythmic activity by serotonin via cyclic AMP in the coelenterate *Renilla köllikeri*, *J. Comp. Physiol.*, in press.

Anctil, M., and Grimmelikhuijzen, C. J. P., 1989, Excitatory action of the native neuropeptide antho-RFamide on muscles in the pennatulid *Renilla köllikeri*, *Gen. Pharmac.* **20**:381-384.

Anctil, M., Boulay, D., and LaRiviere, L., 1982, Monoaminergic mechanisms associated with control of luminescence and contractile activities in the coelenterate, *Renilla köllikeri*, *J. Exp. Zool.* **223**:11-24.

Anctil, M., Germain, G., and LaRiviere, L., 1984, Catecholamines in the coelenterate *Renilla köllikeri*. Uptake and radioautographic localization, *Cell. Tiss. Res.* **238**:69-80.

Anderson, P. A. V., and Case, J. F., 1975, Electrical activity associated with bioluminescence and other colonial behavior in the pennatulid *Renilla köllikeri*, *Biol. Bull.* **149**:80-95.

Anderson, P. A. V., and Schwab, W. E., 1982, Recent advances and model systems in coelenterate neurobiology, *Progr. Neurobiol.* **19**:213-236.

Brizzi, G., and Blum, J. J., 1970, Effect of growth conditions on serotonin content of *Tetrahymena pyriformis*, *J. Protozool.* **17**:553-555.

Carlberg, M., 1983, Evidence of dopa in the nerves of sea anemones, *J. Neural Transmission* **57**:75-84.

Carlberg, M., 1988, Localization and identification of catechol compounds in the ctenophore *Mnemiopsis leidyi*, *Comp. Biochem. Physiol.* 91C:69-74.

Carlberg, M., and Rosengren, E., 1985, Biochemical basis for adrenergic neurotransmission in coelenterates, *J. Comp. Physiol.* **B155**:251-255.

Chung, J. M., Spencer, A. N., Gahm, K. H., 1989, Dopamine in tissues of the hydrozoan jellyfish *Polyorchis penniciIatus* as revealed by HPLC and GC/MS. *J. Comp. Physiol. B.* **159**:173-181.

Dahl, E., Falck, B., von Mecklengurg, C., and Myhrberg, H., 1963, An adrenergic nervous system in sea anemones, *Quart. J. Micr. Sci.* **104**:531-534.

Dahlstrom, A., and Carlsson, A., 1986, Making visible the invisible, in: *Discoveries in Pharmacology, Vol. 3: Pharmacological Methods, Receptors and Chemotherapy*, pp. 97-125, (M. J. Parnham and J. Bruinvels, eds.), Elsevier, Amsterdam.

David, C. N., 1983, Stem cell proliferation and differentiation in hydra, in: *Stem Cells, their Identification and Characterization*, pp. 12-27 (C. S. Potten, ed.), Churchill-Livingstone, Edinburgh-London.

De Waele, J.-P., Anctil, M., and Carlberg, M., 1987, Biogenic catecholamines in the cnidarian *Renilla köllikeri*: radioenzymatic and chromatographic detection, *Can. J. Zool.* **65**:2458-2465.

Edwards, N. C., Thomas, M. B., Long. B. A., and Amyotte, S. J., 1987, Catecholamines induce metamorphosis in the hydrozoan *Halocordyle disticha* but not in *Hydractinia echinata*, *Roux's Arch. Dev. Biol.* **196**:381-384.

Elofsson, R., and Carlberg, M., 1989, Gland cells in the tentacles of the jellyfish *Cyanea lamarcki* reactive with an antibody against 5-hydroxytryptamine, *Cell Tiss. Res.* **255**:419-422.

Elofsson, R., Falck, B., Lindvall, O., and Myhrberg, H., 1977, Evidence for new catecholamines or related amino acids in some invertebrate sensory neurons, *Cell. Tiss. Res.* **182**:525-536.

Falck, B., Hillarp, N. A., Thieme, G., and Thorp, A., 1962, Fluorescence of catecholamines and related compounds condensed with formaldehyde, *J. Histochem. Cytochem.* **10**:348-354.

Field, K. G., Olsen, G. J., Lane, D. J., Giovannoni, S. J., Ghiselin, M. T., Raff, E. C., Pace, N. R., Raff, R. A., 1988, Molecular phylogeny of the animal kingdom, *Science* **239**:748-753.

Geffard, M., Kah, O., Onteniente, B., Seguela, P., Le Moal, M., and Delaage, M., 1984, Antibodies to dopamine: radioimmunological study of specificity in relation to immunocytochemistry, *J. Neurochem.* **42:**1593-1599.

Germain, G., Anctil, M., 1988, Luminescent activity and ultrastructural characterization of photocytes dissociated from the coelenterate *Renilla köllikeri*, *Tissue and Cell* **20:**701-720.

Grimmelikhuijzen, C. J. P., Dockray, G. J., and Schot, L. P. C., 1982, FMRFamide-like immunoreactivity in the nervous system of hydra, *Histochemistry* **73:**499-508.

Hamon, M., Gozlan, H., El Mestikawi, S., Emerit, M. B., Cossery, J. M., and Lutz, O., 1988, Biochemical properties of central serotonin receptors, in: *Neuronal Serotonin*, pp. 393-422 (N. N. Osborne and M. Hamon, eds.), John Wiley & Sons, New York, London.

Hanai, K., and Kitajima, M., 1984, Two types of surface amine receptors modulating the feeding response in *Hydra japonica*: the depressing action of dopamine and related amines, *Chem. Senses* **9:**355-367.

Hille, B., 1984, *Ionic Channels of Excitable Membranes*, Sinauer Associates, Sunderland, MA.

Janakidevi, K., Dewey, V. C., and Kidder, G. W., 1966a, The biosynthesis of catecholamines in two genera of Protozoa, *J. Biol. Chem.* **241:**2576-2578.

Janakidevi, K., Tewey, V. C., and Kidder, G. W., 1966b, Serotonin in Protozoa, *Arch. Biochem. Biophys.* **113:**758-759.

Juorio, A. V., and Killick, S. W., 1973, The distribution of monoamines and some of their acid metabolites in the posterior salivary glands and viscera of some cephalopods, *Comp. Biochem. Physiol.* **44A:**1059-1067.

Klemm, N., 1985, The distribution of biogenic monoamines in invertebrates, in: *Neurobiology. Current Comparative Approaches*, pp. 280-296 (R. Gilles and J. Balthazart, eds.), Springer-Verlag, Berlin-Heidelberg.

Lesh-Laurie, G. E., 1988, Coelenterate endocrinology, in: *Invertebrate Endocrinology, Vol. 2: Endocrinology of Selected Invertebrate Types*, pp. 3-29 (H. Laufer and R. G. H. Downer, eds.), Alan R. Liss, New York.

Mackie, G. O., 1984, Introduction to the diploblastic level, in: *Biology of the Integument, Vol. 1: Invertebrates*, pp. 43-46 (J. Bereiter-Hahn, A. G. Matoltsy and K. S. Richards, eds.), Springer-Verlag, Heidelberg.

Martin, S. M., and Spencer, A. N., 1983, Neurotransmitters in coelenterates, *Comp. Biochem. Physiol.* **74C:**1-14.

Matus, A. I., 1971, Histochemical localization of biogenic amines in the posterior salivary glands of octopods, *Tissue and Cell* **3:**389-394.

Ness, J. C., and Morse, D. E., 1985a, Regulation of galactokinase gene expression in *Tetrahymena thermophila*. I. Intracellular catecholamine control of galactokinase expression, *J. Biol. Chem.* **260:**10001-10012.

Ness, J. C., and Morse, D. E., 1985b, Regulation of galactokinase gene expression in *Tetrahymena thermophila*. II. Identification of 3,4-dihydroxyphenylalanine as a primary effector of adrenergic control of galactokinase expression, *J. Biol. Chem.* **260:**10013-10018.

Osborne, N. N., Hiripi, L., and Neuhoff, V., 1975, The in vitro uptake of biogenic amines by snail (*Helix pomatia*) nervous tissue, *Biochem. Pharmacol.* **24:**2141-2148.

Paton, D. M., 1976, Characteristics of uptake of noradrenaline by adrenergic neurons, in: *The Mechanism of Neuronal and Extraneuronal Transport of Catecholamines*, pp. 49-66 (D. M. Paton, ed.), Raven Press, New York.

Pazos, A., Engel, G., and Palacios, J. M., 1985, Beta-adrenoceptor blocking agents recognize a subpopulation of serotonin receptors in brain, *Brain Res.* **343:**403-408.

Reuter, M., Wikgren, M., and Lehtonen, M., 1986, Immunocytochemical demonstration of 5-HT-like and FMRFamide-like substances in whole mounts of *Microstomum lineare* (Turbellaria), *Cell. Tiss. Res.* **246:**7-12.

Robson, E. A., 1975, The nervous system of coelenterates, in: *Simple Nervous Systems*, pp. 169-209 (P. N. R. Usherwood and D. R. Newth, eds.), Edward Arnold, London.

Rosenzweig, Z., and Kindler, S. H., 1972, Epinephrine and serotonin activation of adenylate cyclase from *Tetrahymena pyriformis*, *FEBS Lett.* **25:**221-223.

Ross, D. M., 1960a, The effects of ions and drugs on neuromuscular preparations of sea anemones. I. On preparations of the column of *Calliactis* and *Metridium*, *J. exp. Biol.* **37:**732-752.

Ross, D. M., 1960b, The effects of ions and drugs on neuromuscular preparations of sea anemones. II. On sphincter preparations of *Calliactis* and *Metridium*, *J. exp. Biol.* **37**:753-774.

Satterlie, R. A., Anderson, P. A. V., and Case, J. F., 1980, Colonial coordination in anthozoans: Pennatulacea, *Mar. Behav. Physiol.* **7**:25-46.

Soghomonian, J.-J., Beaudet, A., and Descarries, L., 1988, Ultrastructural relationships of central serotonin neurons, in: *Neuronal Serotonin*, pp. 57-92 (N. N. Osborne and M. Hamon, eds.), John Wiley & Sons, New York, London.

Sukhdeo, S. C., and Sukhdeo, M. V. K., 1988, Immunohistochemical and electrochemical detection of serotonin in the nervous system of *Fasciola hepatica*, a parasitic flatworm, *Brain Res.* 463:57-62.

Takeuchi, Y., 1988, Distribution of serotonin neurons in the mammalian brain, in: *Neuronal Serotonin*, pp. 25-56 (N. N. Osborne and M. Hamon, eds.), John Wiley & Sons, New York, London.

Umbriaco, D., Anctil, M., and Descarries, L., 1989, Serotonin-immunoreactive neurons in the cnidarian *Renilla köllikeri*, *J. Comp. Neurol.*, in press.

Umbriaco, D., and Anctil, M., 1988, Immunohistochemical evidence of catecholaminergic cells in the coelenterate *Renilla köllikeri*, *Bull. Can. Soc. Zool.* **19(2)**:23.

Vaughan, P. F. T., 1988, Amine transmitters and their associated second messenger systems, in: *Comparative Invertebrate Neurochemistry*, pp. 124-174 (G. G. Lunt and R. W. Olsen, eds., Cornell Univ. Press, Ithaca.

Venturini, G., Silei, O., Palladini, G., Carolei, A., and Margotta, V., 1984, Aminergic neurotransmitters and adenylate cyclase in *Hydra*, *Comp. Biochem. Physiol.* **78C**:345-348.

Verhofstad, A. A. J., Steinbusch, H. W. M., Joosten, H. W. J., Penke, B., Varga, J., and Goldstein, M., 1983, Immunocytochemical localization of noradrenaline, adrenaline and serotonin, in: *Immunohistochemistry. Practical Applications in Pathology and Biology*, pp. 143-168 (J. M. Polak and S. Van Noorden, eds.), Wright PBS, Bristol-London.

Walker, R. J., 1984, 5-Hydroxytryptamine in invertebrates, *Comp. Biochem. Physiol.* **79C**:231-235.

Webb, R. A., Mizukawa, K., 1985, Serotonin-like immunoreactivity in the cestode *Hymenolepis diminuta*, *J. Comp. Neurol.* **234**:431-440.

Wilson, E. B., 1883, The development of *Renilla*, *Phil. Trans. R. Soc. Lond.*, **B174**:723-815.

Chapter 11

Rethinking the Role of Cholinergic Neurotransmission in the Cnidaria

ELIANA SCEMES

1. Introduction

After Otto Loewi's confirmation of the concept of chemical synaptic transmission, a major emphasis in the field of comparative physiology of neuromuscular systems concerned the question of which invertebrates employ adrenergic or cholinergic transmission. This problem was of particular interest to zoologists attempting to establish phylogenetic relationships among invertebrates. Thus, Bacq and Pantin began a series of studies around 1935 to verify the presence of adrenaline, acetylcholine (ACh) and cholinesterase in invertebrates ranging from the Coelenterata to the Tunicata. Following their finding of a widespread sensitivity of animals to the application of ACh, adrenaline and related substances, these authors proposed that chemical transmission occcurred in most animals, but that the problem of neurotransmission in the coelenterates had been solved by employing chemical mechanisms other than those described for the vertebrate neuromuscular junction.

Since then, certain criteria have been developed to identify putative neurotransmitters. The first is that the candidate transmitter should be demonstrated to be released from the presynaptic nerve terminals when the nerve is stimulated. This

ELIANA SCEMES • Departamento de Fisiologia Geral, Instituto de Biociências, Universidade de São Paulo, SP, Brasil criterion is easily satisfied in isolated organs, the main problem being the need for a method for detecting release that does not itself destroy the functional and structural integrity of the tissue being analyzed.

The second criterion is to document the presence of the neurotransmitter in the tissue in question, and numerous specific cytochemical methods for both light microscopy and electron microscopy have now been developed. However, the presence of a transmitter, per se, indicates neither releasability nor neuroeffectivness (e.g., ACh in nerve-free placenta).

The third criterion is that the candidate substance should mimic the action of transmitter released by nerve stimulation. However, application of exogenous transmitters is frequently very problematic. For instance, diffusional barriers may selectively retard the entrance of many types of molecules, and, in addition, catabolic enzymes may destroy the transmitter as it diffuses to its supposed site of action. A further complication of this method of administration is that the interval between application of the agent and the onset of the response can be quite long (seconds to minutes) compared to the millisecond intervals required for junctional transmission. Long delays make determination of the primary site of action of a transmitter extremely difficult when the observed response involves chains of neurons.

A fourth criterion for the identification of a synaptic transmitter requires that the pharmacological effect of drugs which potentiate or block postsynaptic responses to both neurally released and administered samples be the same. Because the pharmacological effects are often not identical (e.g., the effects of blocking agents applied to the CNS are extrapolated from their effects on peripheral autonomic organs), this criterion is often satisfied only indirectly, with circumstantial evidence.

The question to be considered here is, have we satisfied the criteria necessary to consider ACh as a putative neurotransmitter in the Cnidaria? In spite of the apparent absence of ACh and cholinesterase in coelenterates (Bacq, 1935, 1937, 1946; Pantin, 1935; Bacq and Nachmansohn, 1937; Bullock and Nachmansohn, 1942), it has long been known that these animals are sensitive to cholinergic drugs. Atropine causes convulsive swimming movements in *Sarsia* (Romanes, 1885), increased pulsation rate in *Gonionemus* and longitudinal contractions of whole *Metridium* preparations (Moore, 1917). However, an extensive study of the action of cholinergic drugs on the facilitated responses of sea anemones (Ross, 1946) showed that the responses were not analogous to those described for the vertebrate junctions. Instead, the evidence was against transmission through the direct action of certain substances on muscle and on facilitated processes.

Since then, a variety of cnidarian preparations have been examined for the action and presence of neurotransmitters. As will become evident during this brief review of cholinergic transmission in the Cnidaria, the interpretation of whether or not cholinergic mechanisms are present in Cnidaria remains a difficult challenge, principally owing to the conflicting data available, and the great diversity of biological responses analyzed.

2. Scyphozoa

Very little is known about the effects of ACh and related substances on scyphozoans. ACh, physostigmine and curare have no effect on the swimming rhythm of *Cyanea* (Horridge, 1959) and neither do other related substances (Anderson, personal communication). Furthermore, acetylcholinesterase cannot be demonstrated histochemically in the swimming motor neurons of *Cyanea* (Scemes and Anderson, unpublished) and homogenates of *Chrysaora isocella* do not exhibit acetylcholinesterase activity (Scemes, unpublished). Thus, ACh is probably not employed as a neurotransmitter in the Scyphozoa, at least in the *Cyanea* swimming pathway. Whether cholinergic transmission is involved in other conducting pathways in this class of animals remains an open question.

3. Anthozoa

The data on cholinergic transmission in the Anthozoa, particularly that concerning the effectivness of ACh, are rather conflicting. ACh (0.1 and 1 mM), has no effect on preparations of the sea anemones *Metridium* and *Calliactis* (Bacq, 1946; Ross, 1946, 1960a,b), even in the presence of eserine, and in *Renilla kölliken*, ACh and curare fail to induce luminescence (Anctil et al., 1982). However, other anthozoan preparations respond to applied cholinergic agonists and antagonists. Pieces of tissue from the body wall of the sea anemone *Anthopleura xanthogrammica* are sensitive to ACh (1%), as is the mesenteric retractor muscle, although to a lesser extent. The pharynx, however, is totally insensitive (Winkler and Tilton, 1962).

Strips from the body wall of the sea anemone *Bunodosoma caissarum* contract in response to single doses of ACh, propyonylcholine, benzoylcholine, acetyl-β-methylcholine, carbamylcholine (0.1 mM), and are highly sensitive to nicotine and butyrylcholine (0.1 mM) (Mendes and Freitas, 1984). Surprisingly, eserine potentiates this effect (Mendes, 1976) even though cholinesterase activity is absent in body wall homogenates of the same anemone (Scemes et al., 1982), and neither acetylcholinesterase nor cholineacetyltransferase have been found in *Tealia felina*, *Anemonia sulcata*, *Metridium senile* or *Cerianthus membranaceus* (Van Marle, 1977). Unexpectedly, atropine induces dose-dependent responses of body wall strips of *Bunodosoma caissarum* that are unaffected by ACh, propyonylcholine, benzoylcholine, acetyl-β-methylcholine, butyryltrimethilamonium, hexamethonium, DMPP and succinylcholine (Scemes and Jurkiewicz, 1984). Furthermore, Ross (1946) found that atropine enhanced electrically evoked contractions of intact *Metridium* and *Calliactis*, but failed to do so in isolated preparations (Ross, 1960a,b), whereas Winkler and Tilton (1962) observed that different parts of the sea anemone *Anthopleura xanthogrammica* displayed differential sensitivity to ACh.

These conflicting data have led many investigators (Horridge, 1959; Ross, 1960a) to state that the Anthozoa, and perhaps the Phylum Cnidaria as a whole, do not employ ACh as a neurotransmitter. Alternatively, these results may indicate either a nonspecific action of cholinergic substances in the anthozoan or reflect different levels of neural and muscular organization.

Electrophysiological studies have shown that different sea anemones possess the same conducting systems (TCNN, SS1, SS2), even though their behaviors are very different, and controlled in different ways (Lawn and McFarlane, 1976; McFarlane, 1976, 1982). Even at the anatomical level, there are variations in the degree of nervous development in different species (Robson, 1961, 1963, 1965). Furthermore, a single conducting system, such as the TCNN, can induce both excitatory and inhibitory action in circular and sphincter muscle preparations (Ewer, 1960; Lawn, 1976a, b). Thus, the anatomical and physiological complexity of the nervous systems of sea anemones, in combination with their limitations as experimental preparations, must be taken into account when analyzing pharmacological data. However, a systematic analysis of the action of cholinergic drugs on anthozoan preparations may yet reveal a role for these agents in these animals.

4. Hydrozoa

Although most studies of cholinergic transmission in the Hydrozoa have been made on *Hydra*, the effects of the applied drugs have been analyzed on a great variety of biological responses. ACh (0.5 mM) and physostigmine (0.1 mM) inhibit spawning of *Hydractina echinata* (Yoshida, 1959) and regeneration of *Hydra litorallis* is inhibited by decamethonium (0.46 mM), d-tubocurarine (0.29 mM), atropine (0.30 mM) and physostigmine (0.24 mM) (Lentz and Barrnett, 1963). Nematocyst discharge is induced by ACh, physostigmine, decamethonium and atropine, and blocked by hexamethonium and d-tubocurarine (Lentz and Barrnett, 1962) while phototaxis in *Hydra piradi* is enhanced by acetyl-β-methylcholine (0.01 mM), physostigmine (0.01 mM) and neostigmine (0.001 mM) and blocked by atropine (Singer, 1964). Nicotinic antagonists (d-tubocurarine, decamethonium and hexamethonium) reduce ectodermal contraction pulse bursts in *Hydra attenuata*, and muscarinic antagonist (atropine) enhanced them (Kass-Simon and Passano, 1978). Finally, *Hydra viridis* and *Pelmatohydra oligactis* are relaxed by atropine, while hexamethonium and decamethonium cause contraction.

Although insensitive to ACh, these various species do possess an enzyme that hydrolyzes ACh, but it is not blocked by eserine, BW-284051 or iso-OMPA. The histochemical and cytochemical localization of this enzyme has been problematic, and cholineacetyltransferase activity has not been detected (Van Marle, 1977; Erzen and Brzin, 1978).

Thus, in spite of, or perhaps because of, the variety of biological responses analyzed, the data do suggest the presence of a cholinergic sensitivity in which muscarinic and nicotinic agents may play opposing roles. Further evidence for the presence of a cholinergic mechanism in the Hydrozoa comes from the trachymedusa *Liriope tetraphylla*, a widespread planktonic jellyfish.

Like other jellyfishes, the swimming pattern of *Liriope tetraphylla* is characterized by periods of activity interruped by periods of quiescence; six to seven swimming contractions are followed by about 5 sec of quiescence (Scemes and Freitas, 1989). *Liriope* does not crumple when mechanically or electrically stimulated, but does so in the presence of atropine (Scemes and Mendes, 1986, 1988). Since crumpling behavior, described as radial muscle contraction leading to margin infolding, influences the swimming pattern of jellyfish (Hyman, 1940; Mackie, 1975; Mackie and Singla, 1975) and also changes the pattern of both radial and marginal potentials (Scemes and Freitas, 1989), the swimming behavior of *Liriope* was used to assay the effects of drugs on the conducting pathways underlying such behavior. The experiments essentially consist of measuring the number of pulses and the duration of swimming activity during a 5 min observation period. Known concentrations of cholinergic agonists and antagonists were then added to the medium and the same parameters measured.

A reduction of pulsation rate and radial muscle contraction were induced by butyrylcholine, nicotine and atropine, whereas propyonylcholine, carbamylcholine, pilocarpine, DMPP and gallamine affected only the pulsation rate. No effect was observable when ACh, acetylthiocholine, hexametonium or succinylcholine were applied (Scemes and Mendes, 1986; Scemes, unpublished).

The effect of atropine on *Liriope* was dose-dependent. At a concentration of 0.01 mM no significant behavioral effect was seen, at 0.1 mM there was a reduction in the swimming frequency and at 1 and 10 mM, atropine induced radial muscle contraction and altered the pulsation rate (Scemes and Mendes, 1986; Scemes and Freitas, 1989). The other above-mentioned cholinergic agonists and antagonists were effective only at high concentrations (1 and 10 mM).

It is arguable that the drug concentrations employed were overly high and that the observed responses may, therefore, be nonspecific. However, whole animal preparations were used, and both the epithelia and mesogloea provide a barrier to drug diffusion, as evidenced by the increased response delay observed upon decreasing the concentration of applied atropine (see Fig. 1 in Scemes and Freitas, 1989).

In an attempt to determine whether the site of action of atropine was pre- or post-synaptic, jellyfish were anesthetized in isosmotic (0.2 M) $MgCl_2$, to block chemical transmission. Under these conditions, atropine failed to induce radial muscle contraction. The same result occurred when animals were bathed in Ca^{++}-free, artificial sea water (ASW). On returning the animals to normal sea water (SW), radial muscle contraction resumed (Scemes and Mendes, 1986, 1988).

This result, together with the fact that ACh, even in the presence of neostigmine, (known to block acetylcholinesterase in *Liriope*, Scemes and Mendes, 1986), did not induce radial muscle contraction, while the nicotinic agonists nicotine and butyrylcholine, and the muscarinic antagonist atropine do, suggests that both muscarinic and nicotinic receptors may modulate the release of an as yet unknown neurotransmitter at the neuromuscular junction, antagonistically. Thus, the effect may be analogous to that in the mammalian heart (Lindmar et al., 1968; Westfall and Brasted, 1972). In *Liriope*, this cholinoceptive site may not reside on the muscle, but rather lie presynaptic to the neuromuscular junction in such a way that excitatory nicotinic and inhibitory muscarinic presynaptic receptors may modulate the release of the transmitter responsible for radial muscle contraction.

This model predicts that if the inhibitory muscarinic receptors are blocked, further stimulation of the excitatory nicotinic receptors would enhance the observed response. When *Liriope* was pre-exposed to atropine at a concentration (0.01 mM) that did not affect observable behavior and then exposed to different concentrations (0.01-1.0 mM) of a nicotinic agonist (ACh, nicotine, butyrylcholine, carbamylcholine, DMPP, pilocarpine), only ACh, up to 0.01 mM, induced a reduction in the swimming rate and a radial muscle contraction (Scemes and Mendes, 1988). However, when the jellyfish were pre-exposed to atropine at concentrations (0.1 mM) that do cause a reduction in swimming rate, radial muscle contraction was potentiated by nicotinic agonists in the following sequence of effectiveness: ACh > DMPP > nicotine > carbacol > pilocarpine > butyrylcholine (Scemes and Mendes, 1988; Scemes, unpublished).

These nicotinic receptors cannot be blocked by either hexamethonium (1 mM) or succinylcholine (1 mM); in the presence of these nicotinic antagonists, atropine still induces radial muscle contraction (Scemes, unpublished). However, the muscarinic agonist, PCh (1 mM), does inhibit the radial muscle contraction induced by atropine (1 mM), but does not affect the reduction in swimming rate caused by the latter (Scemes and Mendes, 1986).

The presence of cholinesterse activity in *Liriope* is further evidence for the role of a cholinergic mechanism. Whole body homogenates of *Liriope* hydrolyze acetylthiocholine at a rate of 0.9 μMol/hr/μg protein but do not hydrolyze butyrylthiocholine. The K_m (0.14 ±0.02 mM) and Vmax (0.98 ±0.09 μMol/hr/μg protein) indicate a moderate to high affinity for the substrate. Neostigmine (0.033-0.167 mM) totally blocked acetylthiocholine (0.033-0.167 mM) hydrolysis (Scemes and Mendes, 1986).

Histochemical localization of acetylcholinesterase activity in *Liriope* has so far been difficult, mainly due to the structural damage the available techniques cause. However, acetylcholinesterase has been detected in the ectoderm and endoderm of the manubrium of the leptomedusa *Eucheilota ventricularis* (Scemes and Leite, unpublished), thus confirming previous biochemical data on the presence of cholinesterase activity (1.2 μmol/h/μg protein) in whole body homogenates (Scemes, unpublished). *Eucheilota* also crumples in the presence of atropine, as does *Obelia* spp. However, further histochemistry with *Liriope* is still required to confirm that the above-mentioned effects of cholinergic substances are, indeed, a reflection of a cholinergic pathway in this animal.

Ultrastructural studies on *Liriope* have shown that the ectodermal neuromuscular system is the most likely morphological substrate for the control of crumpling behavior. The ectodermal neuromuscular system extends from the mouth to the margin, and consists of smooth muscle cells and neurons which contain either clear or dense-cored vesicles. This radial nerve net may contact the inner nerve ring responsible for the swimming pulses (Scemes and McNamara, 1989). Activation of the radial muscle, thus, is probably brought about by cholinergic synapses within the radial nerve net. Whether or not the marginal pathway is also involved in inducing radial muscle contraction by way of the radial nerve net and whether the activity of the latter can alter the bursting activity in the marginal pathway, remains to be demonstrated.

5. Discussion

In any given line of animal evolution, the higher representatives are believed to possess more elaborate and effective mechanisms of synaptic transmission than the less evolved members. Structurally, cnidarian neurons resemble those of higher animals. Furthermore, the biophysical basis of electrogenesis in neurons, muscle and epithelial cells is conventional, and intercellular junctions such as chemical and electrical synapses are similar to those found in all higher animals, although the existence of bidirectional synapses is somewhat unusual (Anderson, 1985).

The "simplicity" of the cnidarian nervous system does not, therefore, appear to lie at the level of individual cells, but rather in the organization of such cells into conducting systems. The manner of neuronal organization seen in this group may reflect primitiveness.

The role of a cholinergic transmission in the Cnidaria has been discussed for a number of years now. Biochemical investigations of some cnidarian nervous systems have revealed the presence of a few components of the cholinergic system (AChE) but not others (ACh, ChAt), while physiological studies have yielded rather confusing evidence for a role of ACh. This may reflect the fact that the structural and morphological peculiarities of the Cnidaria impede the use of conventional pharmacological methods for identification of neurotransmitters.

The presence of an inactivating enzyme is not often regarded as a necessary criterion for the identification of a neurotransmitter, but this need not be the case. There are two ways by which transmitter might be removed in cholinergic synapses: one, the classic scheme, involves junctional cholinesterase, the other relies on diffusion of the transmitter. Diffusion of ACh into extrajunctional spaces, and away from the receptors, may be the simplest way of removing the transmitter from the junctional area and, indeed, diffusion may have been the first inactivating mechanism for many neurotransmitters and other chemical regulators, that later developed specific degradative pathways. Thus, the absence of cholinesterase activity in sea anemones and its presence in hydrozoans may be an indication of junctional evolution towards more efficient transmission, achieved by adding a factor that shortens the duration of the transmitter-receptor complex, thus rendering transmission more phasic, although at the cost of stability.

To comprehend the relevance of molecules such as ACh, it is important to ascertain their value as neurotransmitters. The ultimate test of the significance of a molecule is its degree of biological activity, but this does not imply that a molecule should be discarded because biological activity is not observable. It does follow, however, that such activity is indicative of a functional purpose. Such may be the case for the action of ACh on cnidarian preparations. Cholinergic antagonists such as atropine cause many different biological responses. This does not mean that chemical synaptic transmission in the Cnidaria is different from that of vertebrates. Instead, the data described here may be explained in terms of the modulatory activity of neuronal transmission, as described for *Liriope*. Therefore, the lack of response to applied ACh may reflect its simultaneous action on both nicotinic excitatory and muscarinic inhibitory receptors. It should be remembered, however, that this model of synaptic interaction (Scemes and Mendes, 1986, 1988) is based on an analysis of the behavioral responses of the intact jellyfish and that the location of the precise site and mechanism of action of cholinergic agents must be examined at the cellular level. Recent techniques permiting the production of primary cultures of identifiable neurons from Hydromedusae (Przysiezniak and Spencer, 1989) should be useful in determining the nature of the chemical transmitters in cnidarian nervous systems.

ACKNOWLEDGEMENTS. The author is greatful to Dr. John C. McNamara for his comments and for reviewing the manuscript. The studies mentioned in this work were financed by FAPESP (79/282; 82/0143-6; 82/1672-2; 85/1406-9; 85/1407-5) and CNPq (303983/86.8; 500338/88.4).

References

Anderson, P. A. V., 1985, Physiology of a bidirectional, excitatory chemical synapse, *J. Neurophysiol.* **53**:821-835.

Anctil, M., Boulay, D., and Lariviere, L., 1982, Monoaminergic mechanisms associated with control of luminescence and contractile activities in the coelenterate, *Renilla köllikeri*, *J. exp. Biol.* **223**:11-24.

Bacq, Z. M., 1935, Recherches sur la physiologie et la pharmacologie du système nerveux autonome. XVII. Les esters de la choline dans les extraits de tissue des invertébrés, *Arch. Int. Physiol.* **42**:24-46.

Bacq, Z. M., 1937, Nouvelles observations sur l'acétylcholine et la choline-estérase chez les invertébrés, *Arch. Int. Physiol.* **44**:174-189.

Bacq, Z. M., 1946, L'acétylcholine et l'adrénaline chez les invertébrés, *Biol. Rev.* **22**:73-91.

Bacq, Z. M., and Nachmansohn, D., 1937, Cholinesterase in invertebrate muscles, *J. Physiol. (Lond.)* **89**:368-371.

Bullock, T. H., and Nachmansohn, D., 1942, Choline esterase in primitive nervous systems, *J. Cell. Comp. Physiol.* **20**:239-242.

Ewer, D. W., 1960, Inhibition and rhythmic activity of the circular muscles of *Calliactis parasitica* (Couch), *J. exp. Biol.* **37**:812-831.

Erzen, I., and Brzin, M., 1978, Cholinergic mechanisms in *Hydra*, *Comp. Biochem. Physiol.* **59C**:39-43.

Horridge, G. A., 1959, The nerves and muscles of medusae. VI. The rhythm, *J. exp. Biol.* **36**:72-91.

Hyman, L. H., 1940, Observation and experiments on the physiology of medusae, *Biol. Bull.* **79**:282-296.

Kass-Simon, G., and Passano, L. M., 1978, A neuropharmacological analysis of the pacemakers and conducting tissue of *Hydra attenuata*, *J. Comp. Physiol.* **128:**71-79.

Lawn, I. D., and McFarlane, I. D., 1976, Control of shell settling in the swimming sea anemone *Stomphia coccinea*, *J. exp. Biol.* **64:**419-429.

Lentz, T. L., and Barrnett, R. J., 1962, Enzyme histochemistry of *Hydra*, *J. exp. Zool.* **147:**125-149.

Lentz, T. L., and Barrnett, R. J., 1963, The role of nervous system in regenerating *Hydra*. The effect of neuropharmacological agents, *J. exp. Biol.* **154:**305-327.

Lindmar, R. R., Loffelholz, K., and Muscholl, E., 1968, A muscarinic mechanism inhibiting the release of noradrenaline from peripheral adrenergic nerve fibres by nicotinic agents, *Br. J. Pharmac. Chemother.* **32:**280-294.

Mackie, G.O., 1975, Neurobiology of *Stomotoca*. II. Pacemakers and conduction pathways, *J. Neurobiol.* **6:**357-378.

Mackie, G. O., and Singla, C. L., 1975, Neurobiology of *Stomotoca*. I. Action systems, *J. Neurobiol.* **6:**339-356.

McFarlane, I. D., 1976, Two slow conducting systems coordinate shell-climbing behaviour in the sea anemone *Calliactis parasitica*, *J. exp. Biol.* **64:**431-446.

McFarlane, I. D., 1982, *Calliactis parasitica*, in: *Electrical Conduction and Behaviour in "Simple" Invertebrates* pp. 243-265 (G. A. B. Shelton, ed.), Clarendon Press, Oxford.

Mendes, E. G., 1976, Chemical mediation in Coelenterata, *An. Acad. Bras. Cienc.* **47(supl.):**101-104.

Mendes, E. G., and Freitas, J. C., 1984, The responses of isolated preparations of *Bunodosoma caissarum* (Cora, 1964) (Cnidaria, Anthozoa) to drugs, *Comp. Biochem. Physiol.* **79C:**375-382.

Moore, A. R., 1917, *Proc. Nat. Acad. Sci.* **3:**598. Appud ROSS, D.M. 1946.

Pantin, C. F. A., 1935, Responses of the leech to acetylcholine, *Nature* **135:**875.

Przisiezniak, J., and Spencer, A. N., 1989, Primary culture of identified neurons from a cnidarian, *J. exp. Biol.* **142:** 97-113.

Robson, E. A., 1961, A comparision of the nervous systems of two sea anemones, *Calliactis parasitica* and *Metridium senile*, *Q. J. microsc. Sci.* **102:**319-326.

Robson, E. A.. 1963, The nerve-net of a swimming anemone, *Stomphia coccinea*, *Q. J. microsc. Sci.* **104:**535-549.

Robson, E. A., 1965, Some aspects of the structure of nervous system in the anemone *Calliactis*, *Am. Zool.* **5:** 403-410.

Romanes, G. J., 1885, *Jelly-fish, Star-fish and Sea-urchins, Being a Research on Primitive Nervous Systems*, Appleton, New York.

Ross, D. M., 1946, Facilitation in sea anemones. I. The action of drugs, *J. exp. Biol.* **22:**21-31.

Ross, D. M.. 1960a, The effects of ions and drugs on neuromuscular preparations of sea anemones. I. On preparations of the column of *Calliactis* and *Metridium*, *J. exp. Biol.* **37:**732-752.

Ross, D. M., 1960b, The effects of ions and drugs on neuromuscular preparations of sea anemones. II. On sphincter preparations of *Calliactis* and *Metridium*, *J. exp. Biol.* **37:**753- 774.

Scemes, E., and Freitas, J. C., 1989, Electrophysiology of the swimming system of hydromedusae and the effects of atropine-induced crumpling, *Braz. J. Med. Biol. Res.* **22:**189-198.

Scemes, E., and Jurkiewicz, A., 1984, O efeito de agonistas e antagonistas colinérgicos sobre a contractilidade de tiras de parede do corpo da anêmona do mar *Bunodosoma caissarum*, in: *Resumos do II Congresso Brasileiro de Farmacologia e Terapêptica Experimental*, p. 264.

Scemes, E., and McNamara, J. C., 1989, Ultrastructure of the radial neuromuscular system of *Liriope tetraphylla* (Trachymedusae), in: *5th International Conference on Coelenterate Biology*.

Scemes, E., and Mendes, E. G., 1986, Cholinergic mechanism in *Liriope tetraphylla* (Cnidaria, Hydrozoa), *Comp. Biochem. Physiol.* **83C:**171-178.

Scemes, E., and Mendes, E. G., 1988, Pharmacology of the radial neuromuscular system of *Liriope tetraphylla* (Hydrozoa, Trachymedusae), *Comp. Biochem. Physiol.* **90C:**385-389.

Scemes, E., Mendes, E. G., and Chaimovich, H., 1982, Absence of cholinesterase activity in the body wall homogenates from the sea anemone *Bunodosoma caissarum* Correa, *Comp. Biochem. Physiol.* **73C:**415-418.

Singer, R. H., 1964, The effect of neuropharmacological drugs on the light response of *Hydra piradi*, *Anat. Rec.* **148:**402-403.

Westfall, T. C., and Brasted, M., 1972, The mechanism of action of nicotine on adrenergic neurons in the perfused guinea-pig heart, *J. Pharmac. exp. Ther.* **182**:409-418.

Winkler, L. R., and Tilton, B. E., 1962, Predation on the sea hare, *Aplysia californica*, by the great green sea anemone, *Anthopleura xanthogrammica* (Brandt), and the effect of sea hare toxin and acetylcholine on anemone muscle, *Pacific Sci.* **16**:286-290.

Van Marle, J., 1977, Contribution to the knowledge of the nervous system in the tentacle of some coelenterates, *Bijd. Dierk.* **46**:219-260.

Yoshida, M., 1959, Effect of acetylcholine and eserine on spawing of *Hydractina echinata*, *Nature* **184**:1151.

Chapter 12

Wide Range Transmitter Sensitivities of a Crustacean Chloride Channel

HANNS HATT and CH. FRANKE

1. Introduction

Crustacean neuromuscular junctions use glutamate as the excitatory transmitter and GABA as the inhibitory transmitter. Excitatory, glutamate-activated channels have been studied in locust and crayfish by recording single channel currents (Patlack et al., 1979; Cull-Candy et al., 1980; Franke et al., 1983). Recently, the recording technique has been improved to the point at which GΩ-seals can be obtained, and these have been used to characterize the excitatory, glutamate-activated channels in muscle fibers of crayfish (Franke et al., 1987; Hatt et al., 1988a) and locust (Dudel et al., 1988). The channel has a high conductance of about 100 pS and single openings are very short: on the average 0.2 to 0.3 ms. At high glutamate concentrations, channel openings are grouped in bursts, the number of openings per burst and burst duration increasing when glutamate concentration rises from 0.2 to 10 mM. The channel described above is located at the excitatory synapses of muscle fibers of crayfish.

In the course of our study of these glutamate-activated channels, an additional, small amplitude channel activity was occasionally observed in hyperpolarized cell-attached patches. Under defined recording conditions, using excised patches, we learned that this glutamate-triggered activity marked the opening of chloride (Cl^-) channels. It is well known that in crayfish muscle, like in many invertebrate and vertebrate cells, the inhibitory synaptic transmitter γ-amino-butyric acid (GABA), triggers a Cl^- conductance in the membrane (Fatt and Katz, 1953; Dudel and Kuffler,

HANNS HATT and CH. FRANKE • Physiologisches Institut der Technischen Universität München, Biedersteiner Straße 29, 8000 München 40, Federal Republic of Germany.

Evolution of the First Nervous Systems
Edited by P.A.V. Anderson
Plenum Press, New York

1961; Takeuchi and Takeuchi, 1971a,b; Otsuka et al., 1966) and the action of GABA and the effects of stimulation of the inhibitory axon, have been studied extensively in crayfish muscle by voltage clamp current and current noise measurements (Onodera and Takeuchi, 1976, 1979; Dudel, 1977; Dudel et al., 1980). To our surprise, the glutamate-activated Cl^- channels introduced above were also activated by GABA, and by acetylcholine (ACh). Here, we review the properties of this channel. For additional information, readers are referred to the original literature on the subject (Franke et al., 1986b; Zufall et al., 1988, 1989).

2. Methods

Muscles from freshwater crayfish *Austropotamobius torrentium* were isolated and pinned down in the chamber described in Franke et al. (1986a). The following muscles were used: an opener and a closer of the first walking leg, the superficial and the deep abdominal extensor, M. contractor epimeralis and the intrinsic stomach muscles gm 6a and 5b. Cell-attached and outside-out patches of muscle membrane were established as described in Franke et al. (1986a). Preparations were bathed in a modified van Harrevelt solution containing (in mM): 200 NaCl, 5.4 KCl, 13.5 $CaCl_2$, 2.5 $MgCl_2$ and 10 Tris-maleate buffer (pH 7.6). The same solution was used to bathe the external face of outside-out patches. The intracellular side of excised patches contained (in mM): 150 KCl, 5 NaCl, 2 $MgCl_2$, 1 $CaCl_2$ and EGTA (effective Ca^{++}-concentration of 10^{-8} M) and 10 mM Tris-maleat buffer which was adjusted to pH 7.2 by the addition of 40 mM KOH.

The patch-clamp data were recorded and evaluated as described in Frank et al. (1986a) and Hatt et al. (1988a).

3. Results and Discussion

As mentioned in the introduction, in some cell-attached patch clamp recordings with glutamate in the pipette, low-amplitude channel activity, which differed from the activity of the glutamate-activated excitatory channel, was seen. The left hand row of traces of figure 1 show an example. These low-amplitude channels were not observed in excised outside-out patches, if a low Cl^- "intracellular solution" was used. However, if the Cl^--concentration in the "intracellular solution" in the pipette was raised to between 6 mM and the physiological value of 161 mM, channel current increased rapidly then saturated.

In outside-out excised patches, 1 μM glutamate elicited many, relatively long duration channel openings. These openings have a basic amplitude of 22 pS, but in addition to this basic state 1, openings also occurred to a state 2 of 44 pS, and rarely,

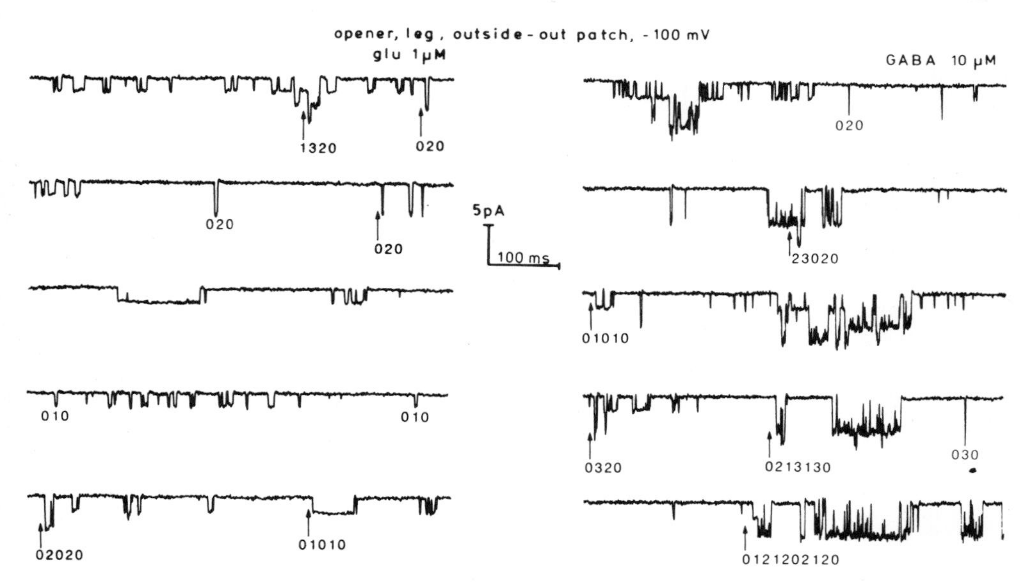

Figure 1. Patch clamp recordings from a single outside-out patch polarized to -100 mV, showing Cl^--channel openings elicited by 1 μM glutamate (left panel) and 10 μM GABA (right panel). The channel opens to 3 substates I_1, I_2, I_3, with amplitudes of about -2, -4 and -6 pA. Some sequences of substates achieved during opening are marked by 0, 1, 2, 3 which refer to the closed state, and open states I_1, I_2, I_3, respectively. When necessary, the starting points of the respective sequences are marked by arrows. Patch from opener muscle of the first walking leg. Bandwidth 0 to 1 kHz (from Franke et al., 1986b).

to state 3 of 66 pS. These higher "states" are not superimposed openings of several channels, but rather openings of single channels, as is evident when records are viewed at higher time resolution. The occurrence of state I_1, I_2 and I_3 is demonstrated also in the channel amplitude distribution of figure 2A. The right hand column of figure 1 shows channel activity recorded in the same patch, in response to 10 mM GABA. With GABA, state 2 was predominant, and state 3 occurred more often than in the presence of glutamate. Activation by both glutamate and GABA can be suppressed by picrotoxin (Fig. 2), the typical blocker of inhibitory currents (Takeuchi and Takeuchi, 1969; Smart and Constanti, 1986).

A comparison of the effects of glutamate or GABA alone with those produced by simultaneous application of both show that these substances really activate the same channel (Franke et al., 1986b; Zufall et al., 1988). Furthermore, quisqualate acts like glutamate and ß-guanidino-propionic-acid or muscimol, acts like GABA.

Average open times were determined for the different states of channels. They are characteristic for the state, but apparently largely independent of the agonist used. The burst length contained components of about 3 ms and about 35 ms, for activation by both glutamate and GABA (Franke et al., 1986b).

Further experiments (Zufall et al., 1988; Zufall et al., 1989, and submitted) showed, rather surprisingly, that nicotinic agonists also opened this Cl^- channel. Figure 3 shows,

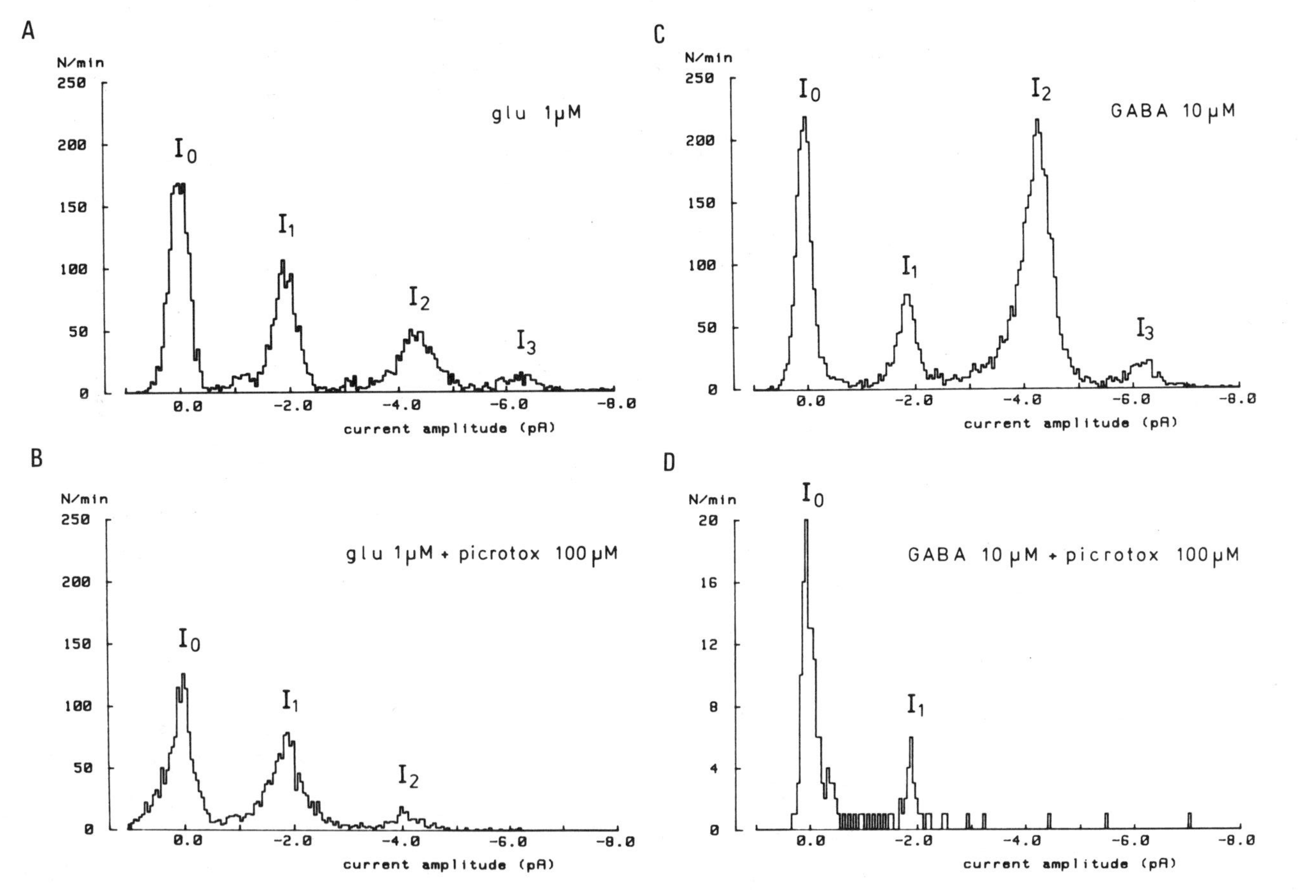

Figure 2. Amplitude distributions for single chloride channel currents from the same patch that yielded the results in Figure 6. **(A).** channels opened by 1 μM glutamate. **(B).** channels opened by 1 μM glutamate in the presence of 100 μM picrotoxin. **(C).** channels opened by 10 μM GABA. **(D).** channels opened by 10 μM GABA in the presence of 100 μM picrotoxin. Polarization -100 mV (from Franke et al., 1986b).

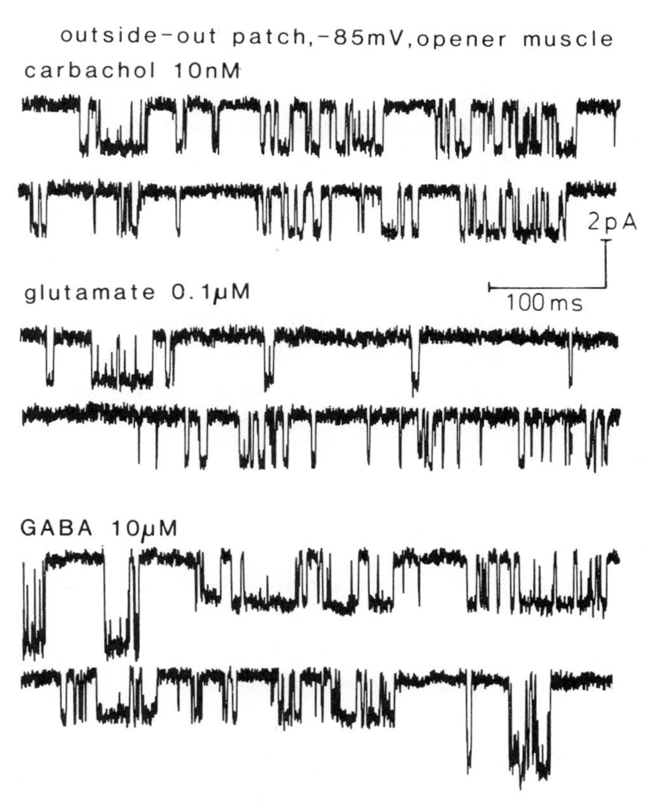

Figure 3. Single-channel Cl⁻ currents recorded in the same patch, in the presence of 10 nM carbachol (upper two traces), 0.1 μM glutamate (middle traces) and 10 μM GABA (bottom traces). While carbachol and glutamate activate only state 1 (see Figure 6), GABA also activates state 2.

for one patch, activation of the same channel type by carbachol, glutamate and GABA. Carbachol, like glutamate, elicited predominately state 1 channel openings. The distribution of life times of openings (> 1 ms) elicited by ACh, glutamate and GABA could be fitted by a single exponential with a time constant of about 2.5 ms, corresponding to the mean open time. Distribution of the closed times could be fitted with the sum of two exponentials with time constants of 3.8 ms and 60 ms for all the agonists. The short time constants represent the short closing separating the individual openings in a burst. The closed times between the bursts are represented by the longer time constant.

We also evaluated the voltage-dependence of the single channel currents activated by ACh. In figure 4, the amplitudes of single channel currents activated by 10 nM ACh are shown at different membrane potentials. The reversal potential of the channel, determined after replacing intracellular K^+ with Cs^+, was -18 mV and corresponded to the Nernst potential for Cl^-.

The activation of Cl^- channels by GABA and nicotinic agonists depends very much on the extracellular Ca^{++}-concentration. In the experiment shown in figure 5, relatively high concentrations of agonists were applied by means of the liquid filament switch (Franke et al., 1987), a means of achieving a near-instantaneous change in the composition of the medium bathing the patch. The maximum currents elicited reached -700 pA with GABA, indicating simultaneous activation of several hundred channels. The presence of such large numbers of Cl^- channels, compared to the few glutamate-activated cation channels in the same patch, was a general finding. Figure 5 shows that the effect of glutamate was little affected by lowering of the extracellular Ca^{++}-concentration. Glutamate produced strong desensitization, with a time constant of several hundred ms. In contrast, 10^{-5} M GABA opened relatively few channels at the normal Ca^{++}-concentration (13.5 mM), but its effectivity increased more than 50-fold in 0.1 mM Ca^{++}. Carbachol 10^{-4} M activated very few channel openings at 13.5 mM Ca^{++}, but had a several hundred-fold stronger effect at 0.1 mM Ca^{++}. This Ca^{++} dependence of the effects of GABA and of nicotinic agonists is in the opposite direction to the Ca^{++}-dependence of the excitatory channels (Hatt et al., 1988b). It should be noted that the rise and decay of channel activity at the onset and at the end of the application of agonist pulse were much slower than for the excitatory channels recorded under the same conditions. Further, only activation by glutamate (or quisqualate) caused a relatively rapid desensitization of the Cl^- receptors/channels. Activation by GABA led to desensitization with a time constant of minutes.

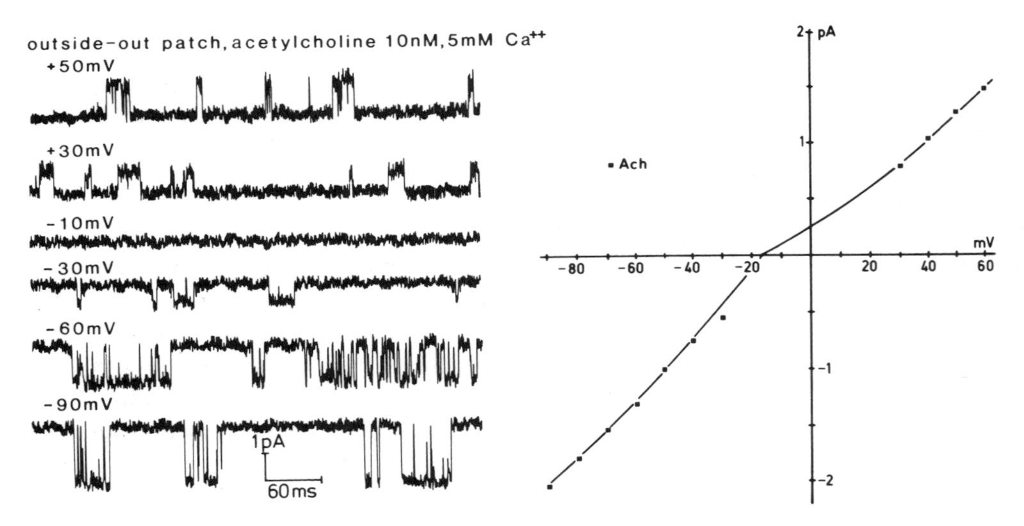

Figure 4. Recordings and evaluations from an outside-out patch from the gm 6b stomach muscle. Left side: single traces of recordings of single channel currents activated by 10 nM ACh, measured at different membrane potentials. Extracellular Ca^{++} concentration was 5 mM. Right side: measurements of amplitudes of single channel currents elicited by ACh. The fit curves were calculated according to the GHK equation with single channel permeabilities of $\pi_1 = 1.4 \times 10^{-14}$ cm^3 s^{-1} and $\pi_2 = 3.9 \times 10^{-14}$ cm^3 s^{-1}.

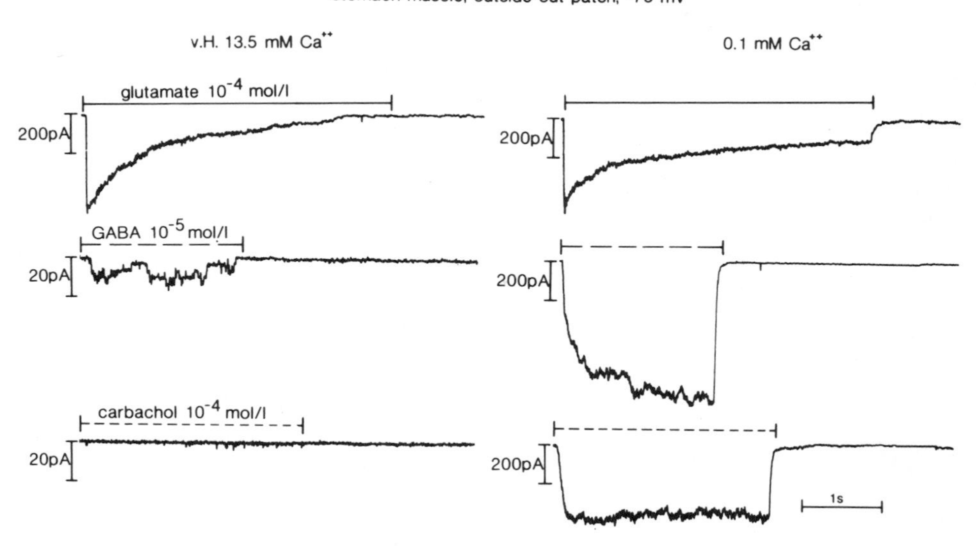

Figure 5. Chloride currents elicited in the same patch by one pulse each of glutamate, GABA and carbachol, as indicated. Pulses were applied via a liquid filament switch. In the recordings in the left column, the Ca^{++} concentration in the background superfusion of the patch, as well as in the liquid filament was 13.5 mM. In the right column, the background perfusion contained 13.5 mM Ca^{++}, but in the liquid filament stream that contained the agonists, the Ca^{++} concentration was 0.1 mM. Note the different amplitude calibrations. Isolated single 2 pA channel openings are visible only in the presence of carbachol 10^{-4} M, Ca^{++} 13.5 M; the other recordings are superpositions of up to several hundred channels.

The fact that carbachol has a very low probability of opening Cl^- channels in normal Ca^{++} can be used to demonstrate competition of the different agonists for the same receptor. In the experiment shown in figure 6, the agonists glutamate and GABA were applied to the same patch with the rapid switch either alone or in the presence of carbachol. In the left hand column, 100 μM glutamate elicited a large current. When glutamate was added to 10 μM carbachol, channel opening was suppressed. When glutamate alone was applied again, the same large number of channel openings was evoked as in the control before. The analogous experiment is shown for GABA and carbachol in the right hand column. Evidently, carbachol in the presence of glutamate or GABA, occupies most of the receptors and prevents the action of the agonist. These results show that carbachol, glutamate and GABA activate the same population of channels, and further, that in high Ca^{++}, carbachol reacts with the receptors, but does not open most channels.

The usual reactivity of the Cl^- receptor/channels to three classes of transmitters raises the suspicion that the opening of chloride channels is mediated by way of different receptors on the cell surface, which act on intracellular second messengers via G-proteins. However, chloride channel openings could also be triggered in inside-out

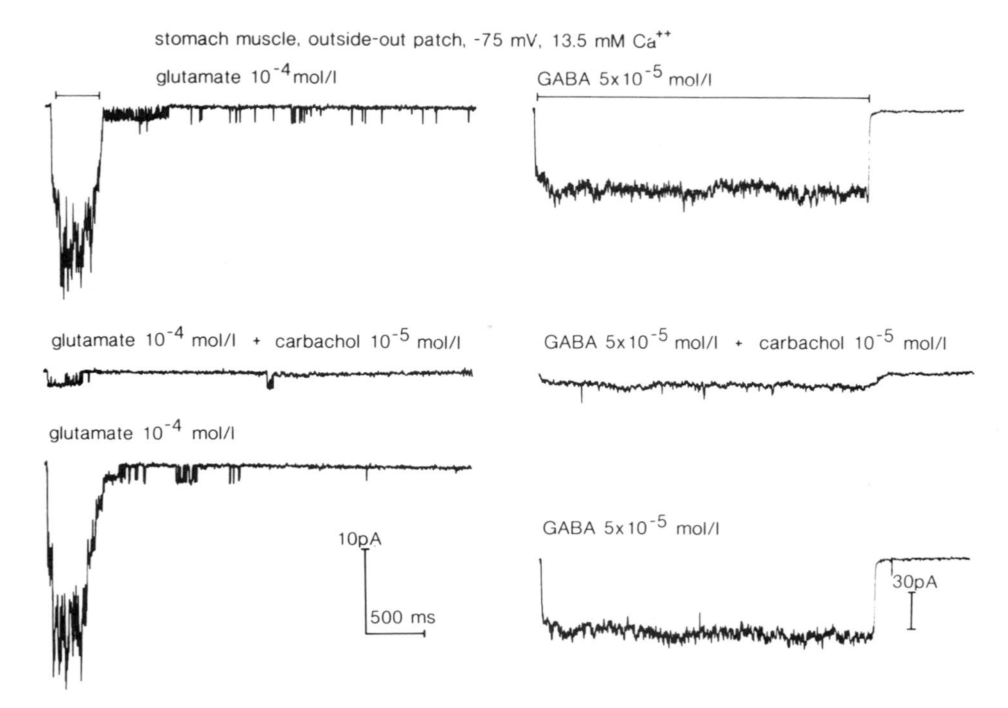

Figure 6. Chloride currents elicited in one patch, as in figure 5. The top and the bottom traces are controls, obtained before and after the record in the middle trace. In the latter, the background perfusion of the patch, as well as the liquid filament stream, contained 10^{-5} M carbachol, and the agonists glutamate or GABA were added to the filament stream. Note the different amplitude calibrations in the left and right columns.

patches which were superfused rapidly. Variation of the intracellular Ca^{++} concentration did not affect channel opening, neither did addition of adenosine triphosphate (ATP), cyclic nucleotides or guanosine triphosphate (GTP). Also, addition of GTP-ß-S or of pertussis toxin had no effect on the gating of Cl^- channels. It seems very improbable, therefore, that a second messenger system is involved in the opening of the Cl^- channels on crayfish muscle by agonists of glutamate, GABA and ACh. Biochemical studies have shown that at least nicotinic, GABA and glycine receptor channel complexes are closely related proteins (Barnard et al., 1987). The channel molecules activated by glutamate, GABA and ACh probably form a closely related family. We suggest, therefore, that the Cl^- channels found on crayfish muscle are primitive or non-selective members of this family, and that they accept different classes of the transmitters to achieve binding to and/or opening of the channel, but are quite selective in allowing only anions to pass the channel. Other examples of this class of channels have been seen (King and Carpenter, 1987).

In most crayfish and other crustaceans, excitation is mediated synaptically by glutamate and inhibition by GABA. The excitatory channels activated by glutamate should be the synaptic channels. However, it is difficult to see, how the Cl^- channels studied here could be synaptic ones. The Ch^- channels were found on the opener muscle which has rich inhibitory innervation, but also on intrinsic stomach muscles which have no inhibitory innervation. At least some of the chloride channels, therefore, are extrasynaptic. In the opener muscle, inhibitory synaptic currents were measured by voltage clamping muscle fibers while stimulating the inhibitory axons (Dudel, 1977). The inhibitory postsynaptic currents (IPSCs) were not depressed by applied carbachol or glutamate, as expected for the Cl^- channel studied here. However, the IPSCs were blocked rapidly and reversibly by picrotoxin, indicating that the inhibitory synapses were readily accessible to superfused drugs. On the other hand, low glutamate concentrations and also carbachol in low Ca^{++}, elicited a small current with the same equilibrium potential as the IPSCs. Chloride channels with the same characteristics as those observed in patches, thus, seem to be present in the intact muscle. We assumed, tentatively, that treatment with collagenase and isolation in a patch either directly affects the GABA-ergic synaptic or extrasynaptic channel molecules, or may modify the membrane microenivornment of the receptors to make these channels less selective towards the agonist than they are in the synaptic complexes.

References

Barnard, E. A., Darlison, M. G., and Seeburg, P., 1987, Molecular biology of $GABA_A$ receptor: the receptor/channel superfamily. *Trends Neurosci.* **10:**502-509.

Cull-Candy, S. G., Miledi, R., and Parker, I., 1980, Single glutamate-activated channels recorded from locus muscle fibres with perfused patch-clamp electrodes. *J. Physiol. (Lond.)* **321:**195-210.

Dudel, J., 1977, Dose-response curve of glutamate applied by superfision to crayfish muscle synapses. *Pflügers Arch.* **368:**49-54.

Dudel, J., and Kuffler, S. W., 1961, Presynaptic inhibition at the crayfish neuromuscular junction. *J. Physiol. (Lond.)* **155:**543-562.

Dudel, J., Finger, W., and Stettmeier, H., 1980, Inhibitory synaptic channels activated by γ-aminobutyric acid (GABA) in crayfish muscle. *Pflügers Arch.* **387:**143-151.

Dudel, L., Franke, Ch., Hatt, H., Ramsey, R. L., and Usherwood, P. N. R., 1988, Rapid activation and desensitization by glutamate or excitatory, cation-selective channels in locust muscle. *Neurosci. Lett.* **88:**33-38.

Fatt, P., and Katz, B., 1953, The effect of inhibitory nerve impulses on a crustacean muscle fibre. *J. Physiol. (Lond.)* **121:**374-389.

Franke, Ch., Dudel, J., and Finger, W., 1983, Single synaptic channels recorded at glutamate sensitive patches on a crayfish muscle. *Neurosci. Lett.* **42:**7-12.

Franke, Ch., Hatt, H., and Dudel, J., 1986a, The excitatory glutamate-activated channel recorded in cell-attached and excised patches from the membranes of tail, leg and stomach muscles of crayfish. *J. Comp. Physiol. A.* **159:**579-589.

Franke, Ch., Hatt, H., and Dudel, J., 1986b, The inhibitory chloride channel activated by glutamate as well as -amino-butyric acid (GABA). Single channel recordings from crayfish muscle. *J. Comp. Physiol. A.* **159:**591-609.

Franke, Ch., Hatt, H., and Dudel, J., 1987, Liquid filament switch for ultra-fash exchanges of solutions at excised patches of synaptic membrane of crayfish muscle. *Neurosci. Lett.* **77:**199-204.

Hatt, H., Franke, Ch., and Dudel, J., 1988a, Ionic permeabilities of L-glutamate activated, excitatory synaptic channel in crayfish muscle. *Pflügers Arch.* **411**:8-16.

Hatt, H., Franke, Ch., and Dudel, J., 1988b, Calcium dependent gating of the L-glutamate activated, excitatory synaptic channel on crayfish muscle. *Pflügers Arch.* **411**:17-26.

King, W., and Carpenter, D. O., 1987, Distinct GABA and glutamate receptors may share a common channel in Aplysia neurons. *Neurosci. Lett.* **82**:343-348.

Onodera, K., and Takeuchi, A., 1976, Inhibitory postsynaptic current in voltage-clamped crayfish muscle. *Nature* **263**:153-154.

Onodera, K., and Takeuchi, A., 1979, An analysis of the inhibitory postsynaptic current in the voltage-clamped crayfish muscle. *J. Physiol. (Lond.)* **286**:265-282.

Otsuka, M., Iversen, L. L., Hall, Z. W., and Kravitz, E. A., 1966, Release of gamma-amino butyric acid from inhibitory nerves of lobster. *Proc. Natl. Acad. Sci. U.S.A.* **56**:1110-1115.

Patlak, J. B., Gration, K. A. F., and Usherwood, P. N. R., 1979, Single glutamate activated channels in locust muscle. *Nature* **278**:643-645.

Smart, T. G., and Constanti, A., 1986, Studies on the mechanism of action of picrotoxin and other convulsants at the crustacean muscle GABA receptor. *Proc. R. Soc. Lond. B* **227**:191-216.

Takeuchi, A., and Takeuchi, N., 1969, A study of the action of picrotoxin on the inhibitory neuromuscular junction of the crayfish. *J. Physiol. (Lond.)* **205**:377-391.

Takeuchi, A., and Takeuchi, N., 1971a, Anion interaction at the inhibitory postsynaptic membrane of the crayfish neuromuscular junction. *J. Physiol. (Lond.)* **212**:337-351.

Takeuchi, A., and Takeuchi, N., 1971b, Variations in the permeability properties of the inhibitory post-synaptic membrane of the crayfish neuromuscular junction when activated by different concentrations of GABA. *J. Physiol. (Lond.)* **217**:341-358.

Zufall, F., Franke, Ch., and Hatt, H., 1988, Acetylcholine activates a chloride channel as well as glutamate and GABA. Single channel recordings from crayfish stomach and opener muscles. *J. Comp. Physiol. A* **163**:609-620.

Zufall, F., Franke, Ch., and Hatt, H., 1989, The insecticide avermectin B_{1a} activates a chloride channel in crayfish muscle membrane. *J. exp. Biol.* **142**:191-205.

Chapter 13

Two Pathways of Evolution of Neurotransmitters-Modulators

C. LADD PROSSER

1. Introduction

Nature provides many examples of parallel and convergent evolution. For example, two large categories of contractile proteins have evolved in parallel - myosins and tubulins. My thesis is that there have been two parallel pathways of evolution of neurotransmitters and neuromodulators: (1) Amino acids and amines which are formed by simple reactions from amino acids, phospholipid bases and nucleotides; (2) Neuropeptides, which are encoded as long precursors that are later cleaved and may be converted further, post-translationally. In some cells, the two types of neuro-agents co-exist.

2. Amino Acids and Biogenic Amines; Purines

We can start with prokaryotes. For motile bacteria such as *E. coli*, a few amino acids, particularly the acidic (negatively charged) glutamate (Glu) and aspartate, are attractants, while others, mainly non-polar (hydrophobic) ones, such as leucine and

C. LADD PROSSER • Department of Physiology and Biophysics, University of Illinois, Urbana, Illinois 61801, USA.

Evolution of the First Nervous Systems
Edited by P.A.V. Anderson
Plenum Press, New York

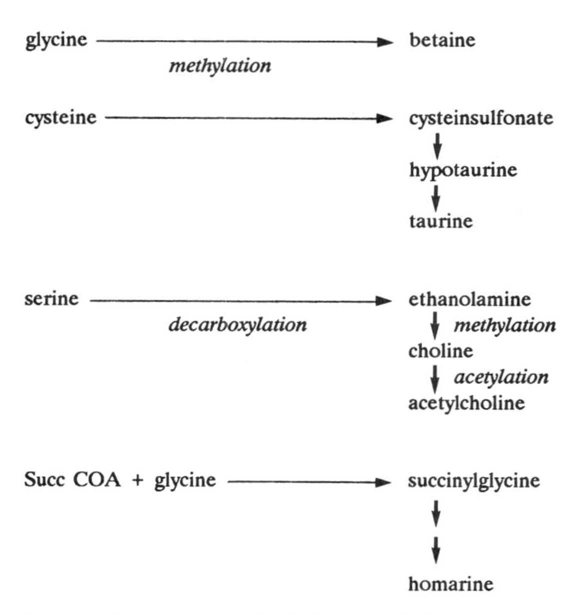

Figure 1. Pathways of synthesis for several biogenic amines.

isoleucine, are repellants (Adler, 1975). Some amino acids are feeding stimulants in some cnidarians (Lenhoff, 1976), as is the short chain peptide glutathione (-Glu-Cys-Gly), which is synthesized by the action of glutamate-cystine synthetase, followed by the addition of glycine by glutathione synthetase.

Several amino acids are known to be neurotransmitters in animals that have complex nervous systems. Glutamate and, to a lesser extent, aspartate, are excitatory transmitters, and open cationic channels. Gamma-amino butyric acid (GABA) and, in some synapses, glycine are inhibitory and open chloride (Cl^-) channels (McGeer, 1989).

GABA is synthesized from Glu by decarboxylation and is inactivated by oxidative deamination to succinate. A further variation on the use of single amino acids is the role of the dipeptide carnosine (Ala-His) as a transmitter in olfactory receptor cells, in the olfactory bulb (Margolis, 1980). Although carnosine is a neuropeptide, it is synthesized directly by the action of carnosine synthetase, rather than being formed as a long chain precursor, as is the case for other neuropeptides (see below). Thus, its synthesis is similar to that of other amino acid neurotransmitters.

There are usually multiple receptors for single amino acid transmitters, and this feature may have evolved very early. A good example of multiple receptors are the four known receptors for glutamate, especially the clinically important NMDA receptor.

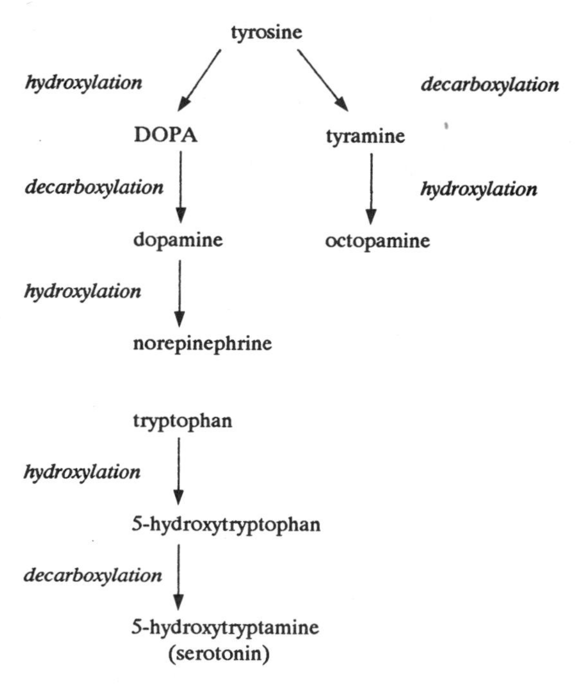

Figure 2. Pathways of synthesis of several catecholamines.

The next step in the elaboration of neurotransmitters was the conversion of amino acids to amines. This occurs through decarboxylation, methylation, oxidation, hydroxylation and acetylation (Fig. 1), using enzymes that evolved in prokaryotes and appear to be little changed in animals. Choline is formed by two reactions: decarboxylation of serine to ethanolamine which is then methylated to choline. The choline is then acetylated by choline acetyl transferase resulting in the important neurotransmitter acetylcholine (ACh). The enzyme for hydrolysis of ACh (ACh esterase) is well characterized and widely distributed.

Several amines which are formed by direct reactions of amino acids, however, have no apparent synaptic function; betaines, taurines, and, possibly, homarine are osmolytes (Huxtable, 1982) and homarine is also a methyl donor (Netherton and Gurin, 1982). A further enigma is that the amine bases of phosphatidyllic compounds (phosphatidyl choline, phosphatidyl serine, phosphatidyl ethanolamine and phosphoinositol) are present in all eukaryotic cell membranes where they function to regulate of membrane permeability but, with the exception of choline, these amines have not apprently been used as neurotransmitters. Why? There are several possible reasons. For instance, ethanolamine is highly lipid soluble and could not, therefore, be easily retained in vesicles; serine, on the other hand, is readily incorporated into proteins, while inositol is a component of the important second messenger IP_3. The adaptive features of ACh which may underlie its selection as a ubiquitous neurotransmitter are that it is a long molecule with a strong dipole, it has molecular flexibility, especially at the quaternary nitrogen, and has a low lipid solubility (Greenberg, 1970). ACh occurs in yeast, plants and all animals.

Similar questions can be raised in the case of the monoamines (Fig. 2). Tyrosine is the precursor for the known neurotransmitters, dopamine, norepinephrine, epinephrine and octopamine. Dopamine, an intermediate in the synthetic pathway leading to epinephrine and norepinephrine (Fig. 2) has only a limited function as a neurotransmitter, whereas norepinephrine and epinephrine are used far more extensively. Tyramine, on the other hand, the intermediate in the synethesis of octopamine, a neurotransmitter in arthrodpods but not in vertebrates, is apparently not used as a neurotransmitter at all. Similarly, 5-hydroxytryptamine (serotonin or 5HT), a common neurotransmitter in many phyla, is formed from trytophan (Fig. 2). The intermediate in its synthesis, 5-hydroxytryptophan, is apparently not functional at synapses. The reasons for these various selections are not clear at this time.

Another class of transmitters is the purines, specifically, adenosine, ATP. No pyrimidines are known to act as transmitters. The reason for this is unclear, particularly so since cyclized derivatives (cAMP and cGMP) are used extensively by cells as 2nd messengers. Once again, there are multiple receptors for purinergic transmitters (Burnstock, 1981).

Most active biogenic amines act at several kinds of receptors. ACh, for instance, acts via nicotinic and muscarinic receptors. The gene for the nicotinic ACh receptor has been sequenced (Noda et al., 1983) and consists of four transmembrane loops, with the N and C terminals located outside the membrane (DiPaola et al., 1989). The muscarinic receptor is very dfferent, being composed of seven transmembrane loops. Norepinephrine acts via α and β receptors, and several 5-HT receptors are known (Dixon, et al., 1986; Hall, 1987).

Several biogenic amine transmitters can coexist in the same neuron (ACh and 5-HT are found together in some molluscan neurons) and neurons can express different transmitters at different developmental stages. For instance, the sympathetic neurons that innervate the sweat glands of rats change from adrenergic transmission to cholinergic transmission during development (Landis and O'Keefe, 1983) and other sympathetic neurons do the same when in culture with non-neuronal tissue (Potter et al., 1983). Finally, there is also variability in the mode of action of biogenic amines; some alter ion permeability directly (ligand-gated channels), others use 2nd messengers.

It is thus concluded that many amino acids and biogenic amines exist but that relatively few of those available to the cells are apparently used as transmitters. In all three classes, amino acids, biogenic amines and purines, the neurotransmitter substances apparently coevolved with multiple receptor proteins. Some active di- and tri-peptides are formed directly from amino acids; these are present in prokaryotes and probably antedated the development of nervous systems.

3. Neuropeptides

Neuropeptides probably appeared as transmitters and modulators later than amino acids and amines, since although some amino acid chains do occur in prokaryotes, none resemble known neuropeptides. Among simple eukaryotes, yeast cells synthesize polypeptide pheromones that function as mating factors (Betz, 1979). One of these has 13 amino acids, another has 10 (Stoltzer, 1976; Whiteway, et al., 1989). However, once again, none resemble known neuropeptides.

All known neuropeptides are synthesized on ribosomes associated with endoplasmic reticulum (ER), and are transferred to Golgi apparatus and transported to the cell membrane for exocytosis. During transportation, they are cleaved and processed to produce the specific peptide(s). The synthesis of peptides in the form of long precursors may have evolved in prokaryotes, when secretion became essential for life functions. Some bacteria, for instance, synthesize proteins by way of precursors, and the proteins are later released. Examples of post-translational processing in *E. coli* are maltase and alkaline phosphatase (Randall, et al., 1987). The bacterial precursor has a leader sequence that is cleaved off by protease, prior to secretion by transport sites in the cell membrane. Similarly, the mating factors of yeast are synthesized as precursors and subsequently cleaved prior to secretion. In eukaryotes in general, precursor synthesis is general for proteins that are to be secreted, much less for proteins that are retained in cells. This distinction is typified by neuropeptides.

The mRNAs for several neuropeptides have been isolated and the corresponding cDNAs cloned and sequenced. In general, mRNAs for eukaryote peptides have long leader sequences, while prokaryote mRNAs correspond more closely in length to the secreted proteins. Pro- and pre-pro-precursor compounds are genetically encoded; the precursor proteins are cleaved post-translationally at pairs of basic amino acids by proteinases. It is uncertain how the genomes came to encode long sequences that are later cleaved (and portions eliminated) before secretion, but the precursors may be an adaptation to ER function.

Various approaches have been used to reveal the presence of particular peptides and confirm their function. These include:

(1) Localization and identification of putative peptides using antibody staining. However, classification based on antibody staining is not definitive since the antibodies are invariably raised against small terminal sequences or other parts of the molecule, and not against the entire peptide molecule. Thus, they are of limited specificity (Grimmelikhuijzen, et al., 1988). A new technique of *in situ* cDNA hybridization may be a more specific means of localizing neuropeptides (Valentino et al., 1987).

(2) Isolation and sequencing of peptides. Comparison of different sequences is used for classification of families. However, often the members of a family occur in such diverse and unrelated animals that speculation regarding phylogenetic relations is limited (see below).

FMRFamides

Mollusc prototypes	F-M-R-F-NH_2 F-L-R-F-NH_2
Cnidarians	pE-G-R-F-NH_2 pE-S-L-R-W-NH_2 pE-L-L-G-G-R-F-NH_2
Pulmonate	S-D-P-F-L-R-F-NH_2
Cephalopod	A-F-M-R-F-NH_2
Homarus	S-D-R-N-F-L-R-F-NH_2
Homarus	T-N-R-N-F-L-R-F-NH_2
Drosophila	D-P-L-Q-D-F-M-R-F-NH_2
Ascaris	K-N-E-F-I-R-F-NH_2
Chick brain	L-P-L-R-F-NH_2

Figure 3. Amino acid sequences of several FMRFamides and related compounds (assembled from Greenberg and Price, 1988).

(3) Examination of the action of a peptide or related compounds. Most neuropeptides have an excitatory or inhibitory action on muscles or neurons so this appraoch may not be definitive. Furthermore, many unrelated compounds have similar actions. Why so many peptides have similar actions is unclear, but redundancy is common in nervous systems.

(4) Attempting to block the action of the putative peptide. A limited number of pharmacological agents (i.e. the opioid blocker naloxone) can be used but their specificity is often limited.

(5) Sequencing precursor molecules within families of related compounds. The mRNAs for most precursors have long signal and terminating sequences which are apparently used for binding to transport endoplasmic reticulum. Cleavage of these precursors occurs after pairs of basic amino acids, and in many of the resulting peptide chains, the C-terminal is amidated. Furthermore, a Glu at the N-terminal becomes cyclized so that the two terminals lose their charges. Thus, this approach provides little useful information about the functioning of a neuropeptide.

(6) Receptor identification. Few receptors for neuropeptides have been identified, largely because their low concentrations in membranes makes isolation difficult. Sequencing of receptor proteins and identification of binding sites has been frequently used for describing transmembrane and steric organization of biogenic amine receptors.

MAMMALIAN TACHYKININS

	1 11
Substance P	-Arg-Pro-Lys-Pro-Gln-Gln-Phe-Phe-Gly-Leu-Met (NH_2)
	11
Substance K	-H-Lys-Thr-Asp-Ser-Phe-Val-Gly-Leu-Met (NH_2)

AMPHIBIAN TACHYKININS

	1 11
Physalaemin	<pGlu-Ala-Asp-Pro-Asn-Lys-Phe-Try-Gly-Leu-Met (NH_2)
	1 11
Uperolein	<pGlu-Pro-Asp-Pro-Asn-Ala-Phe-Tyr-Gly-Leu-Met (NH_2)
	1 11
Phyllomedusin	<pGlu-----Asn-Pro-Asn-Arg-Phe-Ile-gly-Leu-Met (NH_2)

MOLLUSCAN TACHYKININS

	1 11
Eledoisin	<pGlu-Pro-Ser-Lys-Asp-Ala-Phe-Ile-Gly-leu-Met (NH_2)

Figure 4. Amino acid sequences in several members of tachykinin family (from Maggio, 1988 and others).

However, analysis of receptors for neuropeptides has, to date, given little information regarding peptide function.

It is concluded that present information does not permit construction of a phylogenetic picture of neuropeptide evolution. It seems more profitable, therefore, to discuss diversity and general principles, specifically the basis for selection of particular sequences.

The vertebrate nervous system evolved as a neural tube which expanded at the anterior end. The number of peptides produced during this process multiplied along with the development of hypothalamus and pituitary but many of the peptides produced by this region are unrelated to any in invertebrate animals. The following discussion attempts to give reasons for the selection of particular amino acid sequences within existing neuropeptides.

(1) Chain length. Many neuropeptides of invertebrate animals are short and many of vertebrates are long. However, this is not a general condition. Many neuropeptides of vertebrates are 10 to 100 amino acids long but the egg-laying peptide of the mollusc *Aplysia* are also long, consisting of 36 amino acids (Stuart et al., 1980). With the exception of carnosine, which is synthesized in a completely different manner (see above), the shortest vertebrate neuropeptide is thyroid releasing hormone (TRH) which consists of only three amino acids (GHP). This is shorter than the shortest invertebrate peptide; even though cnidarian neurons are stimulated by the dipeptide, RFamide and are stained by antibodies raised against this sequence, the native cnidarian peptides are considerably longer (see Chapter 7, by Grimmelikhuijzen et al., in this book). Nevertheless, in the hydromedusa *Polyorchis*, the common requirement of the several longer

Bio-opiates

α-endorphin	Y-G-G-F-L-T-S-E-K-S-Q-T-P-L-V-T
ß-endorphin	Y-G-G-F-L-T-S-E-K-S-Q-T-P-L-V-T-L-F-K-N-A-I-I-K-N-A-Y-K-K-G-E
dynorphin A	Y-G-G-F-L-R-R-I-R-P-K-L-K-W-D-N-Q
met-enkephalin	Y-G-G-F-M
leu-enkephalin	Y-G-G-F-L

Figure 5. Sequences of several bio-opioids (from biochemistry textbooks).

peptides which have a similar action, is the presence of RF at the C-terminal (Spencer, 1988).

(2) Sequence similarities. Most families of neuropeptides are classified by similarities at the C-terminal, a few at the N-terminal. The recurrence of similar terminal sequences could result from selective advantage or from genetic drift. Some selective advantage of a sequence is suggested by the frequent occurrence of similar sequences within and between families. This need not result from phylogenetic relationships but could be due to covergence - repeated appearance of certain sequences that have selective value.

The tetrapeptide family of FMRFamides was discovered in molluscs and members occur in many invertebrate animals with relatively few substitutions (Fig. 3). A neuropeptide of less abundance in molluscs is FLRFamide. Others occur in arthropods. A similar pentapeptide has been found in chick brain, LPLRFamide even though there is no phylogenetic relationship between chick and mollusc (Greenberg et al., 1988; Greenberg and Price, 1988; Price et al., 1987).

The tachykinin family of peptides has members in diverse phyla (Fig. 4). The prototype C-terminal is FXGLMamide, where X is an aromatic or branched aliphatic amino acid. In mammals, substance P (SP) and substance K (SK) are functional tachykinin neuropeptides. The skin of toads contains at least seven different tachykinins of unknown function (Erspamer and Melchiorri, 1980). Another tachykinin is found in a cephalopod *Eledone*; this is eledoisin. It appears, therefore, that tachykinins occur in unrelated groups of animals. The same gene codes for three tachykinins. The precursor of SP has 130 amino acids, that for SK has 115 (Krause et al., 1987).

There is some evidence for occurrence of bio-opioids in molluscs as well as in vertebrates. The fact that the similarities between different bio-opioids occur at the N-terminal (Fig. 5) rather than at the C-terminal suggests different selective pressures from similarities at the C-terminal, and implies separate evolution.

(3) Internal sequences. Selection based on internal amino acids may have occurred also. For example, recent evidence obtained with the two radular muscles (protractor

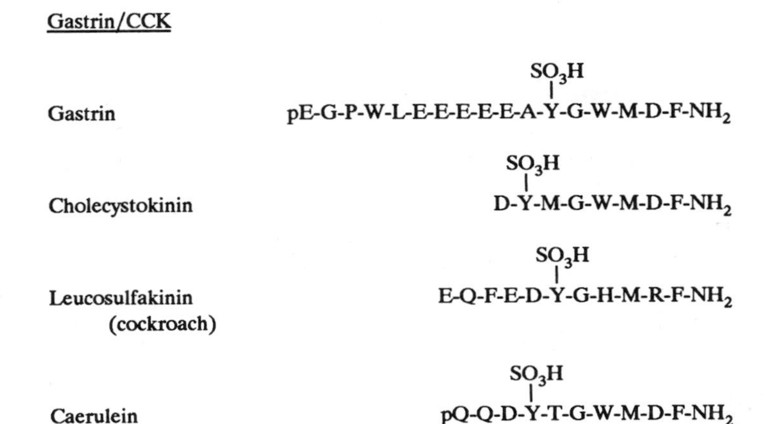

Figure 6. Amino acid sequence of members of the gastrin-leucosulfakinin -CCK family (from Rehfeld, 1981 and Andries et al., 1989).

and retractor) of the mollusc *Rapana* indicates that FMRFamide and FLRFamide have opposite actions (Yanagowa and Kobayashi, 1988). Several analogs of FMRF differ in effectiveness in stimulating the heart of *Rapana*. The RF at the C-terminal is critical and longer chains, also substitutions, are less effective (Kobayashi and Muneoka, 1986).

All members of the gastrin-CCK family have an internal tyrosine with an attached SO_3H, which is presumably adapted for specific receptors (Fig. 6). In some animals, the non-sulfated CCK is as effective as the sulfated forms (Andries et al., 1989). From immunohistochemistry it is argued that a caerulein-like peptide is primitive in that coelenterates stain for it (Rehfeld, 1981).

Some similarities of mid-regions may be fortuitous. In the mid-region of the precursor of enkephalin (Met and Leu) is a sequence of four amino acids that is similar to FMRFamide; this is unlikely to be due to a relationship between enkephalins and FMRFamide. Two hepatopeptides of opioid action occur in skin of a frog, one has high affinity for μ receptors, the other for δ receptors. Both have a D-amino acid at the second position; this gives flexibility to the molecules (Erspamer et al., 1989). The amino acid sequences in mid-regions may, therefore, be related to tertiary structure. The tertiary structure of neuropeptides has not been much investigated, but it is probable that long neuropeptides show some coiling. A minimum of six to eight amino acids is essential for molecular folding to occur.

(4) Precursors. Analysis of precursor molecules has given little information concerning the significance of peptide sequences. The mRNAs of all neuropeptide precursors have long leader and terminal sequences that are not translated. Also in the precursor proteins, long sequences of amino acids precede and are interposed between active neuropeptides. These non-neuronal sequences may function in the liberation of a neuropeptide from the cell membrane.

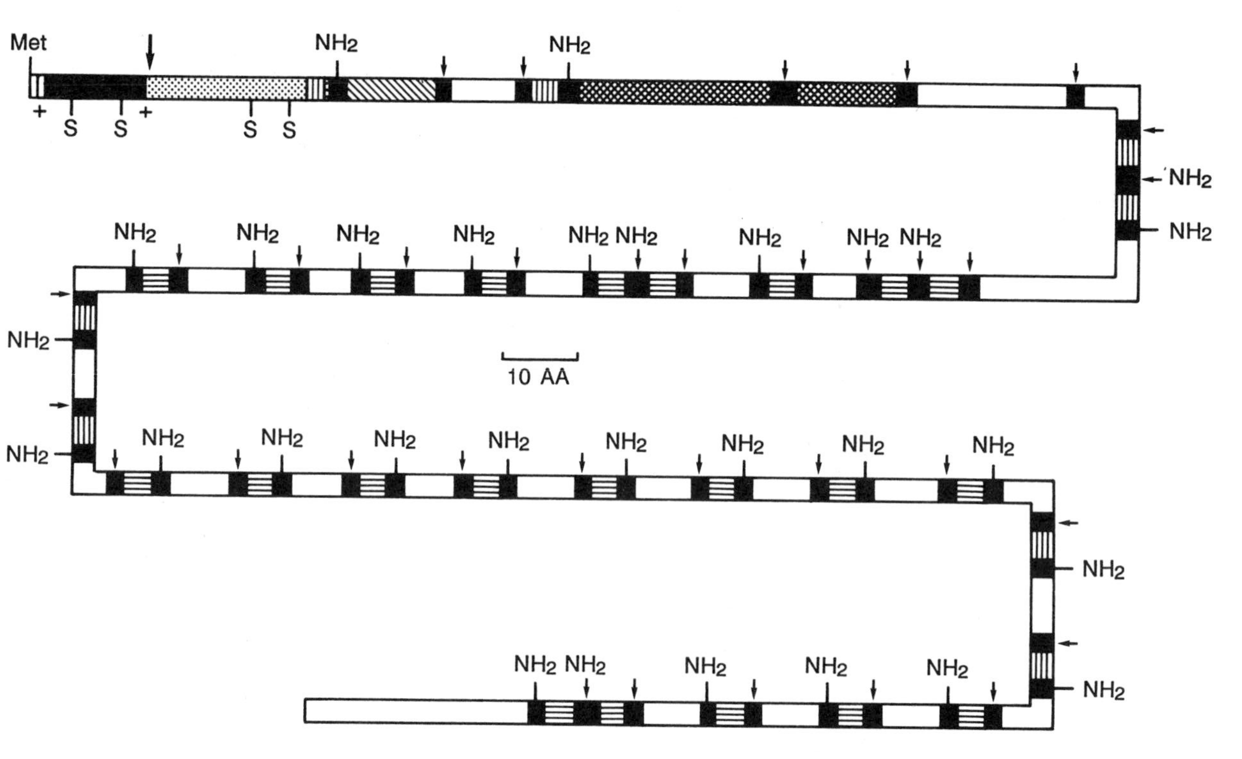

Figure 7. Schematic of the FMRFamide precursor protein. A solid region flanked by positively charged amino acids (+) indicates hydrophobic residues. The 28 FMRF tetrapeptides are denoted by the parallel lines oriented parallel to the long axis of the molecule, and the single FLRF and the carboxy terminal LRF are indicated by parallel oriented perpendicular to the same axis (after Taussig et al., 1988).

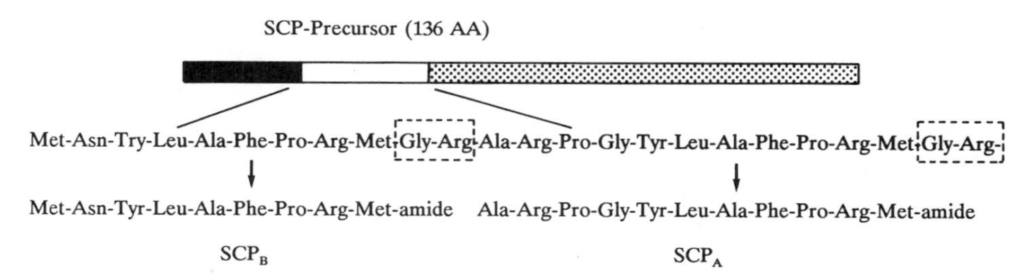

Figure 8. Diagram illustrating the cleavage of the precursor of SCP to form SCP_A and SCP_B (from Lloyd, personal communication).

In precursors, post-translational cleavage occurs at pairs of basic amino acids - usually lysine and arginine (Fisher and Scheller, 1988; Nagel et al., 1988). This is a property of precursors of all families, hence is unrelated to terminal sequences. Rather, this pattern of cleavage reflects the conservation of alkaline proteases (trypsin-like) that occur in prokaryotes and most eukaryotes.

Most known precursor proteins contain multiple copies of a neuropeptide; some precursors code for several different polypeptides. The precursor for FMRFamide contains 38 copies of FMRF and one of FLRF (Fig. 7) (Taussig et al., 1988). The

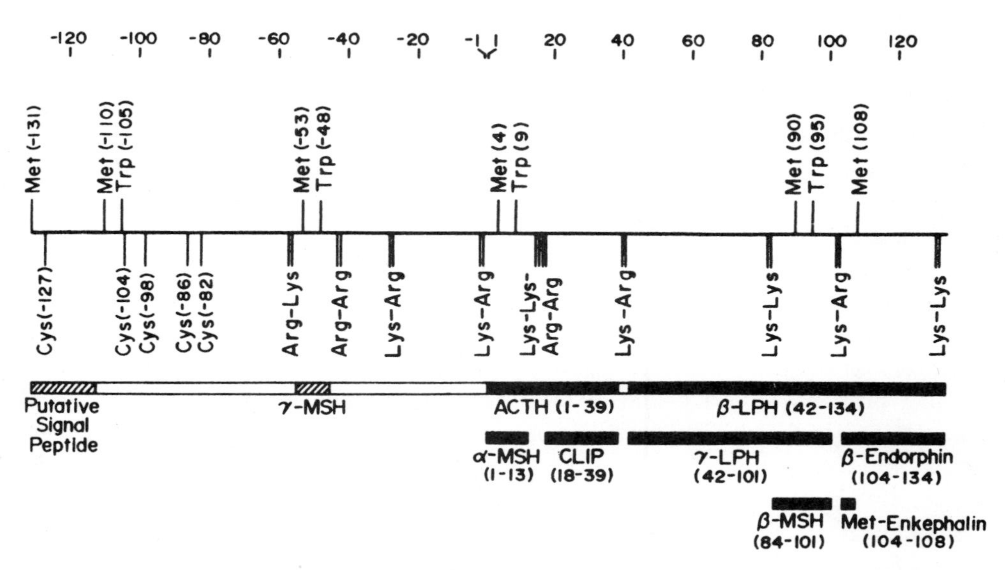

Figure 9. Diagram of precursor of bovine ACTH, ßLPH. Translation initiation at Met residue. Solid bars represent regions for which amino acid sequences are known. Shaded bars locate α MSH and putative signal peptide (from Nakanishi et al., 1979).

mRNA for CCK contains 150 nucleotide bases. This corresponds to a precursor of 39 amino acids which is cleaved first to CCK 33, and then to CCK 12 and CCK 8 in intestine. Progastrin, which encodes for 34 amino acids, is cleaved in stomach and brain to 17 and 14 sequences, respectively. Cleavage into two related but different neuropeptides occurs with the two small cardiac peptides (SCP) of *Aplysia*, SCP_A and SCP_B (Fig. 8) (Lloyd, 1982; Mahon and Lloyd, 1985; Morris et al., 1982).

The long precursor pro-opio-melanotropic-cortin (POMC) is an example of a precursor for several peptides of different families. The cDNA has 1091 base pairs. The precursor has a terminal sequence which includes the sequence for melanocyte stimulating hormone (γMSH). Next is the adrenal corticotrophic hormone (ACTH) region, which encodes for 39 amino acids that includes α MSH and corticotropin-like intermediate lobe peptide (CLIP). Next is a long precursor sequence for β lipotropin (β-LPT), which consists of message for 91 amino acids and contains sequences for γ-LPT and β endorphin. γLPT contains a sequence for β MSH; β endorphin (30 amino acids) contains the sequence for met-enkephalin (four amino acids) (Fig. 9) (Nakanishi et al., 1979). However, met-enkephalin (Met-enk) is formed by a different route; pre-pro-enkephalin has 4 copies of Met-enk, one of Leu-enk, and one each of two metenkaphalins with two additional amino acids (Fig. 10). (Gubler et al., 1982; Noda et al., 1982). The occurrence of the short enkephalin sequence in long β endorphin is another example of replication of the same sequence in two precursors, similar to Phe-Met in enkephalins. POMC probably evolved by a series of gene duplications of common ancestral DNA fragments followed by substitutions, additions and deletions of some regions.

(5) Receptors. Receptor proteins are specific for various ligands. The numerous and different actions of the members of a peptide family, suggest that there are multiple receptor sites, e.g., between FMRFamide and FLRFamide, and there are probably multiple receptors for each neuropeptide. Unlike the amines, however, very few receptors for neuropeptides have been isolated and sequenced. Knowledge of binding sites and configuration of receptors is likely to provide evidence for selection of particular sequences.

Multiple receptors for bio-opioids have been identified by differences in sensitivity to the various neuropeptides and by blocking agents; these receptors are classified μ_1, μ_2, κ and δ. In rat brain locus coeruleus, it appears that opioid receptors, α-adrenergic receptors and somatostatin receptors are all coupled to a K^+ channel via a GTP binding protein (Mikaki et al., 1989). The one peptide receptor that has been completely sequenced is for the tachykinin substance K (SK) (Masu et al., 1987). The amino acid composition, deduced from the cDNA for SK, consists of 384 amino acids which are arranged in seven hydrophobic membrane-spanning segments thus resembling the muscarinic and β-adrenergic receptors, and rhodopsin. The binding sites for SK on this molecule are not known.

(6) Coexistence. Two or more neuropeptides or a neuropeptide and a biogenic amine may occur together in the same cell. In the case of peptides, each is encoded by its own gene. SCP coexists with ACh or 5-HT in some *Aplysia* neurons. Serotonin and

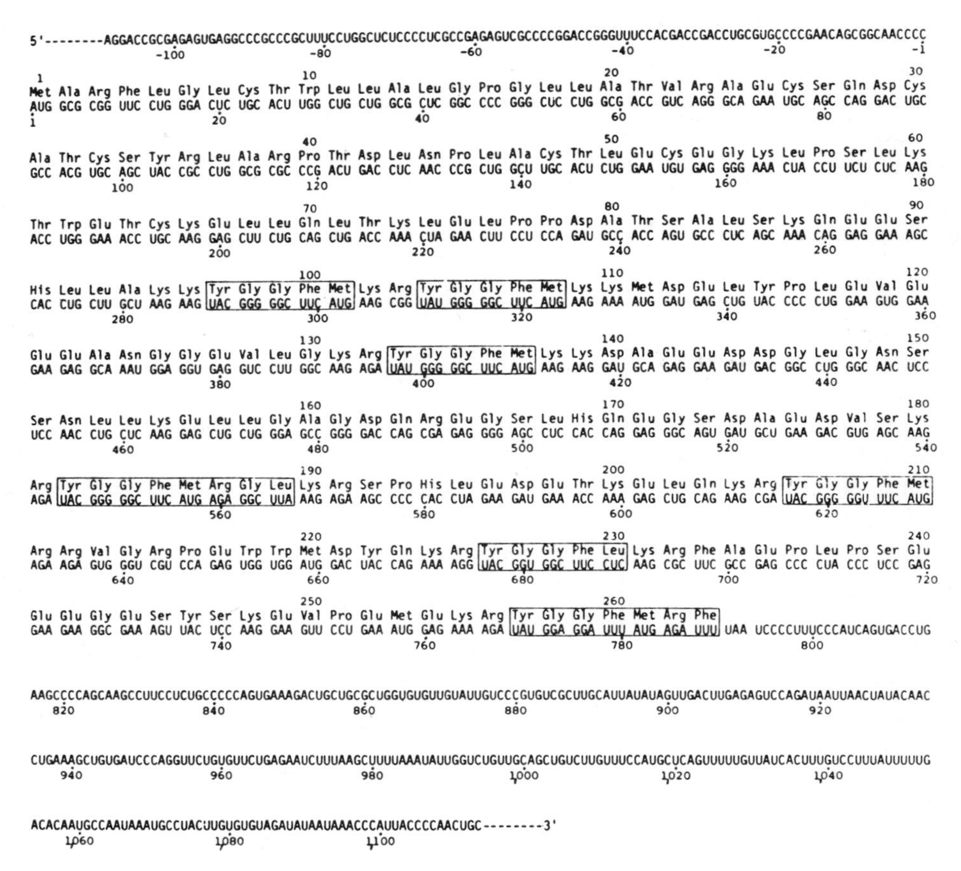

Figure 10. Primary structure of bovine preproenkephalin in RNA. Amino acid sequences above nucleotide sequences. Sequence of Met-enk, Leu-enk and Met-enk with carboxyl extensions bounded by paired basic amino acids (from Noda et al., 1982).

met-enk, neuropeptide Y and norepinephrine (NE), and oxytocin with CCK are examples of coexistence in brain cells. In some sensory neurons of *Aplysia*, the S current of a K^+ channel is modulated by two coexisting transmitters (Siegelbaum et al., 1982). FMRFamide or dopamine act via arachidonic acid to open the K (S) channel; SCP_B or 5-HT act via cAMP to close the channel; the neuro-agents occur in separate afferent terminals. In mammalian dorsal root ganglion cells, NE and Enk act to close a Ca^{++} channel.

(7) Biological actions. Many neuropeptides have been discovered by their excitatory or inhibitory actions on smooth muscles. Members of a family frequently have similar actions but differ quantitatively. For example the various bio-opioids are all analgesic but differ in effectiveness. The effective series differs with species. The tachykinins and CCK are excitatory for intestinal smooth muscle but differ in potency. Many

neuropeptides act at several sites. For example, CCK is excitatory to the intestine and in the brain (hypothalamus) it causes the sensation of satiety. The redundancy or similarity of action of many peptides is unexplained.

Degradation of neuropeptides, unlike that of biogenic amines, is by non-specific proteinases (O'Shea and Schaffer, 1985). The affinity of proteinases for different neuropeptides has not been much examined but may give evidence of selection based on terminal sequences.

4. Conclusions

Known neurotransmitters can be divided into two broad groups on the basis of their synthesis. The simplest transmitters are amino acids, biogenic amines and purines, which are either existing molecules, or are produced by relatively simple reactions of existing cellular components; biogenic amines from amino acids and purines from nucleotides. The other group of transmitters is the neuropeptides, which are gentically coded and post-translationally modified. A survey of the distribution and biochemical properties of these various types of molecule provides evidence that the two basic classes of neurotransmitter probably evolved separately and in parallel. The basis for this conclusion can be summarized as follows.

Some of the amino acids that serve as excitatory or inhibitory transmitters at synapses in the nervous systems of multicellular animals are sensed by the membranes of certain prokaryotes. Modification of those amino acids to amines or, in the case of carnosine, a dipeptide, involve relatively simple reactions such as decarboxylation and methylation. Curiously, only a limited number of available amines are used as neurotransmitters; possible reasons for this selection have been discussed.

In contrast, neuropeptides are invariably synethized as long-chain precursors that are encoded in the mRNA. A precursor chain may include multiple copies of a single peptide or several separate peptides, but in all case, the precursor is subsequently cleaved very specifically, to produce the respective neuropeptides. This form of synthesis of peptides that are to be secreted by cells, appears to have evolved early inasmuch as peptides that are secreted from bacterial cells are first synthesized as precursors and are subsequently cleaved. Cleavage of precursors is between flanking pairs of basic amino acids and this presumably reflects evolution of the proteases, rather than that of the peptides. Many cleaved peptides become amidated at the C-terminal post-translationally; many with glutamine at the N-terminal become cyclized, thus charge at the two ends is neutralized.

Similarities in the amino acid sequences of neurotransmitters, hormones and modulators suggest that many may have common origins and peptides are commonly grouped into families on the basis of similarities of amino acid sequences, usually at the C-terminal, but at the N-terminal in bio-opiods. However, since peptides with similar sequences are found in animals that have little or no phylogenetic relationship, it is possible that the reoccurrence of specific sequences in distantly related animals indicates convergence rather than phylogenetic relationships, similarities not homologies. If so, the basis for this recurrence of similar sequences presumably reflects a selective advantage, possibly at the level of precursor or receptors structure, rather than genetic drift.

In some neurons of advanced nervous systems, one finds coexistence of peptides with amines and of peptides with peptides, but in each case synthesis of each is independent, and follows the two parallel lines discussed here.

References

Adler, J., 1975, Chemotaxes in bacteria, *Ann. Rev. Biochem.* **44**:341-356.

Andries, J. C., Belemtougri, G., Croix, D., and Tramu, G., 1989, Gastrin -CCK like immunoreactivity in nervous system of *Alschna* (Odonata), *Cell Tissue Res.* **257**:105-113.

Betz, R., 1979, a-Factor arrests division of opposite mating types in yeast, *Europ. J. Bioch.* **95**:469-475.

Burnstock, G., 1981, Neurotransmitters and trophic factors in the autonomic nervous system, *J. Physiol. (Lond.)* **313**:1-35.

Di Paoloa, M., Czajakowski, C. and Karlin, A., 1989, The sidedness of the COOH terminus of the acetylcholine receptor subunit, *J. Biol. Chem.* **264**: 15457-15463.

Dixon, R., Kobilka, B., Strader, D., Benovic, J., Dohlman, H., Frielle, T., Bolanowski, M., Bennett, C., Rando, E., Diehl, R., Mumford, R., Slater, E., Sigal, F., Caron, M., Lefkowitz, R., and Strader, X. C., 1986, Cloning of gene and cDNA for mammalian ß-adrenergic receptor and homology with rhodopsin, *Nature* **321**:75-79.

Erspamer, V., and Melchiorri, P., 1980, Tachykinins in frog skin, *Trends in Pharmacol. Sci.* **1**:391-395.

Erspanmer, V., Melchiorri, P., Falconieric-Ersrpamer, G., Negri, L., Corsi, R., Severine, C., Barra, D., Simmaco, M., and Kobil, G., 1989, Deltorphins - A family of naturally occuring peptides with high affinity and selectivity for δ opioid binding sites, *Proc. Nat. Acad. Sci. USA* **86**:5188-51921.

Fisher, J., and Scheller, R., 1988, Prohormone processing and the secretory pathway, *Jour. Biol. Chem.* **263**:16515-16518.

Greenberg, M., 1970, A comparison of acetylcholine structure-activity relations on the hearts of bivalve molluscs, *Comp. Bioch. Phys.* **33**:259-294.

Greenberg, M. J., and Price, D., 1988, The phylogenetic and biomedical significance of extended neuropeptide families, in: *Biomedical Importance of Marine Organisms* (Fautin, D. G., Fenical, W. and Kem, W. R. eds.) *Mem. Cal. Acad. Sci. : 85-96.*

Greenberg, M. J., Payza, K., Nachman, R., Holman, G., and Price, D., 1988, Relationships between FMRF amide-related peptides and other peptide families, *Peptides* **9**:125-135.

Grimmelikhuijzen, C. J., Graff, G., and Spencer, A. N., 1988, Stucture, location and possible actions of Arg-Phe-amide peptides in coelenterates, in: *Neurohormones in Invertebrates*, pp. 199-217 (M. T. Thorndyke and G. Goldsworthy, eds.), Cambridge Univ. Press.

Gubler, U., Seeburg, P., Hoffman, B., Gage, P., and Udenfriend, S., 1982, Molecular cloning establishes proenkephalon as precursory enkephalin-containing peptides, *Nature* **295**:206-208.

Hall, Z. W., 1987, Three of a kind; adrenergic receptor, muscarinic receptor and rhodopsin, *Trends in Neuroscience* **10**:99-101.

Huxtable, R. J., 1982, *Taurine in Nutrition and Neurology*, Plenum Press, New York.

Kobayashi, M., and Muneoka, Y., 1986, Structural requirements for FMRFamide-like activity on heart of prosobranch *Rapana thomasiana*, *Comp. Bioch. Phys.* **84C**:349-352.

Krause, R. M., Chirgivin, J. M., Carter, M., Xu Z., and Hershey, A., 1987, Preprotachykinin mRNAs, substance P and neurokinin (SK), *Proc. Nat. Acad. Sci. USA* **84**:881-885.

Kravitz, E. A., 1986, Serotonin, octopamine and proctolin; two amines and a peptide, aspects of lobster behavior, in: *Fast and Slow Chemical Signalling in the Nervous System*, pp. 244-259 (L. Iverson and F. Goodman, eds.), Oxford Univ. Press.

Landis, S. C., and O'Keefe, D., 1983, Evidence for neurotransmitter plasticity *in vivo*, *Developmental Biology* **98:**349-372.

Lenhoff, H., 1976, Chemical stimulation of feeding reactions in Hydra, in: *Coelenterate Ecology and Behavior*, pp. 571-579 (G. O. Mackie, ed.), Plenum Press, New York.

Lloyd, P. E., 1984, Evidence for parallel actions of a molluscan neuropeptide and serotonin in mediating arousal in *Aplysia*, *Proc. Natl. Acad. Sci. USA* **81:**2434-2937.

Lloyd, D., 1986, Small cardiac peptides of *Aplysia*, *Proc. Natl. Acad. Sci. USA* **83:**9794-9798.

Loh, Y., and Gainer, H., 1983, Biosynthesis and processing of brain peptides, in: *Brain Peptides*, pp. 79-116 (D. Krieger et al., eds.), Wiley.

Maggio, J.E., 1988, Tachykinins, *Ann. Rev. Neurosci.* **11:**13-28.

Mahon, A., Lloyd, P. et al., 1985, Small cardioactive peptides A and B of *Aplysia* are derived from common precursor molecule, *Proc. Nat. Acad. Sci. USA* **82:**3925-3929.

Margolis, S. L., 1980, Carnosine: An olfactory neuropeptide, in: *Role of Peptides in Neuronal Function*, pp. 545-572 (J. L. Baker and T. Smith, eds.), Dekker.

Masu, Y. K., Nakayama, H., Amaki, T., Harada, Y., Kuno, M., and Nakanishi, S., 1987, DNA cloning of bovine substance-K receptor through oocyte expression system, *Nature* **329:**836-838.

McGeer, P. L., and McGeer, E. G., 1989, Amino acid transmitters, in: *Basic Neurochemistry, Molecular and Cellular Aspects*, 4th edition, pp. 311-332 (G. J. Siegel, ed.), Raven Press, New York.

Miyaki, M., North, A., and Christie, M., 1989, Single potassium channel opened by opioids in rat locus coeruleus, *Proc. Natl. Acad. Sci. USA* **86:**3419-3422.

Morris, H. R., Pancio, M., Kaplus, A., Lloyd, P., and Riniker, B., 1982, Elucidation by FAB-MS of structure near cardioactive peptide from *Aplysia*, *Nature* **300:**643-645.

Nagle, G., Painter, S., Blankenship, J., and Kurosky, A., 1988, Proteolytic processing of egg-laying hormone-related precursor in *Aplysia*, *J. Biol. Chem.* **263:**9223-9237.

Nakanishi, S., Inoue, A., Kita, T., Nakamura, M., Chang, S., Cohen, S., and Numa, S., 1979, Nucleotide sequence of cloned cDNA for bovine corticotropin-beta-lipotropin precursor, *Nature* **278:**423-427.

Netherton, J., and Gurin, S., 1982, Biosynthesis and physiological role of homarine in marine shrimp, *J. Biol. Chem.* **257:**11971-11975.

Noda, M., Furutani, Y., Takahashi, H., Toyosato, M., Hirose, T., Inayama, S., Nakanish, S., and Numa, S., 1982, Cloning and sequence analysis of cDNA for bovine adrenal preproenkephalin, *Nature* **295:**202-208.

Noda, M., Takashi, H., Tanube, T., Toyosato, M., Kikyotani, S., Furutani, Y., Hirose, T., Takashima, H., Inayama, S., Miyata, T., and Numa, S., 1983, Structural homology of *Torpedo californica* acetycholine receptor subunits, *Nature* **302:**528-537.

O'Shea, M., and Schaffer, M., 1985, Neuropeptide function. The invertebrate contribution, *Ann. Rev. Neurosci.* **8:**171-198.

Potter, D. D., Furshpan, E. J., and Landis, S. C., 1983, Transmitter status in cultured sympathetic neurons, *Feder. Proc.* **42:**1628-1634.

Price, D., Davies, N., Doble, K., and Greenberg, M., 1987, Variety and distribution of the FMRFamide-related peptides in molluscs, *Zool. Sci.* **4:**395-410.

Randall, L., Hardy, S. J., and Thom, J., 1987, Export of protein: a biochemical view, *Ann. Rev. Microbiol.* **41:**507-541.

Rehfield, J. F., 1981, Four basic characteristics of the gastrin - cholecystokinin system, *Amer. Jour. Physiol.* **240:**G255-G266.

Siegelbaum, S. A., et al., 1982, Serotonin and cAMP close K^+ channels in *Aplysia* sensory neurons, *Nature* **299:**413-417.

Spencer, A. N., 1988, Effect of Arg-Phe amide peptides on identified motor neurons in the hydromedusa *Polyorchis penicillatus*, *Canad. J. Zool.* **66:**639-645.

Stotzler, D., 1976, Sequences of mating factors in yeast, *Europ. J. Bioch.* **69:**397-400.

Stuart, D., Chiu, A., and Strumwasser, F., 1980, Neurosecretion of egg-laying hormone and other peptides from electrically active bag cell neurons of *Aplysia*, *J. Neurophysiol.* **43:**488-498.

Taussig, R., Nambu, J. R., and Scheller, R. H., 1988, in: *Neurohormones in Invertebrates*, pp. 299-309 (M. Thorndyke and G. Goldsworthy, eds.), Cambridge Univ. Press.

Valentino, K. L., Eberwine, J. H., and Barchas, J. (eds.), 1987, *In situ Hybridization: Applications to Neurobiology*, Oxford Univ. Press, New York.

Whiteway, M., Hougan, L., Diegnard, D., Thomas, D., Bell, L., Saari G., Grant, F., O'Hara, P., MacKay, V., 1989, STE 4 and STE 18 genes of yeast encode ß and subunits of mating factor receptor coupled G proteins, *D. Cell* **56**:467-477.

Yanagawa, M., and Kobayashi, M., 1988, Potentiating effects of some invertebrate neuropeptides on twitch contraction of the radula muscles of a mollusc *Rapana thomasiana*, *Comp. Bioch. Phys.* **90C**:73-77.

Chapter 14

Summary of Session and Discussion on Intercellular Communication

MICHAEL J. GREENBERG

Cells can communicate in three ways: by direct exchange between their cytoplasms, by diffusable chemical signals, and by chemical signals (e.g., adhesion molecules) fixed to cell membranes. The morphological manifestation of the first method is the gap junction, which appears in the nervous system as the electrical synapse. Diffusable chemical messengers and their receptive mechanisms are much more diverse, and their most obvious (but not exclusive) embodiment in the nervous system is the chemical synapse. Specific cell contacts guided by membrane macromolecules are important in developmental and immune phenomena. The first two days of this symposium were devoted to characterizing the gap junctions, synapses, and chemical signals in simple animals and, by comparing them with those of more complex organisms, to extract some generalizations about the evolution of the nervous system.

Gap junctions, or related structures, appear in virtually all metazoan phyla, including even sponges. In his discussion of these organelles, Colin Green (Chapter 1) compared the structure, chemistry, and functions of mammalian gap junctions (which are best known) with those in *Hydra*. Three points seemed critical to any evolutionary consideration. First, cross-immunoreactivity suggests that the intrinsic membrane proteins constituting the connexons in *Hydra* and rat liver are homologous, and thus members of the same large family. This family is, however, quite heterogeneous; i.e., the proteins vary significantly with tissue and species. Second, gap junctions subserve a variety of functions; e.g., metabolic and ionic coupling, patterning in development and regeneration, as well as neural transmission. Moreover, although chemical signalling is certainly an ancient trick (as W. E. S. Carr shows, Chapter 6),

MICHAEL J. GREENBERG • Whitney Laboratory, University of Florida, 9505 Ocean Shore Blvd., St. Augustine, Florida 32086, USA.

low molecular weight signals, such as the head inhibitor of *Hydra*, can clearly be transmitted by gap junctions, as Green describes. Finally, gap junctions are not mere conduits; the opening is regulated, and the electrical synapses in nervous systems can be rectified and even inhibitory. Their utilization in conducting pathways should not, therefore, be viewed as necessarily simple or primitive.

Synaptic transmission in the Cnidaria (reviewed by A. N. Spencer, Chapter 2) can be chemical or electrical, but these mechanisms are not uniformly distributed throughout the phylum. Electrical synapses are limited to the Class Hydrozoa, where they couple all of the cells of a particular neural network or tissue layer. Chemical synapses, in contrast, occur in all classes; in the Hydrozoa they serve to link each network with its effectors, or two networks.

The absence of gap junctions, from the Anthozoa and Scyphozoa (reported by I. D. McFarlane et al., Chapter 8, as well as by Spencer) is surprising but has its counterpart in the non-uniform tissue distribution in mammals. These discontinuities are potentially instructive. Gap junctions in scyphozoans and anthozoans have only been sought by ultrastructural and electrophysiological methods; the homologous proteins may yet be found biochemically, or by cloning, in a structure not recognizable as a conventional gap junction and performing a function unrelated to neurons. A related problem is posed by the novel apical punctate junctions of ctenophores, described in this symposium by Marie-Luz Hernandez-Nicaise and her associates (Chapter 3). These structures may, or may not, contain proteins homologous to those in *Hydra* and (presumably) in the classical gap junctions of ctenophores. That is, gap-junction-like structures might have originated more than once, and with different proteins.

The structure and physiology of both chemical and electrical synapses are generally conventional. But there are surprises; e.g., the strong electrical coupling of the *Polyorchis* swimming motor neuron network, and the bidirectional synapses in the motor network of the schyphozoan jellyfish *Cyanea*. These might, as Spencer suggests, reflect some special cnidarian feature, such as the radial symmetry; yet almost all of these peculiarities -including the bidirectional synapse and its proposed quantile, non-vesicular release of transmitter - also occur in "higher" animals.

Notwithstanding the constraints of two dimensional construction, radial symmetry, and sheer simplicity, the nervous coelenterate nervous system is well- and predictably organized. In *Hydra*, morphogenisis is regulated by a set of factors derived from its nervous system. These are (according to Chica Schaller and her associates): a pair of peptides (head and foot activators), and a pair of low molecular weight substances (head and foot inhibitors). The head activator (HA), at least, is not restricted in its occurrence to hydrozoans, or even cnidarians; i.e., this peptide has been also been sequenced from ox intestines. The current model, proposed by Sabine A. H. Hoffmeister and S. Dubel (from Schaller's lab) (Chapter 4), requires that the interstitial stem cells be stimulated twice during the cell cycle in order to produce head directed or foot-directed nerves. In contrast, E. Hobmayer and Charles N. David (Chapter 5) find that, although HA stimulates the formation of tentacle nerve and epithelial cells, the final differentiation of innervated battery cells is dependent on cell-cell interactions during tentacle formation. These views were not reconciled during this symposium.

Interneuronal transmission by means of chemical signals, whether at long range or at synapses, requires the participation of several processes or mechanisms, each specific to either the sender or the receiver. W. E. S. Carr proposed that the entire repertoire had already evolved in unicellular organisms. Comparing slime mold and yeast with higher metazoans, he drew parallels with regard to: the synthesis and storage of the signal (either a small molecule or a peptide); the pulsatile presentation of the signal; the structure of the receptor; the degradation and removal of excess signal; and the contributions of G-proteins and second messenger. The hypothesis is that signalling between cells within metazoans evolved from signalling through the external milieux between unicellular organisms; the internalization of cAMP by slime molds provided a model that suggested how this might have occurred.

The nature and actions of the chemical signals used in cnidarian nervous systems, and their implications for the evolution of neurochemical signalling in general, were major foci of discussion. Peptides, as well as classical monoamine transmitters, were considered.

A set of endogenous, authentically cnidarian peptides, identified by C. J. P. Grimmelikhuijzen and his co-workers (Chapter 7), were discussed by him and by I. D. McFarlane. Actual sequences have so far been determined to date in one hydrozoan (*Polyorchis*) and two anthozoans (*Anthopleura* and *Renilla*), and fall into two groups on the basis of their C-terminals: -Gly-Arg-Phe-NH_2 (two Antho-RFamides and two Pol-RFamides), and -Leu-Arg-Trp-NH_2 (two Antho-RWamides). The pairs of congeners may be processed from common precursors; time (and molecular genetics) will tell. Immunochemical studies suggest that these peptides are ubiquitous in the coelenterates, but taxonomic range and variability, and family relationships, will become clearer as sequences from more species are acquired. In previous studies, Grimmelikhuijzen has stained hydras with antibodies to several of the known vertebrate neuropeptides. The neuronal patterns were distinctive (Grimmelikhuijzen et al., 1982); so although the antigens are completely unknown, there are clearly many regulatory peptides still to be identified in coelenterates.

The biology of the sequenced cnidarian peptides seems to be straightforward. Immunochemical evidence is consistent with the view that their synthesis, intravesicular packaging and transport, and release are similar to those of other metazoan peptides. Moreover, immunoreactivity has been observed only within nerves, particularly those innervating muscles (i.e., in hydrozoans and anthozoans). Synthetic peptides, applied to preparations of endodermal (sphincters and circular muscle rings) and ectodermal (isolated tentacles and oral disc radials) muscle of anthozoans, affect both tension and rhythmicity, although the actions of the Antho-RFamides and -RWamides differ qualitatively. Neither the sites nor modes of action of the peptides is known, a matter to be discussed below. Nevertheless, the broad proposal, that these are neuropeptides acting as myoneural transmitters or modulators, seems reasonable.

The concept of a peptide superfamily, originating in the coelenterates and extending to all Metazoa, characterized by the general sequence <Glu...Arg-X-NH_2, and interacting with a broad class of membrane receptors requiring the penultimate arginyl residue for binding, is a grand, and ultimately testable, one.

The case for biogenic amines and amino acids having a role in chemical signalling in coelenterate nervous systems is less substantial than that for the peptides. To start with, Grimmelikhiujzen detailed the evidence indicating that catecholamines, serotonin, and acetylcholine are neither present nor functional in the nervous system of *Hydra*. In contrast, Michel Anctil (Chapter 10) drew data from several approaches to show that adrenaline (as well as noradrenaline and dopamine) is present in neurons and non-neural cells in all of the tissue layers of *Renilla*. [J. van Marle (Chapter 9) also observed catecholamine-like immunofluorescence in *Tealia*, but only in the ectoderm of this sea anemone.] Adrenaline also induces bioluminescence (in *Renilla*) and anthozoan muscle contractions. Anctil also reports the presence of serotonin in *Renilla* and suggests its involvment in the rhythmic response to water flow. Finally, Eliana Scemes (Chapter 11) recounted the patchy, mostly negative evidence about the role of ACh in coelenterates -- and then suggested, on pharmacological and biochemical evidence, that some sort of cholinergic mechanisms might be active in "crumpling" behavior of (at least) the trachymedusan, *Liriope*.

The difficulties involved in identifying potential transmitters in coelenterates are yielding to new techniques, but coelenterate pharmacology --whether of peptides or monoamines -- remains thorny. To achieve a credible threshold, isolated muscles must be tediously trimmed. The study of isolated tentacles or even whole organisms is technically much simpler, but the responses must be extraordinarily complex, including even sensory effects; thus, sites and mechanisms of action cannot be analyzed. The solution is to isolate nerves and muscles in culture as, for example, Peter Anderson has done with the motor nerve net of *Cyanea*.

Illustrating the reductionist aproach, Hanns Hatt (Chapter 12) described, in diverse crayfish muscles, a glutamate-activated "inhibitory" Cl^- channel that could also be activated by GABA and ACh. The channels seemed not to be at synapses; Hatt suggested that they be thought of as primitive and non-selective with regard to agonist. This finding does bring to mind the observation of McFarlane, that glutamate, at rather high doses, inhibits electrically invoked contractions of circular muscles in sea anemones.

C. Ladd Prosser (Chapter 13) summarized by suggesting that the use of biogenic amines and peptides as transmitter agents arose separately and evolved in parallel, along with their biosynthetic processes, their receptors, and their intracellular mechanisms of action. This view is well accepted; the questions seem to be whether the two evolutionary pathways started at about the same time, and whether they proceeded at the same rate in all taxa. The discontinuous distribution of the biogenic amines, described above, led Grimmelikhuijzen to argue that, since these amines are not present in all cnidarian neurones, they cannot be the most ancient transmitters in animals; this "honor" he assigns to peptides.

Finally, there was a general discussion of a paper comparing the sequences of 18S ribosomal RNA from 10 animal phyla (Field et al., 1988). The authors of this paper proposed that the Cnidaria arose from a protistan ancestry different from that of the Bilateria, and that the Bilateria diverged early into the flatworms and the coelomate phyla. The argument was made that, even if the Cnidaria are primitive, if they are not on the same branch of the phylogenetic tree as the rest of the Eumetazoa, then they can tell us very little about the evolution of the higher animals. The Platyhelminthes, in contrast, are genuinely primitive with respect to the Bilateria, and might be quite informative (as the poster by M. Reuter and I. Palmberg on microturbellarians was), but they are rarely studied.

The responses to Field et al. (1988) were spirited and along the lines of those written to *Science* (see Various authors, 1989). The nervous systems of Cnidaria may not have any evolutionary relics; i.e., characteristics or mechanisms that are basically different from those found in so-called higher animals. They probably have, instead, a rather different distribution of mechanisms than do higher animals, but taken from the same rich assortment that was available very early on in evolution. As for flatworm nervous systems, all agreed that they probably represent the organization extant at the origin of bilaterality, and that more should be done to understand them.

References

Field, K. G., Olsen, G. J., Lane, D. J., Giovannonia, S. J., Ghiselin, M. T., Raff, E. C., Pace, N. R., and Raff, R. A., 1988, Molecular phylogeny of the animal kingdom, *Science* **239**:748-753.

Grimmelikhuijzen, C. J. P., Dockray, G. J., and Schot, L. P. C., 1982, FMRFamide-like immunoreactivity in the nervous system of hydra, *Histochemistry* **73**:499-508.

Various authors, 1989, Letters on phylogeny and molecular data, *Science* **243**:548-551.

Part 2

ELECTRICAL EXCITABILITY

Chapter 15

Ion Channels of Unicellular Microbes

CHING KUNG

1. Introduction

Before the first "nervous system," there must have been "nervous molecules". The quintessential nervous molecules are the ion channels. These integral membrane proteins enclose the hydrophilic pathways across the hydrophobic membrane that would be otherwise impenetrable for charged or polar molecules. Since a room without a door is but a tomb, one could argue on first principles that channels probably evolved soon after cell membranes. Arguments and speculations aside, we have now shown that protozoa, yeast, and even bacteria all have ion channels. It appears that all cellular forms of life have ion channels. We are also forced to conclude that they must have evolved very early.

Ion channels are "nervous" because they readily respond to stimuli. A certain stimulus can increase the open probability of a given channel. Such a stimulus (gating principle) can be (1) an external ligand, e.g., acetylcholine for the nicotinic acetylcholine receptor/channel, (2) an internal second messenger, e.g., Ca^{++} for Ca^{++}-gated K^+ channel or cGMP for the cGMP-gated Na^+ channel in the rod outer segment, or (3) cross membrane voltage, e.g., the voltage-gated Na^+ channel and delayed rectifier K^+ channel of nerves. These classes of channels have been extensively studied and reviewed (Hille, 1984). More recently, other ion channels have been found to be gated by (4) GTP-binding proteins (G- proteins) (Yatani et al., 1988; Kim et al., 1989), (5) by arachidonic acid (Kim and Clapham, 1989; Ordway et

CHING KUNG • Laboratory of Molecular Biology and Department of Genetics, University of Wisconsin, Madison, Wisconsin 53706, USA.

al., 1989) and by (6) mechanical forces on the membrane (Guharay and Sachs, 1985, Sachs, 1988). Once open, channels allow ions to flow down their electrochemical gradients. Such ion flux changes membrane potential and ion concentrations in the cytoplasm. Therefore, ion channels are transducers that are nervously waiting to relay various chemical, electric or mechanical stimuli into ionic or voltage signals for the cells.

I will first review briefly the ion channels found in *Paramecium*, yeast and *Escherichia coli* and then speculate on evolutionary matters. Reviews of microbial channels can be found in Saimi et al. (1988a,b) and Martinac et al. (1988).

2. Ion Channels of *Paramecium*

Paramecium is the most thoroughly studied protozoan in terms of electrophysiology. Work by Naitoh, Eckert, Machemer and many others has been reviewed often (see Machemer, 1988; Preston and Saimi, 1989, for recent reviews), and will be only briefly summarized here.

Paramecium is not a simple cell. Membrane depolarization opens two voltage-gated channels: (1) a Ca^{++} channel (Oertel et al., 1977) and (2) a K^{+} channel (Machemer and Ogura, 1979; Satow and Kung, 1980a). These channels activate upon depolarization within milliseconds. The Ca^{++} that entered through the Ca^{++} channel opens two other Ca^{++}-gatedchannels: (3) a Na^{+} channel (Saimi and Kung, 1980) and (4) a second K^{+} channel (Satow and Kung, 1980b). These Ca^{++}-gatedchannels are slower, activating in tens to hundreds of milliseconds after the depolarization that open Ca^{++} channel. Recent investigation of currents activated by membrane hyperpolarizations reveals more complexity. Hyperpolarization opens two different, voltage-gated channels: (5) a second Ca^{++} channel (Saimi, 1986; Hennessey, 1987) and a (6) third K^{+} channel (Oertel et al., 1978; Preston et al., in preparation). The Ca^{++} entering through the second, hyperpolarization-activated Ca^{++} channel then opens (7) a fourth K^{+} channel (Preston et al., in preparation). Again, the voltage-gated channels activate faster than the Ca^{++}-gated one. Besides these voltage- or Ca^{++}-gated channels, *Paramecium* is also equipped with two channels gated by mechanical forces: anterior touch opens (8) a Ca^{++}-passingchannel and a posterior touch opens (9) a fifth K^{+}-specific channel (Ogura and Machemer, 1980).

The currents through these nine channels have all been recorded from whole cells, under two-electrode voltage clamp. Such experiments are successful because *Paramecium* is a very large cell, measuring 150 μm in length and 50 μm in diameter. Furthermore, whole-cell membrane currents are much larger than the current leaking through the points of puncture. Individual membrane currents triggered by step depolarizations or hyperpolarizations were sorted by established methods, and their voltage- or Ca^{++}-dependencies,ion specificities, and time courses distinguished

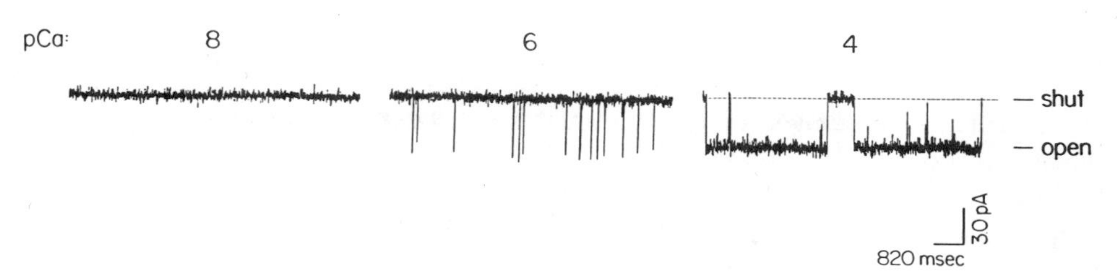

Figure 1. Single-channel activity of a Ca^{++}-dependent K^+ channel in an excised patch of membrane from *Paramecium tetraurelia.* The patch was bathed in a 100 mM K^+ solution with three different Ca^{++} concentrations. **Left**: Little activity was observed at 10^{-8} M Ca^{++}. Middle: Only some short openings were detected at 10^{-6} M Ca^{++}. **Right**: The channel was very active at 10^{-4} M Ca^{++} showing long openings. The patch was held at -50 mV throughout. Ca^{++} concentration was changed in the order of 10^{-6}, 10^{-8}, and then 10^{-4} (from Saimi and Martinac, 1989).

one from another. Ionic or organic blockers specific to certain channels were also used in current separations. In *Paramecium*, mutations were also used to separate currents, since some mutations affect only specific types of channels and not others. Mutational dissection of membrane currents has been periodically reviewed (Saimi and Kung, 1987; Hinrichsen and Schultz, 1988; Ramanathan et al., 1988; Preston and Saimi, 1989; Hennessey, in this book).

Recently, the activities of single channels of *Paramecium* have been recorded using the patch-clamp technique (Hamill et al., 1981). Blisters are induced on paramecia by incubating them in a Na^+ glutamate solution. Patches can be excised from these blisters and channel activities are readily observed (Saimi and Martinac, 1989; Kubalski et al. 1989). Figure 1 shows the activity of a Ca^{++}-gated K^+ (I_{KCa}) channel in such a patch.

The functions of some of these currents are understood. The anterior mechanosensitive channel (Channel 8) is for the generation of a depolarizing receptor potential (Eckert et al., 1972) which triggers an action potential. The Ca^{++} action potential and how it relates to ciliary reversal and, therefore, the "avoiding reaction" has been extensively reviewed over the years (Eckert 1972, Machemer 1988). Conceptually, the Ca^{++} action potential of *Paramecium* is modeled after the Na^+ action potential of axon with a Ca^{++} upstroke (by Channel 1) and a K^+ downstroke (Channel 2). Such action potentials can be recorded when *Paramecium* is bathed in a simple Ca^{++} or Ca^{++}/K^+ solution. Cells bathed in culture media or similar solutions containing Ca^{++}, K^+ and Na^+ have more complicated forms of electrical discharge (Satow and Kung, 1974). Such discharges correspond to the stops, turns, retreats, or spurts exhibited by cells in such solutions. Ca^{++}-gated channels (Channels 3 and 4) appear to be used during excitations in the more natural situations (Saimi et al., 1983). Membrane hyperpolarization corresponds to ciliary augmentation and rapid forward swimming (Naitoh et al., 1973; Machemer, 1988). The posterior mechanosensitive channel (Channel 9) appears to be a trigger for hyperpolarization. The functions of the other channels in *Paramecium* are not clearly understood. They may participate in setting the resting membrane potential.

While the *Paramecium* plasma membrane, which encloses the cell body as well as the cilia, is certainly complex, it should be noted that most metazoan neurons, when thoroughly investigated, are also found to have many channel types strategically located in the cell body, axon, dendrites and the synaptic surfaces. *Paramecium* is more than a "swimming neuron". Its membrane has to sense, excite, and respond, besides serving other housekeeping functions. Beyond locomotive behavior, this single-cell animal also feeds, pumps water, secretes, excretes, mates and autogamizes. Whether or not ion channels are used in these other aspects of their biology has not been investigated.

From the viewpoint of channel evolution, it is clear that by the time protozoa emerged, most of the basic types of channels and their gating mechanisms had probably already been operating. At least three gating principles, voltage, Ca^{++}, and mechanical force, are extensively used by these unicells.

3. Ion Channels of Yeast

The budding yeast, *Saccharomyces cerevisiae*, is much smaller than *Paramecium*. It would be difficult to perform classical voltage clamp experiments with it. It also has a wall of glucan which prevents contact between the cell membrane and an electrode. Removal of this wall with zymolyase or similar enzymes yields spheroplasts some 5-7 μm in diameter. Patch-clamp experiments can readily be performed with them. Gigaohm seals can be formed between the spheroplast membrane and the patch-clamp pipette. Ion-channel activities are routinely observed and recorded most commonly in the cell-attached or whole-cell mode.

Two types of ion channels are always found in the spheroplast membrane. One is a voltage-gated K^+-channel (Fig. 2, top) (Gustin et al. 1986). It has a unit conductance of about 20 pS in 100 mM K^+ solution. It is strongly selective for K^+ over Na^+ and is blocked by the usual K^+-channel blockers such as tetraethyl-ammonium (TEA), quinidine, Ba^{++}, or Cs^+. The open probability increases with membrane depolarization. Its kinetics include rapid millisecond flickers between the open and closed states. Flickers are clustered in bursts each lasting for seconds, separated by silent interbursts which also last for seconds. Thus, the behavior of this K^+ channel is rather similar to depolarization-sensitive K^+ channels found in animals. We estimated that there are only about 100 such channels in each yeast cell.

The second channel we discovered in yeast spheroplasts is a stretch-activated channel (Fig. 2, bottom) (Gustin et al., 1988). In the whole-cell configuration, this channel is activated upon application of pressure equivalent to a few centimeter mercury column exerted through the pipette. The applied forces are physiological. The pressure of a 5 cm Hg column is equivalent to the pressure generated by a less

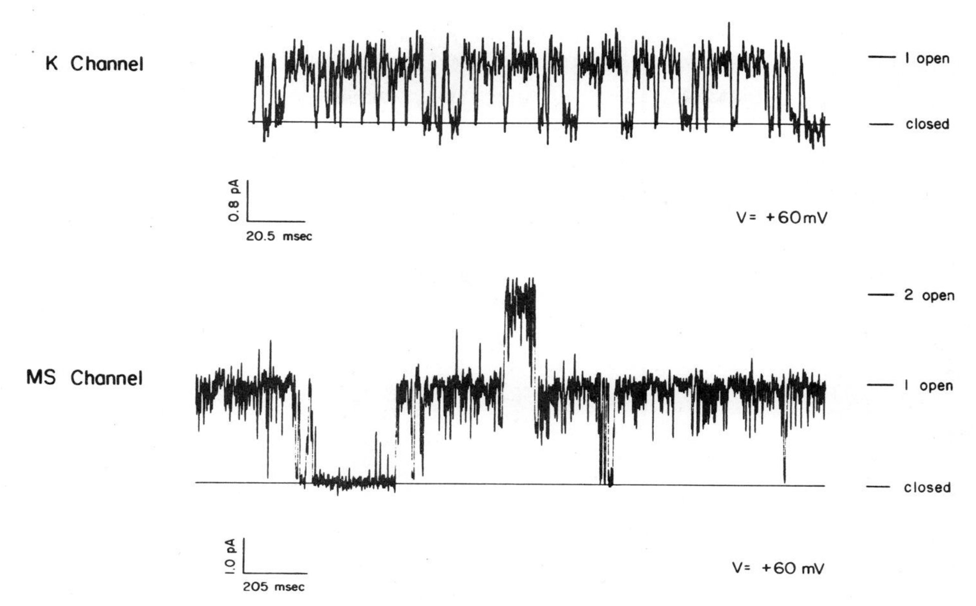

Figure 2. Activities of two types of channels in spheroplasts of the budding yeast, *Saccharomyces cerevisiae*. **Top**: The activity of a single depolarization-activated K^+ channel registered in the whole-cell recording mode. The pipette solution was 180 mM KCl; the bath was 30 mM KCl, 120 mM NaCl, 50 mM $MgCl_2$, at 6.5 °C. Pipette voltage was at +60 mV. See Gustin et al. (1986) for detail. **Bottom**: The activity of two mechanosensitive mechanosensitive (MS) channels in an excised outside-out patch. Pipette voltage was at +60 mV, corresponding to membrane depolarization. For details see Gustin et al. (1988).

than 3 milliosmolar difference across the membrane. A dilution of the bath solution, equivalent to increasing an outward osmotic pressure, also open these channels, monitored in the whole-cell mode. These channels have a unit conductance of about 40 pS. They are poorly ion selective, passing both cations and anions. We found that less pressure is needed to open the channels in larger cells. Laplace's law states that tension on the thin wall of a sphere is directly proportional to the product of pressure and diameter of the sphere. Our results shows that the channel molecule is gated by tension along the plane of the membrane and not by pressure perpendicular to it. This channel shows an adaptation to mechanical force. When the membrane is at negative voltages, the application of high pressure causes a transiently high level of channel activity, which soon adapts to a low steady state during sustained pressure. Interestingly, adaptation is dependent on voltage. Positive voltages prevent adaptation.

Yeast has the body plan of a plant cell. Within the cytoplasm is a large vacuole. Osmotically lysing spheroplasts releases their vacuoles. These vacuoles can also be examined by patch-clamp electrodes. There are several types of channels on this membrane. One of them appears to be a K^+ channel of about 130 pS conductance in symmetric 150 mM K^+ solution. Channel opening is dependent upon millimolar concentrations of free vacuolar Ca^{++}. Like the K^{+} channel of the sarcoplasmic reticulum, this channel can be blocked by 400 μM decamethonium (Minorsky et al., 1989; Minorsky, personal communication).

4. Ion Channels of Bacteria

A typical bacillus is too small for electrophysiology, even with a patch clamp. *Escherichia coli*, for example, is only 0.2 μm in diameter and 1 to 2 μm in length. However, we have found ways to generate giant cells or giant spheroplasts, some 5 to 10 μm in diameter, sufficient in size for patch-clamp experiments. For example, *E. coli* can be cultured in cephalexin, a penicillin analog, which prevents septation. The cells, which cannot septate, grow into filaments 50-150 μm long. Lysozyme, together with EDTA, can then be applied to nick the peptidoglycan wall, thereby converting the filaments into giant spheroplasts (Martinac et al., 1987). The activities of ion channels that we found in these spheroplasts are not artefacts of cephalexin or EDTA-lysozyme, since four other methods were used successfully. Some of these methods exclude the use of one, the other, or both of these agents.

There are at least two types of ion channels found in *E. coli* spheroplasts. One type of ion channel turned out to also be mechanosensitive. Suction of a few cm Hg applied to a patch in an on-cell mode recording opens these channels. Depolarization also facilitates pressure-sensitive gating. The unit conductance is very large, being 900 pS in 300 mM salt solution. It has very little selectivity, although it prefers anions slightly. Large organic ions, such as glutamate, can pass through this channel (Martinac et al., 1987).

The second type of channel in the spheroplast is voltage-sensitive. It is open most of the time and it displays frequent, brief transitions to closed levels. The most frequent transitions have a unit conductance of 90 pS. Depolarizations induce what appear to be cooperative closure of several units. This channel prefers cations (Delcour et al., 1989). A gram-negative bacterium, such as *E. coli*, has two membranes. From a variety of considerations, we came to the conclusion that the two channels described above belong to the outer membrane.

Bacterial membranes can also be reconstituted into artificial liposomes. The ion channels so reconstituted can be examined with patch-clamp pipettes (Delcour et al., 1989). We have modified the method of Criado and Keller (1987) for this purpose. *E. coli* inner and outer membrane fractions are partially separated through sucrose-gradient centrifugation (Osborn et al., 1972). Azolectin (lipid mixture extracted from soybean) is sonicated in a buffer and then frozen and thawed twice. The bacterial membranes are then mixed with the azolectin to a desired protein-to-lipid ratio. The mixture is then dried into a film on glass and rehydrated before a Mg^{++} solution is added to induce blisters. Patch-clamp electrodes can then be used to repeatedly sample such blisters. Figure 3 shows the

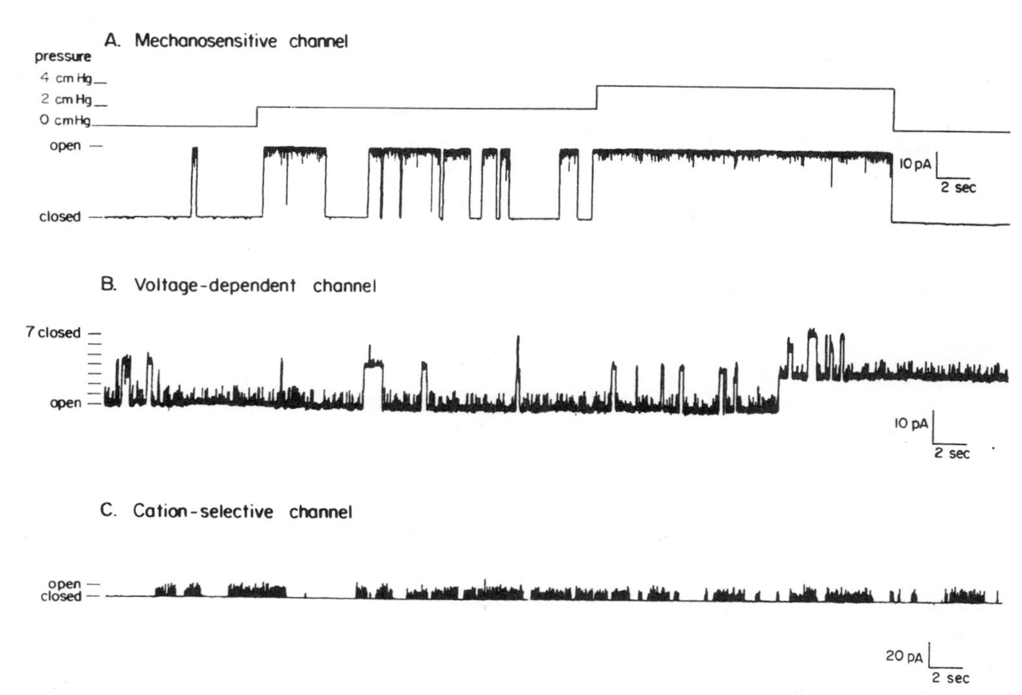

Figure 3. Activities of three types of ion channels from the bacterium *Escherichia coli*. Membrane fractions were reconstituted into liposomes. Blisters were induced from the liposomes and sampled with patch-clamp pipettes. In all cases both the pipette and the bath solutions were 150 mM KCl, 0.1 mM EDTA, 10^{-5} M $CaCl_2$, 5 mM HEPES, pH 7.2., at room temperature. **Top:** A mechanosensitive ion channel. Activity was observed when suction was applied to the patch as shown by the upper trace. Open probability increased with suction. Pipette voltage was +40 mV. **Middle:** A voltage-sensitive channel. Only the open level and maximally closed level, which corresponds to the closures of seven units of conductance, are labeled; other levels are indicated by tick marks. The pipette voltage was -50 mV. **Bottom:** A cation-selective channel of small conductance. The pipette voltage was +120 mV (from Delcour et al., 1989).

types of channel activities we found in such blisters. Both the mechanosensitive channel (Fig. 3A) and the voltage-sensitive channel (Fig. 3B) can be funtionally reconstituted. In addition, a cation-sensitive channel is also observed (Fig. 3C). It has a unit conductance of 30 pS in 150 mM KCl and seems more active when the pipette is made negative in voltage. This channel has yet to be studied in detail and we are not certain whether it comes from the outer or the inner membrane.

5. Solute Senses vs. Solvent Senses

Mechanosensitive ion channels must have evolved early since they are found in all these very different microbes, including a lower eucaryote (yeast) and a procaryote (*E. coli*). They seem to be touch receptors in *Paramecium*. What functions do they serve in yeast and *E. coli*? One possibility is that they are used to monitor water concentration in the environment by measuring the osmotic pressure.

Water is of obvious concern to life. Most animals and plants have to deal with dehydration or overhydration. Microbes are no exception. When confronted with water, *E. coli* immediately discards its small molecules and ions through certain "holes," but retains its macromolecules (Britten and McClure, 1962; Epstein and Schultz, 1965). Operons are turned on later in the low-osmotic environment to synthesize the membrane-derived oligosaccharides in the periplasm (Kennedy, 1982) and to reproportion the major porin species (Hall and Silhavy, 1981). When confronted with high osmolarity (the threat of dehydration), *E. coli* immediately pumps more K^+ inward, using the constitutive uptake system, Trk (Epstein and Schultz, 1965). Operons are turned on later to scavenge K^+ (the Kdp system in low external K^+, Rhoads et al., 1976; Laimins et al., 1981), to synthesize the osmoprotectants: proline and glycine betaine (le Rudulier et al. 1984), and to reproportion the major porins (Hall and Silhavy, 1981). Several species of motile bacteria were known to avoid high or low osmolarity, a behavior termed "osmotaxis" by Massart (1891). We have recently demonstrated osmotaxis in *E. coli* with modern techniques (Li et al., 1988). Despite the wealth of information on the various responses to hydrating or dehydrating stresses, the molecular mechanism(s) that senses the changes of osmolarity remains unknown. Whether or not the stretch-activated channels in yeast or *E. coli* are actually the sensors of osmotic stress has yet to be proven.

Biophysically, water has to be measured by a mechanism different from those that detect other molecules in solution. Recall that pure water is 55.56 Molar in concentration. Yet, cells seem to be able to detect and respond to millimolar changes in water concentrations. It stands to reason, therefore, that our ingrained notion on how chemicals (solutes) are detected by cells does not apply to water (the solvent). A water-binding pocket with a K_m of nM to mM simply would not do. A consequence of the high concentration of water and the thinness of the lipid bilayer is that water permeates the lipid bilayer (Finkelstein and Cass, 1968) without having to go through protein channels. Water equilibrates across proteinless liposome membranes rapidly. If a device measures the difference in total bombardment of particles (osmotic pressure) inside and outside the membrane, it is, in effect, measuring the water concentration outside. Stretch-activated ion channels can indeed be opened by changing osmolarity (Gustin et al., 1988; Martinac et al., unpublished result). I can imagine the following scenario of overhydration. Water partitions into the cell when these microbes find themselves in rain. The swelling cells open their stretch-activated channels which then allow small osmolites (solutes), including charged solutes, to escape and, thereby, relieve the osmotic tension. *E. coli* is not a rigid rod. The whole cell body, including its peptidoglycan wall, can swell or shrink (Stock et al., 1977).

I would like to speculate that early cells evolved molecular devices, including channels and other receptors, to measure water concentration and that these devices

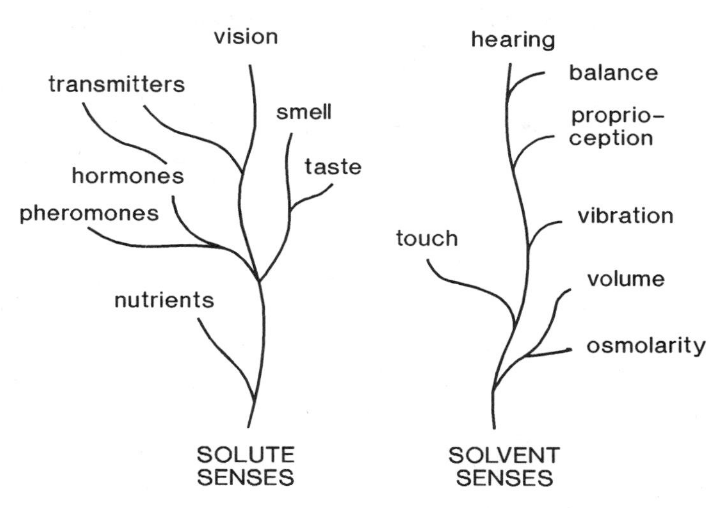

Figure 4. A scheme of evolution of senses. The scheme emphasizes the completely separate origins of solutes senses and solvent senses. The order and timing of the emergence of different senses are less critical here.

are categorically different from devices evolved to detect other chemicals. Because water detectors are transducers of mechanical forces, it is not difficult to imagine that they eventually became the molecular mechanisms for touch, vibrations, hearing, balance, and proprioception. I propose to call these senses collectively the "solvent senses." Solute detectors, perhaps first used to detect nutrients and wastes, could be remodeled to serve in taste, smell, pheromone or hormone reception, neurotransmission, and vision. They can be collectively called the "solute senses." Figure 4 represents this speculation. So categorized, it is interesting to note that many of the molecular mechanisms of solute senses are worked out, including the now famous G-protein circuitry. On the other hand, the molecular mechanisms for the solvent senses are completely obscure. There is presently no structural information, from primary sequence to 3-D modeling, on any mechanosensitive proteins.

Two caveats need to be pointed out. First, mechanosensitive channels may have another function, that of morphogenesis. They may well be devices for the cell to measure its own size and shape (Christensen, 1987; Falke and Misler, 1989). However, osmosensation and morphogenesis are not necessarily mutually exclusive. The same channels may serve both functions. Second, there might well be convergent evolution resulting in channels of different molecular structures serving the same function. Note that sequence information of the cloned eucaryotic plasmamembrane channels indicates that they are giant multimeric proteins with similar subunits or giant monomeric proteins with similar repeated internal domains (Noda et al., 1982, 1984; Tanabe et al., 1987; Papazian et al., 1987; Schwarz et al., 1988).

Models of these channels have the ion pathways lined by transmembrane alpha helices, each contributed by one subunit or repeat. On the other hand, sequences of bacterial porins (channels in the outer membrane of Gram-negative bacteria) and VDAC (voltage-dependent anion channels in the outer membrane of mitochondria) suggest that they are made of beta-pleated sheets (Paul and Rosenbusch, 1985; Rosenbusch, 1986; Forte et al., 1987). The stretch-activated channel we found in *E. coli* may be structurally very different from that in yeast, for example. We have yet to learn the structural detail of the microbial channels reviewed here.

ACKNOWLEDGEMENTS. I thank Yoshiro Saimi, Boris Martinac, Robin Preston and Beverly Seavey for comments on the manuscript. Work in our laboratory and collaborating laboratories headed by Prof. Julius Adler and Prof. Michael Culbertson is supported in part by NIH GM22714, GM36386, GM37925, DK29121, and a grant from the Lucille P. Markey Trust.

References

Britten, R. J., and McClure, F. T., 1962, The amino acid pool of *Escherichia coli*, *Bact. Rev.* **26:**292-335.

Christensen, O., 1987, Mediation of cell volume regulation by Ca^{2+} influx through stretch-activated channels, *Nature* **330:**66-68.

Criado, M., and Keller, B. U., 1987, A membrane fusion strategy for single-channel recordings of membranes usually non-accessible to patch-clamp pipette electrodes, *FEBS Letters* **224:**172-176.

Delcour, A. H., Martinac, B., Adler, J., and Kung, C., 1989, A modified reconstitution method used in patch-clamp studies of *Escherichia coli* ion channels, *Biophysical J.* **56:**631-636.

Eckert, R., 1972, Bioelectrical control of ciliary activity, *Science* **176:**473-381.

Eckert, R., Naitoh, Y., and Friedman, K., 1972, Sensory mechanisms in *Paramecium*. I. Two components of the electric response to mechanical stimulation of the anterior surface, *J. exp. Biol.* **56:**683-694.

Epstein, W., and Schultz, S. G., 1965, Cation transport in *Escherichia coli*. V. Regulation of cation content, *J. Gen. Physiol.* **49:**221-234.

Falke, L. C., and Misler, S., 1989, Activity of ion channels during volume regulation by clonal N1E115 neuroblastoma cells, *Proc. Natl. Acad. Sci.* **86:**3919-3923.

Finkelstein, A., and Cass, A., 1968, Permeability and electrical properties of thin lipid membranes, *J. Gen. Physiol.* **52:**145s-172s.

Forte, M., Guy, H. R., and Mannella, C. A., 1987, Molecular genetics of the VDAC ion channel: structural model and sequence analysis, *J. Bioenergetics Biomembranes* **19:**341-349.

Guharay, F., and Sachs, F., 1985, Mechanotransduction ion channels in chick skeletal muscle: The effects of external pH, *J. Physiol. (Lond.)* **363:**119-134.

Gustin, M. C., Martinac, B., Saimi, Y., Culbertson, M. R., and Kung, C., 1986, Ion channels in yeast, *Science* **233:**1195-1197.

Gustin, M. C., Zhou, X.-L., Martinac, B., and Kung, C., 1988, A mechanosensitive ion channel in the yeast plasma membrane, *Science* **242:**762-765.

Hall, M. N., and Silhavy, T. J., 1981, Genetic analysis of the major membrane proteins in *Escherichia coli*, *Ann. Rev. Genetics.* **15:**91-142.

Hamill, O. P., Marty, A., Neher, E., Sakmann, B., and Sigworth, F. J., 1981, Improved patch-clamp technique for high-resolution current recording from cell and cell-free membrane patches, *Plugers Arch.* **391:**83-100.

Hennessey, T. M., 1987, A novel calcium current is activated by hyperpolarization of *Paramecium tetraurelia*, *Soc. Neurosci. Abs.* **13:**108.

Hille, B., 1984, *Ion Channels of Excitable Membranes*, Sinauer Assoc. Inc., Sunderland, MA.

Hinrichsen, R. D., and Schultz, E. J., 1988, *Paramecium*: a model system for the study of excitable cells, *Trends in Neurosciences* **11:**27-32.

Kennedy, E. P., 1982, Osmotic regulation and the biosynthesis of membrane-derived oliogosaccharides in *Escherichia coli*, *Proc. Natl. Acad. Sci. U.S.A.* **79:**1092-1095.

Kim, D., and Clapham, D. E., 1989, Potassium channels in cardiac cells activated by arachidonic acid and phospholipids, *Science* **242:**1174-1176.

Kim, D., Lewis, D. L., Graziadei, L., Neer, E. J., Bar-Sagi, D., and Clapham, D. E., 1989, G-protein beta gamma-subunits activate three cardiac muscarinic K^+-channel via phospholipase A_2, *Nature* **337:**557-560.

Kubalski, A., Martinac, B., and Saimi, Y., 1989, Proteolytic activation of a hyperpolarization- and Ca^{2+}- activated K channel in *Paramecium*, *J. Membrane Biol.*, in press.

Laimins, L. A., Rhoads, D. B., and Epstein, W., 1981, Osmotic control of kpd operon expression in *Escherichia coli*, *Proc. Natl. Acad. Sci. U.S.A.* **78:**464-468.

le Rudulier, D., Strom, A. R., Dandekar, A. M., Smith, L. T., and Valentine, R. C., 1984, Molecular biology of osmoregulation, *Science* **224:**1064-1068.

Li, C.-Y., Boileau, A. J., Kung, C., and Adler, J., 1988, Osmotaxis in *Escherichia coli*, *Proc. Natl. Acad. Sci. U.S.A.* **85:**9451-9455.

Machemer, H., 1988, in: *Paramecium*, pp. 186-215 (H.D. Gortz, ed.), Springer-Verlag, Heidelberg.

Machemer, H., Ogura, A., 1979, Ionic conductances of membranes in ciliated and deciliated *Paramecium*, *J. Physiol. (Lond.)* **296:**49-60.

Martinac, B., Buechner, M., Delcour, A. H., Adler, J., and Kung, C., 1987, Pressure-sensitive ion channel in *Escherichia coli*, *Proc. Natl. Acad. Sci. U.S.A.* **84:**2297-2301.

Martinac, B., Saimi, Y., Gustin, M. C., Culbertson, M. R., Adler, J., and Kung, C., 1988, Ion channels in microbes, *Periodicum Biologorum* **90:**375-384.

Massart, J., 1891, Recherches sur les organismes inferieurs, *Acad. Roy. de Med. de Belgique* **22:**148-163.

Minorsky, P. V., Zhou, X.-L., Culbertson, M. R., and Kung, C., 1989, A patch clamp analysis of a cation-current in the vacuolar membrane of the yeast *Saccharomyces*, *Plant Physiology* **89:**S-882.

Naitoh, Y., and Eckert, R., 1973, Sensory mechanism in *Paramecium*. II. Ionic basis of the hyperpolarizing mechanoreceptor potential, *J. exp. Biol.* **59:**53-65.

Noda, M., Takahashi, H., Tanabe, T., Toyosato, M., Furutani, Y., Hirose, T., Asai, M., Inayama, S., Miyata, T., and Numa, S., 1982, Primary structure of alpha-subunit precursor of *Torpedo californica* acetylcholine receptor deduced from cDNA sequence, *Nature* **299:**793-797.

Noda, M., Shimizu, S., Tanabe, T., Takai, T., Kayano, T., Ikeda, T., Takahashi, H., Nakayama, H., Kanaoka, Y., Minamino, N., Kangawa, K., Matsuo, H., Raferty, M. A., Hirose, T., Inayama, S., Hayashida, H., Miyata, T., and Numa, S., 1984, Primary structure of *Electrophorus electricus* sodium channel deduced from cDNA sequence, *Nature* **312:**121-127.

Oertel, D., Schein, S. J., and Kung, C., 1977, Separation of membrane currents using a *Paramecium* mutant, *Nature* **268:**120-124.

Oertel, D., Schein, S. J., and Kung, C., 1978, A potassium channel activated by hyperpolarization in *Paramecium*, *J. Membrane Biol.* **43:**169-185.

Ogura, A., and Machemer, H., 1980, Distribution of mechanoreceptor channels in the *Paramecium* surface membrane, *J. Comp. Physiol.* **135(A):**233-242.

Ordway, R. W., Walsh, J. V. Jr., and Singer, J. S., 1989, Arachidonic acid and other fatty acids directly activate potassium channels in smooth muscle cells, *Science* **244:**1176-1179.

Osborn, M. J., Gander, J. E., Parisi, E., and Carson, J., 1972, Mechanism of assembly of the outer membrane of *Salmonella typhimurium*, *J. Biol. Chem.* **247:**3962-3972.

Papazian, D. M., Schwarz, D. L., Tempel, B. L., Jan, Y. N., and Jan, L. Y., 1987, Sequence of a probable potassium channel component encoded a Shaker locus of *Drosophilia*, *Science* **237:**749-753.

Paul, C., and Rosenbusch, J. P., 1985, Folding patterns of porin and bacteriorhodopsin, *EMBO J.* **4:**1593-1597.

Preston, R. R., and Saimi, Y., 1989, Calcium ions and the regulation of motility in *Paramecium*, in: *The Structure and Function of Cilary and Flagellar Surfaces* (R. Bloodgood, ed.), Plenum Press, in press.

Ramanathan, R., Saimi, Y., Hinrichsen, R., Burgess-Cassler, A., and Kung, C., 1988, A genetic dissection of ion-channel functions in *Paramecium*, in: *Paramecium*, pp. 236-253 (H. D. Gortz, ed.), Springer-Verlag, Heidelberg.

Rhoads, D. B., Waters, F. B., and Epstein, W., 1976, Cation transport in *Escherichia coli*. VIII. Potassium transport mutants, *J. Gen. Physiol.* **67**:325-341.

Rosenbusch, J. P., 1986, Three-dimensional structure of membrane proteins, in: *Bacterial Outer Membranes as Model Systems*, pp. 141-162 (M. Inouye, ed.), Wiley, N.Y.

Sachs, F., 1988, Mechanical transduction in biological systems, *CRC Critical Review Biomedical Engineering* **16**:141-169.

Saimi, Y., 1986, Calcium-dependent sodium currents in *Paramecium*: mutational manipulations and effects of hyperdepolarization, *J. Membrane. Biol.* **92**:227-236.

Saimi Y., and Kung, C., 1980, A Ca-induced Na-current in *Paramecium*, *J. exp. Biol.* **88**:305-325.

Saimi Y., and Kung, C., 1987, Behavioral genetics of *Paramecium*, *Ann. Rev. Genetics* **21**:47-65.

Saimi Y., and Martinac, B., 1989, A calcium-dependent potassium channel in *Paramecium* studied under patch-clamp, *J. Membrane Biol.*, in press.

Saimi, Y., Hinrichsen, R. D., Forte, M., and Kung, C., 1983, Mutant analysis shows that the Ca^{2+}-induced K^+ current shuts off one type of excitation in *Paramecium*, *Proc. Natl. Acad. Sci. USA* **80**:5112-5116.

Saimi Y., Martinac, B., Gustin, M. C., Culbertson, M. R., Adler, J., and Kung, C., 1988a, Ion channels in *Paramecium*, yeast, and *Eschrichia coli*, *Trends in Biochemical Sciences* **13**:304-309.

Saimi Y., Martinac, B., Gustin, M. C., Culbertson, M. J., Adler, J., and Kung, C., 1988b, Ion channels in *Paramecium*, yeast, and *Escherichia coli*, *Cold Spring Harbor Symposia on Quantitative Biology* **53**:667-673.

Satow Y., and Kung, C., 1974, Genetic dissection of active electrogenesis in *Paramecium aurelia*, *Nature* **247**:69-71.

Satow Y., and Kung, C., 1980a, Membrane currents of pawn mutants of the pwA group in *Paramecium tetraurelia*, *J. exp. Biol.* **84**:57-71.

Satow Y., and Kung, C., 1980b, Ca-induced K^+-outward current in *Paramecium tetraurelia*, *J. exp. Biol.* **88**:293-303.

Schwarz, T. L., Tempel, B. L., Papazian, D. M., Jan, Y. N., and Jan, L. Y., 1988, Multiple potassium-channel components are produced by alternative splicing at the Shaker locus in *Drosophilia*, *Nature* **331**:137-142.

Stock, J. B., Rauch, B., and Roseman, S., 1977, Periplasmic space in *Salmonella typhimurium* and *Escherichia coli*, *J. Bacteriol.* **262**:7850-7861.

Tanabe T., Takashima, H., Mikami, A., Flockerzi, V., Takahashi, H., Kangawa, K., Kojima, M., Matsuo, H., Hirose, T., and Numa, S., 1987, Primary structure of the receptor for calcium channel blockers from skeletal muscle, *Nature* **328**:313-318.

Yatani, A., Mattera, R., Codina, J., Graf, R., Okabe, K., Padrell, E., Iyengar, R., Brown, A. M., and Birnbaumer, L., 1988, The G protein-gated atrial K^+ channel is stimulated by three distince G_i alpha-subunits, *Nature* **336**:680-682.

Chapter 16

Ion Currents of *Paramecium*: Effects of Mutations and Drugs

TODD M. HENNESSEY

1. Introduction

Paramecium is a eukaryotic, single-celled organism which is used as a simple model system for studying the regulatory mechanisms governing excitable and sensory cell functions. In terms of ion channel functions, some strategies may have been so efficient in the primative protozoans that they may remain as universal mechanisms in most sensory and excitable cells while other specialized functions may have been refined and further evolved in higher organisms, leading to diversity in the structure, function, and regulation of ion channels.

Paramecium offers the unique advantage of a "Genetic Dissection" approach for ion channel studies because: 1. Cells are large (about 150 μm) so that intracellular electrophysiology and microinjection can be done easily, 2. Electrophysiological procedures are well described. Both voltage-clamp and patch-clamp analysis (Saimi et al., 1988) can be done and most of the membrane ion currents have been well characterized, 3. Swimming behaviors have been so well studied that they can be used as simple bioassays to estimate specific ion channel activities and to identify "mutant curing factors", 4. Mutant selection and genetic analysis is easily done. Over 300 lines of behavioral mutants exist. Swimming behavior and viability selections are used to isolate behavioral mutants with ion channel defects. Molecular biological techniques, such as cloning and transformation, are also applicable for such studies, 5. Axenic cultures can be grown for large amounts of one type of genetically identical cell and, 6. Biochemical analyses are easily done because large amounts of starting material can

TODD M. HENNESSEY • Department of Biological Sciences, State University of New York at Buffalo, Buffalo, New York 14260, USA.

Evolution of the First Nervous Systems
Edited by P.A.V. Anderson
Plenum Press, New York

be obtained. Both the cilary membrane (which is the excitable membrane) and the body membrane can be easily purified and analyzed (Mortillaro, White and Hennessey, in preparation).

Paramecium can be considered "free-swimming sensory cells" because they respond to environmental stimuli by changing their membrane potential. Changes in membrane potential are translated into alterations in swimming behavior because the ciliary beat direction and frequency are both correlated with the electrophysiological state of the cell. This provides a mechanism for them to be attracted towards favorable conditions or to avoid noxious stimuli. The major stimuli that *Paramecium* respond to are ionic (Machemer, 1989), chemical (Van Houten, 1988), mechanical (Machemer, 1988), and thermal (Hennessey et al., 1983). Some species also respond to light (Iwatsuki and Naitoh, 1981).

2. Membrane Ion Currents

2.1. Resting currents

It is generally assumed that the major contributors to the resting membrane potential (V_m) of *Paramecium* are passive Ca^{++} and K^+ currents but other resting conductances have not been ruled out (Machemer, 1988). Contributions of electrogenic pumps have been suggested (Connoly and Kerkut, 1983) but none have yet been described. At rest, the internal Ca^{++} is assumed to be about 10^{-7} M (Kung and Naitoh, 1973) and internal K^+ ranges from 17-18 mM (Oertel et al., 1978) to 34 mM (Ogura and Machemer, 1980). The resting membrane potential of wild type in 4.0 mM K^+ and 1.0 mM Ca^{++} is -35 mV, reflecting the relative contributions of both Ca^{++} and K^+ to resting V_m. There may also be ion currents associated with thermoreception (Hennessey et al., 1983), chemoreception (Van Houten, 1988) and trichocyst release (Satir et al., 1988) but these currents have not yet been characterized. Although electrogenic exchangers may also contribute to membrane ion currents, none have been systematically described.

2.2. Mechanosensory Currents

Due to assymetric distribution of the ion channels involved, different mechanosensory receptor potentials are generated by stimulating different parts of the cell. These currents are located on the body membrane because they exist in both intact and deciliated cells (Ogura and Machemer, 1980).

Posterior stimulation causes an increase in a mechanosensory K^+ conductance and a membrane hyperpolarization. This mechanosensory K^+ current is partially isolated by posterior mechanosensory stimulation under voltage clamp conditions. Under free-running membrane potential recording conditions, posterior stimulation in low K^+ media produces a large "regenerative hyperpolarization" which prolongs the hyperpolarization of free swimming cells (Satow and Kung, 1977), but the ion currents involved have not been characterized.

Anterior mechanical stimulation causes an increase in a mechanosensory Ca^{++} current and depolarization of V_m. If the membrane is sufficiently depolarized, an action potential can be elicited (Machemer, 1988). The mechanosensory Ca^{++} current is also isolated by voltage clamping the cell and stimulating the anterior end mechanically. This current does not involve the ciliary I_{Ca} because mutants which lack I_{Ca} (pawns) have normal anterior mechanoreceptor currents (Ogura and Machemer, 1980).

2.3. Hyperpolarization-induced Currents

These currents have not been studied as extensively as the depolarization-induced currents and, therefore, are not well understood at the present time. Further characterization of these currents, along with resting currents, will not only contribute to a complete electrophysiological description of the ion currents but will also provide missing information as to the ionic mechanisms involved in hyperpolarization-dependent behavioral responses (see below).

When cells are voltage clamped in a solution containing 4.0 mM K^+ and 1.0 mM Ca^{++}, a complex set of currents are seen during hyperpolarization. During a 500 msec hyperpolarization from the holding level of -40 mV to -100 mV, the predominant inward current seen within the first 50 msec is carried by K^+ (Oertel et al., 1978), but the majority of these currents (and the later currents) are actually Ca^{++}-dependent K^+ currents (Richard et al., 1986; and personal observations). A Ca^{++}-dependent Na^+ current is also seen during hyperpolarization in high Na^+ solutions. Ca^{++} entry is apparently necessary to activate these Ca^{++}-dependent currents because they are not seen when external Mg^{++} is substituted for Ca^{++} (Saimi, 1986). The Ca^{++} source for these two different Ca^{++}-dependent ion currents has not been identified (for hyperpolarization-induced responses), but a possible hyperpolarization-induced inward Ca^{++} current has been suggested (Richard et al., 1986; Saimi, 1986). Since the Ca^{++}-dependent Na^+ channel can also carry Ca^{++} (Saimi, 1986), it provides one possible route for Ca^{++} entry during hyperpolarization.

Observations on the hyperpolarization-induced currents of various mutants (see below) suggest the possibility that there may be more than one type of hyperpolarization-induced Ca^{++}-dependent K^+ channels (Richard et al., 1986) as well as other possible carriers for unidentified hyperpolarization-induced currents.

2.4. Depolarization-induced Currents

Under free-running membrane potential conditions (in a Ca^{++}/K^+ solution) current injection produces a transient Ca^{++}-spike, followed by a slow repolarization. Following termination of stimulation, an indicative slow afterhyperpolarization is seen (see Fig. 5). The upstroke of this action potential is due to the voltage-dependent opening of the ciliary Ca^{++} channels and the downstroke is due to activation of the delayed, voltage-dependent, K^+ conductance (Satow and Kung, 1980a). The slow repolarization and afterhyperpolarization are due to an increase in the Ca^{++}-dependent K^+ conductance (Satow and Kung, 1980b).

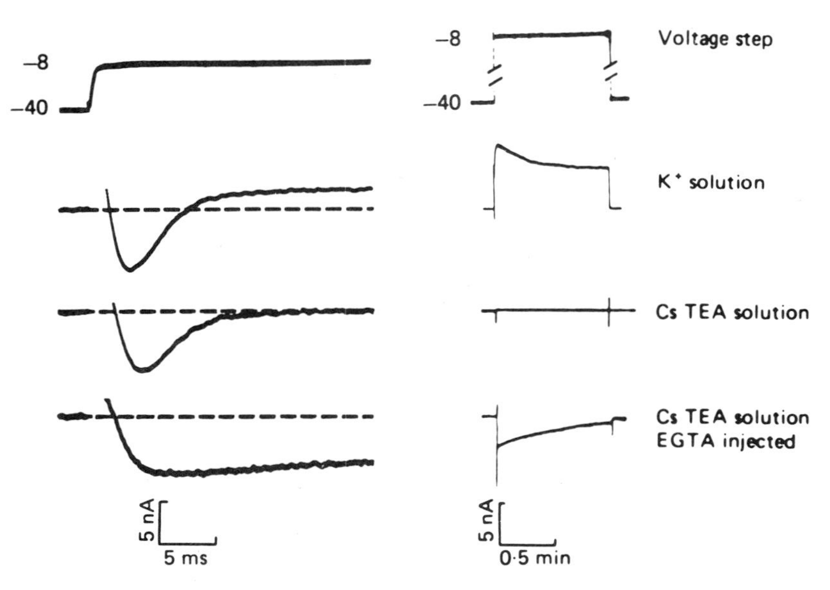

Figure 1. Fast (left) and slow (right) currents of *P. caudatum* induced by step depolarizations were analyzed under voltage clamp. Voltage steps from -40 to -8 mV were applied to cells bathed in three different solutions. In K^+ solution (with 2 M KCl electrodes) a transient inward current was seen in the first 10 ms followed by a large outward current which relaxed over 1 min. In Cs TEA solution (with 2 M CsCl electrodes) the outward K^+ current was blocked, unmasking the inward Ca^{++} current in isolation. Only this fast current was seen during the first 20 ms; any small outward current seen during a 1 min depolarization without activation or inactivation was assigned to leakage. In Cs TEA solution (with 0.5 M EGTA added to the 2 M CsCl in the electrodes) the Ca^{++}-dependent Ca^{++}-channel inactivation (fast inactivation) was blocked by EGTA. The isolated inward current relaxed gradually over the 1 min depolarization, suggesting a different, slow inactivation (from Hennessey and Kung, 1985).

When cells are voltage clamped (at -40 mV) in solutions containing both Ca^{++} and K^+, depolarization produces a transient inward current which peaks at -10 to 0 mV within the first 5 msec. This is followed by a delayed outward K^+ current which peaks within the next few tens of milliseconds (Satow and Kung, 1980a). If the depolarization is prolonged for more than 50 msec, a Ca^{++}-dependent K^+ current is activated which takes up to 500 msec to develop. Following repolarization, a slow outward tail current with a time constant of about 70 msec is seen in wild type (Satow and Kung, 1980b; see Figs. 3B and 7).

If Na^+ is included in the bath, a Ca^{++}-dependent Na^+ inward current can also be seen, but it develops even more slowly than the Ca^{++}-dependent K^+ current. The inward tail current which follows termination of this current is even slower than the outward tail; its time constant is up to 400 msec (Saimi and Kung, 1980; Fig. 3C). In both cases the tail current kinetics most likely represent the rates of Ca^{++} removal following termination of the depolarization-induced inward currents, rather than the rates of channel closure (Evans et al., 1987).

Four different depolarization-induced ion currents have been isolated and characterized:

1. The voltage-dependent inward Ca^{++} current (I_{Ca}). It is effectively isolated for study by blocking the outward K^+ currents with internal Cs^+ and external TEA^+. This procedure also unveils a sustained, inward Ca^{++} current, which is believed to involve the same Ca^{++} channels (Hinrichsen and Saimi, 1984).

The isolated I_{Ca} shows a fast, Ca^{++}-dependent inactivation and the remaining sustained inward Ca^{++} current shows slow, voltage-dependent inactivation (Hennessey and Kung, 1985; Fig. 1). The rate of return from Ca^{++}-dependent Ca^{++} channel inactivation can be measured by eliciting two isolated I_{Ca} currents with twin depolarizing pulses. When the duration between the pulses is too short for complete Ca^{++} removal (less than 100 msec), the amplitude of the second current is reduced due to Ca^{++}-dependent inactivation (see Fig. 6). When the interstimulus interval is long enough, the two currents become indistinguishable (Brehm and Eckert, 1980). The time constant for this process is, therefore, an estimate of the rate of Ca^{++} removal (Evans et al., 1987). EGTA injection eliminates this process so that the amplitudes of the two currents are similar, even with short interstimulus intervals. This also produces a significant broadening of the transient inward Ca^{++} current because of the loss of Ca^{++}- dependent inactivation. Since EGTA eliminates the fast, Ca^{++}-dependent inactivation, the slow relaxation of the remaining inward Ca^{++} current represents the rate of voltage-dependent Ca^{++} current inactivation (Hennessey and Kung, 1985; Fig. 1).

2. The outward voltage-dependent K^+ current (I_K). It is isolated by removal of the I_{Ca} by mutations (Saimi and Kung, 1987), deciliation (Dunlap, 1977) or drugs (Hennessey and Kung, 1984). The Ca^{++}-dependent K^+ currents do not normally contaminate this current because they activate so slowly that they do not begin to develop until after I_K has inactivated. They are also eliminated by conditions which remove the I_{Ca} (Oertel et al., 1977).

3. The Ca^{++}-dependent K^+ current (I_{KCa}). It is partially isolated by electronic subtraction of digitized current traces before and after EGTA injection (Saimi et al., 1983). This is not an optimal procedure however because EGTA also increases the inward sustained I_{Ca}, producing an underestimate and a change in the kinetics of the Ca^{++}-dependent K^+ current.

4. The Ca^{++}-dependent Na^+ current (I_{NaCa}). It is isolated by subtraction of the depolarization-induced currents in a Ca^{++}/choline solution (choline is not permeable) from those in a Ca^{++}/Na^+ solution. Since the Ca^{++}-dependent Na^+ current is also seen during hyperpolarization, this same isolation procedure can be done with hyperpolarizing steps (Saimi, 1986). Partial isolation of this current can be be seen by eliminating the K^+ currents with internal Cs^+ and external TEA^+ but, since both the transient and sustained I_{Ca} remain, this does not yield a complete isolation of this current.

3. Contributions of Ion Currents to Swimming Behavior

In most excitable and sensory cells a stimulus elicits a mechanical response by way of membrane potential-dependent changes in internal Ca^{++} concentrations. In neurons (and most secretory cells), the mechanical response is vesicular release, while in muscle the response is contraction. In *Paramecium* the response is a change in swimming behavior. Although the mechanical responses differ, there may be many aspects of sensory transduction that are common to all of these cell types and can be studied in the simplest system possible.

Four different types of behavioral responses can be seen when *Paramecium* encounter a depolarizing stimulus, depending upon the type of stimulus and duration of exposure to it. Since each has specific electrophysiological and behavioral correlates, these swimming behaviors can be used as behavioral bioassays for predicting the electrophysiological state of the cell. These are just estimates, however, and must always be verified with intracellular recordings. The responses are:

3.1. Changes in Swim Speed

There is generally a correlation between swimming speed and membrane potential; fast swimming indicates hyperpolarization and slow swimming suggests depolarization. Although this is primarily due to the voltage dependence of ciliary beat frequency, the correlation is not absolute because membrane potential also affects the angle of ciliary inclination and the ciliary orientation, factors which also contribute to forward swim speed (Machemer, 1988). The regulating factor(s) which transduce changes in membrane potential into ciliary beat frequency changes may involve Ca^{++} (Machemer, 1989), cyclic nucleotides (Bonini, et al., 1986), or other uncharacterized factors (Hennessey et al., 1985).

Hyperpolarization, increased beat frequency, and increased forward swim speed occur in response to posterior mechanical stimulation. In this manner, cells can burst forward to avoid approaching posterior stimuli.

Membrane potential and swim speed changes are also seen in response to changes in external ion concentrations, temperature, and chemicals but the ionic conductances involved (g_x) have not been characterized. Chemoattractants cause faster forward swimming and hyperpolarization of the membrane potential while repellents cause depolarizations which slow the swim speed. These chemosensory responses are transient; normal swimming returns after tens of seconds, even though the membrane potential is still altered (Van Houten, 1988). The mechanism of this chemosensory adaptation is unknown.

3.2. Avoiding Reactions (A.R.)

If a depolarizing stimulus is sufficient, voltage-dependent Ca^{++} channels open and an inward Ca^{++} current ($I_{Ca}{}^{++}$) results. This Ca^{++} current is exclusively localized to the ciliary membrane (Dunlap, 1977), causing intraciliary Ca^{++} to rise above 10^{-6} M. This rise in Ca^{++} causes the cilia to reverse their direction of beating by a Ca^{++}-dependent mechanism. Electrophysiologically, this is seen as either a transient action potential under current clamp or as a transient inward current under voltage-clamp conditions. The upstroke of the action potential is due to the opening of voltage-dependent Ca^{++} channels. The down stroke is due to a combination of Ca^{++}-dependent Ca^{++} channel inactivation (Brehm, et al., 1980) and activation of the delayed-rectifying K^{+} current (I_K). Unlike the Na^{+} action potentials of nerve, this response is graded with the strength of the stimulus. All-or-none action potentials can be seen with strong stimuli, however. Behaviorally, this is seen as a brief (1-2 body lengths) backward excursion followed by normal forward swimming. This has been called Type I excitation (Saimi et al., 1983).

Some stimuli, such as Ba^{++}, Na^{+}, and TEA^{+}, cause repetitive A.R. in what is referred to as Type II excitation. A typical example is the "Ba^{++} dance", or repetitive A.R. This represents trains of action potentials due to prolonged depolarization and inadequate Ca^{++} channel inactivation. It is terminated by the repolarizing action of a delayed, Ca^{++}-dependent K^{+} current. If the depolarization is strong or the Ca^{++}-dependent K^{+} current is inhibited, cells show more prolonged responses or Type III excitation (Saimi et al., 1983). This is typically assayed as the duration of backward swimming (continuous ciliary reversal) in either high K^{+} (20 to 40 mM) or in TEA^{+}/Na^{+} solutions.

Anterior mechanosensory stimulation causes depolarization, ciliary reversal and backward swimming. Therefore, cells jerk backwards if they should happen to run headlong into a physical disturbance. Since cells reorient themselves after a bout of such backward swimming, they swim off in a new, random direction to avoid continually running into the same stimulus. Since the avoiding reaction requires both a depolarizing generator potential and functioning voltage-dependent Ca^{++} channels, the absence of A.R. usually indicates either a decrease in some g_x or a decreased Ca^{++} channel activity. This is why behavioral selections are used to isolate mutants with decreased I_{Ca} or decreased g_x to depolarizing stimuli (Kung et al., 1975). However, electrophysiological verification of the implied defect is necessary because conditions exist where A.R. are eliminated due to either a loss of Ca^{++}-dependent axonemal responses (Hinrichsen et al., 1984) or an abnormally rapid repolarization mechanism (Hennessey and Kung, 1987).

3.3. Continuous Ciliary Reversal (CCR)

When a depolarization is strong and prolonged, backward swimming can last tens of seconds due to continued, high internal Ca^{++}. Although the majority of the I_{Ca} inactivates within the first few milliseconds, by Ca^{++}-dependent Ca^{++} channel inactivation, a sustained inward Ca^{++} current persists up to minutes (Hinrichsen and Saimi, 1984; Hennessey and Kung, 1985; Fig. 1). This provides a persistant Ca^{++} influx to keep internal Ca^{++} elevated and to cause CCR. Forward swimming can resume by one of two mechanisms, membrane repolarization and voltage-dependent Ca^{++} channel inactivation.

Since there is generally a correlation between the duration of high K^+ stimulated CCR and the amplitude of the voltage-dependent inward Ca^{++} current, the duration of CCR is often used as a bioassay for purifying "curing factors" which can restore Ca^{++} channel activity to mutants which lack this current (Haga et al., 1984) and as an assay for Ca^{++} channel blockers (Hennessey and Kung, 1984; Gustin and Hennessey, 1988). In general, a decrease in high K^+ CCR suggests a decreased inward Ca^{++} current.

The actual measurement of the duration of response to high K^+ has two components, the length of time which the cell swims backwards (CCR) and the time it takes to recover forward swimming after the CCR (recovery time or RT). Since the orientation of the ciliary power stroke is Ca^{++}-dependent (Machemer, 1988) and defective Ca^{++} removal causes an increase in the recovery time (Evans et al., 1987, Evans and Nelson, 1989), the recovery time is an estimate of how long Ca^{++} remains high after CCR. Behaviorally, the recovery time is seen as a whirling or rolling in place. For this reason, it is likely that the duration of backward swimming (CCR) is an accurate estimate of the Ca^{++} channel activity while the recovery time (RT) indicates the rate of removal of the Ca^{++} which accumulated during the CCR.

Any suggested Ca^{++} current decrease must always be confirmed electrophysiologically because conditions can exist where the amplitude of the inward Ca^{++} current is decreased but the duration of the high K^+ stimulated response is increased (Evans et al., 1987). A diagnostic for this condition is that the I_{Ca} returns to normal amplitudes when EGTA is injected (Evans et al., 1987), suggesting that the reduced I_{Ca} is due to Ca^{++}-dependent Ca^{++} channel inactivation. Recovery of forward swimming is not due to axonemal adaptation because when permeabilized cells are supplied with Ca^{++} and ATP they continue to swim backwards as long as the ATP is available (Kung and Naitoh, 1973). Both of these processes represent forms of sensory cell adaptation to a prolonged depolarizing stimuli.

Prolonged CCR in solutions containing a mixture of 5.0 mM TEA^+ and 10.0 mM Na^+ is used as a bioassay for estimating the activity of the Ca^{++}-dependent K^+ current because this is the primary mechanism for repolarization from prolonged depolarizations (Saimi et al., 1985). However, increased CCR in TEA^+/Na^+ solutions can also be due to decreased Ca^{++} channel inactivation (Hinrichsen and Saimi, 1984), an increased Ca^{++}-dependent Na^+ current (Saimi and Kung, 1980), or a defect in Ca^{++} removal (Evans et al., 1987). Obviously, conditions which reduce the ability of the cell to generate action potentials will also reduce this response.

3.4. Cellular Adaptation

Cellular adaptation is a time-dependent decay of responsiveness of a receptor cell to a sustained stimulus (Machemer, 1989). Since this necessitates that only one stimulus be used, previous studies of adaptation may not be comparable to this discussion because cells were exposed to high K^+ and responses to a different stimulus (Ba^{++}) were assayed (Schusterman et al., 1978).

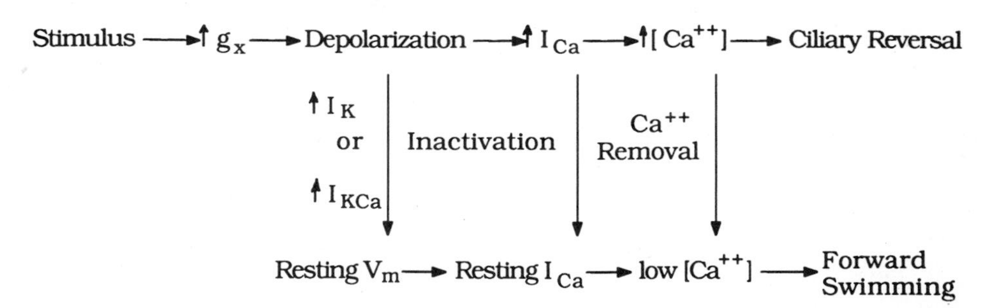

Figure 2. Sensory transduction pathway for depolarizing stimuli. Upward arrows indicate increases. Definitions: g_x, ionic conductance; I_{Ca}, inward calcium current; I_K, delayed outward potassium current; I_{KCa}, calcium-activated potassium conductance.

An example of cellular adaptation in *Paramecium* is that cells which swim backwards in high K^+ solutions for tens of seconds can resume forward swimming even though the membrane potential remains depolarized. This is because the sustained depolarization causes voltage- dependent Ca^{++} channel inactivation to decrease the inward Ca^{++} current (Hennessey and Kung, 1985; Fig. 1). In this adaptation response, CCR is initially seen because the sustained inward Ca^{++} current keeps the internal Ca^{++} high, but when the rate of Ca^{++} removal exceeds the rate of influx, the internal Ca^{++} concentration falls until a low enough level is reached to permit forward swimming. Therefore, the extents of sensory responses to prolonged depolarizing stimuli are determined by the relative rates of Ca^{++} entry and removal. This adaptation mechanism may also contribute to termination of Type III excitation as well as chemosensory adaptation to repellents.

The contributions of the various ion currents to swimming behavior are summarized in figure 2. This sensory transduction pathway shows the responses to depolarizing stimuli, such as temperature, chemorepellents, anterior mechanosensory stimulation, or depolarizing ions. These stimuli produce a depolarizing generator (or receptor) potential due to stimulus-specific changes in conductance (g_x). If this depolarization is relatively small, the response is slower swimming. If the depolarization is stronger, it can cause increased I_{Ca} and a Ca^{++} action potential. The Ca^{++} which accumulates during this response causes ciliary reversal, which continues as long as the internal Ca^{++} is high.

The termination of ciliary reversal can occur at any point along this stimulus transduction pathway. If the receptor potential is transient, the V_m will repolarize and the cell will swim forward at normal speed. A stronger stimulus-induced depolarization can be terminated (or at least attenuated) by the repolarizing action of an increased outward K^+ current. This K^+ current can be provided by either the delayed, depolarization-induced K^+ current (I_K) or, if the internal Ca^{++} remains high long enough, by the Ca^{++}-dependent K^+ current (I_{KCa}). Under conditions where the V_m cannot be efficiently brought back to rest, the I_{Ca} will normally decrease even though the depolarization persists. This is due initially to the fast, Ca^{++}- dependent Ca^{++}

channel inactivation, but during continued depolarization, the remaining sustained I_{Ca} is decreased by voltage-dependent, Ca^{++} channel inactivation. Even when all of these mechanisms are functioning normally, ciliary reversal may continue if internal Ca^{++} remains high. Therefore, a Ca^{++} removal process is necessary to permit forward swimming, especially in the presence of a sustained depolarization. It can be seen from figure 2 that a change in swimming behavior can, therefore, result from electrophysiological defects in many different steps along this pathway.

A similar pathway for the response to hyperpolarizing stimuli (faster forward swimming) will not be discussed because the hyperpolarization responses have not been as well characterized.

4. Effects of Mutations on Membrane Ion Currents

The rationale behind behavioral selections is that mutants with altered swimming behaviors may have altered electrophysiological properties. Genetic analysis is done on cultured cells derived from single cells (to assure clonal lines). These diploid cells can undergo autogamy, a process that makes each cell homozygous at all loci. Mating between cells occurs by sexual conjugation and the phenotype of the offspring gives information on dominance, complementation and linkage.

While no mutants have yet been described with changes in either the mechanosensory Ca^{++} or K^{+} currents, there are mutations which decrease (and others which increase) the I_{Ca}, I_{KCa}, I_{NaCa} and resting currents. Several pleiotrophic mutants affect I_{K}, but no mutant has yet been described with a primary defect in this current. The described mutant defects are:

4.1. Decreased I_{Ca}

The first behavioral mutant of *Paramecium tetraurelia* was named pawn because, like the chess piece, it cannot move backwards (Kung, 1971). The phenotype of this mutant is that it does not show backward swimming or avoiding reactions in response to any stimulus as long as the membrane is intact. These mutants have been selected by both behavioral (Kung, 1971) and survival screens (Schein, 1976). Pawn-like mutants have also been selected in *Paramecium caudatum*, designated as cnr mutants (caudatum non-reversal) (Takahshi et al., 1985).

Both the pawn and cnr phenotypes are due to missing or decreased voltage-dependent inward Ca^{++} currents (I_{Ca}) (Kung and Eckert, 1972; Takahashi and Naitoh, 1978). Therefore, these mutants show that the I_{Ca} is necessary for ciliary reversal. These cells are not defective in their ciliary reversal machinery because permeabilized pawns swim backwards the same as wild type (Kung and Naitoh, 1973). The I_{Ca} is not completely lacking in all of these mutants, some are "leaky" and will show some A.R. under specific conditions, depending upon growth phase, culturing conditions and growth temperature. As shown in figure 3A, the cnrD appears to completely lack the I_{Ca} (Takahshi et al., 1985).

Microinjection of wild type cytoplasm into the pawn and cnr mutants causes them to be able to swim backwards and, therefore, temporarily "cures" them. Using duration of high K^+ CCR as a bioassay for Ca^{++} channel activity, a "curing factor" has been partially purified which restores both I_{Ca} and CCR to the cnrC mutant when microinjected (Fig. 4A). This curing is not a genetic transformation because the mutant phenotype returns in a few days and curing is resistant to DNAse and RNAse. The soluble curing factor is a protein (or proteins) which is apparently necessary for the proper function of existing Ca^{++} channels (Haga et al., 1984). Injection of RNA can also produce this curing, apparently by providing the message for *in vivo* translation of the wild type gene within the mutant (Nock et al., 1982). Cytoplasmic microinjection has also been used to define the seven different complementation groups for the pawn and cnrD phenotypes, suggesting at least seven diffent genes for regulation of the Ca^{++} channel (Haga et al., 1982).

4.2. Increased I_{Ca}

An overresponsive mutant, called dancer, shows repeated A.R. to weak depolarizing stimuli which elicit little response in wild type. The phenotype is longer backward swimming in K^+ and Ba^{++} solutions (Hinrichsen and Saimi, 1984). Since the Ca^{++}-dependent Ca^{++} channel inactivation is slowed in this mutant (see Fig. 3A), more Ca^{++} enters during stimulation. This causes exaggerated responses to even weak stimuli. This mutant is not cured by cytoplasmic microinjection, and it may be that the dancer mutation affects either the Ca^{++} channel itself or its immediate surrounding (Hinrichsen and Saimi, 1984).

4.3. Decreased I_{KCa}

The pantophobiac mutant, like kshy and dancer, is an overresponsive mutant. Pantophobiac swims backwards longer in response to most depolarizing stimuli because it has a decreased Ca^{++}- dependent K^+ current (Fig. 3B). The decreased I_{KCa} can be seen both in response to depolarizations (Saimi et al., 1983) and hyperpolarizations (personal observation). In wild type, this current is normally used to repolarize the cell during a strong depolarization to most depolarizing stimuli but it cannot repolarize the V_m in high external K^+ because of the reduced driving force for K^+. Therefore, the pantophobiac mutant swims backwards longer in TEA^+/Na^+ solutions because it cannot repolarize as well, but its duration of CCR in high K^+ is normal. Behaviorally, this categorizes the pantophobiac mutant as an adaptation mutant. It cannot adapt to most depolarizing stimuli as well as wild type, but it does show behavioral adaptation to high K^+. This is in contrast to the dancer and kshy mutants which are also adaptation mutants, but which show prolonged CCR in virtually all depolarizing stimuli (including high K^+), due to altered relative rates of Ca^{++} entry and removal. None of

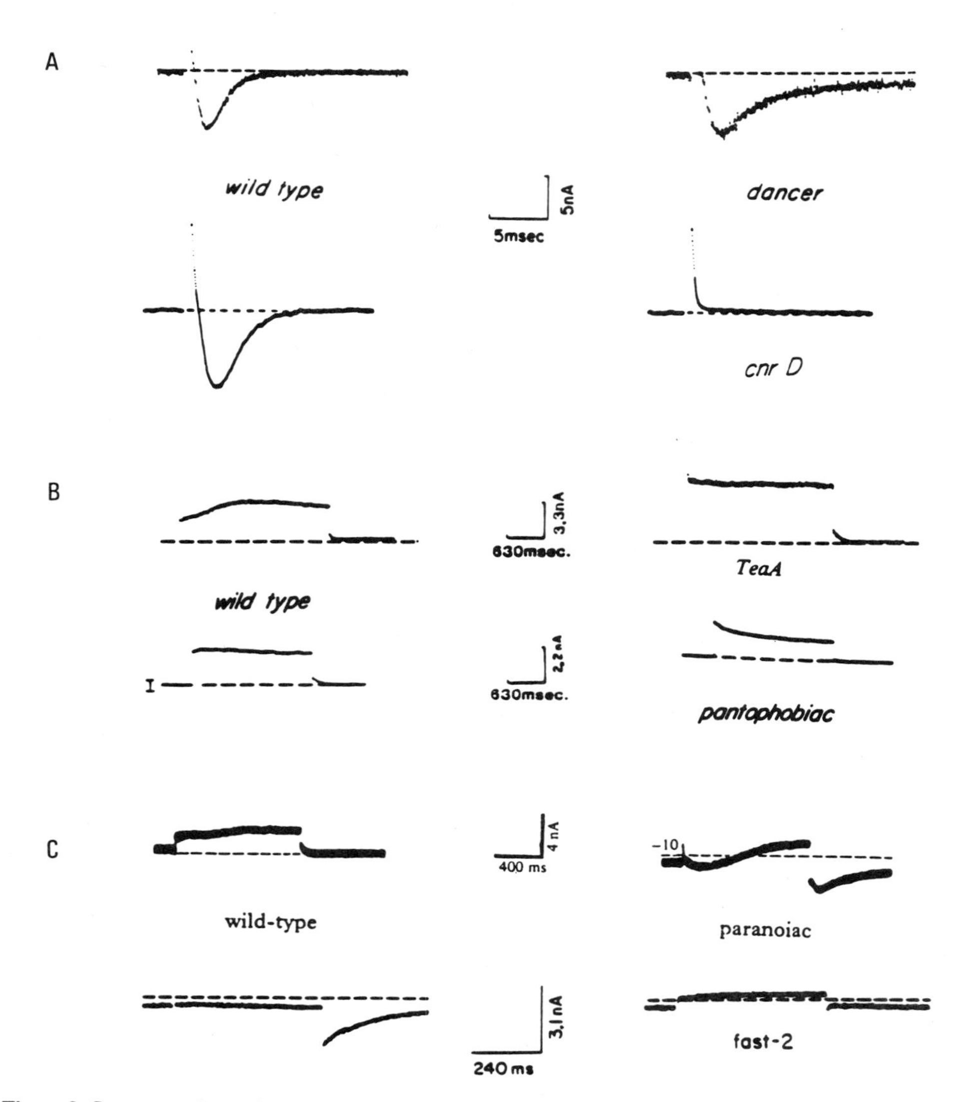

Figure 3. Summary of membrane ion current changes in different behavioral mutants. In each case, the appropriate wild type control is shown on the left and both "up" and "down" mutations in the same current are shown. **(A)**. The isolated I_{Ca} is increased in dancer, due to less inactivation of the Ca^{++} channel (from Hinrichsen and Saimi, 1984), while the cnrD is completely lacking the I_{Ca} (from Takahashi et al., 1985). **(B)**. The I_{KCa} is increased in the teaA mutant (from Hennessey and Kung, 1987) and decreased in the pantophobiac mutant (from Hinrichsen et al., 1986). **(C)**. The I_{NaCa} is increased in paranoiac (from Saimi and Kung, 1980) and eliminated in fast-2 (Saimi, 1986). Since 3 M KCl electrodes were used with paranoiac (and the wild type control), outward K^+ currents contaminate I_{NaCa}. The fast-2 currents were, however, isolated with internal Cs^+ and external K^+.

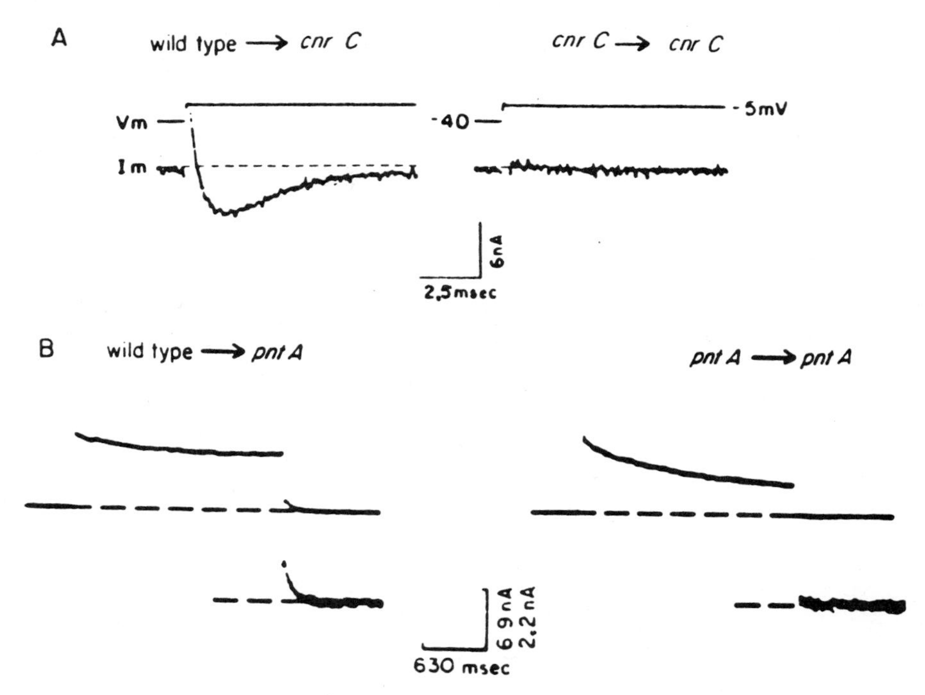

Figure 4. Restoration of ion currents in behavioral mutants by microinjection of wild type "curing factors." The voltage step is from -40 mV to -5 mV in all cases. **(A).** A protein fraction was purified from either wild type or cncC cells and injected into cnrC mutants. The fraction from wild type restored the I_{Ca} in the mutant while the mutant fraction was ineffective. **(B).** Calmodulin was purified from either wild type or the pantophobiac mutant and injected into pantophobiac. Only the wild type calmodulin restored the I_{KCa} to the pantophobiac mutant. Tail currents are enlarged in the insets (adapted from Haga et al., 1984 and Hinrichsen et al., 1986).

these three mutants are adaptation-minus mutants because all do eventually regain forward swimming.

Biochemically, pantophobiac is the only behavioral mutant in which the specific altered gene product has been identified. It is a calmodulin mutant. The pantophobiac calmodulin has a phenylalanine substituted for serine at position 101 (Schaefer et al., 1987). Microinjection of purified wild type calmodulin into pantophobiac causes renormalization of both the swimming behavior and the I_{KCa} (see Fig. 4B). Therefore, it appears that pantophobiac swims backwards longer in response to depolarizing stimuli because a functional calmodulin is necessary for proper Ca^{++}-dependent K^+ channel function.

4.4. Increased I_{KCa}

The teaA mutant was selected as a "TEA-insensitive" mutant because it does not show A.R. in TEA solutions. The reason for this phenotype is that its I_{KCa} current is larger (Fig. 3B) and activates too rapidly, effectively short-circuiting the action potential

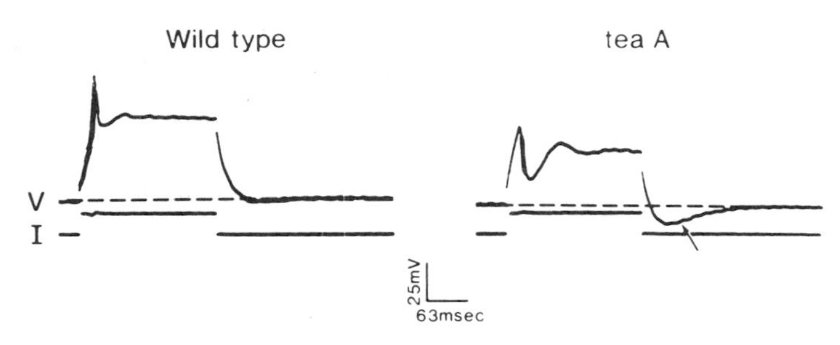

Figure 5. Responses of the free-running membrane potential to outward current injected in wild type (left) and the mutant (teaA (right). Cells were bathed in the Ca^{++}-K^{+} solution and recorded with electrodes filled with 500 mM KCl. The lower traces show the 250-msec, 1.0-nA outward current injected. The upper traces show the responses of the membrane potentials. Note that the mutant has a smaller action potential, a lower plateau depolarization during the current pulse, and a prominent hyperpolarization after the current pulse (arrow) (from Hennessey and Kung, 1987).

(Fig. 5). Since this I_{KCa} is not blocked by TEA, the overactive I_{KCa} can eliminate TEA-induced action potentials. This mutant also shows reduced A.R. in Na^{+} solutions because of this short circuiting.

4.5. Decreased I_{NaCa}

Fast-2, a mutant which lacks the I_{NaCa} (Fig. 3C), swims forward in Na^{+} solutions. In wild type, an I_{NaCa} is seen in response to both depolarization and hyperpolarization. Since the single gene mutant fast-2 eliminates both currents, it appears that they both involve the same gene product. Although the fast-2 mutation does not affect any of the other characterized ion currents, it does appear to decrease an uncharacterized Ca^{++} current seen during hyperpolarization, suggesting that the I_{NaCa} conductance pathway is also permeable to Ca^{++} (Saimi, 1986). Such a current would be a Ca^{++}-dependent Ca^{++} current because it is eliminated by EGTA and by removal of external Mg^{++}.

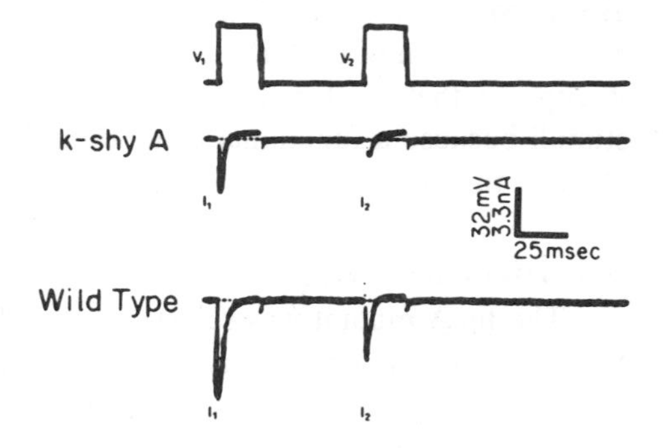

Figure 6. Recovery from the Ca^{++}-dependent "fast" inactivation of the Ca^{++} current. For both wild type and kshy A the twin voltage steps (V_1 and V_2) were 25 msec in duration, +34 mV in amplitude, and separated by 50 msec. I^1 and I^2 refer to the two I^{Ca} produced by the voltage steps. The Cs-TEA procedure was used to isolate the I^{Ca} for this analysis.

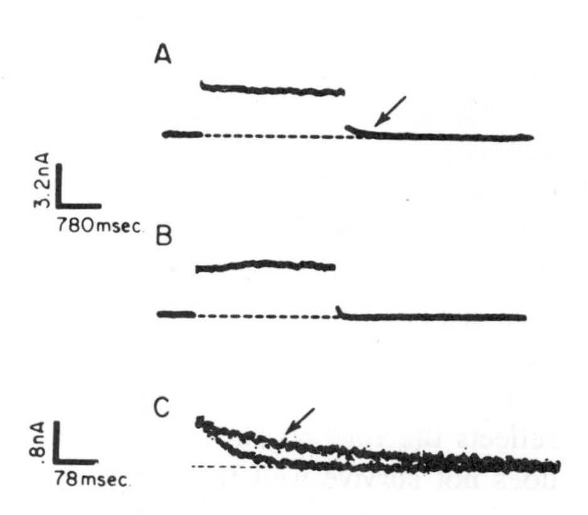

Figure 7. Tail currents of the Ca^{++}-dependent K^+ current. The Ca^{++}-dependent K^+ current and its indicative tail (see arrow) are shown for the kshy A mutant (A) and for wild type (B) with a 2.5 sec voltage step to +34 mV. The tail currents are expanded and superimposed in (C). The arrow marks the kshy A tail current. The K^+-Ca^{++} solution and KCl electrodes were used.

4.6. Increased I_{NaCa}

A mutant which overresponds to high external Na^+, called paranoiac, has an increased I_{NaCa}. The isolated I_{NaCa} has not been described in this mutant but an increased inward current has been seen in Na^+ solution (Fig. 3C). An ultrastructural change in the distribution of intramembrane particles in the "plaque" region of the cilia has been reported (Byrne and Byrne, 1978), but the functional significance of this is yet to be elucidated.

4.7. Resting Current

A behavioral mutant called restless is hyperpolarized, relative to wild type, in most solutions. It has an increased I_{KCa}, but this mutant is not allelic to teaA or pantophobiac, suggesting that it affects a different Ca^{++}-dependent K^+ channel (Richard et al., 1986). The characterization and identification of multiple Ca^{++}-dependent K^+ currents has not yet been described, however. It is not known whether this, or any other Ca^{++}-dependent K^+ current, contributes to the resting V_m of wild type. This mutant was originally selected as a fna (fast forward swimming in Na^+ solutions), and its phenotype is best represented as no A.R. in Na^+ solutions. Like teaA, restless shows a decreased response to most depolarizing stimuli, but this is because it is more hyperpolarized at rest. Unlike the teaA mutation, the relative resting K^+ conductance of restless is also greatly increased. This causes restless to swim faster because it is unable to hold its V_m as depolarized as wild type does, hence its restless nature. Since its relative K^+ conductance is so high, the V_m acts in a near Nernstian manner to K^+, making restless a free-swimming K^+ electrode. All of these alterations are normalized by EGTA injection.

A different behavioral mutant, teaB, is depolarized with respect to wild type in most solutions. Its phenotype is forward swimming in TEA^+ solutions. It has an apparent reduction of membrane surface charge which also causes a shift in the voltage sensitivities of I_{Ca} and I_K (Satow and Kung, 1981). This phenotype can be mimicked in wild type by addition of external Ca^{++} or Mg^{++}. TeaB also has a reduced heat-induced depolarization (Hennessey et al., 1983).

4.8. Other Behavioral Mutants

The kshy mutants also have a decreased I_{Ca} but this is due to high internal Ca^{++} and, consequently, Ca^{++}-dependent Ca^{++} channel inactivation. This mutant has high internal Ca^{++} for a longer time because its rate of Ca^{++} removal is defective. Behaviorally, this is seen as an overresponse (increased recovery time) to most stimuli and an inability to show correct high K^+ adaptation. Electrophysiologically, this phenotype is represented by increased I_{Ca} inactivation (Fig. 6) and a decreased rate of decay of the I_{KCa} tail current (Fig. 7). The slow recovery of the I_{KCa} tail current is the easiest diagnostic for kshy and suggests that the kinetics of this tail current decay reflects the rate of Ca^{++} removal and not the rate of channel closure. This mutant does not survive well in high K^+ solutions (Evans et al., 1987), suggesting that the mechanism involved in behavior and adaptation to high K^+ may also be important for high K^+ survival. Recently, it has been shown that this mutant can be cured by microinjection, providing a bioassay for identifying "curing factors" (Evans and Nelson, 1989). Assays for Ca^{++} pumping and Ca^{++}-ATPase activities in various membrane domains is currently underway in an effort to define the biochemical defect of kshy (Hennesey, unpublished).

The baA mutant also shows decreases in both I_{Ca} and I_K. Since there is a correlation between I_{Ca} and membrane lipid composition, it was proposed that the Ca^{++} and K^+ channels are normal, but that the currents are reduced because of the alteration in the channels' lipid environment. The indicative phenotype of this mutant is increased duration of CCR in Ba^{++} solutions (Forte et al., 1981).

Like baA, the amp^{res} mutant has both a decreased I_{Ca} and an increased CCR in Ba^{++}. The major amp^{res} mutant phenotype is that it can survive in concentrations of amphotericin B which kill wild type. Since amphotericin B kills cells by interacting with membrane lipids, this mutant was selected as a possible membrane lipid mutant (Forte et al., 1986). However, the biochemical basis of its defect has not yet been resolved.

Several chemosensory defective mutants, called ace and fol, have alterations in their responses to acetate and folate and are very useful for the study of the mechanisms involved in chemosensory transduction (Van Houten, 1988). Mutants have also been described which are resistant to high K^+ killing and which may have defects in K^+ adaptation (Schusterman et al., 1978). Other behavioral mutants have also been described (Kung et al., 1975) but their electrophysiological characteristics have not been documented.

5. Effects of Drugs on Membrane Ion Currents

The I_{Ca} is not inhibited by dihydropyridines, verapamil, or any of the traditional Ca^{++} channel blockers. Omega-conotoxin is without effect on I_{Ca} (personal observations). Although this was initially a problem, the pharmacological sensitivities of these currents is now an interesting topic in itself. Since it is now known that there are many different kinds of Ca^{++} channels (Bean, 1989) and Ca^{++}-dependent K^+ channels (Lattore et al., 1989), their pharmacological specificities can give information as to the similarities to other organisms as well as their evolutionary lineage. If it is found that a *Paramecium* ion channel has the same pharmacological specificity as a higher organism, it can be used as a more convenient and ethical model system in which to study the properties of that channel.

The I_{Ca} can be completely inhibited by the external addition of W-7 (Hennessey and Kung, 1984) or neomycin (Gustin and Hennessey, 1988). In both cases, the duration of high K^+ CCR varies with drug concentration in parallel with I_{Ca} values. The concentration of W-7 which inhibits CCR and I_{Ca} by 50% is about 20 μM and a similar value for neomycin was 80 μM. Both of these drugs are immediately effective and reversible. The effects of W-7 do not require it to enter the cell because a non-permeant W-7 (quarternary W-7) also inhibits high K^+ CCR and I_{Ca}, at concentrations similar to W-7 (Hennessey, 1989). W-7 (and other more effective W-7 analogs) have also been shown to block voltage-sensitive Ca^{++} channels from *Paramecium* when these channels were reconstituted into artificial lipid membranes (Ehrlich et al., 1988). I_{Ca} is also inhibited by an antibody which was raised against ciliary membranes (Ramanathan et al., 1983).

The I_{Ca} can also be altered by drugs and conditions which change the membrane lipid composition. Cerulenin, a fatty acid desaturase inhibitor, causes wild type to phenocopy kshy both behaviorally and electrophysiologically (Hennessey and Scolese, in preparation). After exposure to this drug, wild type shows extended high K^+ CCR and the same kinds of electrophysiological changes seen in figures 6 and 7. Substitution of sitosterol for stigmasterol as the growth-requiring sterol causes both I_{Ca} and high K^+ CCR to increase (Weglar et al., 1989). These observations, along with those of Forte et al. (1981) support the hypothesis that proper functions of ion channels (and possibly Ca^{++} removal systems) are dependent upon membrane lipid composition.

Effective organic blockers of the other ion currents have yet to be identified. The I_{NaCa} is not affected by tetrodotoxin (TTX) or saxitoxin (STX) (Saimi and Kung, 1980) and the depolarization-induced I_{KCa} is not effectively blocked by charybdotoxin, apamin, or even external TEA (personal observation).

Unlike the depolarization-induced Ca^{++}-dependent K^+ current (I_{KCa}) of wild type, the abnormally large restless I_{KCa} is blocked by external TEA (Richard et al., 1986). The increased I_{KCa} current of teaA is not blocked by external TEA. Both Ca^{++}-dependent currents are eliminated by either EGTA injection or internal Cs^+, however. These two observations suggest that either there are multiple types of these channels in wild type (which have escaped characterization) or that new types of channel have "evolved" in restless and teaA.

A survey of several eukaryotic phyla has suggested that Ca^{++} channels of protozoa, porifera, coelenterates, and ctenophores are all blocked by W-7, while all of these Ca^{++} channels are insensitive to dihydropyridines (see Table 1). This not only provides interesting information about the commonality of Ca^{++} channels across phyla, but also keeps in perspective the extent to which analogies can be drawn in comparative studies.

Table 1. Comparative pharmacology of the Ca channel

Phylum	Genus	Ref. or source	Drug sensitivity W-7	Drug sensitivity DHP
Fungi	*Saccharomyces*	28	NP	NP
Protozoa	*Paramecium*	14	Yes	No
Porifera[1]	*Microciona*	29	Yes	No
Coelenterata	*Obelia*	K. Takeda K. Dunlap, and P. Brehm[2]	Yes	No
Ctenophora[1]	*Beröe*	30[3]	Yes	No
Nematoda	NT		?	?
Echinodermata	NT		?	?
Mollusca	*Loligo*	R. Chow[2]	No	No
	Helix	D. Swandulla[2]	No	No
Anelida	*Hirudo*	A. Kleinhaus[2]	Yes[4]	No
Arthropoda	*Drosophilia*	L. Byerly and H.-T. Leung[2]	Yes[4]	No
Chordata	*Rana*			
	L-type	B. Bean[2]	No	Yes
	T-type		No	No
	Rattus			
	L-type	G. Cota[2]	No	Yes
	T-type		No	No

Abbreviations: DHP, dihydropyridine; NT, none tested; NP, channels not present.
[1]Tested behaviorally.
[2]Personal communication.
[3]In another ctenophore, *Pleurobrachia*, the block is as found for annelids and arthropods.
[4]Block in these organisms is irreversible and often leads to cell death.
(after Ehrlich et al., 1988)

Since ion current mutants are viable, it is just a question of natural selection as to whether such a spontaneously arising mutation would be retained as an evolutionary advantage. It would not be suprising, therefore, that an ion channel seen in a mutant would be more analogous to one in a "higher" organism. *Paramecium* is, therefore, an interesting model system in terms of not only existing ion currents which may be conserved, but also the potential for these ion currents to be precursors for evolutionary diversity.

ACKNOWLEDGEMENTS. This was supported by NSF grant BNS 8607542 and BRSG SO RR 07066

References

Bean, B., 1989, Classes of calcium channels in vertebrate cells, *Ann. Rev. Physiol.* **51**:367-384.

Bonini, N., Gustin, M., and Nelson, D., 1986, Regulation of ciliary motility by membrane potential in *Paramecium*: A role for cAMP. *Cell Motil. Cytoskel.* **6**:256-272.

Brehm, P., Eckert, R., and Tillotson, D., 1980, Calcium-mediated inactivation of calcium currents in *Paramecium. J. Physiol. (Lond.)* **306**:193-203.

Byrne, B., and Byrne, B., 1978, An ultrastructural correlate of the membrane mutant "paranoiac" in *Paramecium*, *Science* **199**:1091-1093.

Connoly, J., and Kerkut, G., 1983, Ion regulation and membrane potential in *Tetrahymena* and *Paramecium*, *Comp. Biochem. Physiol.* **76A**:1-16.

Dunlap, K., 1977, Localization of calcium channels in *Paramecium caudatum*, *J. Physiol. (Lond.)* **271**:119-133.

Ehrlich, B. E., Jacobson, A. R., Hinrichsen, R., Sayre, L. M., and Forte, M. A., 1988, *Paramecium* calcium channels are blocked by a family of calmodulin antagonists, *Proc. Natl. Acad. Sci.* **85**:5718-5722.

Evans, T., and Nelson, D., 1989, New mutants of *Paramecium tetraurelia* defective in a calcium control mechanism: Genetic and behavioral characterizations, *Genetics* (in press).

Evans, T., Hennessey, T., and Nelson, D., 1987, Electrophysiological evidence suggests a defective Ca^{++} control mechanism in a new *Paramecium* mutant, *J. Membr. Biol.* **98**:275-283.

Forte, M., Satow, Y., Nelson, D., and Kung, C., 1981, Mutational alteration of membrane phospholipid composition and voltage-sensitive ion channel function in *Paramecium*, *Proc. Natl. Acad. Sci. (USA)* **78**:7195-7199.

Forte, M., Hennessey, T., and Kung, C., 1986, Mutations resulting in resistance to polyene antibiotics decrease voltage-sensitive calcium channel activity in *Paramecium*, *J. Neurogenet.* **3**:75-85.

Gustin, M., and Hennessey, T., 1988, Neomycin inhibits the calcium current of *Paramecium*, *Biochim. Biophys. Acta* **940**:99-104.

Haga, N., Forte, M., Saimi, Y., and Kung, C., 1982, Microinjection of cytoplasm as a test of complementation in *Paramecium*, *J. Cell Biol.* **82**:559-564.

Haga, N., Forte, M., Ramanathan, R., Hennessey, T., Takahashi, M., and Kung, C., 1984, Characterization and purification of a soluble protein controlling Ca^{++}-channel activity in *Paramecium*, *Cell* **39**:71-78.

Hennessey, T. M., 1989, Calcium currents of *Paramecium* are blocked by a non-permeant analog of W-7, Quaternary W-7, *J. Cell Biol.* **109**:355a.

Hennessey, T. M., and Kung, C., 1984, An anticalmodulin drug, W-7, inhibits the voltage-dependent calcium current in *Paramecium caudatum*, *J. exp. Biol.* **110**:169-181.

Hennessey, T. M., and Kung, C., 1985, Slow inactivation of the calcium current of *Paramecium* is dependent on voltage and not internal calcium, *J. Physiol. (Lond.)* **365**:165-179.

Hennessey, T. M., and Kung, C., 1987, A calcium-dependent potassium current is increased by a single-gene mutation in *Paramecium*, *J. Membr. Biol.* **98**:145-155.

Hennessey, T., Saimi, Y., and Kung, C., 1983, A heat-induced depolarization of *Paramecium* and its relationship to thermal avoidance, *J. Comp. Physiol.* **153**:39-46.

Hennessey, T., Machemer, H., and Nelson, D., 1985, Injected cyclic AMP increases ciliary beat frequency in conjunction with membrane hyperpolarization, *Eur. J. Cell Biol.* **36**:153-156.

Hinrichsen, R., and Saimi, Y., 1984, A mutation that alters properties of the Ca^{++} channel in *Paramecium tetraurelia*, *J. Physiol. (Lond.)* **351**:397-410.

Hinrichsen, R., Saimi, Y., Hennessey, T., and Kung, C., 1984, Mutants in *Paramecium tetraurelia* defective in their axonemal response to calcium, *Cell Motil.* **4**:283-295.

Hinrichsen, R., Burgess-Cassler, A., Soltvedt, B., Hennessey, T., and Kung, C., 1986, Restoration by calmodulin of a Ca^{++}-dependent K^+ current missing in a mutant of *Paramecium*, *Science* **232**:503-506.

Kung, C., 1971, Genic mutations with altered system of excitation in *Paramecium teraurelia*: II. Mutagenesis, screening and genetic analysis of the mutations, *Genetics* **69**:29-45.

Kung, C., and Eckert, R., 1972, Genetic modifications of electric properties in an excitable membrane, *Proc. Natl. Acad. Sci. (USA)* **69:**93-97.

Kung, C., and Naitoh, Y., 1973, Calcium-induced ciliary reversal in the extracted models of pawns, a behavioral mutant of *Paramecium*, *Science* **179:**195-196.

Kung, C., Chang, S.-Y., Satow, Y., VanHouten, J., and Hansma, H., 1975, Genetic dissection of behavior in *Paramecium*, *Science* **188:**898-904.

Iwatsuki, K., and Naitoh, Y., 1983, Behavioral responses of *Paramecium multinucleatum* to visible light, *Photochem. Photobiol.* **37:**415-419.

Lattore, R., Oberhauser, A., Labarca, P., and Alvarez, O., 1989, Varieties of calcium-activated potassium channels. *Ann. Rev. Physiol.* **51:**385- 399.

Machemer, H., 1988, Electrophysiology, in: *Paramecium*, pp. 186-215 (H.-D. Gortz, ed.), Springer Verlag, Berlin, Heidelberg.

Machemer, H., 1989, Ions in modulating cellular behavior: electrophysiological consideration, in: *Signal Transduction in Chemotaxis* (W. Stoekenius, ed.), Alan R. Liss, New York.

Nock, A. H., Kretschar, M., Lipps and Schultz, J. E., 1982, Restoration of membrane excitability by microinjection of cytoplasmic wild-type RNA into *Paramecium tetraurelia* pawn C mutants. *FEMS Microbiol. Lett.* **13**:275-277.

Oertel, D., Schein, S., and Kung, C., 1977, Separation of membrane currents using a *Paramecium* mutant, *Nature* **268:**120-124.

Oertel, D., Schein, S., and Kung, C., 1978, A potassium conductance activated by hyperpolarization in *Paramecium*, *J. Membr. Biol.* **43:**169-185.

Ogura, A., and Machemer, H., 1980, Distribution of mechanoreceptor channels in the *Paramecium* surface membrane, *J. Comp. Physiol.* **135:**233-242.

Ramanathan, R., Saimi, Y., Peterson, J., Nelson, D., and Kung, C., 1983, Antibodies to the excitable membrane of *Paramecium tetraurelia* alter membrane excitability, *J. Cell Biol.* **97:**1421-1428.

Richard, E., Saimi, Y., and Kung, C., 1986, A mutation that increases a novel calcium-activated potassium conductance of *Paramecium tetraurelia*, *J. Membr. Biol.* **91:**173-181.

Saimi, Y., 1986, Calcium-dependent sodium currents in *Paramecium*: Manipulations and effects of hyper- and depolarization, *J. Membr. Biol.* **92:**227-236.

Saimi, Y., and Kung, C., 1987, Behavioral genetics of *Paramecium*, *Ann. Rev. Genet.* **21:**47-65.

Saimi, Y., Hinrichsen, R., Forte, M., and Kung, C., 1983, Mutant analysis shows that the Ca^{++}-induced K^{+} current shuts off one type of excitation in *Paramecium*, *Proc. Natl. Acad. Sci. (USA)* **80:**5112-5116.

Saimi, Y., and Kung, C., 1980, A Ca^{++}-induced Na-current in *Paramecium*, *J. exp. Biol.* **88:**305-325.

Saimi, Y., Martinac, B., Gustin, M., Culbertson, M., Adler, J., and Kung, C., 1988, Ion channels of *Paramecium*, yeast and *Escherichia coli*, *TIBS* **13:**304-309.

Satir, B.H., Busch, G., Vuoso, A., and Murtaugh, T. J., 1988, Aspects of signal transduction in stimulus exocytosis-coupling in *Paramecium*, *J. Cell. Biochem.* **36:**429-443.

Satow, Y., and Kung, C., 1977, A regenerative hyperpolarization in *Paramecium*, *J. Comp. Physiol.* **119:**99-110.

Satow, Y., and Kung, C., 1980a, Membrane currents of the pwnA group in *Paramecium teraurelia*, *J. exp. Biol.* **84:**57-71.

Satow, Y., and Kung, C., 1980b, Ca^{++}-induced K^{+} outward current in *Paramecium tetraurelia*, *J. exp. Biol.* **88:**293-303.

Satow, Y., and Kung, C., 1981, Possible reduction of surface charge by a mutation in *Paramecium tetraurelia*, *J. Membr. Biol.* **59:**179-190.

Schaefer, W., Hinrichsen, R., Burgess-Cassler, A., Kung, C., Blair, I., and Watterson, D., 1987, A mutant *Paramecium* with a defective calcium dependent potassium conductance has an altered calmodulin: a nonlethal selection alteration in calmodulin regulation, *Proc. Natl. Acad. Sci. (USA)* **84:**3931-3935.

Schein, S., 1976, Nonbehavioral selection for pawns, mutants of *Paramecium aurelia* with decreased excitability, *Genetics* **84:**453-468.

Shusterman, S., Theide, E., and Kung, C., 1978, K^{+}-resistant mutants and "adaptations" in *Paramecium*, *Proc. Natl. Acad. Sci. (USA)* **75:**5645-5649.

Takahashi, M., and Naitoh, Y., 1978, Behavioral mutants of *Paramecium caudatum* with defective membrane electrogenesis, *Nature* **271**:656-658.

Takahashi, M., Haga, N., Hennessey, T., Hinrichsen, R., and Hara, R., 1985, A gamma-ray induced non-excitable membrane mutant in *Paramecium caudatum*: A behavioral and genetic analysis, *Genet. Res.* **46**:1-10.

Van Houten, J., and Preston, R., 1988, Chemokinesis in *Paramecium*, in: *Paramecium* (H.D. Gortz, ed.), Springer Verlag.

Weglar, D., Howe-McDonald, S., and Hennessey, T., 1989, The inward calcium current is increased by sterol supplementation in *Paramecium*, *Comp. Biochem. Physiol.* **94A**:25-32.

Chapter 17

Membrane Excitability and Motile Responses in the Protozoa, with Particular Attention to the Heliozoan *Actinocoryne contractilis*

COLETTE FEBVRE-CHEVALIER, ANDRÉ BILBAUT, JEAN FEBVRE, and QUENTIN BONE

1. Introduction

Unicellular microorganisms are regarded as primitive eukaryotes. However, they are extremely adaptable and display a remarkable diversity of shape, behavior and mode of life. They are sensitive to various stimuli (Naitoh, 1982; Machemer and Deitmer, 1987) and the response to external signals is generally a transient modification in motile activity. Some protists remain fixed to their substrate and perform contraction-relaxation cycles (some heliozoans and ciliates), the others are free-living, capable of swiming or "walking" (flagellates, amoebae, actinopods, most of the ciliates).

Motility closely depends on three kinds of dynamic cytoskeletal structures: 1) acto-myosin microfilaments (MFs) that are involved in cytoplasmic streaming, amoeboid movements and cytokinesis (Edds, 1981; Cohen et al., 1984; Detmers et al., 1985; see also Pollard and Weihing, 1974; Schliwa, 1986), 2) nonactin nanofilaments (NFs) that drive cell contraction in flagellates, actinopods and ciliates (Routledge et al., 1976; Febvre-Chevalier and Febvre, 1986; Roberts, 1987), 3) Microtubules (Mts) that can be stable or unstable, according to their microtubule associated proteins (MAPs) (Dustin, 1984). Stable Mts serve

COLETTE FEBVRE-CHEVALIER[1], ANDRÉ BILBAUT[2], JEAN FEBVRE[1], and QUENTIN BONE[3] • [1]Laboratoire de Biologie Cellulaire Marine, U.R.A 671, 06230, Villefranche-sur-Mer, France. [2]Laboratoire de Cytologie Experimentale, URA 651, Parc Valrose, 06034, Nice-Cedex, France. [3]The Marine Laboratory, Citadel Hill, Plymouth, PL1 2PB, United Kingdom.

Evolution of the First Nervous Systems
Edited by P.A.V. Anderson
Plenum Press, New York

that can be stable or unstable, according to their microtubule associated proteins (MAPs) (Dustin, 1984). Stable Mts serve as a scaffold for maintaining cell shape and guiding intracellular traffic. They are also responsible for bending movements of the cell body, ciliary and flagellar beatings (Satir, 1984; Huitorel, 1988; Gibbons, 1989). In contrast, unstable Mts form the cytoskeleton of sensitive cytoplasmic extensions (named axopods in actinopods) that are instantly destabilized upon stimulation (Davidson, 1975; Febvre-Chevalier and Febvre, 1980, 1986).

Over the last 20 years, the relationships between external stimulus and motile behavior have been extensively studied in free-living ciliates and in some stalked species. Calcium plays a pivotal role in protists, as it does in all other cells (Naitoh, 1982; Machemer and Deitmer, 1987). In amoebae for instance, amoeboid movement is controlled by calcium. Likewise in ciliates, reversed ciliary beating during avoidance reaction is dependent upon a transient Ca^{++}influx (Eckert et al., 1976). Information regarding secondary messengers (cAMP, cGMP), internal Ca^{++} levels and Ca^{++} regulatory proteins is important for a complete understanding of reception, transmission and motile responses (Satir et al., 1980; Fukui and Yumura, 1986; Andrivon, 1988; Otter, 1989; Stephens and Stommel, 1989).

The present paper reviews correlations between motility and membrane excitability in a very sensitive contractile heliozoan *Actinocoryne contractilis* (Febvre-Chevalier, 1981, 1987). Until now, this is the only protozoan in which 1) rapid, overshooting depolarizing action potentials have been recorded, and 2) the depolarizing phase of the action potential is Na^{+} and Ca^{++}-dependent (Febvre-Chevalier et al., 1986). The bioelectric activity associated with cell motility in some other protists will be reviewed to illustrate the distinctiveness of *Actinocoryne*.

2. Cytology of *Actinocoryne contractilis* and the Kinetics of Contraction-relaxation

The cell body of free-living heliozoans is spherical and bears hundreds of long axopods that are sensitive to mechanical stimuli (Febvre-Chevalier, 1973; Davidson 1975), high pressure (Tilney et al., 1966), ultrasound (Cachon and Cachon., 1982), electrical stimulation (Nishi et al., 1986) and low temperature (Tilney and Porter, 1967). The first cine-records (50 frames. sec^{-1}) of contraction were made by Davidson (1975), who reported velocities ranging from 30 to 100 lengths per second. The response to the application of diverse ions and inhibitors is a slow retraction of the axopods, which is correlated with disassembly of axopodial microtubules (Tilney, 1968; Shigenaka et al., 1971, 1975; Cachon and Cachon, 1982; Matsuoka and Shigenaka, 1986).

In contrast with other heliozoans, *Actinocoryne* is benthic, stalked and exhibits remarkable sensitivity and contractility. The cell body consists of a flat base which is fixed to a hard substrate and gives rise to a long (50 μm to 250 μm) peduncule bearing

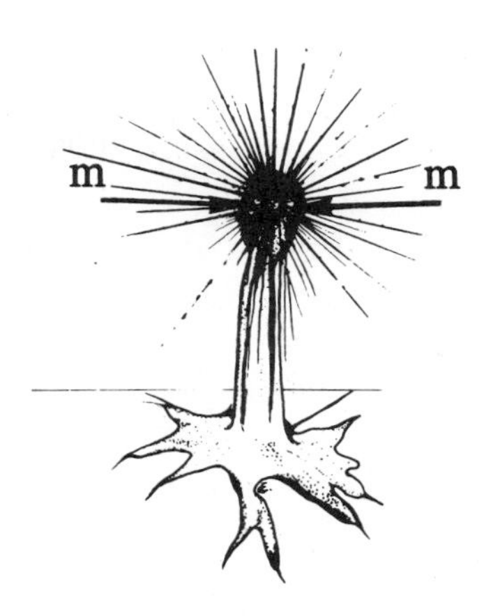

Figure 1. A drawing of *Actinocoryne* attached to substrate (base not visible) and impaled by two microelectrodes (m) in the head.

a head, 50 μm in diameter. Hundreds of long axopods radiate from the head, forming a large trap involved in prey capture (Fig. 1).

Patterned arrays of microtubules stiffen the axopodia and the peduncle. All Mt-bundles (50-800 microtubules in each) arise from a spherical microtubule organizing center (MTOC) (or axoplast), which is located in the center of the head (Febvre-Chevalier, 1980). Mechanical and electrical external stimulation and intracellular current injection elicit rapid contraction of the axopodia, and of the head and stalk (Latency: 7 ms $\pm$3 msec, n=80). It is an exponential process: the first phase is extremely rapid and lasts 2-10 msec (i.e. velocity of between 1 and 5 cm.sec^{-1}); the second is much longer (30-40 msec duration). The animal contracts to 90% of its original length (Febvre-Chevalier, 1981; Febvre-Chevalier and Febvre, 1980). At present, *Actinocoryne contractilis*, is the only cell known to produce such a rapid contraction. During this event all Mt-bundles disassemble. Different kinds of data suggest that contraction is correlated with the fragmentation of Mt-axonemes prior to depolymerization from the ends of newly formed segments (Febvre-Chevalier and Febvre, 1986).

Re-extension of the stalk, head and axopodia is a much slower process (mean velocity: 0.16 μm.sec^{-1} for the peduncle (stalk); 0.58 μm.sec^{-1} for the axopods) (Febvre-Chevalier, 1987).

3. Control of Contraction and Stabilization Assays in *Actinocoryne*

The rapid contraction that follows the action potential is dependent upon external Ca^{++}. Evidence for this is: 1) contraction is abolished in Ca^{++}-free solutions containing 1 mM EGTA or, by the presence of Ca^{++} channel blockers (Co^{++}, Mn^{++}, La^{+++} D600) in a calcium-containing medium, 2) the use of Ca^{++}-ionophore (A23187) or mild detergents in a calcium-containing medium induces immediate rapid contraction. In Ca^{++}-free solutions or in the presence of Ca^{++} inhibitors, however, a delayed and much slower (one to several seconds duration) contraction occurs, suggesting a role for internal Ca^{++} stores. The very slow contraction observed in the presence of inhibitors of oxidative and alternative pathways of ATP production suggests that the mechanisms of rapid contraction are ATP-dependent (Febvre-Chevalier and Febvre, 1986; Febvre-Chevalier, 1987).

Ca^{++}-and Na^{+}-free artificial sea water (ASW) stabilize *Actinocoryne*. Heavy water, which stabilizes unstable Mts (Marsland et al., 1971), allowed us to fix the protist for electron microscopy without triggering contraction (Febvre-Chevalier, 1980). The use of polyethylene glycol (PEG), as an additional stabilizing agent (1 mM in Ca^{++}-and Na^{+}-free ASW), was insufficient to stabilize the cell in sea water (SW). Similarly, taxol, which has been reported to stabilize Mts and promote Mt-lengthening via tubulin polymerization (de Brabander et al., 1981), is not obviously effective in the case of *Actinocoryne*. It induces a very slight lengthening in the first 2 hours, as reported in another heliozoan (Hausmann et al., 1983), but rapid contraction occurs after this delay.

Control of stabilization and reactivation of contraction will be detailed elsewhere.

4. Membrane Excitability in Relation to Contractile Activity in *Actinocoryne*

4.1. Receptor Potential, Action Potential, Contractile Activity

Conventional KCl-filled glass microelectrodes (initial resistance: 8-10 MΩ) were used to perform electrophysiological experiments. Microelectrodes could be inserted into the head without triggering contraction of the stalk. The cells have a remarkably stable resting membrane potential of -78 ±8 mV, and this is mainly potassium-dependent. The relationship between membrane potential and the logarithm of external K^{+} concentration closely approximated the Nernst equation, with a slope of 56 mV per decade.

The cell membrane was mechanically excitable. A series of mechanical stimuli of increasing strength evoked a series of graded depolarizations of the cell membrane. These were regarded as receptor potentials. Fast overshooting action potentials could be triggered by either mechanical stimuli of sufficient strength or summation of sub-threshold stimuli (Fig 2A). Receptor potentials were not observed in Na^{+}-deficient solutions, suggesting that they were mainly Na^{+}-dependent.

The cell membrane was also electrically excitable. Injection of a current pulse across the cell membrane initiated an overshooting action potential, whose characteristics were comparable to those evoked mechanically (Fig. 2B). The threshold was between 58 and 75 mV above the resting potential; the amplitude of the overshoot

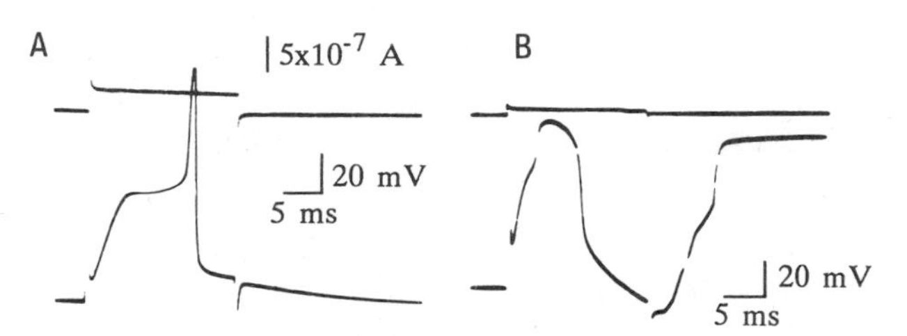

Figure 2. (A) Graded receptor potentials evoked by graded mechanical stimuli, the largest giving rise to an action potential. The return to zero potential at the end of the action potential marks the onset of the contraction that dislodged the recording electrode. (B) Action potential evoked by current injection (current pulse monitored on upper trace). The subsequent return to zero potential results from the loss of the impalement, following contraction.

was between 30 and 50 mV; the duration of the action potential measured at half amplitude was 0.5 ms; there was no undershoot.

When a microelectrode was inserted into the base of a contracted cell, action potentials could be initiated by mechanical stimuli. Simultaneous records from microelectrodes placed in the head and the base of the protist indicated that the action potential propagated at around 3 m sec^{-1}, and confirmed that the entire cell membrane was excitable.

In SW the electrode potential returned to 0 mV a few milliseconds after complete repolarization, giving evidence for cell contraction.

4.2. Ionic Basis of the Action Potential

Study of the ionic basis of the action potential showed that: 1) the depolarizing phase of the spike was Na^+- and Ca^{++}-dependent, and 2) the fast contractile activity occurred only in the presence of extracellular calcium. These conclusions were drawn from the following observations: i) in Ca^{++}-free ASW containing 1 mM EGTA, action potentials were similar to those evoked in normal SW, but contractile activity of the cell was not observed (Fig. 3A). When the external Na^+ concentration was lowered below 150 mM, action potentials could not be evoked. Both action potentials and

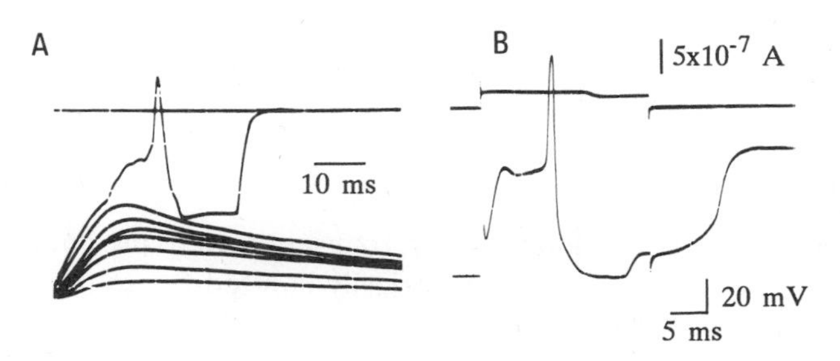

Figure 3. (A) Action potential evoked by current injection (upper trace current pulse) in Ca^{++}-free solution containing 1 mM EGTA. No contraction follows the action potential. (B) Action potential evoked by current injection (upper trace current pulse) in Na^+-free solution containing 10 mM Ca^{++}. Contraction follows the action potential, which bears a plateau.

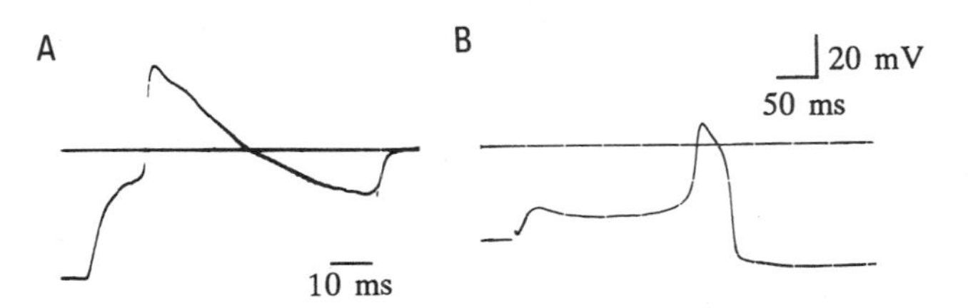

Figure 4. (A) TEA (10 mM) prolongs the repolarizing phase of the action potential (evoked by mechanical stimulation); contraction occurs before repolarization is complete. (B) 4-aminopyridine (10 mM) prolongs the repolarizing phase of the action potential (evoked by mechanical stimulation), but no contraction follows.

contraction were unaffected in the presence of TTX (10^{-6} M) added to ASW. ii) In Na^+-free solutions containing less than 10 mM Ca^{++}, no regenerative responses were seen, but overshooting action potentials could be elicited when the external Ca^{++} concentration was increased. The depolarizing phase of these action potentials was prolonged by a short plateau and repolarization was followed by contraction of the cell (Fig. 3B). The relationship between the amplitude of these action potentials and the external Ca^{++} concentration was 28 mV per decade, as predicted by the Nernst equation. This suggests that Ca^{++} is also involved in membrane electrogenesis and in the initiation of rapid contraction. The effect of Ca^{++} channel blockers such as Mn^{++} or Co^{++} added to ASW was the same as that of Ca^{++}-free ASW containing EGTA; the depolarizing phase of the action potentials was unaffected, but the repolarizing phase was prolonged and the rapid contraction blocked. Tetraethylammonium (TEA) and 4-aminopyridine (4-AP) (10 mM in SW) prolonged the repolarizing phase of the action potential (Fig. 4A). Cell contraction was not observed in the presence of the 4-AP (Fig. 4B).

5. Correlation Between Membrane Excitablity and Contraction in Other Protists

Besides heliozoans, contractile activity is found in the dinoflagellate *Noctiluca*, acantharians and in ciliates.

5.1. Dinoflagellates

In the dinoflagellate *Noctiluca miliaris*, transient membrane potential changes can be recorded during spontaneous or electrically-induced extension and flexion of the food-gathering tentacle. The resting membrane potential of the cells is -45 to -60mV and changes in membrane potential were found to be involved in the control of tentacle movement. The tentacle-regulating potentials appear to consist of a pre-spike pace-maker and an action potential which is prolonged by a plateau (Sibaoka and

Eckert, 1967). External Ca^{++} was shown to be needed for tentacle movements (Eckert and Sibaoka, 1967). Recently, the bioelectrical conditions and ionic mechanisms that control initiation of feeding were re-examined in intact and detergent-treated specimens, and it was shown that Ca^{++}, Cl^- and H^+ are involved in tentacle-regulating potentials and in tentacle movements (Nawata and Sibaoka, 1986; Oami et al., 1988; Naitoh and Oami., 1989).

5.2. Acantharians

In acantharians, changes in cell shape are caused by rapid coordinated contraction of myonemes, nanofilament bundles along which the axopodial axonemes lie (Febvre, 1971, 1981; Febvre and Febvre-Chevalier, 1982). Rapid contraction of the myonemes is a Ca^{++}-dependent process which is abolished in Ca^{++}-free solutions or in the presence of Ca^{++} channel blockers (Co^{++}, Mn^{++}, La^{+++}) in normal SW. At high extracellular Ca^{++} (40 mM instead of the 10 mM in normal SW), myonemes exhibit a series of fast, spontaneous contraction-relaxation cycles. Although no electrophysiological data are available, these facts suggest that myoneme contractility is dependent upon external Ca^{++} and that it enters via a selective membrane conductance during myoneme contraction (Febvre and Febvre-Chevalier, 1989).

5.3. Contractile Ciliates

Heterotrich and peritrich ciliates can respond to external stimuli by contraction. They use special nanofilament-based organelles called spasmonemes and myonemes.

Heterotrich ciliates are sensitive to mechanical, electrical and photic stimuli (Wood, 1970, 1982; Kim et al., 1984; Matsuoka and Shigenaka, 1986; Fabczac and Fabczac, 1988). In *Stentor*, the negative phototactic response is correlated with ciliary reversal. In *Blepharisma*, photic stimulation elicits elongation of the cell body, while mechanical or electrical stimulation evokes rapid contraction. Contraction-relaxation cycles involve two antagonistic cytoskeletal systems: bundles of Mts, called "Km fibers" and spasmin-like bundles of nanofilaments named myonemes (Huang and Pitelka, 1973; Matsuoka and Shigenaka, 1986). Membrane excitability has been mainly studied in *Stentor* (Wood, 1970, 1982) (Table 1). The resting membrane potential is between -49 mV and -56 mV, is K^+-dependent, and involves a minor Cl^--dependence. Depolarizing receptor potentials can be generated by mechanical stimulation. Mechanical or electrical stimuli evoke Ca^{++} and Cl^--dependent action potentials (amplitude: 65-75 mV). Substitution of Ca^{++} by Ba^{++} evokes an all-or-none depolarization, indicating that Ba^{++} mimics and emphasizes the effects of Ca^{++}. No hyperpolarizing receptor potential has been reported (Table 2).

Peritrich ciliates (*Vorticella*, *Carchesium*, *Zoothamnium*) are marine or fresh-water, solitary or colonial ciliates. They are fixed on hard substrates by a long, contractile peduncle. The oral area of the zooid has a prominent peristome, bordered by membranelles. In fresh-water species, the cell body includes a contractile vacuole. The contractile organelle (spasmoneme) in the peduncle is made of 2-5 nm nonactin

Table 1. Properties of resting and mechanoreceptor potentials in protists

	MEMBRANE POTENTIAL		MECHANORECEPTOR POTENTIAL	
	Value (mV)	Ionic Dependence	Depolarizing	Hyperpolarizing
CILATES				
Nassophorida				
Paramecium	-20 to -40[1]	[1]K^+/Ca^{++}	Ca^{++}-dependent[2]	K^+-dependent[3]
Oxytricha				
Stylonychia	-51 ±1.6[4]	[5]K^+/Ca^{++}	$Ca^{++}(Mg^{++})$-dependent[6]	voltage- and K^+-dependent[4]
Heterotricha				
Stentor	-48 to -56[7]	[7]$K^+ > Ca^{++} > Cl^-$	$Ca^{++} > Cl^-$ volt-dependent[7] habituation[6,9]	not present[7]
Blepharisma	4.1 ±1.7[10]	[10]K^+/Na^+		
Fabrea	50 to 60[11]			
Peritricha				
Vorticella	-10 to -50[12]			not present[12]
Zoothamnium	-30 ±12[13]			not present[13]
HELIOZOANS				
Echinosphaerium	-20 to -50[14]		Ca^{++} ?	
Actinocoryne	-78 ±6[15]	[15]K^+	$Na^+ > Ca^{++}$ [15]	not present[15]

[1]Eckert and Brehm, 1979.
[2]Eckert et al., 1972.
[3]Naitoh and Eckert, 1973.
[4]Deitmer, 1981.
[5]de Peyer and Machemer, 1977.
[6]de Peyer and Deitmer, 1960.
[7]Wood, 1982.
[8]Wood, 1988a.
[9]Wood, 1988b.
[10]Fabczak and Fabczak, 1988.
[11]Dryl et al., 1982.
[12]Shiono et al., 1980.
[13]Moreton and Amos, 1979.
[14]Nishi et al., 1986.
[15]Febvre-Chevalier et al., 1986.

filaments (i.e. spasmin, a 20 kD molecular weight, Ca^{++}-binding protein; Amos et al., 1975). Contraction occurs spontaneously or after mechanical, electrical, photic or chemical stimulus. Contraction velocity is up to 200 lengths.sec^{-1} (Routledge et al., 1976). In *Vorticella*, the resting membrane potential is unstable (-10 to -50 mV) and K^+-dependent with a minor Ca^{++} component (Table 1). An all-or-none, transient, Ca^{++}-dependent membrane depolarization (amplitude: 38 ±13 mV) can occur spontaneously or following mechanical or electrical stimulation. Small depolarizing and hyperpolarizing potential changes have been recorded during contraction of the contractile vacuole (Moreton and Amos, 1979; Shiono et al., 1980) (Table 2).

In Suctorians, the resting membrane potential lies between -20 and -30 mV. Spontaneous, brief hyperpolarizing or depolarizing potentials (0.2 - 2.0 sec duration, 8 mV amplitude) were recorded that correspond to the discharge of contractile

Table 2. Properties and function of electrogenic responses in protists

	Membrane Responses	Trigger	Ionic Basis	Behavioral Change
CILIATES				
Nassophorida				
Paramecium	Regenerative in normal medium[1]; All-or-none in Ba^{++} [2] or EGTA injection[3]	Current injection	Ca^{++}	Ciliary reversal
Oxytricha				
Stylonychia	Graded, two-peak action potential[4]	Current injection	Ca^{++}/Ba^{++} Sr^{++}	Cilary reversal
Heterotricha				
Stentor	Action potential[5]	Mechanical, current injection	Ca^{++}	Cilary reversal, contraction
Blepharisma	Action potential[7]	Current injection	?	Cilary reversal
Fabrea	Action potential in Ca^{++}, Ba^{++} [8] and TEA[9]	Spontaneous, current injection	? $Ca^{++}(Ba^{++})$	"Periodic ciliary reversal"
Peritricha				
Vorticella	"Large pulse"[10]	Spontaneous	?	Zoid contraction
Zoothamnium	"Action potential"[11]	Mechanical	?	Contraction
HELIOZOANS				
Echinosphaerium	Action potential[12]	Spontaneous, mechanical, current injection	Ca^{++}	Stays quiescent
Actinocoryne	Action potential[13]	Current injection, mechanical	Na^{+}/Ca^{++}	Contraction

[1]Naitoh et al., 1972.
[2]Naitoh and Eckert, 1968.
[3]Satow, 1977.
[4]de Peyer and Machemer, 1977.
[5]Wood, 1982.
[6]Wood, 1970.
[7]Fabczac and Fabczac, 1988.
[8]Dryl et al., 1982.
[9]Kubalsky, 1987.
[10]Shiono et al., 1980.
[11]Moreton and Amos, 1979.
[12]Nishi et al., 1986.
[13]Febvre-Chevalier et al., 1986.

vacuoles, but not to tentacle contraction (Butler and McCrohan, 1987). Calcium, which is needed for the slow contraction of tentacles, may be provided by the IP_3 route (Evans et al., 1988; Evans et al., 1989).

5.4. Free-living Heliozoa

Nishi et al. (1986, 1988) recorded spontaneous action potentials that occur in the absence of apparent cell contractility in the free-living heliozoans, *Echinosphaerium* and *Actinosphaerium*. These action potentials can be evoked by electrically or mechanically (Table 2). They can also be elicited at the end of a long-lasting hyperpolarizing potential that is associated with contraction of the contractile vacuole. As described in other protists, these action potentials are pure Ca^{++}-spikes, thus confirming the uncommon properties of membrane electrogenesis in the heliozoan *Actinocoryne*.

As in most other protists, rapid contraction in *Actinocoryne* is clearly correlated with a Ca^{++} influx during the depolarizing phase of the action potential (Table 2). However, the action potential itself is very unusual among those in the protista: 1) In *Actinocoryne* this action potential is a very fast membrane response which can overshoot the zero potential by 50 mV or more. 2) The depolarizing phase of this action potential is mainly Na^+-dependent, but TTX-insensitive, and also involves a minor Ca^{++}-component, whereas in other protists studied to date, action potentials are pure Ca^{++} spikes. Evidence for a small Na^+ conductance has been found in some *Paramecium* mutants (Hinrichsen and Saimi, 1984). However, activation of this Na^+ conductance was a consequence of Ca^{++} entry. Such a possibility can be discarded in *Actinocoryne*, since the Na^+-dependent action potential was observed at an external Ca^{++} concentration below 10^{-9} M.

Although voltage clamp experiments are lacking in *Actinocoryne*, examination of the ionic bases of the action potentials suggests the existence of at least 5 separate membrane conductances: 1) a mechano-sensitive Na^+ conductance, 2) a TTX-insensitive, voltage-activated Na^+ conductance, 3) a voltage-activated, Co^{++}-sensitive Ca^{++} conductance, 4) a Ca^{++}-activated K^+ conductance, 5) a voltage-activated K^+ conductance, sensitive to TEA and 4-AP.

6. Correlation Between Membrane Excitability and Locomotion in Protists

6.1. Flagellates

Flagellates and unicellular green algae respond to stimulation by modifying the form, amplitude and frequency of the flagellar beating. This process partly results from sliding interactions between Mts and dynein-ATPase (Goodenough and Heuser, 1989; Satir, 1989), and partly from bundles of nonactin filaments (flagellar rootlets, flagellar links made of a calcium-binding protein, centrin) whose contraction-relaxation cycles change the angle between flagella (Salisbury et al., 1984; Mcfadden et al., 1987). The resting membrane potential and correlations between Ca^{++}-dependent membrane depolarization and "ciliary" reversal were first studied in the parasitic flagellate *Opalina ranarum* (considered until recently as belonging to the ciliates, Small and Lynn, 1985) (Kinosita, 1954). It was shown in *Chlamydomonas* and *Euglena* that initiation and directional control of flagellar beating depended on changes in the internal Ca^{++} concentration (Schmidt and Eckert, 1976; Nichols and Rickmenspoel, 1978) and, more recently, that any responses to Ca^{++} may be mediated by calmodulin (Otter, 1989). One can presume that rapid reorientation of the basal bodies is caused by transient Ca^{++} influx across the cell membrane, in reponse to stimulation.

6.2. Amoebae and Slime Molds

Amoebae are relatively insensitive to electrical stimulation (Tasaki and Kamiya, 1964) while mechanical stimuli induce rapid contraction of the stimulated pseudopod. The locomotor pattern is widely modified by chemicals and pH (Braatz-Schade, 1978) and, recently, cAMP was shown to induce a rapid chemotactic response (Fukui and Yumura, 1986).

Basic electrophysiological data were mostly known about 30 years ago (Naitoh, 1982). The resting membrane potential is around -80 mV and is mainly K^+-dependent. Current pulses induce depolarizing responses that last as long as the stimulus. Spontaneous membrane potential changes called spike potentials (amplitude: 20 mV, duration 1-20 sec) were found to be correlated with inward Ca^{++} currents (Nuccitelli, et al., 1977).

Amoeboid movement is a complex process that involves pseudopod formation and extension, and cell substrate adhesion. Actin polymerization that accompanies the motile response was shown to be one of the major driving forces for pseudopod formation (Taylor, 1976; Taylor and Condeelis, 1979; Schliwa, 1986). Locomotion is stopped by inhibitors of actin microfilaments (cytochalasin, NBD-phallacidin) and monoclonal antibodies that dissociate the actin filaments (Condeelis et al., 1988; Kiehart and Pollard, 1984). Since regulation of any movement that depends on the actomyosin system requires Ca^{++}, locomotion is inhibited by Ca^{++} channel blockers and drugs that modify the Ca^{++} storing capacity of membranes (Braatz-Schade, 1978; Götz von Olenhusen and Wohlfarth-Bottermann, 1979). Amoeboid movement is also inhibited by poisons of the cAMP-dependent protein kinase and the inositol-phosphate pathways (Thiery et al., 1988).

6.3. Free-living Ciliates

Free-living ciliates are very excitable protists that have differentiated various motor organelles (cilia, membranelles, cirri). They respond to stimuli by rapidly changing the direction, amplitude and speed of ciliary beating.

Bioelectric processes have been studied in the following taxa: oxytrichs, (*Stylonychia*), nassophorids (*Paramecium*, *Euplotes*), and hymenostomes (*Tetrahymena*). In these ciliates, the cell body bears rows of somatic cilia and the oral cavity is bordered with polykinetid appendages or so-called membranelles. In *Stylonychia*, *Tetrahymena* and *Euplotes* the ventral face bears cirri that are used to "walk" on the substratum (for nomenclature and taxonomy, see Small and Lynn, 1985).

The resting membrane potential of *Paramecium* was first measured by Kamada (1934). It is between -20 and -50 mV and mainly K^+-dependent (Table 1). Correlations between changes in membrane potential and locomotor behavior were also recorded (Kinosita et al., 1964). In this genus, depolarizing current pulses evoke depolarizing action potentials which trigger ciliary reversal and backward locomotion by way of a transient increase in the Ca^{++} concentration in the cilia (Naitoh, 1968). Repolarization results from an outward K^+ current. Electrical stimulation of this type mimics mechanical stimulation of the anterior region, which evokes a graded Ca^{++}-dependent, depolarizing receptor potential, followed by an action potential that induces backward locomotion (Eckert et al., 1972; Naitoh et al., 1972). Conversely, stimulation of the posterior region elicits a hyperpolarizing receptor potential carried by K^+ ions, that triggers forward locomotion (Naitoh and Eckert, 1973) (Table 2). In *Paramecium* and *Stylonychia*, the repolarization phase of the action potential is prolonged by TEA and 4AP (Eckert et al., 1972; Machemer and Deitmer, 1987). Further analysis of the transmembrane ionic currents in *Paramecium* has been carried out with voltage and patch clamp techniques (Ogura and Machemer, 1980) and with the use of behavioral mutants (Kung et al., 1975; Kung and Saimi, 1982, 1985; Hinrichsen et al., 1985; Saimi and Kung, 1988). Both of these approaches distinguished diverse inward and outward currents mainly carried by Ca^{++}or substitutes (Sr^{++}, Ba^{++}), and by K^+.

Finally, it has been recently shown that the membrane potential can regulate intracellular cAMP in *Paramecium* and that Ca^{++} and cAMP have antagonistic actions. Cyclic AMP levels may regulate beat frequency, while Ca^{++} regulates reversal (Bonini et al., 1986).

7. Conclusion

Although protists are single cell microorganisms, they fulfill most functions allotted to specialized cells in multicellular organisms. Protists are sensory cells that respond to mechanical stimulation with depolarizing or hyperpolarizing receptor potentials. Like nerve cells, their membranes are equipped with voltage-gated ionic channels. Action potentials are elicited when the cell membrane is depolarized up to a threshold level by either receptor potentials or by current injection. Likewise, protists are "effector" cells; action potentials elicit changes in motile behavior or trigger contractile activity.

Even if certain protists have developed very distinct organelles or systems, they all generally use similar elements and metabolic routes. For instance, motile behavior results from interactions between various fibrillar proteins, Ca^{++}, ATP or other related nucleotides, and Ca^{++}-modulated proteins. Calcium, provided by the external medium, enters the cell during the rising phase of the action potential. In most protists, Ca^{++} ions are responsible for carrying ionic currents and coupling membrane excitation to motile response. Bioelectric responses appear generally well adapted to the mode of life. As a matter of fact, the sole response observed in protozoans that are fixed to substrates is a depolarizing response which is closely associated with the only motile response they can perform: cell contraction. In contrast, in free-living ciliates, which need a relatively large pattern of motile activities (avoidance, forward, normal and amplified locomotions, backward swimming, "walking"), the electrical responses are generally correlated with relatively less precise membrane domains. Although posterior stimulation produces a hyperpolarizing response and evokes forward beating, while anterior stimulation produces a depolarizing response and ciliary reversal, there are gradients of mixed conductances between these areas that allow modulation of the response to the stimulus.

Such adaptability in protists is correlated with a very large variety of ionic channels and membrane currents, as exemplified in ciliates (Naitoh, 1982; Machemer and Deitmer, 1987). Furthermore, the same protist can show several distinct currents carried by the same ion for distinct motile activities. In *Stylonychia* for instance, the voltage-dependent inward current underlying the depolarizing mechanoreceptor potential is carried by Ca^{++}(current I) and localized at the membrane cirri, while inward Ca^{++} current (current II) is localized at the oral membranelle surface (Deitmer, 1986).

With the exception of *Actinocoryne*, where the action potential is produced by separate Na^{+} and Ca^{++} conductances, action potentials in protists are exclusively Ca^{++}-dependent. Thus, membrane activity in protists appears to be similar to that reported in muscle and nerve membranes of some invertebrates where action potentials are pure Ca^{++} spikes.

Why membrane electrogenesis in *Actinocoryne* displays a strong Na^{+} and TTX-insensitive component in addition to a minor Ca^{++}-dependent component is puzzling. In general, the kinetics of activation for Ca^{++} channels are slower than for Na^{+} channels. On the other hand, the activation threshold for Ca^{++} channels is higher than for the Na^{+} ones. Thus, membrane depolarization is faster in cells equipped with both types of channels than in those having only Ca^{++} channels, and it is possible that in *Actinocoryne* the Na^{+} conductance serves to rapidly depolarize the cell, thereby facilitating Ca^{++} entry.

In this paper, we have examined the biological and electrophysiological bases of membrane electrogenesis and rapid contraction. However, *Actinocoryne* displays other patterns of movements (Febvre-Chevalier, 1987); for instance, the slow contraction that takes place in some conditions is presently being studied and should give useful information on unstable microtubule-based contractility.

ACKNOWLEDGEMENTS. We are grateful to Dr. M. Suffness from Dept. of Health and Human Services, Bethesda, Maryland, USA for the generous gift of taxol.

References

Amos, W. B., Routledge, L. M., and Yew, F. F., 1975, Calcium-binding proteins in a vorticellid contractile organelle, *J. Cell Sci.* **19**:203-213.

Andrivon, C., 1988, Membrane control of ciliary movement in ciliates, *Biol. Cell.* **63**:133-142.

Braatz-Shade, K., 1978, Effects of various substances and cell shape, motile activity and membrane potential in *Amoeba proteus*, *Acta Protozool.* **17**:163-176.

Bonini, N. M., Gustin, M. C., and Nelson, D. L., 1986, Regulation of ciliary motility by membrane potential in *Paramecium*: A role for cyclic AMP, *Cell Motility and the Cytoskeleton* **6**:256-272.

Butler, R. D., and McCrohan, C. R., 1987, Stimulus-response coupling in the contraction of tentacles of the suctorian protozoon *Heliophrya erhardi*, *J. Cell Sci.* **88:**121-127.

Cachon, J., and Cachon, M., 1982, Induction of unusual structures after a microtubule disassembly in the protozoan *Sticholonche zanclea*, in: *Microtubules in Microorganisms*, pp. 325-339 (P. Cappuccinelli and N. R. Morris, eds.), Marcel Dekker, Inc., New York, Basel.

Cohen, J., Garreau de Loubresse, N., and Beisson, J., 1984, Actin microfilaments in *Paramecium*: localization and role in intracellular movements, *Cell Motility* **4:**443-468.

Davidson, L. A., 1975, *Studies of the actinopods* Heterophrys marina *and* Ciliophrys marina: *Energetics and structural analysis of their contractile axopodia, general ultrastructure and phylogenetic relationships*, Ph.D. thesis, University of California at Berkeley, pp. 1-163.

de Brabander, M. J., Geuens, G., Nuiden, R., Willebords, R., and de Mey, J., 1981, Taxol induces the assembly of free microtubules in living cells and blocks the organizing capacity of the centrosomes and kinetochores, *Proc. Natl. Acad. Sci. USA* **78:**5608-5612.

de Peyer J. E., and Deitmer, J. W., 1980, Divalent cations as charge carriers during two functionally different membrane currents in the ciliates *Stylonychia*, *J. exp. Biol.* **1988:**73-89.

de Peyer J. E., and Machemer, H., 1977, Membrane excitability in *Stylonychia*. Properties of the two-peak regenerative calcium response, *J. Comp. Physiol.* **121:**15-32.

Deitmer, J. W., 1981, Voltage and time characteristics of the potassium mechanoreceptor current in the ciliate *Stylonychia*, *J. Comp. Physiol.* **141:**173-182.

Deitmer, J. W., 1986, Voltage dependence of two inward currents carried by calcium and barium in the ciliate *Stylonychia mytilus*, *J. Physiol. (Lond.)* **380:**551-574.

Detmers, P. A., Carboni, J. M., and Condeelis, J., 1985, Localization of actin in *Chlamydomonas* using antiactin and NBD phallacidin, *Cell Motility* **5:**415-430.

Dryl, S., Demar-Gervais, C., and Kubalski, A., 1982, Contribution to studies on the role of external cations in excitability of marine ciliate *Fabrea salina*, *Acta Protozoologica* **21:**55-59.

Dustin, P., 1984, in: *Microtubules* (P. Dustin, ed.), Springer Verlag, Berlin, Heidelberg, New York.

Eckert, R., and Brehm, P., 1979, Ionic mechanisms of excitation in *Paramecium*, *Ann. Rev. Biophys. Bioeng.* **8:**353-383.

Eckert, R., Naitoh, Y., and Friedman, K., 1972, Sensory mechanisms in *Paramecium*. I. Two components of the electric response to mechanical stimulation of the anterior surface, *J. exp. Biol.* **56:**683-694.

Eckert, R., Naitoh, Y., and Machemer, H., 1976, Calcium in the bioelectric and motor functions of *Paramecium*, in: *Calcium in Biological Systems*, pp. 233-255 (Duncan, ed.), Cambridge University Press, London.

Eckert, R., and Sibaoka, T., 1967, Bioelectric regulation of tentacle movement in a dinoflagellate, *J. exp. Biol.* **47:**433-446.

Edds, K. T., 1981, Cytoplasmic streaming in a heliozoan, *Biosystems* **14:**371-376.

Evans, R. L., McCrohan, C. R., and Butler, R. D., 1988, Tentacle contraction in *Heliophrya erhardi* (Suctoria): the role of inositol phospholipid metabolites and cyclic nucleotids in stimulus-response coupling, *Exp. Cell Res.* **177:** 382-390.

Evans, R. L., Butler, R. D., McCrohan, C. R., and Cuthbertson, K. S. R., 1989, Control of tentacle contraction in suctorian protozoa, *Intern. Congr. Protozool.* 10-17 July 1989, Tsukuba, Japan. S5-5, p. 53.

Fabczac, S. and Fabczac, H. (1988), The resting and action membrane potentials of the ciliate *Blepharisma japonicum*, *Acta Protozool.* **27:**117-124.

Febvre, J., 1971, Le myonème d'acanthaire: Essai d'interprétation ultrastructurale et cinétique, *Protistologica* **7:**379-391.

Febvre, J., 1981, The myoneme of the acantharia (Protozoa): A new model of cellular motility, *Biosystems* **14:**327-336.

Febvre, J., and Febvre-Chevalier, C., 1982, Motility processes in acantharia (Protozoa). I. Cinematographic and cytological study of the myonemes. Evidence for a helix-coil of the constituent filaments, *Biol. Cell* **44:**283-304.

Febvre, J., and Febvre-Chevalier, C., 1989, Motility processes in acantharia (Protozoa). III. Calcium regulation of the contraction-relaxation cycles of in vivo myonemes, *Biol. Cell* **67**(2), in press.

Febvre-Chevalier, C., 1973, Un nouveau type d'association des microtubules axopodiaux chez les héliozoaires, *Protistologica* **9**:35-43.

Febvre-Chevalier, C., 1980, Behaviour and cytology of *Actinocoryne contractilis* nov. gen. nov. sp. A new stalked heliozoan (Centrohelidia). Comparison with the other related genera, *J. Mar. Biol. Ass. U.K.* **60**:909-928.

Febvre-Chevalier, C., 1981, Preliminary study of the motility processes in the stalked heliozoan *Actinocoryne contractilis*, *Biosystems* **14**:337-343.

Febvre-Chevalier, C., 1987, *Ultrastructure et critères taxonomiques. Excitabilité et contractilité cellulaire d'héliozoaires marins (Protozoa, Actinopoda)*, Thèse Doctorat Es-Sciences, Office National des Thèses, Grenoble.

Febvre-Chevalier, C., Bilbaut, A., Bone, Q., and Febvre, J., 1986, Sodium-calcium action potential associated with contraction in the heliozoan *Actinocoryne contractilis*, *J. exp. Biol.* **122**:177-192.

Febvre-Chevalier, C., and Febvre, J., 1980, Cytophysiology of motility in a pedunculated heliozoan, *Film, SFRS*, Paris.

Febvre-Chevalier, C., and Febvre, J., 1986, Motility mechanisms in the actinopods (Protozoa): A review with particular attention to axopodial contraction-extension and movement of nonactin filament systems, *Cell Motility and the Cytoskeleton* **6**:198-208.

Fukui, Y., and Yumura, S., 1986, Acto-myosin dynamics in chemotactic amoeboid movement of *Dictyostelium*, *Cell Motility and the Cytoskeleton* **6**:662-673.

Gibbons, I. R., 1989, Microtubule-based motility: An overview of a fast-moving field, in: *Cell Movement*, vol. 1, The dynein ATPases, pp. 3-22 (F. D. Warner, P. Satir, and I. R. Gibbons, eds.), Alan R. Liss, New York.

Goodenough, U. W., and Heuser, J. E., 1989, Structure of the soluble and in situ ciliary dyneins visualized by quick-freeze deep-etch microscopy, in: *Cell Movement*, vol. 1, The dynein ATPases, pp. 121-140 (F. D. Warner, P. Satir, and I. R. Gibbons, eds.), Alan R. Liss, New York.

Götz von Olenhusen, K., and Wohlfarth-Bottermann, K. E., 1979, Effects of caffeine and D_2O on persistence and *de novo* generation of intrinsic oscillatory contraction automaticity in *Physarum*, *Cell Tissue. Res.* **197**:479-499.

Hausmann, K., Linnenbach, M., and Patterson, D. J., 1983, The effect of taxol on microtubular arrays. *In vivo* effects on heliozoan axonemes, *J. Ultrastr. Res.* **82**:212-220.

Hinrichsen, R. D. and Saimi, Y., 1984, A mutation that alters properties of Ca^{2+} channels in *Paramecium tetraurelia*, *J. Physiol. (Lond.)* **351**:397-410.

Hinrichsen, R. D., Saimi, Y., Ramanathan, R., Burgess-Cassler, A., and Kung, C., 1985, A genetic and biochemical analysis of behavior, in: *Sensing and Response in Microorganisms*, pp. 147-157 (M. Eisenbach, and M. Balaban, eds.), Elsevier, New York.

Huang, B., and Pitelka, D. R., 1973, The contractile process in the ciliate *Stentor coeruleus*. The role of microtubules and filaments, *J. Cell Biol.* **57**:704-728.

Huitorel, P., 1988, From cilia and flagella to ultracellular motility and back again: A review of a few aspects of microtubule-based motility, *Biol. Cell.* **63**:249-258.

Kamada, T., 1934, Some observations on potential differences across the ectoplsm membrane of *Paramecium*, *J. exp. Biol.* **11**:94-102.

Kiehart, D. P., and Pollard, T. D., 1984, Inhibition of *Acanthomoeba* actomyosin-II ATPase activity and mechanochemical function by monoclonal antibodies, *J. Cell Biol.* **99**:1024-1033.

Kim, I. H., Prusti, R. K., Song, P. S., Haeder, D. P., and Haeder, M., 1984, Phototaxis and photophobic response in *Stentor coeruleus*. Action spectrum and role of calcium fluxes, *Biochim. Biophys. Acta.* **799**:298-304.

Kinosita, H., 1954, Electric potentials and ciliary response in *Opalina*, *J. Fac. Sci. Tokyo Univ. IV* **7**:1-14.

Kinosita, H., Dryl, S., and Naitoh, Y., 1964, Changes in the membrane potential and the response to stimuli in *Paramecium*, *J. Fac. Sci. Tokyo Univ. IV* **10**:291-301.

Kubalski, A., 1987, The effects of tetraethylammonium on the excitability of marine ciliate *Fabrea salina*, *Acta Protozoologica* **26**:135-144.

Kung, C., Chang, S. Y., Satow, Y., Van Houten, J., and Hansma, H., 1975, Genetic dissection of behaviour in *Paramecium*, *Science* **188**:898-904.

Kung, C., and Saimi, Y., 1982, The physiological basis of taxes in *Paramecium*, *Ann. Rev. Physiol.* **44**:519-534.

Kung, C., and Saimi, Y., 1985, Ca^{2+} channnels of *Paramecium*. A multidisciplinary study, in: *Current Topics in Membrane and Transport*, 23, pp. 46-66, Acad. Press.

Machemer, H., and Deitmer, J. W., 1987, From structure to behaviour: *Stylonychia* as a model system for cellular physiology, *Progress in Protistology* **2**:213-330.

Marsland, D., Tilney, L. G., and Hirshfield, M., 1971, Stabilizing effect of heavy water on the microtubular components and needle-like form of the heliozoan axopods. A pressure, temperature analysis, *J. Cell. Physiol.* **77**:187-194.

Matsuoka, T., and Shigenaka, Y., 1986, Elongation and contraction of *Blepharisma* evoked by mechanical or light stimulation, *Arch. Protistenk.* **131**:85-94.

McFadden, G. I., Schulze, D., Surek, B., Salisbury, J. L., and Melkonian, M., 1987, Basal body reorientation mediated by Ca^{2+}-modulated contractile protein, *J. Cell Biol.* **105**:903-912.

Moreton, R. B., and Amos, W. B., 1979, Electrical recording from the contractile ciliate *Zoothamnium geniculatum* Ayrton, *J. exp. Biol.* **83**:159-167.

Naitoh, Y., 1968, Ionic control of the reversal response of cilia in *Paramecium caudatum*. A calcium hypothesis, *J. Gen. Physiol.* **51**:85-103.

Naitoh, Y., 1982, Protozoa, in: *Electric Conduction and Behaviour in "Simple" Invertebrates*, pp. 1-48 (G. A. B. Shelton, ed.), Clarendon Press, Oxford.

Naitoh, Y., Eckert, R., 1968, Electrical properties of *Paramecium caudatum*: all or none electrogenesis, *Z. Vergl. Physiologie* **61**:453-472.

Naitoh, Y., Eckert, R., 1973, Sensory mechanisms in *Paramecium*. II. Ionic basis of the hyperpolarizing mechanoreceptor potential, *J. exp. Biol.* **59**:53-65.

Naitoh, Y., Eckert, R., and Friedman, K., 1972, A regenerative calcium response in *Paramecium*, *J. exp. Biol.* **56**:667-681.

Naitoh, Y., and Oami, K., 1989, Bioelectric control of the tentacle movement in the dinoflagellate *Noctiluca*, *VIIIe Intern. Congr. Protozool.*, July 10-17, 1989, Tsukuba, Japan.

Nawata, T., and Sibaoka, T., 1986, Membrane potential controlling the initiation of feeding in the marine dinoflagellate *Noctiluca*, *Zool., Sci., Tokyo* **3**:49-58.

Nichols, K. M., and Rikmenspoel, R., 1978, Control of flagellar motility in *Euglena* and *Chlamydomonas*. Microinjection of EDTA, EGTA, Mn^{2+} and Zn^{2+}, *Exp. Cell Res.* **116**:33-40.

Nishi, T., Gotow, T., and Kobayashi, M., 1988, Changes in electrical connections during cell fusion in the heliozoan *Echinosphaerium akame*, *J. exp. Biol.* **135**:183-191.

Nishi, T., Kobayashi, M., and Shigenaka, Y., 1986, Membrane activity and its correlation with vacuolar contraction in the heliozoan *Echinosphaerium*, *J. Exp. Zool.* **239**:175-182.

Nuccitelli, R., Poo, M. N., and Jaffe, L. F., 1977, Relations between amoeboid movement and membrane-controlled electrical currents, *J. Gen. Physiol.* **69**:743-763.

Oami, K., Sibaoka, T., Naitoh, Y., 1988, Tentacle regulating potentials in *Noctiluca miliaris*. Their generation sites and ionic mechanisms, *J. Comp. Physiol.* **162**:179-186.

Ogura, A., and Machemer, H., 1980, Distribution of mechanoreceptor channels in the *Paramecium* surface membrane, *J. Comp. Physiol.* **135**:233-242.

Otter, T., 1989, Calmodulin and the control of flagellar movement, in: *Cell Movement*, vol. 1, The dynein ATPases, pp. 281-298 (F. D. Warner, P. Satir, and I. R. Gibbons, eds.), Alan R. Liss, New York.

Pollard, T. D., and Weihing, R. R., 1974, Actin and myosin and cell movement, *CRC Crit. Rev. Biochem.* **2**:1-65.

Roberts, T. M., 1987, Invited review: Fine (2-5 nm) filaments: New types of cytoskeletal structures, *Cell Motility and the Cytoskeleton* **8**:130-142.

Routledge, L. M., Amos, W. B., Yew, F. F., and Weis-Fogh, T., 1976, New calcium-binding contractile proteins, in: *Cell Motility*, pp. 93-114 (B. R. Goldman, T., Pollard, and J. Rosenbaum, eds.), Cold Spring Harbor Laboratory.

Saimi, Y., and Kung, C., 1988, Ion channels of *Paramecium*, yeast and *Escherichia coli*, in: *Current Topics in Membrane and Transport*, 33, pp. 1-11, Acad. Press, New York, London.

Salisbury, J. L., Baron, A., Surek, B., Melkonian, M., 1984, Stiated flagellar roots: Isolation and partial characterization of a calcium modulated contractile organelle, *J. Cell Biol.* **99**:962-970.

Satir, P., 1984, The generation of ciliary motion, *J. Protozool.* **31**:8-12.

Satir, P., 1989, Structural analysis of the dynein cross-bridge cycle, in: *Cell Movement*, vol. 1, The dynein ATPases, pp. 219-234 (F. D. Warner, P. Satir, and I.R. Gibbons, eds.), Alan R. Liss, New York.

Satir, B. H., Garofalo, R. S., Gilligan, D. M., and Maihle, N. J., 1980, Possible functions of calmodulin, in: *Calmodulin and Cell Functions* (D. N. Watterson and F. F. Vicenzi), *Ann. New York Acad. Sci.* **356**:83-91.

Satow, Y., 1978, Internal calcium concentration and potassium permeability in *Paramecium*. *J. Neurobiol.* **9**:81-91.

Schliwa, M., 1986, The cytoskeleton, in: *Cell Biology Monographs*, 13, pp. 1-323, Springer Verlag, Vienne, New York.

Schmidt, J. A., and Eckert, R., 1976, Calcium couples flagellar reversal to photostimulation in *Chlamydomonas reinhardtii*, *Science* **262**:713-715.

Shigenaka, Y., Roth, L. E., and Pihlaja, D. J., 1971, Microtubules in the heliozoan axopodium. III. Degradation and reformation after dilute urea treatment, *J. Cell Sci.* **8**:127-151.

Shigenaka, Y., Tadokoro, Y., Kaneda, M., 1975, Microtubules in protozoan cells. I. Effects of light metal ions on the heliozoan microtubules and their kinetic analysis. *Ann. Zool. Jap.* **48**:227-241.

Shiono, H., Hara, R., and Asai, H., 1980, Spontaneous membrane potential changes associated with the zoid and vacuolar contraction in *Vorticella convallaria*, *J. Protozool.* **27**:83-87.

Sibaoka, T., and Eckert, R., 1967, An electrophysiological study of the tentacle-regulating potential in *Noctiluca*, *J. exp. Biol.* **47**:447-459.

Small, E. B., and Lynn, D. H., 1985, Phylum Ciliophora Doflein, 1901, in: *An Illustrated Guide to the Protozoa*, pp. 393-575 (J. J. Lee, S. H. Hutner, and E. C. Bovee, eds.), Society of Protozoologists, Lawrence.

Stephens, R. E., and Stommel, E. W., 1989, Role of cyclic adenosin monophosphate in ciliary and flagellar motility, in: *Cell Movement, vol. 1, The dynein ATPases*, pp. 299-316 (F. D. Warner, P. Satir, and I. R. Gibbons, eds.), Alan R. Liss, New York.

Tasaki, I., and Kamiya, N., 1964, A study of electrophysiological properties of carnivorous amoebae, *J. Cell Comp. Physiol.* **63**:365-380.

Taylor, D. L., 1976, Motile model systems of amoeboid movement, in: *Cell Motility*, pp. 797-821 (R. Goldman, T. Pollard, and J. Rosenbaum, eds), CSH Conferences on Cell Proliferation. V. III.

Taylor, D. L., and Condeelis, J. S., 1979, Cytoplasmic structure and contractility in amoeboid cells, *Intern. Rev. Cytol.* **56**:57-144.

Thiery, R., Klein, R., and Tatischeff, I., 1988, Phorbol 12-myristate 13-acetate modulates the cAMP-induced light-scattering response of a *Dictyostelium discoideum* cell population, *FEB.* 149-153.

Tilney, L. J., 1968, Studies on the microtubules in heliozoa. IV The effect of colchicine on the formation and maintenance of the axopodia and the redevelopement of pattern in *Actinosphaerium nucleofilum* (Barrett), *J. Cell Sci.* **3**:549-562.

Tilney, L. J., Hiramoto, Y., and Marsland, D., 1966, Studies on the microtubules in Heliozoa. III. A pressure analysis of the role of these structures of the formation and maintenance of the axopodia of *Actinosphaerium nucleofilum* (Barrett), *J. Cell Biol.* **29**:77-86

Tilney, L. J., and Porter, K., 1967, Studies on the microtubules in the heliozoa. II. The effect of low temperature on these structures in the formation and maintenance of the axopodia, *J. Cell Biol.* **34**:327-358.

Wood, D. C., 1970, Electrophysiological studies of the protozoan *Stentor coeruleus*, *J. Neurobiol.* **1**:363-377.

Wood, D. C., 1982, Membrane permeabilities determining resting, action and mechanoreceptor potentials in *Stentor coeruleus*, *J. Comp. Physiol.* **146**:537-540.

Wood, D. C., 1988a, Habituation in *Stentor*. A response-dependent process, *J. Neurosci.* **8**:2248-2253.

Wood, D. C., 1988b, Habituation in *Stentor* produced by mechanoreceptor channel modification, *J. Neurosci.* **8**:2254-2258.

Chapter 18

Ion Channels and the Cellular Behavior of *Stylonychia*

JOACHIM W. DEITMER

1. Introduction

Unicellular organisms require some unique adaptations to their environment, such as the necessity to incorporate all functions needed for survival onto a single cell, and the ability to cope with their direct exposure to the outside world. Their unicellular existence imposes restrictions that multicellular organisms have circumvented by distributing different functions onto different cell types. This is most evident if we compare animal behavior with "cellular behavior" of protozoa. The sequence of events underlying behavior in animals includes sensory perception by sense organs, processing of the sensory information and translating this into motor programs by the nervous system, and movements carried out by activation of muscles. Thus, different types of cells and organs with varying degrees of specialization contribute to the behavioral performance of multicellular animals.

In unicellular organisms, all these principle functions have to be achieved by one cell. Nevertheless, protozoa can perceive stimuli and react to external stimuli by producing the appropriate behavior. These processes are relatively well studied in ciliates, particularly in *Paramecium* and *Stylonychia*. The latter is a hypotrich ciliate and belongs, in evolutionary terms, to the most advanced protozoa. These have a complex morphology, and the ciliary apparatus is more diverse than in any other ciliate order. The structure and location of the compound cilia, cirri and membranelles, provide the means for locomotion and food intake, respectively.

JOACHIM W. DEITMER • Abteilung für Allgemeine Zoologie, FB Biologie, Universität Kaiserslautern, Postfach 3049, D-6750 Kaiserslautern, Federal Republic of Germany.

The behavior of *Stylonychia*, as of other ciliates, is based on bioelectric mechanisms of stimulus detection and integration, electrical excitation, and electromotor coupling. These mechanisms resemble those of sensory receptor cells and of neurons found in animals, which is why ciliates have been called "swimming sensory cells" (Machemer and de Peyer, 1977), "swimming neurons" (Kung and Saimi, 1982) or "walking sensory neurons" (Machemer and Deitmer, 1987), to symbolize their functional abilities.

The key for understanding the cellular behavior of ciliates is located in the cell membrane, and is related to the excitability of this membrane. The structural entities of the electrical excitability are ion channels, which can be gated by mechanical deformation, by chemicals, or by changes in membrane potential. It is the aim of this article to review the central role of some of these ion channels in shaping the behavior of the ciliate *Stylonychia*.

2. Stimulus Reception

Of the many exogenous stimuli, only mechanical stimulation and reception have been studied in some detail in *Stylonychia*. Other stimulus modalities, such as chemical or light, have been investigated only in other ciliates (Doughty and Dryl, 1981; Song, 1981; Van Houten et al., 1981). Therefore, this section will describe exclusively some of the main processes occuring during the reception of mechanical stimuli.

In *Stylonychia*, mechanical stimuli to different parts of the cell produce a dual response, similar to that described for *Paramecium* (Jennings, 1906; Naitoh and Eckert, 1969; Naitoh, Eckert and Friedman, 1972). The cell discriminates between mechanical stimuli applied to the anterior and to the posterior part of the cell body by responding with a depolarizing and a hyperpolarizing receptor potential, respectively (de Peyer and Machemer, 1978a). In both cases the response causes the cell to "escape" from the source of mechanical stimulation, i.e. it moves forward, or accelerates forward swimming, when stimulated at the cell posterior, and moves backward, when receiving stimulus at the cell anterior. These behavioral responses indicate that the cell is able to discriminate between different sites of stimulus reception.

Voltage-clamp experiments, using the two intracellular microelectrode technique, have given further evidence for the bipolar nature of the mechanoelectrical response. An inward current is recorded upon anterior stimulation (Fig. 1 A), and an outward current upon posterior stimulation (Fig. 1 B). The nature of these receptor currents is determined by the ion selectivity of the channels activated by mechanical deformation of the cell membrane. The inward current, which underlies the depolarizing receptor potential, is largely carried by Ca^{++} and Mg^{++}, as well as Ba^{++} or Sr^{++}, if present (Fig. 1 A). This channel is thus selective for divalent cations (de Peyer and Deitmer, 1980). A small fraction of the current is carried by K^{+}, which flows

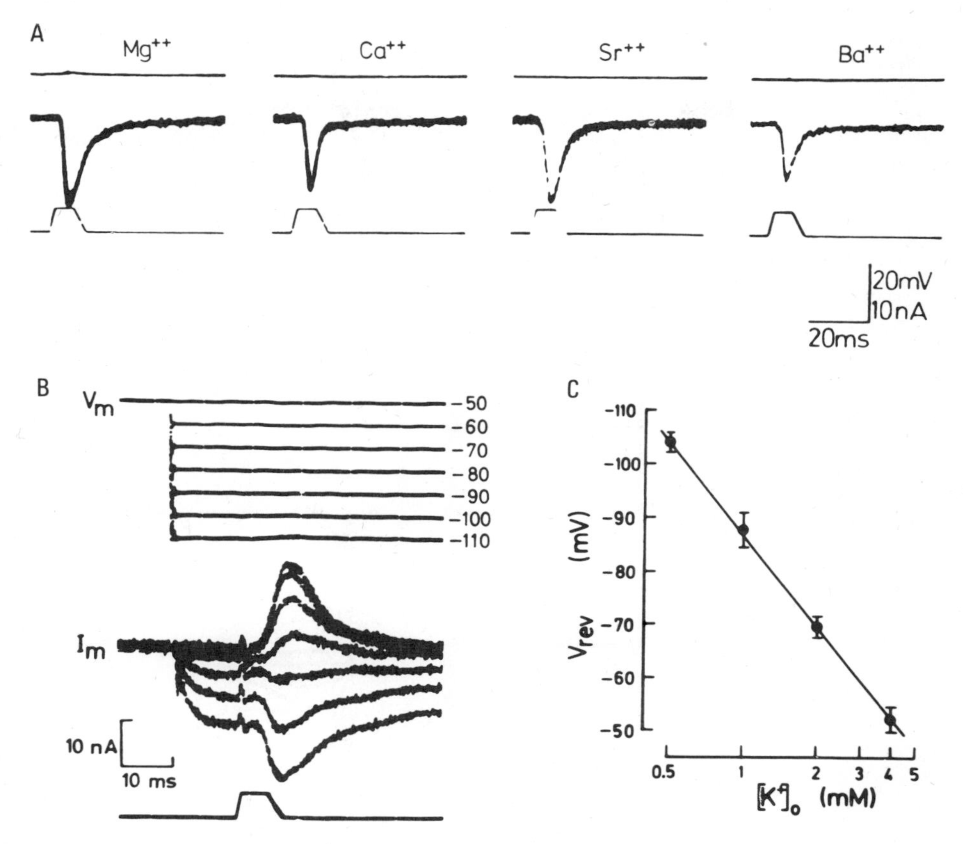

Figure 1. Receptor currents following mechanical stimulation. **(A).** Inward currents recorded during stimulation of the cell anterior are carried by the different divalent cations indicated. **(B).** Outward currents recorded during stimulation of the cell posterior at different membrane potentials. **(C).** The reversal of the outward receptor varies with the external K^+ concentration.

outward and, hence, partly short-circuits the inward current carried by divalent cations. Divalent cations which are known to be Ca^{++}-antagonists, such as Cd^{++}, Mn^{++}, and Co^{++}, inhibit the anterior mechanoreceptor current.

The outward current (Fig. 1 B), which underlies the hyperpolarizing receptor potential, is carried exclusively by K^+ (Deitmer, 1981, 1982). The reversal potential varies by 58 mV per tenfold change in the external K^+ concentration; it is identical to the K^+ equilibrium potential. The channel is impermeable to divalent cations as well as to Na^+, but can be carried by Rb^+. The K^+ channel blocker tetraethylammonium (TEA), but not tetramethylammonium, completely inhibits this current in millimolar concentrations (Deitmer, 1982). Both the amplitude and the kinetics of this K^+ mechanoreceptor current display a voltage dependence (Deitmer, 1981). The non-linearity of the receptor conductance may be of adaptive value for cells exposed

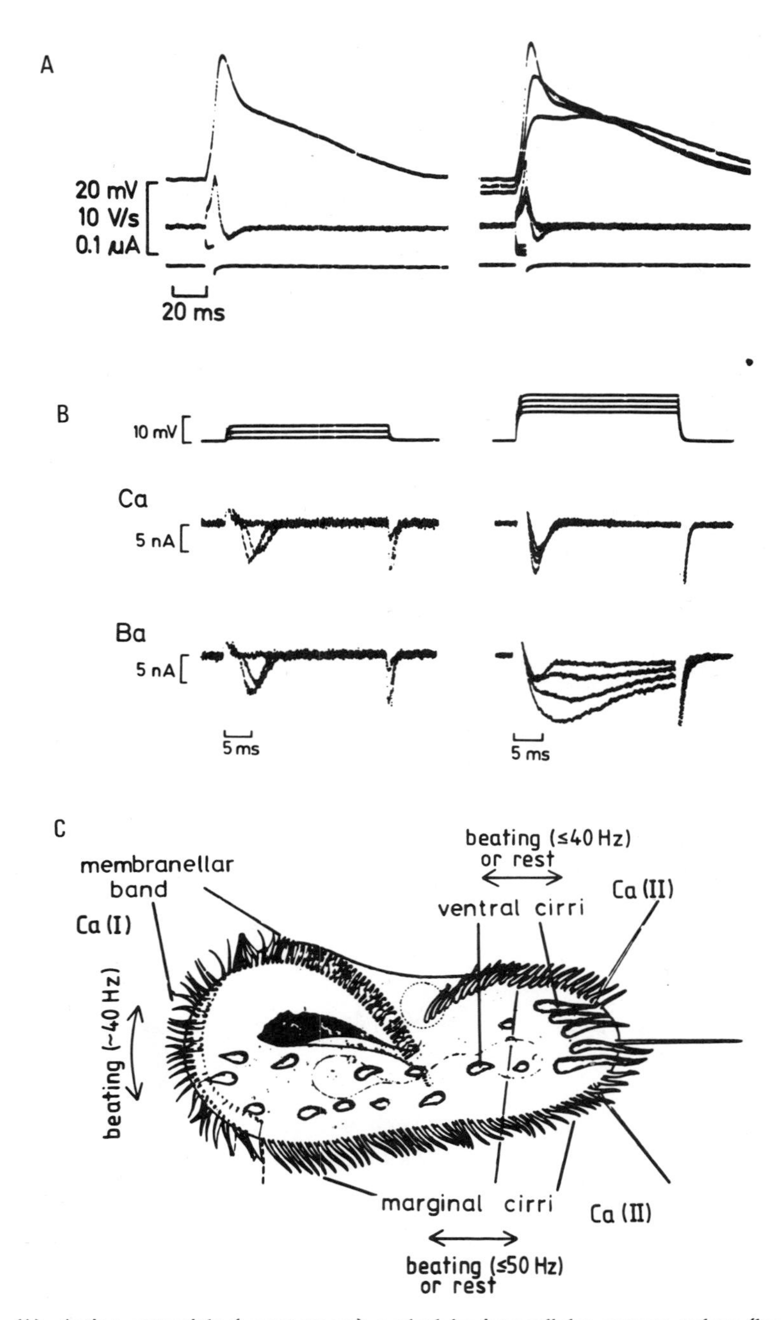

Figure 2. (A). Action potentials (upper traces) evoked by intracellular current pulses (lower traces) showing graded and all-or-none peaks; the first derivative is indicated by the middle traces. **(B).** Ca^{++} and Ba^{++} currents through voltage-dependent channel I (left column) and channel II (right column). **(C).** Ventral view of *Stylonychia* indicating the ciliary organelles, their maximum beating frequency, and the localization of Ca^{++} channels I and II.

to frequent stimulation; a hyperpolarizing posterior stimulus will, thus, be less effective during membrane depolarization.

3. Voltage-dependent Excitability

The action potential of *Stylonychia* has two components, a large, graded peak, and a second, smaller, all-or-none shoulder (Fig. 2 A). It occurs spontaneously, or it may be triggered by exogenous stimuli which depolarize the membrane; the depolarization needed to evoke an action potential is only a few millivolts in *Stylonychia*. In a free-swimming cell an action potential elicits the reversal of ciliary beating and hence a backward movement of the cell.

Both components of the action potential are dependent on extracellular Ca^{++} (de Peyer and Machemer, 1977; de Peyer and Deitmer, 1980). In voltage-clamp experiments, two types of inward current (I and II) carried by Ca^{++} (or Ba^{++}) can be recorded upon membrane depolarization (Fig. 2 B). The characterization of these two currents (Deitmer, 1984, 1986) disclosed a number of different properties of the corresponding voltage-dependent Ca^{++} channels. Among these are the range of membrane potentials for their activation, their mode of inactivation, their kinetics, their pharmacology indicated by different blocking agents, and their location in the membrane. Some of these differences became evident when comparing the currents carried by Ca^{++} and Ba^{++} (Fig. 2 B). The kinetics of activation and inactivation in Ca^{++} and Ba^{++} are similar for current I, but grossly dissimilar for current II when carried by Ba^{++}. In addition, the outward current, presumably carried by K^{+}, is activated when current II is carried by Ca^{++}, but not when carried by Ba^{++}. Injection of the Ca^{++} chelator EGTA into the cell considerably reduces the outward current, suggesting that it is largely Ca^{++}-dependent. Current I, however, never activates an outward current.

4. The Ca^{++} Channels

Stylonychia displays three Ca^{++} channels, one activated by mechanical stimulation, and two activated by membrane depolarization. In order to understand their significance for sensory reception and electromotor coupling in this cell, some of their properties have to be discussed in more detail.

Perhaps the most important feature of these three Ca^{++} channels is their location in the cell membrane (Deitmer, 1988). The mechanoreceptor Ca^{++} channel is located in the anterior cell body membrane. The two voltage-dependent Ca^{++} channels are located in the ciliary membranes (Deitmer, 1984), channel I in the membrane of the membranelles, and channel II in the membrane of the ventral and marginal cirri (Fig.

2 C). Hence, the Ca^{++} influx into the cell is separated, as indicated for the mechanosensory and the voltage-dependent currents. While the Ca^{++} transient produced by the mechanoreceptor current remains local at the inner side of the cell body membrane and has no influence on ciliary beating (de Peyer and Machemer, 1978b), the voltage-dependent Ca^{++} currents produce an intraciliary Ca^{++} transient which directly activates the axoneme. The sequence of events is sketched in figure 3. It is believed that both frequency and direction of ciliary beating is correlated with the rise in intraciliary Ca^{++}. During an action potential, when the voltage-dependent Ca^{++} currents flow across the ciliary membrane, the resulting increase in intracellular Ca^{++} results in a cirral stroke towards the cell anterior and backward locomotion of the cell.

The selectivities of the Ca^{++} channels correspond to their presumptive function. The main, and presumably only function, of the mechanoreceptor Ca^{++} current is to depolarize the membrane to reach the threshold for generating an action potential. Since the anterior mechanoreceptor channel is equally selective for Mg^{++} and for Ca^{++} (de Peyer and Deitmer, 1980), the mechanoreceptor current will be carried as much by Mg^{++} in fresh water, which usually contains similar amounts of Ca^{++} and Mg^{++} (Machemer and Deitmer, 1987). Mg^{++}, however, does not activate the ciliary machine (Naitoh and Kaneko, 1973; de Peyer and Machemer, 1982). Therefore, we believe that the nonselectivity of the anterior mechanoreceptor channel, with respect to divalent cations and the corresponding ion flux, is of no further consequence, because the ions merely act as charge carrier to depolarize the membrane.

This is different for the voltage-dependent Ca^{++} currents across the ciliary membranes, as described above. The selectivity of the voltage-dependent Ca^{++} channel is indeed more specific; Mg^{++} does not permeate the voltage-dependent channels.

All three Ca^{++} channels are also selective for Ba^{++} and Sr^{++} (de Peyer and Deitmer, 1980), but not for Co^{++}, Cd^{++}, or Mn^{++}, which inhibit current flow through the Ca^{++} channels. In the nominal absence of other divalent cations, Mg^{++} and even monovalent cations, such as Na^{+}, were found to permeate voltage-dependent channel I (Ivens, 1986). This behavior resembles that of vertebrate skeletal muscle, where Ca^{++} channels can lose their exclusive selectivity to certain divalent cations, when these are absent (Almers et al., 1984). Apparently, the binding of ions to the channel plays a role in Ca^{++} channel functioning. It seems possible that the different selectivity of the Ca^{++} channels is related to the different binding affinity of various permeant and blocking ions.

The inactivation of currents through the three types of Ca^{++} channels is due to different mechanisms. The decay of the Ca^{++} receptor current is exponetial. For the anterior receptor current, two time constants for inactivation, 1.8 ms and 7.2 ms, were determined when the current was carried by Ca^{++}, and a single time constant of 7.6 ms, when Mg^{++} carried the current (de Peyer and Deitmer, 1980). The current has a similar, single exponential time course when either Ba^{++} or Sr^{++} acted as charge carriers, suggesting that there is a component of the inactivation which is related to Ca^{++}. It is unclear, however, whether or not this phenomenon can be explained by

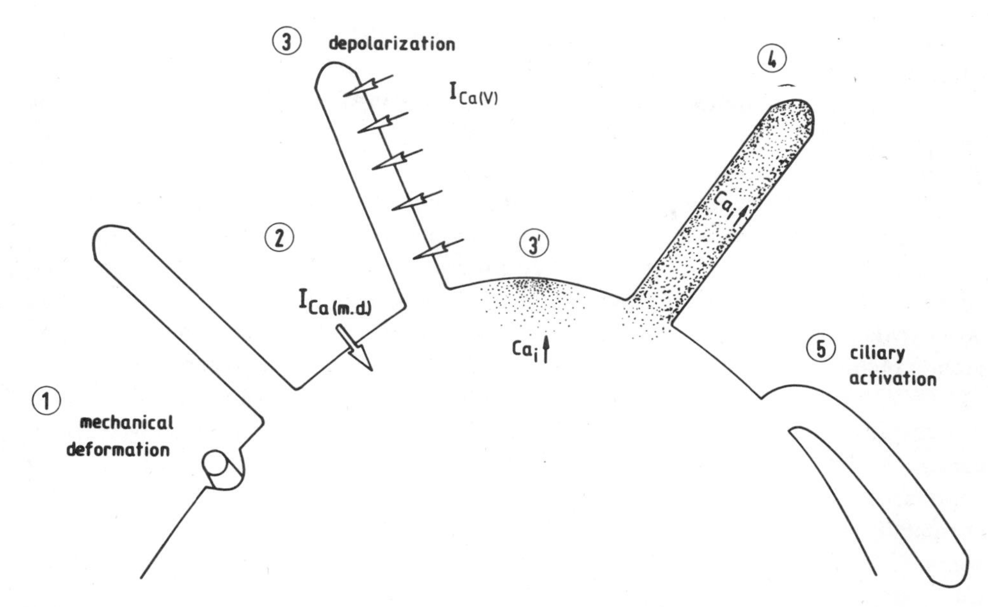

Figure 3. The sequence of events during senso-electro-mechanical coupling: mechanical stimulation of the cell anterior (1) produces a Ca^{++} receptor current (2), which leads to a depolarization and opening of Ca^{++} channels in the ciliary membrane (3), and a local Ca^{++} transient near the somatic membrane (3'). The Ca^{++} influx into the cilia results in a large increase of intraciliary Ca^{++} (4), which activates the ciliary axoneme (5).

Ca^{++}-dependent inactivation, as suggested for voltage-dependent Ca^{++} channels in many tissues (see Eckert and Chad, 1984).

Voltage-dependent current I decays with a similar time course when carried either by Ca^{++}, Sr^{++}, or Ba^{++}. This is in sharp contrast to the voltage-dependent current II, which decays rapidly in Ca^{++}- and in Sr^{++}, even in the presence of K^+ channel blockers, but very slowly and incompletely in Ba^{++} (Fig. 2 B). Inactivation of voltage-dependent current II appears to be due to the influx of Ca^{++}, and is, thus, largely Ca^{++}-dependent, with some voltage dependency; inactivation of current I, on the other hand, only depends on membrane potential.

5. The K^+ Channels

The second group of ion channels to be compared here are the different kinds of K^+ channels, of which there is one mechanoreceptor channel and at least three or four types of voltage-dependent K^+ channels. All these channels appear highly selective for K^+, particularly the mechanoreceptor channel.

The location of these K^+ channels is only partly defined in *Stylonychia* and other ciliates. The mechanoreceptor channel is located mainly in the posterior cell body membrane, but may extend, at much lower density, into the anterior cell body membrane (de Peyer and Machemer, 1978a). The cilia appear to be free of mechanoreceptor channels of either sort.

The voltage-dependent K^+ channels display different properties, which is why they were grouped into separate classes. The importance of these K^+ channels lies in their contributing significantly to the shape and duration of action potentials. It must be admitted, however, that our knowledge of the voltage-dependent K^+ channels in *Stylonychia* is still rather incomplete. The main component of the outward current upon membrane depolarization appears to be a Ca^{++}-dependent K^+ (I_{KCa}) current. Injection of EGTA, or removal of external Ca^{++}, results in the reduction of the major fraction of outward current. It is interesting that a depolarization sufficient to activate Ca^{++} current I (i.e. 2-10 mV from a holding potential of -50 mV), never activates any appreciable outward current. Only more depolarized potentials, which activate Ca^{++} current II, elicits an outward current. This leads to the conclusion that it is the Ca^{++} coming into the cell via Ca^{++} channel II, which is responsible for the activation of the Ca^{++}-dependent K^+ outward current.

The remaining outward currents present during depolarization are presumably the voltage-dependent, delayed rectifier and the rapidly-inactivating, fast outward current, known as the A-current (Deitmer, 1984). The latter is sensitive to diaminopyridines and to predepolarization of the membrane. It is yet unclear to what extent membrane depolarization and Ca^{++} may act cooperatively in activating K^+ currents.

Hyperpolarization of the membrane beyond -80 mV activates the inward rectifier, i.e. K^+ channels which open at rather negative potentials. Ba^{++} blocks this transient K^+ current, as has been shown in frog skeletal muscle (Standen and Stanfield, 1978). Furthermore, in *Stylonychia*, this hyperpolarization-induced K^+ current is blocked by 0.5 nM neomycin.

6. Why Such a Diversity of Ion Channels?

It is amazing to find that this unicellular organism employs so many different ion channels, and there are many more that have not been discussed here (e.g., chemically-activated ion channels. There are two striking similarities between this ciliate and neurons of invertebrate and vertebrate animals which appear to be of utmost functional significance: (1) the presence of a large variety of ion channels in any one cell, and (2) their inhomogenous distribution in the membrane. In *Stylonychia* we have two different types of voltage-dependent Ca^{++} channels in two functionally different sets of ciliary organelles. In neurons, we may also find different kinds of Ca^{++} channels (McCleskey et al., 1986), and they may be located in different neuronal compartments, such as the cell soma, dendrites or presynaptic terminals (Llinas and Yarom, 1981; Penner and Dreyer, 1986). In these different parts of neurons, local Ca^{++} influxes and subsequent local Ca^{++} transients may affect specific ion conductances and, hence, electrical excitability, or the release of neurotransmitter.

The interaction of various inward and outward currents through differently gated ion channels may provide the means for varying the membrane potential. External and internal stimuli seem to be integrated by changes in the membrane potential, which then lead to an appropriate behavioral response. In *Stylonychia*, it is the activation of the ciliary organelles: the initiation, direction and often, also, the frequency of ciliary beating, which are controlled by membrane potential (de Peyer and Machemer, 1982; Deitmer et al., 1984). In addition, changes in membrane potential bring about ion fluxes, such as the Ca^{++} influx, which act as the signal for triggering the electro-motor coupling by activating the axoneme of the cilium. In neurons, variations in membrane potential reflect the state of excitability, which may accelerate or prevent the initiation of an action potential and, hence, influence information processing within the nervous system. In both types of cells, neurons and ciliates, the major form of activity is controlled by changes in membrane potential. In this respect, we may learn about one cell from the other.

7. Concluding Remarks

When comparing ion channels, ion currents and electrical excitability in unicellular organisms such as ciliates with those in neurons of invertebrate or vertebrate animals, we find many similarities between structures which appear to be worlds apart in evolutionary terms. Yet, excitability of cells as a form of signal and information transfer and/or as a means of coordinating and synchronizing biological processes (ciliary activation, exocytosis of synaptic vesicles, or action potential frequency) seems to be under high evolutionary pressure. The speed of behavioral reaction is directly dependent upon these processes, which are initiated by the gating of ion channels in the membrane. Ionic currents passing through these channels and the concomitant membrane potential changes are the fastest signal and information processing in organisms. The mechanisms involved have proven to be extremely important and successful during evolution, as evidenced by its presence in almost all animal cell types. The speed of "flight and fight strategies," such as e.g., ciliary reversal, muscular contraction, or our thinking, may select for survivors.

ACKNOWLEDGEMENTS. The financial support (SFB 114, TP A5 and De 231/4-2) from the Deutsche Forschungsgemeinschaft is gratefully acknowledged.

References

Almers, W., McCleskey, E. W., and Palade, P. T., 1984, A non-selective cation conductance in frog muscle membrane blocked by micromolar external calcium ions, *J. Physiol. (Lond.)* **353:**565.

Deitmer, J. W., 1981, Voltage and time characteristics of the potassium mechanoreceptor current in the ciliate *Stylonychia*, *J. Comp. Physiol. A* **141:**173.

Deitmer, J. W., 1982, The effects of tetraethylammonium and other agents on the potassium mechanoreceptor current in the ciliate *Stylonychia*, *J. exp. Biol.* **96:**239.

Deitmer, J. W., 1984, Evidence for two voltage-dependent calcium currents in the membrane of the ciliate *Stylonychia*, *J. Physiol. (Lond.)* **355:**137.

Deitmer, J. W., 1986, Voltage-dependence of two inward currents carried by calcium and barium in the ciliate *Stylonychia mytilus*, *J. Physiol. (Lond.)* **380:**551.

Deitmer, J. W., 1988, Multiple types of calcium channels: Is their function related to their localization? in: *Calcium and Ion Channel Modulation*, p. 19 (A. D. Grinnell, D. Armstrong, and M. B. Jackson, eds.), Plenum Press, New York.

Deitmer, J. W., Machemer, H., and Martinac, B., 1984, Motor control in three different types of ciliary organelles in the ciliate *Stylonychia*, *J. Comp. Physiol. A* **154:**113.

de Peyer, J. E., and Deitmer, J. W., 1980, Divalent cations as charge carriers during two functionally different membrane currents in the ciliate *Stylonychia*, *J. exp. Biol.* **88:**73.

de Peyer, J. E., and Machemer, H., 1977, Membrane excitability in *Stylonchia*: Properties of the two-peak regenerative Ca-response, *J. Comp. Physiol.* **121:**15.

de Peyer, J. E., and Machemer, H., 1978a, Hyperpolarizing and depolarizing mechanoreceptor potentials in *Stylonychia*, *J. Comp. Physiol.* **127:**255.

de Peyer, J. E., and Machemer, H., 1978b, Are receptor-activated ciliary motor responses mediated through voltage or current? *Nature* **276:**285.

de Peyer, J. E., and Machemer, H., 1982, Electromechanical coupling of cilia. I. Effects of depolarizing voltage steps, *Cell Motil.* **2:**483.

Doughty, M. J., and Dryl, S., 1981, Control of ciliary activity in *Paramecium*. An analysis of chemosensory transduction in a eucaryotic unicellular organism, *Progr. Neurobiol.* **16:**1.

Eckert, R., and Chad, J. E., 1984, Inactivation of Ca channels, *Progr. Biophys. Mol. Biol.* **44:**215.

Ivens, I., 1986, Different properties of two voltage-dependent inward currents of the ciliate *Stylonychia mytilus*, *J. Physiol. (Lond.)* **381:**1.

Jennings, H. S., 1906, *Behavior of the Lower Organisms*, Columbia University Press, New York.

Kung, C., and Saimi, Y., 1982, The physiological basis of taxes in *Paramecium*, *Ann. Rev. Physiol.* **44:**519.

Llinas, R., and Yarom, Y., 1981, Electrophysiology of mammalian inferior olivary neurones *in vitro*. Different types of voltage-dependent ionic conductances, *J. Physiol. (Lond.)* **315:**549.

Machemer, H., and Deitmer, J. W., 1987, From structure to behaviour: *Stylonychia* as a model system for cellular physiology, *Progr. Protistol.* **2:**213.

Machemer, H., and de Peyer, J. E., 1977, Swimming sensory cells: Electrical membrane parameters, receptor properties and motor control in ciliated protozoa, *Verh. Dtsch. Zool. Ges., Erlangen* **1977:**86.

McCleskey, E. M., Fox, A. P., and Tsien, R. W., 1986, Different types of calcium channels, *J. exp. Biol.* **124:**177.

Naitoh, Y., and Eckert, R., 1969, Ionic mechanisms controlling behavioral responses in *Paramecium* to mechanical stimulation, *Science* **164:**963.

Naitoh, Y., Eckert, R., and Friedman, K., 1972, A regererative calcium response in *Paramecium*, *J. exp. Biol.* **56:**667.

Naitoh, Y., and Kaneko, M., 1973, Control of ciliary activities by adenosinotriphosphate and divalent cations in Triton-extracted models of *Paramecium caudatum*, *J. exp. Biol.* **58:**657.

Penner, R., and Dreyer, F., 1986, Two different presynaptic calcium currents in mouse motor nerve terminals. *Pflügers Arch.* **406:**197.

Song, P. S., 1981, Photosensory transduction in *Stentor coeruleus* and related organisms, *Biochem. Biophys. Acta,* **639:**1.

Standen, N. B., and Stanfield, P. R., 1978, A potential- and time-dependent blockade of inward rectification in frog skeletal muscle fibres by barium and strontium ions, *J. Physiol. (Lond.)* **280:**169.
Van Houten, J., Hauser, D. R., and Levandowsky, M., 1981, Chemosensory behavior in Protozoa, in: *Biochemistry and Physiology of Protozoa*, p. 67 (M. Levandowsky and S. H. Hutner, eds.), Academic Press, New York.

Chapter 19

Ionic Currents of the Scyphozoa

PETER A. V. ANDERSON

1. Introduction

Scyphozoans are members of the phylum Cnidaria and, according to Werner (1973), are the second most primitive extant class of this, the most primitive group of animals to possess a recognizable nervous system. Their nervous systems are, for the most part, composed almost entirely of diffuse two-dimensional plexuses of neurons, nerve nets (Passano, 1982), the most prominent of which is the motor nerve net (Anderson and Schwab, 1981), a diffuse plexus of large bipolar neurons that transmits motor activity from marginal pacemakers to the swimming muscle. This nerve net was previously called the Giant Fiber Nerve Net (Horridge, 1954), on account of the unusually large size of its component neurons.

This nerve net provides an excellent preparation for examining the basic processes that underlie nervous function in this class of animals, for two reasons. First, the component neurons are relatively large (cell body diameters can exceed 20 μm) and, more important, in *Cyanea capillata* the myoepithelial cells that envelop the neurons in this species can be removed selectively (Anderson and Schwab, 1983) permitting unrestricted access to the neurons and their synapses. The following is a survey of the variety of ion channels known to exist in these neurons. Additional details may be obtained from the original literature (Anderson, 1987; Anderson and McKay, 1985).

The only other scyphozoan cell type from which voltage clamp recordings have been obtained are cnidocytes (sting cells) from *Chrysaora* (Anderson and McKay, 1987). Data from these cells is included here to provide a wider survey of the variety of ionic channels possessed by these organisms.

PETER A. V. ANDERSON • Whitney Laboratory and Departments of Physiology and Neuroscience, University of Florida, 9505 Ocean Shore Blvd., St. Augustine, Florida 32086, USA.

Evolution of the First Nervous Systems
Edited by P.A.V. Anderson
Plenum Press, New York

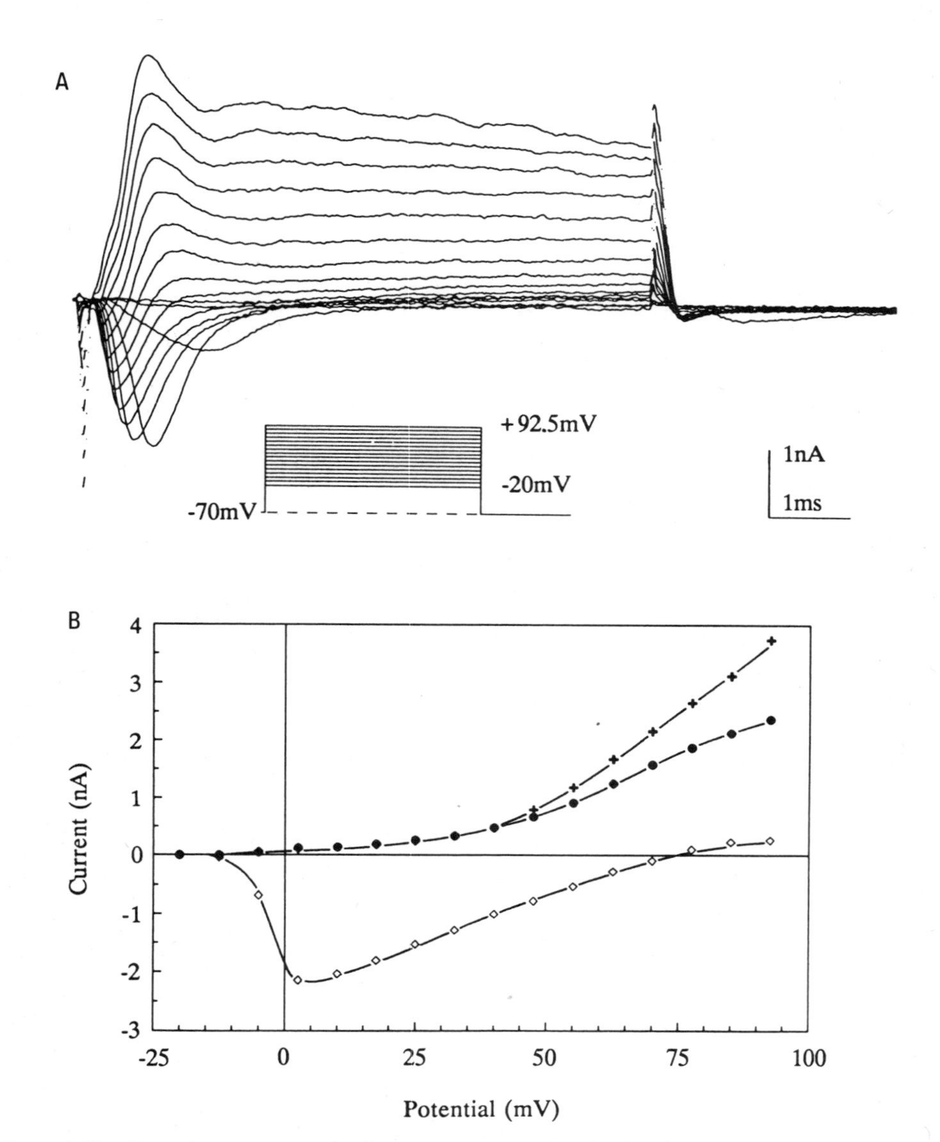

Figure 1. Total membrane currents in *Cyanea* neurons. **(A)** A family of membrane currents generated by the voltage regime shown. **(B)** Current voltage relationships of the inward (o), steady state outward (•) and early transient outward (+) currents. External solution: saline; internal solution: high K^+, low Ca^{++} patch solution.

2. Voltage-Activated Currents

2.1. Total Membrane Currents in *Cyanea* Neurons

Whole-cell, voltage clamp recordings from axotomized cell bodies reveal that these neurons possess a complex family of voltage-activated ionic currents (Fig. 1). When patch pipettes are filled with a mock intracellular solution (high K^+, low Ca^{++}, low Na^+), total membrane currents consist of a fast transient inward current and several outward currents. The transient inward current activates at potentials around -20 mV, as does the first of the outward currents, a relatively steady state outward current. At potentials around +40 mV, an early, transient, outward current appears and gets progressively larger at more positive potentials (Fig. 1). A second, slower, transient outward current activates at +55 mV (not shown). These various currents can be isolated using a variety of pharmacological and physiological approaches to permit examination of many of them in isolation.

2.2. Inward Currents in *Cyanea* Neurons

All outward current can be blocked by intracellular Cs^+ and tetraethylammonium (TEA) and extracellular 4-aminopyridine (4-AP). The inward current that is recorded under these conditions consists of a fast, transient current that activates at -20 mV, reaches peak amplitude at around +10 mV and reverses around +60 mV (Fig. 2). This is a very fast current which reaches peak amplitude in less than 1 ms at depolarized potentials and then inactivates completely over the succeeding 5 ms. Activation of this inward current is clearly voltage-dependent. So is inactivation, as revealed by standard H-infinity experiments in which the currents generated by a voltage step applied after a 50 ms prepulse are compared with those generated by the same test pulse in the absence of a prepulse. Prepulses to potentials more negative than -40 mV have no effect on the amplitude of currents generated by the test pulse, but with more positive prepulses the current amplitude decreases markedly and is essentially completely abolished by prepulses to 0 mV. Half-maximum inactivation is achieved by prepulses to -15 mV.

The ionic basis of this current was determined by ionic substitution. Inward current was abolished by the removal of extracellular Na^+ (tetramethylammonium substitution) and its reversal potential was dependent on the extracellular Na^+ concentration. Thus, this current is a Na^+ current (I_{Na}) whose kinetics and waveform are very similar to those of Na^+ currents in most higher animals.

In Na^+-freesolutions with elevated Ca^{++}, a second inward current is observed (Fig. 3). This current is slower than I_{Na}, taking 2-3 ms to reach peak amplitude and it decays far more slowly. The current/voltage (I/V) relationship of this current (Fig. 3B) is also different inasmuch as the current activates around 0 mV, reaches peak amplitude at +30 mV and reverses around +90 mV. The different kinetics, I/V relationships and ionic dependencies suggest that this is a separate current from the fast, inward current recorded in the presence of extracellular Na^+ (I_{Na}). However, it is arguable that the

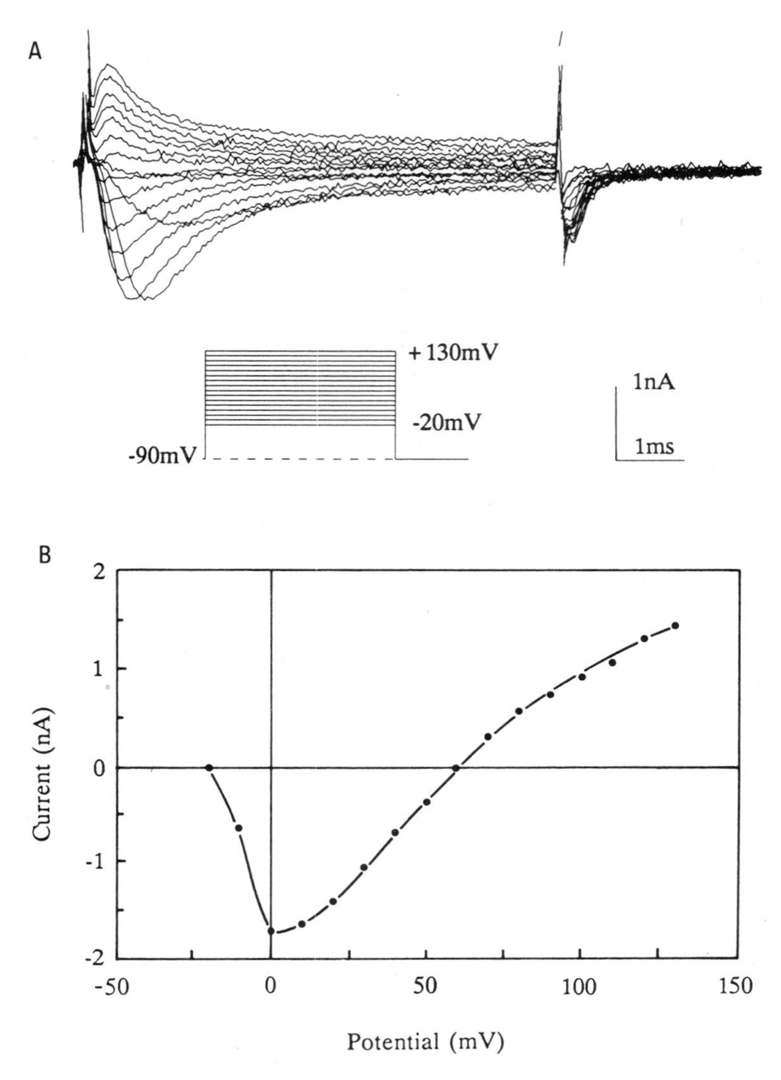

Figure 2. Inward current in *Cyanea* neurons. **(A)** Inward currents generated by the voltage steps shown. Outward current had been blocked by intracellular Cs^+ and TEA and extracellular 4-AP. Under these conditions, inward current is a transient current. **(B)** I/V relationship of the inward current. External medium: saline with 7.5 mM 4-AP; internal solution: Cs^+/TEA patch solution (after Anderson, 1987).

slower current could use the same ionic channels as I_{Na}, with the different kinetics being the result of having Ca^{++} as the charge carrier rather than Na^+ and the different I/V relationships being a result of a shift of the normal I/V relationship of the cell due to the elevated extracellular Ca^{++} concentration (Frankenhaeuser and Hodgkin, 1957). The presence of a fast, transient outward current at the onset of this slower, inward

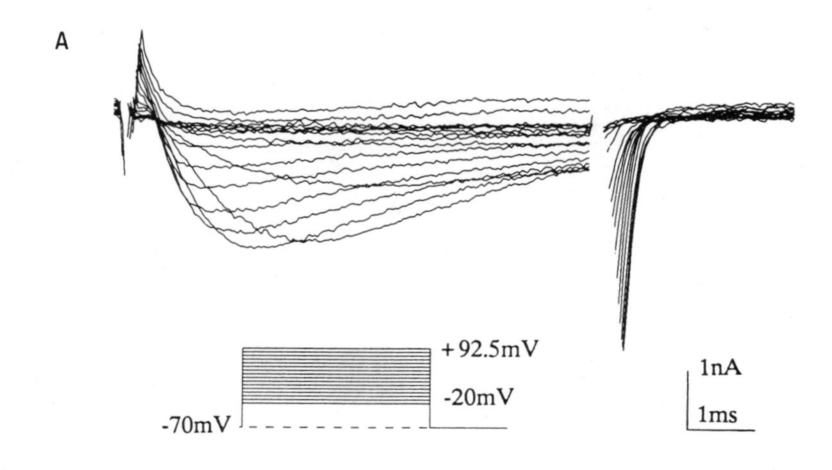

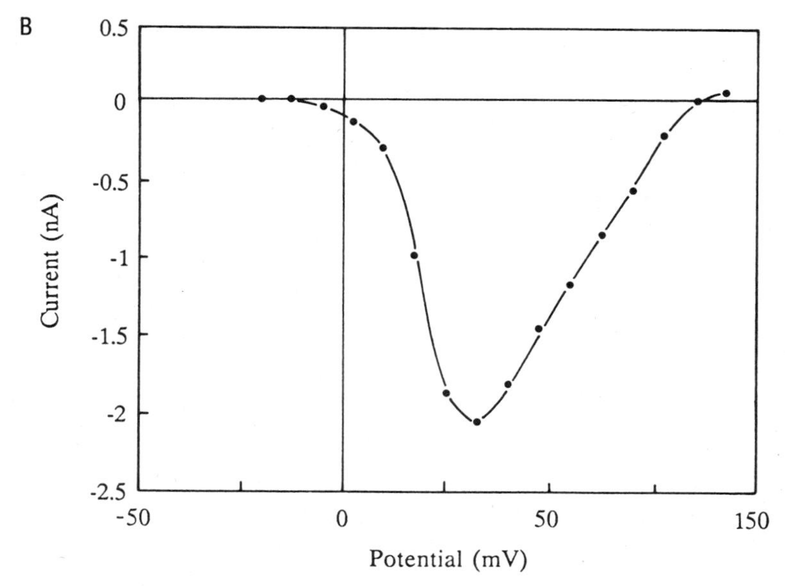

Figure 3. Inward current recorded in Na^+-free saline containing 98 mM Ca^{++}. **(A)** Under these conditions, inward current generated by the voltage steps shown, consists of a slower, transient current. **(B)** I/V relationship of the peak inward current in A. External solution: Na^+-free (TMA substituted) saline with 7.5 mM 4-AP; internal solution: Cs^+/TEA patch solution (after Anderson, 1987).

current was originally (Anderson, 1987) thought to be due to the outward movement of Na^+ through the Na^+ channels. If this were so, then those channels would inactivate and be unavailable for the Ca^{++} ions carrying the slower inward current, providing good evidence that the slower inward Ca^{++} current must have entered through another

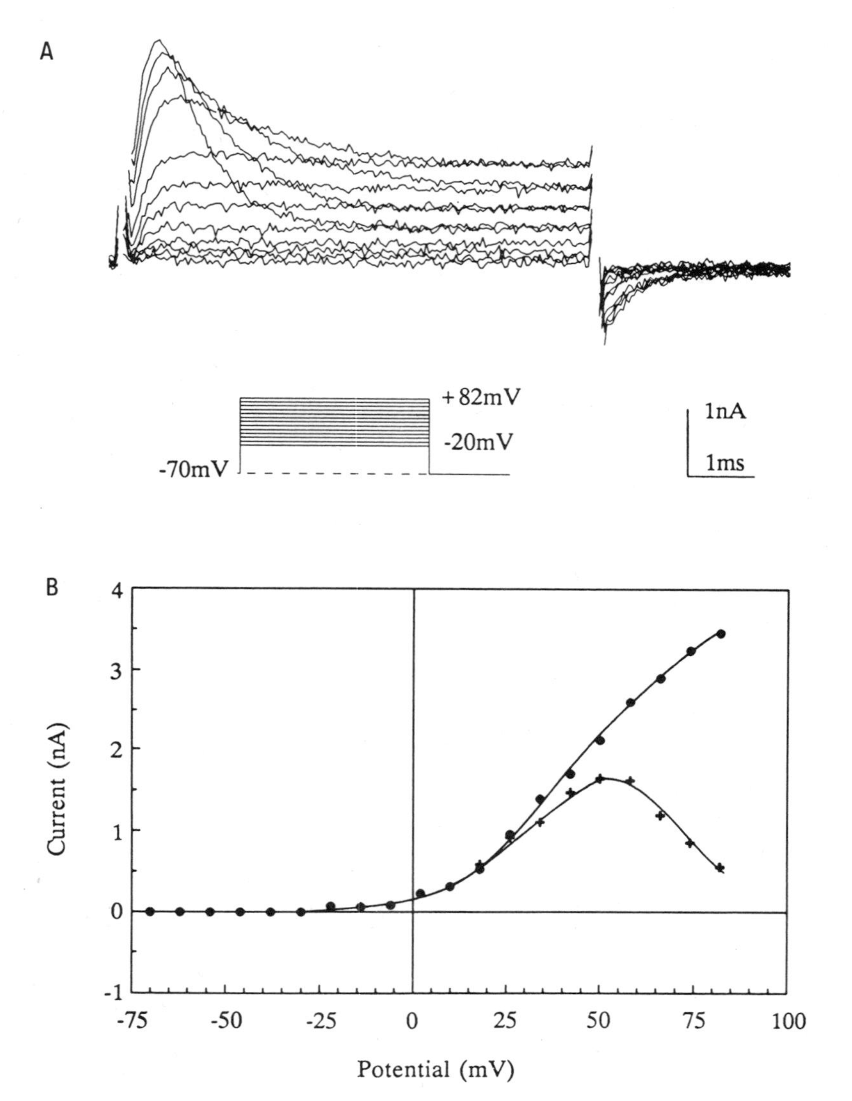

Figure 4. Currents recorded in Na^+-free saline containing 30 mM Ba^{++}. **(A)** Under these conditions, the first current to appear is a fast, steady-state outward current on which is later superimposed an inward-going current. **(B)** Current voltage relationship of the steady-state outward (•) and inward (+) currents shown above. Extracellular solution: Na^+-free saline with 30 mM Ba^{++}; internal solution: Cs^+/TEA patch solution.

population of channels, presmably Ca^{++} channels. However, it now appears that the fast transient outward current recorded at the onset of the voltage step under these conditions is a fast, steady-state outward current that can be carried by a variety of ions including Cs^+, TEA and H^+, and which is not blocked by intracellular Cs^+ and TEA, as are all the other outward currents. This current appears whenever the extracellular divalent cations are manipulated (Fig. 4). For instance, it is present when Ca^{++} concentrations are elevated or when extracellular Ca^{++} is replaced with Ba^{++} or Sr^{++}. As a result, it has prevented detailed examination of the Ca^{++} current in these cells. It is clear, however, that the slower inward current described above is, indeed, different from I_{Na}, since its I/V relationship is more positive than that of I_{Na}, even when the extracellular divalent cation concentration is increased only slightly (Fig. 4A,B).

These results indicate that these cells possess two distinct inward currents. The predominant inward current is a fast, transient current which is carried by Na^+ and is the current responsible for the rising phase of the action potential in these cells. Physiologically, it resembles Na^+ currents recorded from excitable cells in most higher animals. The second, slower inward current is apparently carried by Ca^{++} and its properties, specifically its elevated I/V relationship, are consistent with its being the Ca^{++} current underlying synaptic transmission by these cells (Anderson, 1985).

The most remarkable feature of I_{Na} in these cells is its pharmacology. It has long been recognized that Na^+-dependent action potentials in cnidarians and ctenophores are insensitive to the classical Na^+ channel blocker tetrodotoxin (TTX) (for review see Anderson and Schwab, 1982), but this conclusion was obtained largely with intact tissue where adequate diffusion of the TTX into extracellular spaces is not necessarily ensured. The Na^+ current of motor nerve net neurons is, however, completely insensitive to TTX, at concentrations up to 100 μM. Similarly, this current is also insensitive to STX (10 μM) and a variety of agents such as veratridine (100 μM), batrachotoxin (2 μM), sea anemone toxin (1 μM) and scorpion venom (1 μM), which block Na^+ channel inactivation in higher animals. This Na^+ current can, however, be blocked by lidocaine (1 mM) and the anti-calmodulin drug W7 (20-100 μM), both in a use- and voltage-dependent manner, and by verapamil (2-100 μM), Cd^{++} (2.5-5 mM) and Nicardipine (100 μM). Agents such as lidocaine, W7 and Cd^{++} are fairly non-specific, so their actions here may not be particularly significant. However, the fact that I_{Na} can be blocked by low concentrations of verapamil, a recognized Ca^{++} channel blocker, and by Nicardipine, a member of the highly specific dihydropyridine class of Ca^{++} channel blockers, is very significant. It indicates that while this channel has the physiology and ion selectivity of a Na^+ channel, it has the pharmacology of a Ca^{++} channel. The pharmacology of I_{Ca} in these cells is more conventional, inasmuch as it is blocked by 10 μM verapamil and 2 mM Cd^{++}.

The fact that I_{Na} in these neurons has the physiology of a Na^+ current, but the pharmacology of a Ca^{++} current, raises a major question concerning the evolution of the Na^+ channel. Following a survey of the distribution of voltage-activated ion channels in the animal kingdom, Hille (1984) suggested that the Na^+ channel of higher animals may have evolved from the Ca^{++} channel of protozoans. This step may have occurred to protect neurons from the unacceptably high levels of intracellular Ca^{++} that would otherwise accumulate during repetitive spiking. While the basis for this model may not be entirely valid following the discovery of a Na^+-dependent action potential in the heliozoan *Actinocoryne* (see chapter by Febvre-Chevalier et al. in this

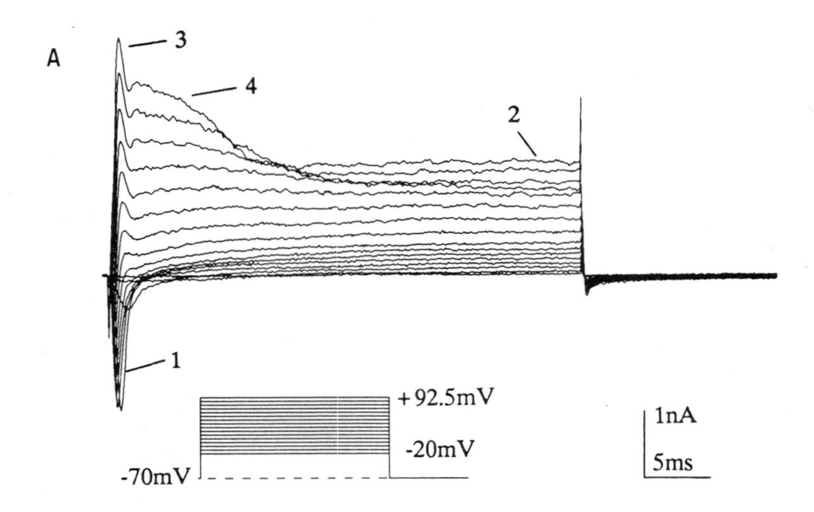

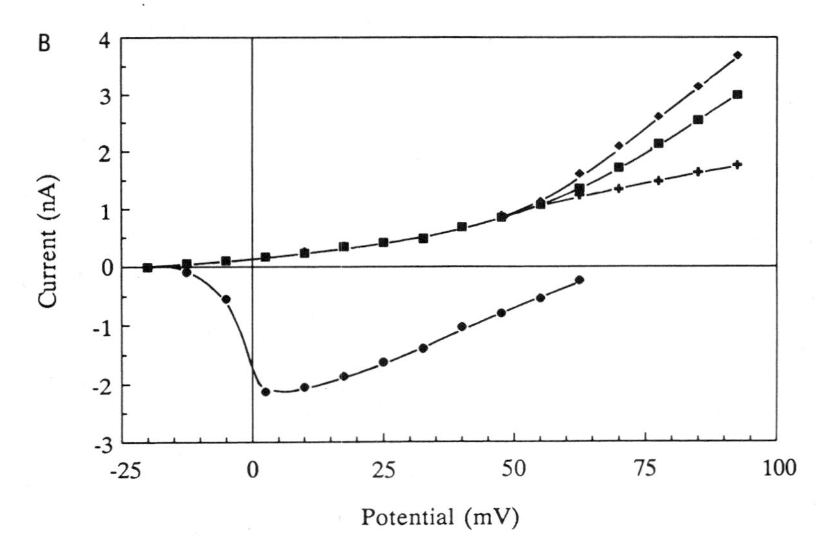

Figure 5. Total membrane currents from *Cyanea* neurons. **(A)** Family of total membrane currents recorded during long (35 ms) voltage steps. Currents consist of an inward current (1), a steady-state outward current (2), an early transient outward current (3) and a slower transient outward current (4). **(B)** I/V relationships of the inward (•), steady state outward (+), early transient outward (▪) and slower transient outward (♦) currents shown in A. External solution: saline; internal solution: high K^+, low Ca^{++} patch solution.

book), the basic concept may still apply, particularly when one considers the marked similarity of the two channel types, as revealed by their respective genes (Noda et al., 1984; Tanabe et al., 1987). If such a model for Na^+ channel evolution is correct, then such a step may well have occurred in an organism such as a primitive cnidarian since these are the first animals known to possess a nervous system. The fact that neurons in a relatively primitive (Werner, 1973) member of this phylum still produce currents with properties intermediate between those of a Na^+ current and a Ca^{++} current suggests that such a step may, indeed, have occurred and that this current in *Cyanea* is a residual intermediate in that step. Final confirmation that the channels through which the Na^+ current in these neurons passes is, indeed, a Na^+ channel, rather than a Ca^{++} channel, or alternatively, a completely different molecule that coincidently gates a current similar to the I_{Na} of higher animals, obviously waits further study at the molecular level.

2.3. Outward Currents in *Cyanea* Neurons

As indicated earlier, these cells possess at least three distinct outward currents, plus the apparently non-specific outward current that appears following modification of the extracellular divalent cation concentration (Fig. 4). The ionic bases of the first three outward currents have not been determined, but it is reasonable to assume that they are K^+ currents. The first outward current to activate during a series of depolarizing voltage steps is a steady-state outward current (Fig. 5). This current activates around -15 mV to -20 mV and remains relatively constant for the duration of 60 ms voltage steps. A transient component of outward current appears on the leading edge of the steady state outward current at +25 mV and gets progressively larger with increasing levels of depolarization and is present even at holding potentials of 0 mV. The third, outward current activates at potentials around +55 mV and appears as a relatively slow hump superimposed on the steady-state outward current. This current gets progressively larger with increasing levels of depolarization. It, too, is present when cells are clamped at a holding potential of 0 mV.

The identity of these three outward currents is unclear. The fact that the two transient outward currents are not abolished by holding potentials of 0 mV suggests that neither can be classical A-currents but, given the very obvious positive shift in the I/V relationships of all currents recorded from these cells, this possibility cannot be ruled out without applying H-infinity protocols that extend to very depolarized potentials. Similarly, one cannot rule out the presence of a Ca^{++}-activatedK^+ current (I_{KCa})merely by the fact that none of the outward currents activates at a potential appropriate for I_{Ca} in these cells and none show a decrease in current amplitude that would correspond to the decline and reversal of the Ca^{++}current shown above (Fig. 3). However, the calculated reversal potential of I_{Ca} in these cells is in excess of +100 mV, so it is very conceivable that the slower transient outward current could be an I_{KCa} with a marked voltage-dependency.

2.4. Total Membrane Currents in *Chrysaora* Cnidocytes

Large numbers of cnidocytes can be isolated from the fishing tentacles of the jellyfish *Chrysaora* and voltage clamp recordings can be obtained fairly easily from one variety, the atrichous isorhizas. These cells are non-spiking cells, unlike stenoteles from the hydrozoan polyp *Cladonema*, and their only active response to injected current is delayed rectification. Not surprisingly, voltage clamp recordings reveal the presence of a strong delayed rectifier current in all cells examined. This current activates at potentials close to 0 mV and reaches peak amplitude 12-13 ms after the onset of the voltage step. In addition to the delayed rectifier, some cells also possess a transient and very rapid outward current that activates at more negative potentials than the steady-state outward current (-10 to -20 mV). Once again, it is tempting to classify the transient, outward current as an A-current, but it is not inactivated by holding potentials up to -20 mV. However, both outward currents in the atrichous isorhizas of *Chrysaora* are very similar to the outward currents in stenoteles from the hydroid *Cladonema* and these have been studied far more extensively. The transient outward current present in *Cladonema* stenoteles is inactivated by holding the cell at 0 mV, so it is quite possible that a similar potential might be required for inactivation of the transient outward current in the atrichous isorhizas from *Chrysaora*. Both currents can be blocked by intracellular Cs^+ and TEA.

3. Ligand-activated Currents

The most obvious ligand-activated current in these cells is the transmitter-activated channel that underlies the EPSP in these cells. Synapses between motor nerve net neurons are unusual inasmuch as they are bidirectional excitatory chemical synapses; they conduct excitatory events in either direction. The synaptic delay at these synapses is of the order of 1 ms and EPSPs attain peak amplitude within 1-1.5 ms. The apparent reversal potential of the EPSP has been measured at +4 mV (Anderson, 1985), suggesting that several ionic species contribute to the underlying current, since no single ionic species has an appropriate reversal potential under the conditions employed. Curiously, however, true reversal of the EPSP has never been observed. This suggests that the transmitter-activated ion channel must have some inherent voltage-sensitivity since, otherwise, an efflux of the outward current-carrying ion should occur and produce a reversed EPSP. The identity of the transmitter at these synapses is not known.

During the search for the transmitter, it was found that applications of acidic saline evoked large depolarizing potentials (Fig. 6A). While it was initially thought that these depolarizations might be EPSPs, the subsequent discovery that they had a different reversal potential (-15 mV), and could be truly reversed, indicated a different

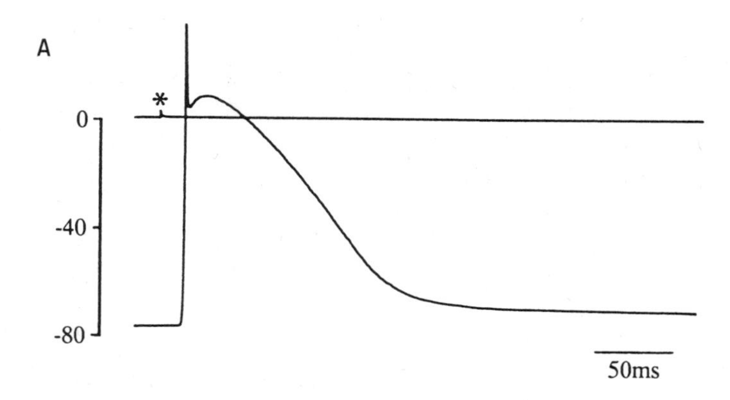

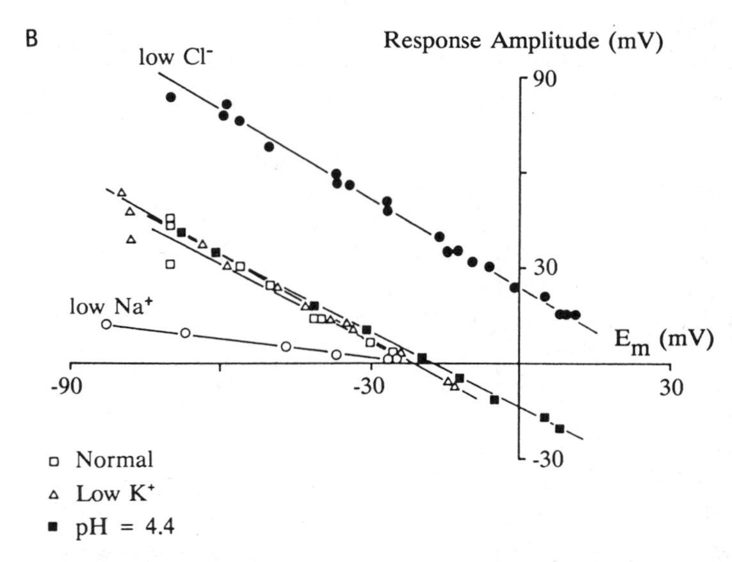

Figure 6. Proton-activated chloride current in *Cyanea* neurons. **(A)** Current clamp recording of the depolarization of a motor nerve net neuron produced by brief application of acidic (pH 5.4) saline. Time of application is indicated by the * on the zero potential trace. **(B)** Relationship between membrane potential (E_m) and amplitude of the proton-activated depolarization, illustrating the reversal potential of the response (-20 to -30 mV) and the fact that only changes in external Cl^- concentration affected that reversal potential (after Anderson and McKay, 1985).

underlying mechanism. The amplitude of these potentials is changed only when the extracellular chloride concentration is altered (Fig. 6B), indicating that the depolarization is carried by Cl^-, specifically a Cl^- efflux. Proton-activated chloride currents have been described in a variety of muscles (Loo et al., 1980; Klein, 1985) and modified muscle cells, electroplax (Hanke and Miller, 1983), but their function remains unclear, as does the equivalent current in *Cyanea* neurons.

4. Conclusions

In many respects, the variety of ionic currents present in scyphozoans closely resemble those found in a variety of higher animals. The Na^+ current in *Cyanea* neurons is very similar physiologically to that recorded from squid axons or vertebrate nerve or muscle preparations. The major difference is a marked positive shift in its I/V relationship and its unusual pharmacology. It is difficult to come to any obvious conclusions about the positive shift in the I/V relationship for both activation and inactivation save to say that it appears to be a consistent feature of scyphozoan cells. It is arguable that this shift may reflect an uncorrected tip potential. However, the measured activation potential for I_{Na} is consistent with spike threshold in these cells (0 mV), suggesting that it may be an accurate measurement, and reflect a primitive feature of voltage-activated currents.

As indicated above, the unusual pharmacology of the Na^+ current in *Cyanea* neurons, specifically its sensitivity to Ca^{++} channel blockers, rather than Na^+ channel blockers poses many questions regarding the evolution of the Na^+ channel. However, while this finding does provide tantalizing evidence that Hille's model of Na^+ channel evolution (Hille, 1984) may be valid, confirmation, or as much as is possible when dealing with evolutionary questions, clearly awaits details on the structure of the Na^+ channel in this organism.

While it has not yet been possible to clearly identify all the outward currents in *Cyanea* neurons and *Chrysaora* cnidocytes, it is clear that both cell types possess a steady-state, outward current of the type usually described as I_K. Furthermore, both cells possess fast, transient outward current which have the appearance of A-currents, voltage-activated and inactivated transient outward currents (Aldrich et al., 1979a,b). Because there are, as yet, no precedents for transient outward currents inactivated by anything other than voltage, it is reasonable to classify these currents as A-currents. However, if they are A-currents, then they must require very depolarized potentials for inactivation since, in the case of *Cyanea* neurons, holding at 0 mV had no obvious effect. The identity of the third outward current in *Cyanea*, the slower transient current, is not known at this time.

The two ligand activated currents described here are important. In the case of the proton activated Cl^- current, the remarkable similarity between the current described in *Cyanea* neurons and that found in higher animals indicates remarkable evolutionary conservation of a basic cell mechanism. While the function of this current in not clear, it is interesting to speculate that it may be involved in regulating the cell's excitability, since we know (Anderson, 1987) that changes in extracellular Cl^- levels produce marked shifts in the I/V relationship of the Na^+ current in *Cyanea* neurons. In the case of the transmitter-activated current responsible for the EPSP at synapses in *Cyanea*, the short latency and short duration of the EPSP argues strongly for the presence of a directly-gated ion channel, rather than one relying on second messenger activation.

Thus, the Scyphozoa possess a variety of ion channels, some of which clearly have counterparts in a variety of higher animals, and others which show tantalizing evidence of being evolutionary precursors or intermediates. However, it should be borne in mind that the various currents described above were all found in only two cell types. Clearly additional work is required to more fully characterize some of the currents already isolated and to pursue other currents in other cell types.

ACKNOWLEDGEMENTS. This work was supported by NSF grant BNS 88-05885.

References

Aldrich, R. W., Getting, P. A., and Thompson, S. H., 1979a, Inactivation of delayed outward current in molluscan neurone somata, *J. Physiol. (Lond.)* **291**:507-530.

Aldrich, R. W., Getting, P. A., and Thompson, S. H., 1979b, Mechanism of frequency-dependent broadening of molluscan neurone soma spikes, *J. Physiol. (Lond.)* **291**:531-544.

Anderson, P. A. V., 1985, Physiology of a bidirectional, excitatory chemical synapse, *J. Neurophysiol.* **53**:821-835.

Anderson, P. A. V., 1987, Properties and pharmacology of a TTX-insensitive Na^+ current in neurones of the jellyfish *Cyanea capillata*, *J. exp. Biol.* **133**:231-248.

Anderson, P. A. V., and McKay, M. C., 1985, Evidence for a proton-activated chloride currents in coelenterate neurons, *Biol. Bull.* **169**:652-660.

Anderson, P. A. V., and McKay, M. C., 1987, The electrophysiology of cnidocytes, *J. exp. Biol.* **133**:215-230.

Anderson, P. A. V., and Schwab, W. E., 1981, The organization and structure of nerve and muscles in the jellyfish *Cyanea capillata* (Coelenterata:Scyphozoa), *J. Morphol.* **170**:383-399.

Anderson, P. A. V., and Schwab, W. E., 1982, Recent advances and model systems in coelenterate neurobiology, *Prog. Neurobiol.* **19**:213-236.

Anderson, P. A. V., and Schwab, W. E., 1983, Action potentials in neurons of the motor nerve net of *Cyanea* (Coelenterata), *J. Neurophysiol.* **50**:671-683.

Anderson, P. A. V., and Schwab, W. E., 1984, An epithelial cell-free preparation of the motor nerve net of *Cyanea* (Coelenterata), *Biol. Bull.* **166**:396-408.

Frankenhaeuser, B., and Hodgkin, A. L., 1957, The action of calcium ions on the electrical properties of squid axons, *J. Physiol. (Lond.)* **137**:218-244.

Hanke, W., and Miller, C., 1983, Single chloride channels from *Torpedo* electroplax, *J. Gen. Physiol.* **82**:25-45.

Hille, B., 1984, *Ionic Channels of Excitable Membranes,* Sinauer, Sunderland, Mass.

Klein, M. G., 1985, Properties of the chloride conductance associated with temperature acclimation in muscles of the green sunfish, *J. exp. Biol.* **114**:581-598.

Loo, D. D. F., McLarnon, A. G., and Vaughan, P. C., 1980, Some observations on the behaviour of chloride current-voltage relationships in *Xenopus* muscle membranes in acid solutions, *Can. J. Physiol. Pharmacol.* **59**:7-13.

Horridge, G. A., 1954, The nerves and muscles of medusae. I. Conduction in the nervous system of *Aurellia aurita* Lamarck, *J. exp. Biol.* **34**:594-600.

Noda, M., Shimizu, S., Tanabe, T., Takai, T., Kayano, T., Ikeda, T., Takahashi, H, Nakayama, H., and Numa, S., 1984, Primary structure of the *Electrophorus electricus* sodium channel deduced from c-DNA sequence, *Nature* **312**:121-127.

Passano, L. M., 1982, Scyphozoa and Cubozoa, in: *Electrical Conduction and Behaviour in "Simple" Invertebrates*, pp. 149-202 (G. A. B. Shelton, ed.), Clarendon Press, Oxford.

Tanabe, T., Takeshima, H., Mikami, A., Flockerzi, V., Takahashi, H., Kangawa, K., Kojima, M., Matsuo, H., Hirose, T., and Numa, S., 1987, Primary structure of the receptor for calcium channel blockers from skeletal muscle, *Nature* **328**:313-318.

Werner, B., 1973, New investigations on the systematics of the class Scyphozoa and the phylum Cnidaria, *Publ. Seto Mar. Lab.* **20**:35-61.

Chapter 20

The Electrophysiology of Swimming in the Jellyfish *Aglantha digitale*

ROBERT W. MEECH

1. Introduction

Many jellyfish are capable of avoiding potentially damaging stimuli, but their responses are generally slow and highly localized (Mackie, 1984). However two members of the Rhopalonematidae, the largest of the five families in the suborder Trachymedusae, have been observed to perform rapid escape swimming. The most well known of these is *Aglantha digitale* (Fig. 1) found in many of the colder waters of the world (Donaldson et al., 1980); the other is *Amphogona apicata* found in the Bahamas (Mills et al., 1985). *Amphogona* closely resembles *Aglantha* and it is to *Aglantha* that the work described in this chapter refers.

In *Aglantha*, the coexistance of excape swimming with a slower form of swimming necessary for feeding (Mackie, 1980a), arises from the capacity of the motor nerves of *Aglantha* to generate two different propagating impulses (Mackie and Meech, 1985). It is these conductance changes in the cell membrane that drive the relatively simple effector system to produce complex behavior, not the presence of structurally different muscle fibers or elaborate patterns of innervation.

We might equate the survival of *Aglantha*, as a species, with evolutionary success, and suppose that it arises from an improved food gathering capability or, possibly, from its unique method of escaping from predators. These are properties that appear to confer an advantage on the animal within the context of a specific set of environmental constraints that we need not immediately identify. We can, therefore, ask questions

ROBERT W. MEECH • Department of Physiology, University of Bristol, University Walk, Bristol BS8 1TD, United Kingdom.

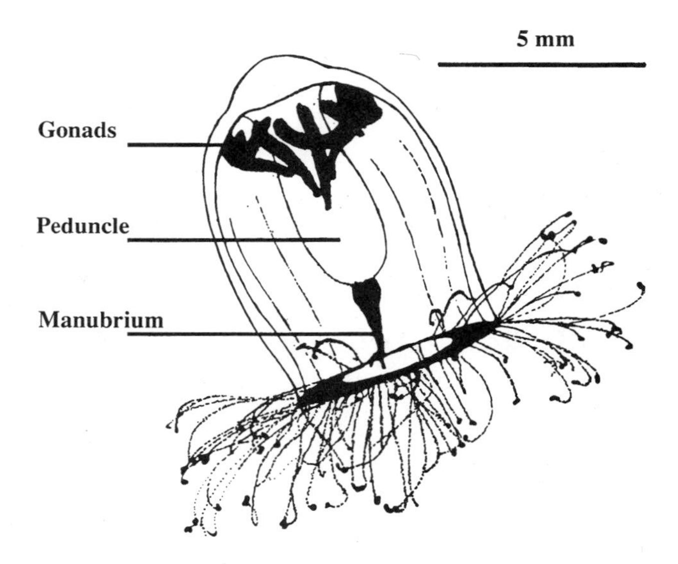

Figure 1. Specimen of *Aglantha digitale*. The animal is in a relaxed state at the end of a slow swim cycle. The gonads, peduncle and manubrium can be seen through the transparent bell-shaped body wall. Many fine tentacles extend from the margin at the base of the bell. The margin also contains the neurons responsible for generating slow swimming. Giant motor axons run radially from the margin to a point just short of the gonads. They are about 40 μm in diameter in adult specimens, and they lie within the sheet of myoepithelium that lines the inner surface of the bell.

The experimental preparation was a single bell, slit up one side, and with the apex removed. In many experiments (see Kerfoot et al., 1985) the margin with its ring giant axon and tentacles was also removed and the preparation became a "strip" of myoepithelium separated into octants by the motor giant axons. The myoepithelium was covered with seawater (12-15 °C), illuminated from below and carefully pinned flat to a layer of Sylgard in a transparent dish with cactus (*Opuntia*) spines. Intracellular recording was carried out using 3M KCl-filled glass pipettes with resistances of 40-50 MΩ and individual axons were stimulated by using brief current pulses from a bipolar stainless steel electrode. For intracellular current injection, thin-walled glass tubing was used to make 3M KCl-filled glass micropipettes with resistances of 10-15 MΩ. In many experiments the preparation was immobilized by adding $MgCl_2$ to the bathing medium. A solution containing 127 mM Mg^{++} was found to prevent contraction in *Aglantha* without significantly affecting the electrical activity recorded in either the motor giant axon or the nearby myoepithelium.

about the physiology of escape or food gathering, as if we were dealing with a problem in design. Are other forms of escape possible or do specific anatomical, physiological or developmental constraints force the solution that we call "*Aglantha*"? The selection of an animal with a simple nervous system and a small genome will aid this analysis, as will its relatively simple pattern of development.

In *Aglantha* the response to environmental pressure has taken the form of a well defined behavior pattern based on equally well defined ion channels. Ion channels are among the most easily identifiable gene products, and we may hope to pick out the key factors that affect their expression. Here too we may expect to detect constraints that limit the complexity of membrane components and the electrical events which they generate.

A constraint set upon written work is that the information must be partitioned and ordered, a process that necessarily introduces bias. On film, it is possible to convey both behavioral and anatomical information simultaneously, but in a book, it is necessary to impose an artificial order. In this chapter the first section gives details of

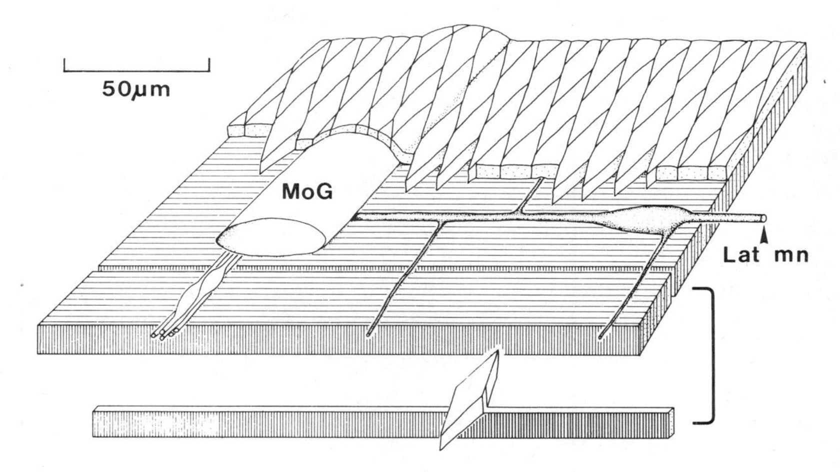

Figure 2. Diagram of part of the subumbrella myoepithelium near a motor giant axon (MoG) which runs from the margin towards the apex of the bell (i.e. radially) over the muscle tails of the myoepithelial cells. The muscle tails run in a circular direction and are shown by the vertical striations. An isolated myoepithelial cell is shown at the bottom of the figure to clarify its structure. The cell body is elongated in the radial direction. A small bundle of fine axons with somas run parallel to the motor giant axon. The lateral motor neuron (lat. mn.) has a principal circumferential axon (which makes contact with the motor giant axon), and finer radial branches. Its cell body is 100-500 μm from the motor giant axon (from Kerfoot et al., 1985).

the anatomy of *Aglantha*. It is followed by details of behavior and important aspects of physiology. Later sections are devoted to the ionic basis of the electrical events that drive the identified behavior patterns. The discussion contains speculations about constraints that may limit complexity at the molecular level.

2. Anatomy

Giant axons in the subumbrella of *Aglantha* were discovered by Singla (1978), who suggested that they mediated escape swimming. *Aglantha* has eight motor giant axons and they run radially up the bell-shaped body wall from the margin at the base of the animal. Neuromuscular synapses are found along their whole length and they make direct contact with the muscle cells of the subumbrella myoepithelium. The myoepithelium is one cell thick and is made up of radially elongated cell bodies with underlying striated muscle tails that are oriented in a circular direction (Fig. 2).

Injection of lucifer yellow into individual giant axons shows that they are dye coupled to laterally-running motor neurons that lie between the cell bodies and the muscle tails of the myoepithelium (Weber et al., 1982). In an animal with a bell height of 9 mm, the lateral motor neurons have cell bodies which are 11-16 μm across and a principal axon 2-4 μm in diameter. Lateral processes are separated by about 350 μm and have radial branches about 400 μm long that run approximately parallel to the motor giant axons (Fig. 2). Sites of visible overlap between adjacent lateral neurons are common near the base of the animal, but nearer the apex, the fields are more distinct. Whether the sites of visible contact are sites of transmission is not known. Synapses between fine axons and myoepithelial cells are found well away from the motor giants, and it seems likely that they occur along the entire length of the lateral neurons (Kerfoot et al., 1985).

Sections of myoepithelium, prepared for electron microscopy and fixed in lanthanum-containing solution, have junctions between their myoepithelial cell bodies with a typical gap junction structure (Kerfoot et al., 1985).

3. Behavior and Physiology

3.1. Swimming

Escape Swimming

Escape swimming in *Aglantha* consists of from one to three violent contractions of the entire bell-shaped, subumbrellar muscle sheet (Donaldson et al., 1980). The first contraction forces water past the velum at the base of the bell and propels the animal through a distance of five body lengths (see figure 3); later contractions begin before the bell has refilled completely with water, and each advances the animal by no more than two or three body lengths. Escape is evoked by touching the bell margin, pinching or gently tugging a tentacle, or creating a pressure wave in the surrounding water. Vibration-sensitive hair cells are present on the velum and at the tentacle bases, but there is no direct evidence for their involvement in initiating the reflex (Arkett et al., 1988). *Aglantha* has been observed from a submersible to escape contact with euphausiids, amphipods, crab larvae, and copepods (Mackie and Mills, 1983). The escape response also prevented capture by a predator (*Aequorea*) in laboratory tests (Donaldson et al., 1980).

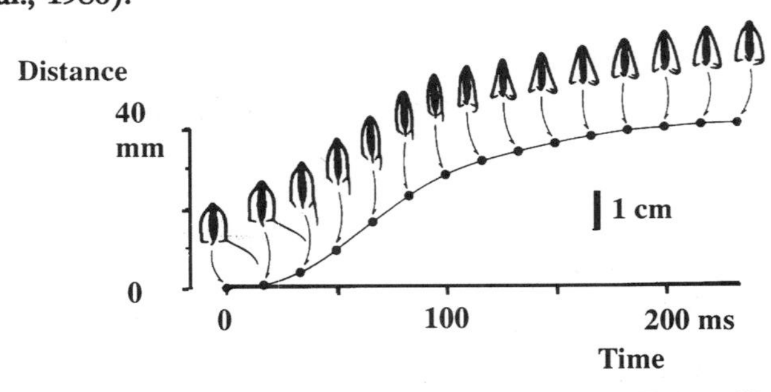

Figure 3. Longitudinal profiles during an escape swim in *Aglantha*. The profiles shown combine stroboscopic data on bell shape changes and overall movements of one animal with cinematographic data on tentacle withdrawal from another. The accompanying graph shows the distance of the movement of the animal through the water (from Donaldson et al., 1980).

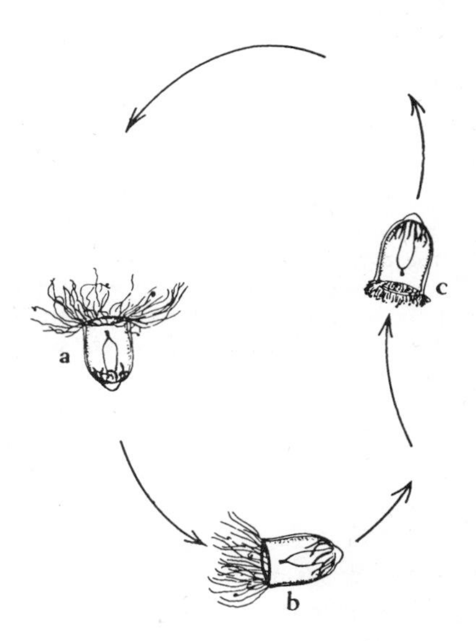

Figure 4. Activity cycle of *Aglantha* during "fishing". **(a)** The animal sinks passively with tentacles extended feeding. **(b)** Asymmetrical swimming ensues until the animal has righted itself. **(c)** The animal swims upward with a series of 30-40 weak contractions (slow swims). The animal then reenters phase a (from Mackie, 1980a).

Slow swimming

Aglantha can also swim slowly and rhythmically like other jellyfish (Gladfelter, 1973), and does so spontaneously in regular activity cycles (Fig. 4). During these "fishing" cycles, *Aglantha* swims upward with its tentacles contracted and then floats down with its tentacles extended. The upward orientation during the swimming phase depends on input from the statocysts (Mackie, 1980a) and several "righting" movements have the effect of bringing the bell into an upright position. Each contraction propels the animal about one body length, and a swimming sequence consists of about 30 contractions and lasts about a minute. As in other medusae, the pacemakers controlling slow swimming are located in the marginal nerve rings (Satterlie and Spencer, 1983).

3.2. Axon Impulses

Stimuli to the base of the animal, which initiate fast swimming movements in an animal pinned for experimentation (see legend, Fig. 1), evoke postsynaptic depolarizing events in the giant motor axon (Roberts and Mackie, 1980). These synaptic potentials summate to give an overshooting action potential that is conducted towards the apex of the bell at a velocity of 1-4 m s^{-1} (Fig. 5). The action potential is tetrodotoxin (TTX)-insensitive, but dependent upon $Na+$ ions in the bathing medium (Kerfoot et al., 1985). Substitution of Ca^{++} for Mg^{++} in the bathing medium has no effect, and the action potential is insensitive to Ca^{++} channel blockers. Its repolarization is delayed by 50 mM tetraethylammonium (TEA) ions.

Much lower amplitude events drive slow swimming. Intracellular recordings from giant motor axons, near their origin at the margin, show sequences of low amplitude spikes, each of which arises from a single postsynaptic potential (Fig. 6). At recording sites half-way along the axon, the slow swim spike is seen alone. Its conduction velocity is about 0.25 m s^{-1} (Mackie and Meech, 1985). The decrementally conducted synaptic

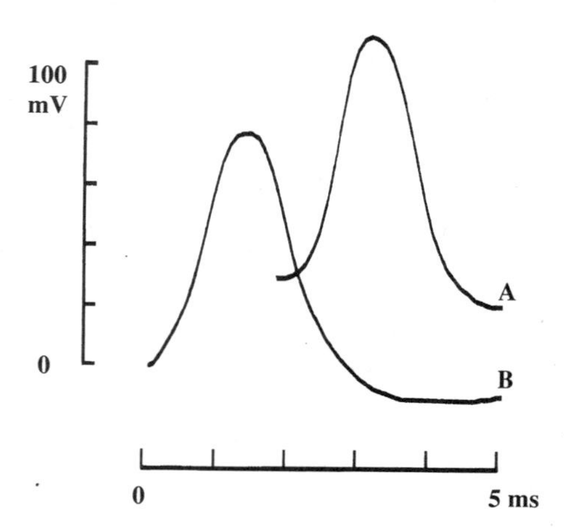

Figure 5. Intracellularly recorded action potential from a giant motor axon during an escape swim. The two recording sites (A and B) were 2.5 mm apart and the action potential was conducted away from the margin. Conduction velocity, 1.4 m s^{-1}. Axon resting potential -58 mV (from Mackie and Meech, 1985).

event is assumed to represent an input from presynaptic pacemaker neurons, as in other hydromedusae.

The amplitude of the slow swim spike is 27 mV (S.E. 4 mV; n=20) compared to 95 mV (S.E. 8 mV; n=23) for the fast swim action potential; the slow swim spike rises more slowly and lasts about 10 times longer. These differences cannot be attributed to different resting membrane potentials because the resting potential of the axon is unchanged during the two forms of activity. Average axon resting potentials were -63 mV (S.E. 6 mV; n=17) for slow swim spiking and -61 mV (S.E. 5 mV; n=10) for the fast swim action potential (Mackie and Meech, 1985).

Slow swim spikes elicited by current injection were blocked by increasing the Mg^{++} concentration in the bathing medium from 50 to 65 mM. They were also blocked by 7 mM Mn^{++}, another Ca^{++} channel blocking agent, and this block could be reversed by increasing the Ca^{++} in the bathing medium to 28 mM. The slow swim spikes were

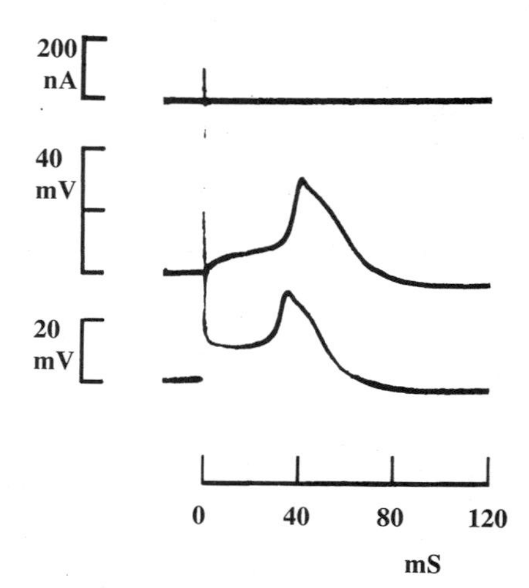

Figure 6. A low amplitude spike recorded from the motor giant axon, in response to a brief depolarizing current pulse (top trace). The current was injected using a bridge circuit so that the membrane potential could be recorded at the site of current injection as well as at a second site 2 mm away. The conduction velocity of the slow swim spike was about 25% of that of the overshooting action potential associated with an escape swim. A small increase in the intensity of the injected current was sufficient to elicit an overshooting action potential in the same axon (not shown). Axon resting potential -72 mV (from Mackie and Meech, 1985).

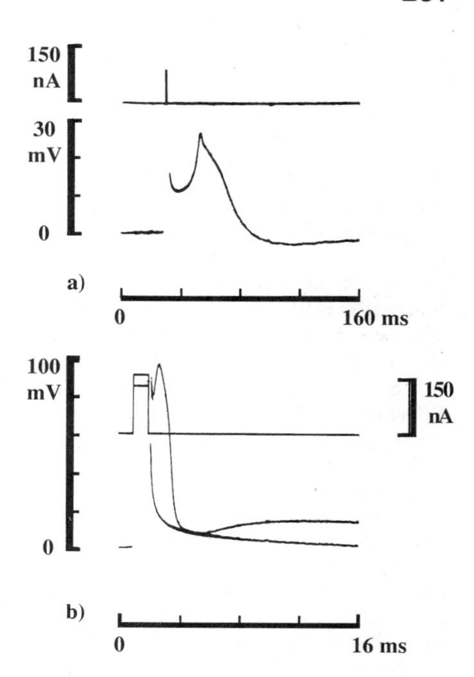

Figure 7. Pharmacological differences between action potentials and slow swim spikes recorded from the giant motor axon. The slow swim spike elicited by a brief depolarizing pulse in normal seawater (**a**) abolished by seawater containing 0.18 mM Nifedipine (**b**). This level of Nifedipine had no effect on the overshooting action potential elicited by a slightly increased current. Axon resting potential -72 mV (from Mackie and Meech, 1985).

relatively insensitive to Nifedipine, but 0.18 mM Nifedipine produced a full blocking effect. Nifedipine at this concentration had no effect on the overshooting Na^+ action potential elicited in the same axon (Fig. 7). TEA ions were found to have little effect on the duration of the spike (Mackie and Meech, 1985).

3.3. Chemical Synapses

During an escape swim, the motor giant axons ensure widespread contraction of the body wall in two ways. They directly excite the myoepithelium at chemical synapses distributed along their length and they electrically excite lateral motor neurons whose branches make further chemical synapses over the whole subumbrella surface (Kerfoot et al., 1985).

Although the transmitter at the *Aglantha* neuromuscular synapse is unknown, measurements of synaptic delay allow us to exclude some of the slower mechanisms such as those based on release of a second messenger. Simultaneous intracellular records from the motor giant axon and from the myoepithelium 40-130 μm away show that the synaptic delay from the peak of the presynaptic action potential to the onset of the postsynaptic response is about 0.7 ms at 12°C (Fig. 8). In the frog neuromuscular junction, the synaptic delay is 1 msec at 12°C (see Katz and Miledi, 1965; Fig. 4), while in the squid giant synapse the synaptic delay measured from the peak of the internally recorded spike is about 1.4 msec at the same temperature (calculated from Lester, 1970; see also Takeuchi and Takeuchi, 1962; Bloedel et al., 1966). At another neuromuscular junction, in the coelenterate *Polyorchis*, Spencer (1982) has reported

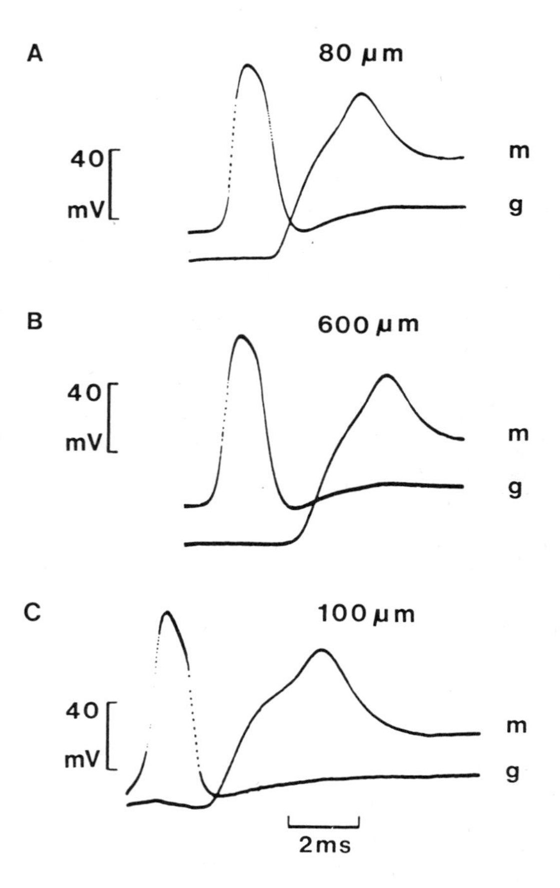

Figure 8. Simultaneous intracellular records from motor giant axon (g) and nearby myoepithelial cell (m) following external stimulation of the giant motor neuron. Records **A** and **B** are from a single penetration of the axon and two separate penetrations of the myoepithelium, 80 μm (**A**) and 600 μm (**B**) from the recording site in the axon. Record **C** is taken from a different experiment: a circle (250 μm diameter) was scored in a region of the myoepithelium which included the motor giant. The axon and a nearby myoepithelial cell (separation: 100 μm) were penetrated within this isolated region. The records were retouched where necessary to show the rising and falling phases of the axon spike. Bathing medium: 90 mM Mg^{++} sea water. Resting potentials: axon A, -62mV; B, -65mV; C, -60mV; myoepithelium: A, -77mV; B, -86mV; C, -67mV (from Kerfoot et al., 1985).

synaptic delays ranging from 0.9 to 7 ms (temperature 14-19°C). And at the bidirectional synapse of *Cyanea* there are delays of 1 ms at room temperature (Anderson, 1985).

3.4. Lateral Neurons

In *Aglantha*, stimulation of a single motor giant axon produces widespread contraction during an escape swim. Intracellular records from the myoepithelium on either side of the axon show that contraction is preceded by a complex electrical response which resembles the end-plate potential at the frog neuromuscular junction (Fatt and Katz, 1951). The response consists of a foot, a sharply rising phase, a shoulder, an overshooting spike-like component and a long phase of repolarization. In frog the muscle response becomes a simple conducted action potential as it moves away from the neuromuscular junction, but in *Aglantha* the myoepithelial response is dominated by the postsynaptic potential at all locations (Kerfoot et al., 1985).

Events which probably represent spontaneous release of transmitter may be recorded at different distances from the motor giant axon, indicating the presence of synaptic sites throughout the myoepithelium, probably associated with the lateral motor neurons. Conduction of excitation through the myoepithelium depends upon the participation of this fine network of lateral motor neurons (Kerfoot et al., 1985) because there is no event in the muscle which behaves like a simple regenerative spike of the kind described in other hydromedusae (Mackie and Passano, 1968; Spencer, 1978).

Stimulation of individual lateral motor neurons elicits a contraction in a well defined area of the myoepithelium (Kerfoot et al., 1985). Escape swimming can be attributed to the concerted action of the entire population of these neurons, each driven by an impulse in the giant motor axon and each producing a contraction in its own field. Slow swimming involves rather more localized contractions (G. O. Mackie, unpublished) but the electrical events appear almost uniform at recording sites up to 500 μm from the giant motor axons, as if driven by Ca^{++} spikes in the lateral neurons.

3.5. Current Spread in the Myoepithelium

Gap junctions are thought to be the site of a low resistance current pathway between cells (Barr et al., 1965; Payton et al., 1969). In *Aglantha* the myoepithelial cells are dye coupled, and there are gap junctions between the cell bodies (Singla, 1978). The cells are also electrically coupled and current injection into one cell produces a voltage change in surrounding cells, which is a linear function of injected current over a wide range (Kerfoot et al., 1985).

In a thin sheet of coupled cells, current flow is restricted to two dimensions, and the steady-state change in membrane potential declines with distance from a point current source as a modified Bessel function (Woodbury and Crill, 1961; Eisenberg and Johnson, 1970; Shiba, 1971; Frömter, 1972; Jack et al., 1975). In *Aglantha*, the intracellularly recorded potential change declines with distance from the current source, as expected for two-dimensional current spread, but unlike the simple epithelium of *Euphysa* (Josephson and Schwab, 1979), the sheet does not behave as if radially symmetrical about the point of injection (Fig. 9). The current spreads further in the circular direction than it does radially. This corresponds to the orientation of the elongated muscle tails, rather than the epithelial cell bodies (Fig. 2), but gap junctions have been found only between the latter (Singla, 1978). Because other characteristics of the sheet, such as its time constant and its internal resistance, are independent of direction, we suppose that the anisotropicity arises from the current path length, and perhaps on the number and distribution of operative gap junctions (see Detwiler and Hodgkin, 1979).

3.6. Muscle Contraction

In *Aglantha*, external Ca^{++} ions are necessary for both neuromuscular transmission and for contraction. When Ca^{++} is added back to a Ca^{++}-free medium, a graded postsynaptic potential is recorded from the muscles, which increases in size with further additions of Ca^{++}. Eventually, it develops a spike-like component. The appearance of

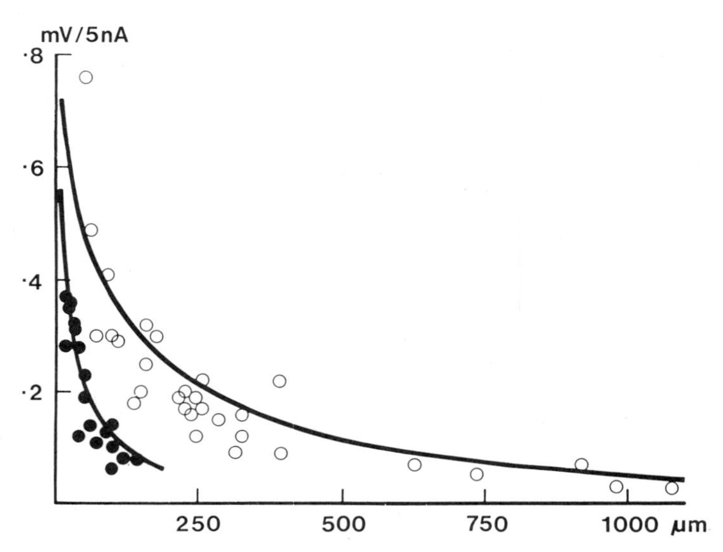

Figure 9. Spread of current in the myoepithelium. Ordinate: mean membrane potential displacement produced by current injected at different distances from the recording site (abscissa). Open circles: measurements in a circular direction; closed circles: measurements in a radial direction. The solid line was calculated using λ_2 = 770 μm and R_i = 201 Ω cm for current flow in a circular direction and λ_2 = 177μm and R_i = 178 Ω cm for current flow in a radial direction. λ_2 is a two dimensional space constant; R_i is the resistivity of the epithelial interior. Bathing medium: 70 mM Mg^{++} sea water (from Kerfoot et al., 1985).

this muscle spike coincides with the appearance of a contraction in the myoepithelium as if an inward Ca^{++} current contributes to the spike but not to the postsynaptic potential. During slow swimming the spike-like component is greatly attenuated.

Although the low amplitude spike associated with slow swimming is very sensitive to increased Mg^{++} in the bathing medium, synaptic transmission during fast escape swimming remains unaffected (Kerfoot et al., 1985). This absence of an effect of Mg^{++} is notable, because at the giant synapse of the squid a doubling of the level of external Mg^{++} (to 108.6 mM) reduces the postsynaptic current to 25% of normal (Takeuchi and Takeuchi, 1962). In fact, the addition of isotonic $MgCl_2$ is often used as a general anaesthetic for invertebrates, because it is assumed that the high levels of Mg^{++} block neuromuscular transmission. High Mg^{++} does, indeed, block muscle contraction in *Aglantha*, but its effect seems to be on the contractile apparatus rather than on transmitter release.

There is a significant difference in the amplitude of the postsynaptic events recorded from the myoepithelium during fast (98 mV, S.E. 8 mV; n = 20) or slow (56 mV, S.E. 6 mV; n = 25) swim movements, and this is matched by a difference in rise-time. The low amplitude and slow rise time presumably produces a reduced Ca^{++} influx and, hence, the reduced contraction observed during slow, as compared to fast, swimming.

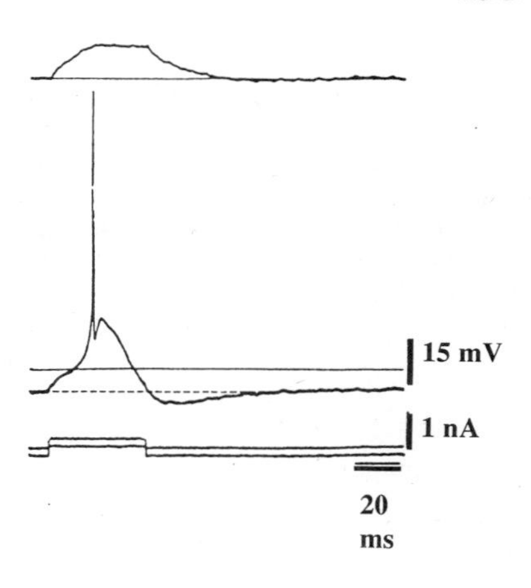

Figure 10. Effect of membrane potential on the excitability of inferior olivary neurons in a brain stem slice preparation. The cells were stimulated by injection of current pulses (bottom trace) either at the resting potential (top) or at 4 mV hyperpolarized to the resting potential (middle). When the cell is hyperpolarized, a depolarizing pulse elicits a multiphasic electrical event that consists of a TTX-sensitive Na^+ action potential riding on top of a low-amplitude Co^{++}-sensitive Ca^{++} spike (from Llinás and Yarom, 1981).

4. Ion Channels

4.1. Sodium and Calcium Currents

As figure 10 shows, low amplitude Ca^{++} spikes, like those in *Aglantha*, are found in areas of the mammalian brain, such as the thalamus (Llinas and Jahnsen, 1984), the inferior olive (Llinás and Yarom, 1981), the pontine reticular formation (Greene et al., 1986), the habenula (Wilcox et al., 1988) and some neocortical areas (Friedman and Gutnick, 1987). However, these cells also have TTX-sensitive Na^+ currents, and as the peak of the Ca^{++} spike is above the threshold of the Na^+ action potential, a complex combination of the two events is generally recorded.

TTX-treated thalamic neurons studied under voltage clamp have inward Ca^{++} currents in the voltage range -60 to +40 mV (Coulter et al., 1989). They are blocked by low levels of Ca^{++} channel blockers, such as Ni^{++} (0.5 mM), Cd^{++} (0.5 mM) or Mg^{++} (8mM) but are unaffected by Nifedipine (25 μM, Crunelli et al., 1989; Lightowler and Pollard, 1989). The currents are fully inactivated at -60 mV, but recover with a time constant of about 240 ms. Inactivation is only fully removed at voltages more negative than -90 mV. In general, the currents resemble the low threshold transient or "T"-type Ca^{++} channels found in a wide range of different preparations (Hagiwara et al., 1975; Fox, 1981; Halliwell, 1983; Deitmer, 1984; Carbone and Lux, 1984; Fox and Krasne, 1984; Armstrong and Matteson, 1985; Bean, 1985; Fedulova et al., 1985; Bossu and Feltz, 1986; Narahashi et al., 1987; Fox et al., 1987).

The slow swim spike in *Aglantha*, like the low amplitude spikes of the vertebrate CNS, has a threshold near the resting potential. During voltage clamp experiments on shortened isolated axons, inward currents appear during steps to voltages in the range -49 to -33 mV. The amplitude of the inward Ca^{++} current is steeply dependent on the holding potential, and if the membrane is held at -58 mV, it is abolished entirely; inactivation is only fully removed at potentials more negative than -70 mV. As in the case of thalamic neurons, the inward current is blocked by an increase in external Mg^{++}.

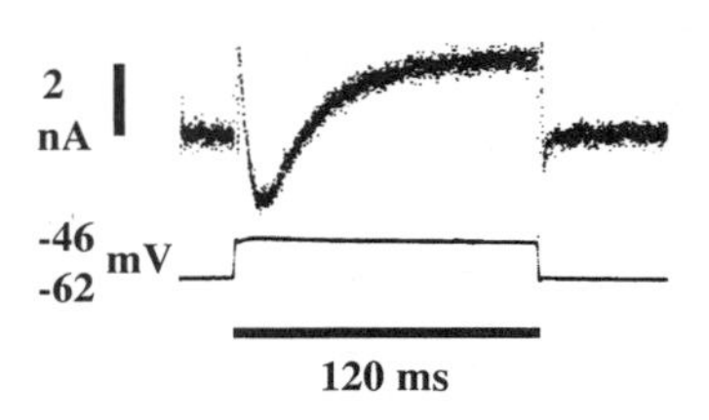

Figure 11. Currents recorded from a shortened giant motor axon under voltage clamp. A conventional, two-electrode voltage clamp was used with micropipettes of the kind described in figure 1. The depolarizing step (bottom trace) was from a holding potential of -62 mV and lasted 120 ms. A transitory inward current, which could be blocked by Mg^{++}, was preceeded and followed by outward currents (G. O. Mackie and R. W. Meech, unpublished).

In *Aglantha*, a step change in membrane potential from -65 to -44 mV produces an inward Ca^{++} current followed 20 ms later by an outward K^+ current (Fig. 11). With steps to potentials near that at which the Ca^{++} current peaks (-33 mV), the outward current turns on sufficiently rapidly to conceal the inward current completely. *In vivo*, this would have the effect of limiting the amplitude of the Ca^{++} spike and keeping it below the threshold of the Na^+ action potential. These K^+ currents appear to be reduced or absent in mammalian neurons.

The separation of the Na^+ action potential from the Ca^{++} spike in *Aglantha* amounts to a division of the signalling process into low and high current systems. The Ca^{++} currents during a Ca^{++} spike are of low amplitude, but they last long enough to increase internal Ca^{++} sufficiently for transmitter release. They are sensitive to external Mg^{++}, and so it is probable that the Ca^{++} influx necessary for transmitter release during the Na^+ action potential (which is insensitive to Mg^{++}) involves a second set of Ca^{++} channels. We have no evidence at present as to whether or not the two channel types are equally distributed.

4.2. Potassium Currents

The evolution of a dual system of impulse propagation has provided *Aglantha* with an escape behavior that appears to be unique among jellyfish. A key feature in the separation of slow and escape swimming is a rapidly developing outward current that prevents the slow swim Ca^{++} spike from reaching the threshold of the escape swim action potential (Mackie and Meech, 1989).

Ensemble currents, derived from patch clamp analysis of intact axons, show that the K^+-selective channels that form the basis of the outward current consist of four different classes which range in their kinetics from a rapidly inactivating A-type (Hagiwara et al., 1961) to a non-inactivating delayed rectifier. All four classes have the same conductance characteristics. Single channel data can be fitted to a theoretical current-voltage relationship derived by Goldman (1943) and Hodgkin and Katz (1949) for a homogenous membrane, with a value of 4-4.6 x 10^{-14} cm^3 s^{-1} for the K^+ permeability constant. This provides a good fit to data from all four channels in concentrations of external K^+ up to 500 mM (Mackie and Meech, 1989).

One possibility that we must consider, is that these findings represent a single channel molecule that exists in kinetically distinct states, rather than different channel molecules. There are reports of abrupt changes in kinetic configuration that take place in certain agonist-activated channels studied in detached patches. However, the change is a relatively simple one (an alteration in a single rate constant), and it affects all of the channels in the patch. In *Aglantha*, the kinetic differences are highly complex; many patches contain a mixed population of channels and, in any case, all the channels we studied remained in the same configuration for the lifetime of the patch, which in some cases was as much as 2 hours.

5. Discussion

5.1. Constraints at a Molecular Level

While it appears that the more demanding environmental niches are colonized by the more highly specialized organisms, some highly specialized organisms may be said to have transformed their environments. A jellyfish with a successful escape reflex has effectively removed from its environment features which still threaten the less sophisticated species that occupy the same (homogeneous) space. Specializations of this kind that extend, rather than limit, the range of habitats that are open to colonization are characteristic of what we may call the more highly evolved species.

The structure and physiology of an organism that comes to occupy a given environmental niche depends not only upon the external constraints that exert what we call evolutionary pressure upon its genetic composition, but also upon internal constraints which arise from the way in which its gene products are organized. The extent to which these factors limit the complexity of any given evolutionary line can be seen most clearly in the less restricted, more highly evolved species. Because of its behavioral complexity, we take *Aglantha* to be a particularly highly evolved jellyfish.

But what sets an upper limit to the degree of sophistication attained by any given evolutionary line? One possibility is that it is the underlying developmental and cellular physiology which, although adequate for less specialized forms, limits the degree of complexity that is attainable. This is not to say that the more complex organisms necessarily utilize *only* the processes present in the simpler ones, because new gene products may well induce changes in the underlying organization. Nevertheless, if constraints exist, they are best uncovered by analyzing those characteristics of an organism that permit it to transform its environment or provide a competitive "edge". In *Aglantha*, the most obvious of these characteristics is its behavior.

It is evident that evolutionary pressures influence the behavioral properties of evolving organisms through changes in molecular entities, such as ion selective membrane channels. Furthermore, it is apparent that the ability of *Aglantha* to escape from predators arises not from the unique properties of any individual ion channel, but from the combined characteristics of a collection of many different channels. Here then, at the molecular level, are possible hidden constraints in organization.

Needless to say, these constraints are not at all well understood, for at this stage the information required can be inferred only from higher level organization. What seems clear is that, because *Aglantha* swims by contracting a single sheet of muscle cells, its ability to generate two forms of swimming depends upon the separation of the two forms of impulse upon which they are based. It is perhaps significant that although the K^+ channels that ensure this separation have different kinetics, they all have the same conductance. Does this mean that they have the same kind of conducting pore? Are they derived from some common ancestor or some common molecular precursor? Or does it indicate a constraint in the mechanism that matches the current density of the different ionic conductances?

In fact, all of the membrane channels we have described in *Aglantha* resemble those found in other organisms. The capacity to make Na^+ and Ca^{++} channels is a property of advanced vertebrates, advanced protozoa, and advanced jellyfish and it seems, at first sight, unlikely that this ability evolved on separate occasions. Of course, channels that have evolved independently may be constrained (by as yet unidentified forces) to take up these specific configurations. Until we can calculate the probability that such structures evolved *once*, it seems fruitless to speculate whether they evolved more than once! At present, we can only note that in the protozoan *Stylonychia*, T-type Ca^{++} channels form a complex Ca^{++} spike with other Ca^{++} channel types whereas, in mammals, there is a combined Na^+ and Ca^{++} spike, and in *Aglantha* the two spike forms are separated. So the channels may be the same, but their organization is different.

Although many of these questions are unanswerable at present, techniques that provide an experimental approach are available. For example, clonal DNA from *Drosophila* (kindly given to us by Y.N. and L.Y. Jan) is being used to probe the *Aglantha* genome, to identify the K^+ channel gene involved in spike separation (Meech et al., 1989). K^+ channels with similar conductances and different kinetics in developing *Drosophila* myotubes are the product of different genes (Zagotta et al., 1988). So, are there multiple K^+ channel genes in *Aglantha*, each one producing a uniform conductance product? *Drosophila* also appears able to generate multiple versions of the A-type K^+ channel by alternative splicing (Schwarz et al., 1988; Iverson et al., 1988). Are there multiple versions of A-channel mRNA in *Aglantha*, or is there a precursor K^+ channel protein that is modified prior to its insertion into the cell membrane? It is the answers to questions like these that will reveal the constraints within the system.

5.2. *Aglantha* as a Design Problem

Although the analysis of the ion channels that form the basis of swimming behavior in *Aglantha* is still in the preliminary stages, considerably more information is available at higher levels of organization. The constraints that exist at these levels may be revealed by a more theoretical approach, such as by treating *Aglantha* as a problem in design and examining alternative ways of generating selected design features.

An important feature that we need to be able to reproduce with this approach is the ability of *Aglantha* to perform an escape swim. The solution to this problem is constrained by the fact that *Aglantha* is a bell-shaped animal and that it swims by contracting a single sheet of muscle. Distribution throughout the muscle sheet of the

signals that elicit these contractions could be accomplished in a limited number of ways:
1. The signal may be mediated by nerve fibers that run parallel to the muscle fibers.
2. The signal may be a mechanical pull from muscle fiber to muscle fiber.
3. The signal may be a chemical transmitter between muscle cells.
4. The signal may be by electrical conduction through gap junctions in a syncytium ("myoid" transmission).

The speed with which contraction spreads throughout the body wall during escape swimming means that mechanisms of propagation that involve chemical transmission between muscle cells or stretch activation are probably too slow. In other hydromedusae (reviewed by Satterlie and Spencer, 1983), conduction within the muscle sheet itself is generally purely myoid, although in *Polyorchis*, an exceptionally large medusa, excitatory nerves extend from the margin up the four radii. Myoid conduction in *Polyorchis* is very slow (about 5cm s^{-1}; Spencer, 1979), and the nerves presumably reduce the conduction time in the radial direction.

When current is injected into a two-dimensional sheet of tissue, the spatial decay of voltage is steeper than the exponential decline observed in an axon, and it is this that probably limits the conduction velocity (Josephson, 1985). Conduction velocities in epithelia are lower than in nerves, and rarely exceed 30 cm s^{-1} (Mackie, 1980b). It would be useful to be able to set a theoretical upper limit, but at present, no such theoretical analysis is available.

With a linear current source, such as might be generated in the myoepithelium of *Aglantha* by the giant motor axons with their multiple synaptic contacts, the voltage change in a thin sheet decays exponentially as the recording site is moved away from the current source. This is the same result as is found in nerve axons with current injected at a point, and so the rate of conduction may be limited by the cross-sectional area of the conducting tissue. A thin epithelium is unlikely to achieve conduction velocities of 4 m s^{-1}, such as occur in the motor giant axons but might approach the much slower rates of propagation attained by the thin lateral neurons.

More information about the role of the lateral neurons is revealed by considering a second design feature: the ability of *Aglantha* to swim in two different ways. Because both movements are generated by electrical events in a single sheet of muscle, the two forms of contraction could be accomplished if:
1. the muscle cells were capable of two kinds of impulse, or
2. the muscles were innervated by two kinds of nerve, or
3. the nerves were capable of two kinds of signal.

Spiking epithelia have a very high threshold to point stimulation, and Schwab and Josephson (1982) have shown that small differences in amplitude produce large differences in conduction velocity. Thus, the two myoid spikes responsible for slow and fast swimming would be much closer in size than the corresponding spikes in an axon. Under these circumstances, it is possible that there would be little difference in the strength of contraction associated with the two different spikes. The situation would be different if the axon acted as a line, rather than a point source of current, but this would demand two pulses in the axon, one of high and one of low amplitude, and as it is inevitable that the low amplitude signal would be conducted relatively slowly, it would fail as a linear source.

The remaining possibilities are either the presence of two kinds of innervation or a single innervation with two kinds of propagating nerve impulse. The important factor here may be the ease of assembling each system. There is a possibility that building multi-impulses into the axons involves fewer commands than building two entirely separate-nerve networks, but a great deal more experimentation is required before we can be sure about that.

6. Conclusion

Aglantha generates two forms of swimming from a single muscle sheet innervated by a single set of motor nerves. Animals with an even simpler anatomy do not have, and are not likely to have more than one form of swimming. Animals with more complex musculature and innervation than *Aglantha* can obviously perform more complex behaviors, but in *Aglantha* the behavioral sophistication (such as it is) is attained by complexity at the level of the nerve membrane and arises from the two electrical pulses in the motor axons. It is possible that such a system can be set up during development by relatively few commands, but the question then arises as to the degree of sophistication attainable. If two impulses are possible, why not an intermediate form for occasions when escape is not appropriate? It is probable that constraints in organization exist at the molecular level.

ACKNOWLEDGEMENTS. I thank George Mackie and Stuart Arkett for their comments, suggestions and discussion.

References

Anderson, P. A. V., 1985, Physiology of a bidirectional, excitatory, chemical synapse, *J. Neurophysiol.* **53**:821-835.

Arkett, S. A., Mackie, G. O., and Meech, R. W., 1988, Hair-cell mechanoreception in the jellyfish, *Aglantha digitale*, *J. exp. Biol.* **135**:329-342.

Armstrong, C. M., and Matteson, D. R., 1985, Two distinct populations of calcium channels in a clonal line of pituitary cells, *Science* **227**:65-67.

Barr, L., Dewey, M. M., and Berger, W., 1965, Propagation of action potentials and the structure of the nexus in cardiac muscle, *J. Gen. Physiol.* **48**:797-823.

Bean, B. P., 1985, Two kinds of calcium channels in canine atrial cells, *J. Gen. Physiol.* **86**:1-30.

Bloedel, J. R., Gage, P. W., Llinás, R., and Quastel, D. M. J., 1966, Transmitter release at the squid giant synapse in the presence of tetrodotoxin, *Nature* **212**:49-50.

Bossu, J.-L., and Feltz, A., 1986, Inactivation of the low-threshold transient calcium current in rat sensory neurones: evidence for a dual process, *J. Physiol. (Lond.)* **376**:341-357.

Carbone, E., and Lux, H. D., 1984, A low voltage-activated, fully inactivating Ca channel in vertebrate sensory neurones, *Nature* **310**:501-502.

Coulter, D. A., Huguenard, J. R., and Prince, D. A., 1989, Calcium currents in rat thalamocortical relay neurones: kinetic properties of the transient, low-threshold current, *J. Physiol. (Lond.)* **414**:587-604.

Crunelli, V., Lightowler, S., and Pollard, C. E., 1989, A T-type Ca^{2+} current underlies low-threshold Ca^{2+} potentials in cells of the cat and rat lateral geniculate nucleus, *J. Physiol. (Lond.)* **413**:543-561.

Deitmer, J. W., 1984, Evidence for two voltage-dependent calcium currents in the membrane of the ciliate *Stylonychia*, *J. Physiol. (Lond.)* **355**:137-159.

Detwiler, P. B., and Hodgkin, A. L., 1979, Electrical coupling between cones in turtle retina, *J. Physiol. (Lond.)* **291**:75-100.

Donaldson, S., Mackie, G. O., and Roberts, A., 1980, Preliminary observations on escape swimming and giant neurones in *Aglantha digitale* (Hydromedusae:Trachylina), *Can. J. Zool.* **58**:549-552.

Eisenberg, R. S., and Johnson, E. A., 1970, Three-dimensional electrical field problems in physiology, *Prog. in Biophys. and Mol. Biol.* **20**:1-65.

Fatt, P., and Katz, B., 1951, An analysis of the end-plate potential recorded with an intracellular electrode, *J. Physiol. (Lond.)* **115**:320-370.

Fedulova, S. A., Kostyuk, P. G., and Veselovksy, N. S., 1985, Two types of calcium channels in the somatic membrane of new-born rat dorsal root ganglion neurones, *J. Physiol. (Lond.)* **359**:431-446.

Fox, A. P., and Krasne, S., 1984, Two calcium currents in *Neanthes arenaceodentatus* egg cell membranes, *J. Physiol. (Lond.)* **356**:491-505.

Fox, A. P., 1981, Voltage-dependent inactivation of a calcium channel, *Proc. Natl. Acad. Sci. USA* **78**:953-956.

Fox, A. P., Nowycky, M. C., and Tsien, R. W., 1987, Kinetic and pharmacological properties distinguishing three types of calcium currents in chick sensory neurones, *J. Physiol. (Lond.)* **394**:149-172.

Friedman, A., and Gutnick, M. J., 1987, Low-threshold calcium electrogenesis in neocortical neurons, *Neuroscience Letters* **81**:117-122.

Frömter, E., 1972, The route of passive ion movement through the epithelium of *Necturus* gallbladder, *J. Memb. Biol.* **8**:259-301.

Gladfelter, W. B., 1973, A comparative analysis of the locomotory systems of medusoid Cnidaria, *Helgol. Wiss. Meeresunters* **25**:228-272.

Goldman, D.E., 1943, Potential, impedance and rectification in membranes, *J. Gen. Physiol.* **27**:37-60.

Greene, R. W., Haas, H. L., and McCarley, R. W., 1986, A low threshold calcium spike mediates firing pattern alterations in pontine reticular neurons, *Science* **234**:738-740.

Hagiwara, S., Kusano, K., and Saito, S., 1961, Membrane changes on onchidium nerve cell in potassium-rich media, *J. Physiol. (Lond.)* **155**:470-489.

Hagiwara, S., Ozawa, S., and Sand, O., 1975, Voltage clamp analysis of two inward current mechanisms in the egg cell membrane of a starfish, *J. Gen. Physiol.* **65**:617-644.

Halliwell, J. V., 1983, Caesium-loading reveals two distinct Ca-currents in voltage-clamped guinea-pig hippocampal neurones *in vitro*, *J. Physiol. (Lond.)* **341**:10P.

Hodgkin, A.L. and Katz, B., 1949, The effect of sodium ions on the electrical activity of the giant axon of the squid, *J. Physiol. (Lond.)* **108**:37-77.

Iverson, L. E., Tanouye, M. A., Lester, H. A., Davidson, N., and Rudy, B., 1988, Potassium channels from *Shaker* RNA expressed in *Xenopus* oocytes, *Proc. Natl. Acad. Sci. USA* **85**:5723-5727.

Jack, J. J. B., Noble, D., and Tsien, R. W., 1975, *Electrical current flow in exitable cells*, pp. 83-97, Clarendon Press, Oxford.

Josephson, R. K., and Schwab, W. E., 1979, Electrical properties of an excitable epithelium, *J. Gen. Physiol.* **74**:213-236.

Josephson, R. K., 1985, Communication by conducting epithelia, in: *Comparitive Neurobiology; Modes of Communication in the Nervous System* (M. J. Cohen and F. Strumwasser, eds.), Wiley Interscience Publications, New York.

Katz, B., and Miledi, R., 1965, The effect of temperature on the synaptic delay at the neuromuscular junction, *J. Physiol. (Lond.)* **181**:656-670.

Kerfoot, P. A. H., Mackie, G. O., Meech, R. W., Roberts, A., and Singla, C. L., 1985, Neuromuscular transmission in the jellyfish *Aglantha digitale*, *J. exp. Biol.* **116**:1-25.

Lester, H. A., 1970, Trasmitter release by presynaptic impulses in the squid stellate ganglion, *Nature* **227**:493-496.

Llinás, R., and Jahnsen, H., 1982, Electrophysiology of mammalian thalamic neurones *in vitro*, *Nature* **297**:406-408.

Llinás, R., and Yarom, Y., 1981, Properties and distribution of ionic conductances generating electroresponsiveness of mammalian inferior olivary neurons *in vitro*, *J. Physiol. (Lond.)* **315**:549-567.

Mackie, G. O., 1980a, Slow swimming and cyclical "fishing" behavior in *Aglantha digitale* (Hydromedusae:Trachylina), *Can. J. Fish. Aquat. Sci.* **37**:1550-1556.

Mackie, G. O., 1980b, *Epithelium*, McGraw Hill Yearbook Science and Technology, McGraw-Hill Book Company, Inc.

Mackie, G. O., 1984, Fast pathways and escape behavior in Cnidaria, in: *Neural Mechanisms of Startle Behavior* (R.C. Eaton, ed.), Plenum Publishing Corp.

Mackie, G. O., and Meech, R. W., 1985, Separate sodium and calcium spikes in the same axon, *Nature* **313**:791-793.

Mackie, G. O., and Meech, R. W., 1989, Potassium channel family in axons of the jellyfish *Aglantha digitale*, *J. Physiol. (Lond.)*, in press.

Mackie, G. O., and Mills, C. E., 1983, Use of the PISCES IV submersible for zooplankton studies in coastal waters of British Columbia, *Can. J. Fish. Aquat. Sci.* **40**:763-776.

Mackie, G. O., and Passano, L. M., 1968, Epithelial conduction in Hydromedusae, *J. Gen. Physiol.* **52**:600-621.

Meech, R. W., Arkett, S. A., Mackie, G. O., and Maitland, N. J., 1989, Potassium channel family in the jellyfish, *Aglantha*, *Soc. Neuroscience Abstracts 1989*.

Mills, C. E., Mackie, G. O., and Singla, C. L., 1985, Giant nerve axons and escape swimming in *Amphogona apicata* with notes on other hydromedusae, *Can J. Zool.* **63**:2221-2224.

Narahashi, T., Tsunoo, A., and Yoshii, M., 1987, Characterization of two types of calcium channels in mouse neuroblastoma cells, *J. Physiol. (Lond.)* **383**:231-249.

Payton, B. W., Bennett, M. V. L., and Pappas, G. D., 1969, Permeability and structure of junctional membranes at an electrotonic synapse, *Science* **166**:1641-1643.

Roberts, A., and Mackie, G. O., 1980, The giant axon escape system of a hydrozoan medusa, *Aglantha digitale*, *J. exp. Biol.* **84**:303-319.

Satterlie, R. A., and Spencer, A. N., 1983, Neuronal control of locomotion in hydrozoan medusae, *J. Comp. Physiol.* **A150**:195-206.

Schwab, W. E., and Josephson, R. K., 1982, Lability of conduction velocity during repetitive activity of an excitable epithelium, *J. exp. Biol.* **98**:175-193.

Schwarz, T. L., Tempel, B. L., Papazian, D. M., Jan, Y. N., and Jan, L. Y., 1988, Multiple potassium channel components are produced by alternative splicing at the *Shaker* locus of *Drosophila*, *Nature* **331**:137-142.

Shiba, H., 1971, Heaviside's "Bessel cable" as an electric model for flat simple epithelial cells with low resistive junctional membranes, *J. Theoretical Biol.* **30**:59-68.

Singla, C. L., 1978, Locomotion and neuromuscular system of *Aglantha digitale*, *Cell Tiss. Res.* **188**:317-327.

Spencer, A. N., 1978, Neurobiology of *Polyorchis*. I. Function of effector systems, *J. Neurobiol.* **9**:143-157.

Spencer, A. N., 1979, Neurobiology of *Polyorchis*. II. Structure of effector systems, *J. Neurobiol.* **10**:95-117.

Spencer, A. N., 1982, The physiology of a coelenterate neuromuscular synapse, *J. Comp. Physiol.* **148**:353-363

Takeuchi, A., and Takeuchi, N., 1962, Electrical changes in the pre- and post-synaptic axons of the giant synapse of *Loligo*, *J. Gen. Physiol.* **45**:1181-1193.

Weber, C., Singla, C. L., and Kerfoot, P. A. H., 1982, Microanatomy of subumbrellar motor innervation in *Aglantha digitale* (Hydromedusae: Trachylina), *Cell Tiss. Res.* **223**:305-312.

Wilcox, K. S., Gutnick, M. J., and Christoph, G. R., 1988, Electrophysiological properties of neurons in the lateral habenula nucleus: an in vitro study, *J. Neurophysiol.* **59**:212-225.

Woodbury, J. W., and Crill, W. E., 1961, On the problem of impulse conduction in the atrium, in: *Nervous Inhibition*, pp. 124-135 (E. Florey, ed.), Pergamon Press, Oxford.

Zagotta, W. N., Brainard, M. S., and Aldrich, R. W., 1988, Single-channel analysis of four distinct classes of potassium channels in *Drosophila* muscle, *J. Neurosci.* **8**:4765-4779.

Chapter 21

Ionic Currents in Ctenophore Muscle Cells

ANDRÉ BILBAUT, MARI-LUZ HERNANDEZ-NICAISE, and ROBERT W. MEECH

1. Introduction

The term Coelenterate includes two phylla, the Cnidaria and the Ctenophora. Typically, they are diploblastic organisms in which an ectoderm is separated from an endoderm by a gelatinous layer, the mesoglea. Ctenophora appear phylogenetically more advanced than Cnidaria. In Cnidaria the mesoglea is generally acellular and the muscle system consists mainly of myoepthelial cells. Ctenophores have differentiated true muscle cells embedded in the mesoglea. The phylogenetic significance of the presence of muscle cells in the mesoglea of Ctenophores is still debated.

In triploblastic organisms, muscle cells are characterized by extreme morphological and physiological diversity. Contraction is initiated by changes in the ionic permeability of the cell membrane, and these often occur in the form of propagated action potentials. According to the type of muscle cell (skeletal, visceral, vascular or cardiac), the action potentials are more or less well-developed and they are coupled to myofilament activation by a variety of complex mechanisms, which will not be considered here.

Voltage-clamp analysis, performed on a great diversity of muscle cell types, has revealed that they, like nerve cells, are equipped with a large collection of different voltage- and cation-gated ionic channels. These membrane conductances have been particularly well studied in vertebrate and upper invertebrate (i.e. arthropod) muscle

ANDRÉ BILBAUT[1], MARI-LUZ HERNANDEZ-NICAISE[1], and ROBERT W. MEECH[2] • [1]Laboratoire de Cytologie Expérimentale, Université de Nice, U.R.A. 651, Parc Valrose, 06034 Nice Cedex, France, and [2]Department of Physiology, The Medical School, University of Bristol, University Walk, Bristol BS8 1TD, United Kingdom.

cells, but in the lower invertebrates the physiological properties of the muscle cell membrane have remained unknown until this last decade.

Ctenophores are among the simplest of the living metazoa with a recognizable muscle system and their muscle cells display unusual morphological characteristics. Their giant size and the fact that they run independantly through the mesoglea has facilitated their isolation and promoted detailed electrophysiological studies.

Giant smooth muscle cells have been described in two ctenophore species, *Beroe ovata* (Hernandez-Nicaise and Amsellem, 1980) and *Mnemiopsis leydii* (Hernandez-Nicaise et al., 1981, 1984). The present report puts into perspective the different morphological and electrophysiological data obtained from *in situ* and isolated muscle cells of both species. The results will be tentatively placed in the context of the evolution of membrane excitability. The reader is referred to the principle articles for more detailed accounts and discussions (for *Beroe*: Hernandez-Nicaise et al., 1980; Bilbaut et al., 1988a,b; for *Mnemiopsis*: Anderson, 1984; Dubas et al., 1988).

2. Morphology of Muscle Cells

Beroe ovata is a beroid ctenophore typically bell-shaped and devoid of tentacles. The largest animals are 10-12 cm in length and 5-6 cm in diameter. The mouth opens into a large, stomodeal cavity which fills the entire body.

The body wall of *Beroe* is soft and consists of a gelatinous ectomesoderm, the mesoglea, in which are embedded endodermic channels and muscle cells. Intramesogleal muscle cells are arranged into three distinct systems: longitudinal, radial and circular sytems. In the longitudinal and radial systems, each muscle cell runs freely through the mesoglea. Longitudinal muscle cells (LMFs) are oriented along the major axis of the animal. Radial muscle cells (RMFs) are oriented perpendicularly to that axis. In the circular sytem, muscle cells are anastomosed and build a monolayer encircling the pharyngeal cavity.

Each muscle cell can be seen clearly through the transparent mesoglea. When they are observed through a stereomicroscope, muscle cells appear like thin filaments of variable diameter (up to 50 μm). Radial cells can be up to 1 cm in length; longitudinal cells can reach several centimeters, justifying the use of the term "giant muscle cells" to describe them. Because LMFs have a tapered end, whereas RMFs are anchored into the epithelia by extensive branching they may be readily identified when isolated.

Intramesogleal muscle cells are also present in the lobate ctenophore *Mnemiopsis*, but the muscle systems are somewhat different from those described in *Beroe*. Most easily identifiable are two bundles of longitudinal giant muscle cells which extend along the stomodeum from the aboral pole to the mouth of the animal. Each bundle consist of 30-50 individual cells 15 μm in diameter and 3-4 cm in length in a medium-sized specimen.

Ultrastructural observations show that each muscle fiber consists of a central core made of a row of nuclei and mitochondria. The central core is surrounded by a sheath of myofilaments devoid of periodic striations. There are two types of filaments. In *Beroe*, thick filaments 16-19 nm in diameter, have irregular profiles in tranverse section. Thin filaments, 6 nm in diameter, are organized around the thick filaments in irregular "rosettes" with a ratio of thin-to-thick filaments varying from 4 to 7 (Hernandez-Nicaise et al., 1980). There are no dense bodies or peripheral attachment plates as in vertebrate muscle cells. Microtubules and sarcoplasmic reticulum, both longitudinally oriented, are scattered in the sheath of myofilaments. No evidence for a peripheral system of junctions between sarcoplasmic reticulum and sarcolemma has been found.

Gap junctions have been ultrastructurally identified as being present between the extremities of muscle cells in *Beroe*, but their distribution is not clear and it is not clear whether or not the cells are electrically coupled. In *Mnemiopsis*, the LMFs are not dye coupled (Anderson, 1984).

In *Beroe*, synapses have been described between nerve endings and sarcolemma of different muscle cells. Synaptic vesicles are present in pre-synaptic terminals where they are associated with mitochondria and reticulum cisternae, but the pharmacology of excitation is unknown.

3. Isolated Muscle Cells

Muscle cells can be isolated from the body wall of *Beroe* (Hernandez-Nicaise et al., 1982) and *Mnemiopsis* (Stein and Anderson, 1984). The method used is basically the same in each case and consists of enzymatically digesting the mesoglea with hyaluronidase and detaching the extremities of the muscle cells from their epithelial insertion with trypsin. In *Beroe*, a piece of body wall may yield several hundred radial and longitudinal cells. Circular muscle cells adhere strongly to the pharyngeal epithelium and need a stronger enzymatic treatment to be released. They have not yet been studied.

Immediately after being released, most isolated muscle cells are in a contracted state. They relax rapidly in a recovery medium containing low levels of Ca^{++} (2 mM in *Beroe*; 1.7 mM in *Mnemiopsis*). Muscle cells which have been cut during dissection or dissociation reseal almost immediately, and retain their membrane excitability for 3-4 days or more (Stein and Anderson, 1984). Relaxed cell segments appear as smooth cylinders, but are sometimes coiled along a short section. Their length is very variable, and may be 2 cm or more. In *Beroe*, isolated RMFs are identified by their extensive terminal branching. In early experiments, LMFs were identified by being longer than an entire radial cell. Once their characteristic electrophysiology had been established, however, shorter lengths could be identified from the waveform of their action potential.

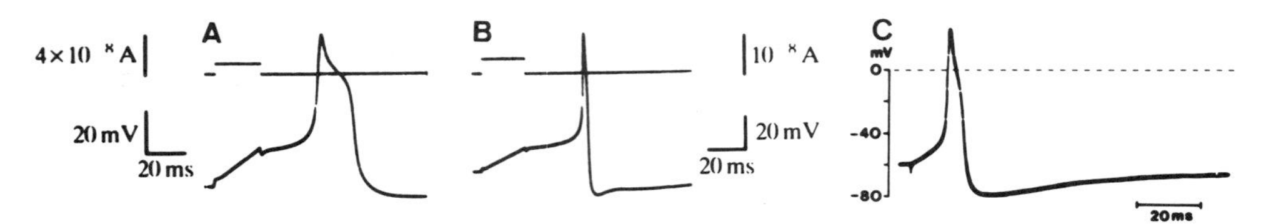

Figure 1. Action potentials from ctenophore isolated smooth muscle cells. (A) LMF from *Beroe*. The depolarizing phase of the action potential (lower trace) has a plateau that is of variable duration in different cells. (B) RMF from *Beroe*. There is no evidence for a plateau phase. Upper trace: current monitor and zero potential (from Bilbaut et al., 1988a). (C) Action potential recorded in a smooth muscle cell isolated from *Mnemiopsis* (from Anderson, 1984).

4. Electrophysiology of Muscle Cells

4.1. Resting Potential

The resting potential of isolated muscle cells of *Beroe ovata* is around -60 mV. The mean resting potential of the radial cells (-66 ±1.4 mV) differs significantly from that of the longitudinal muscle cells (-60 ±1.35 mV; Bilbaut et al., 1988a) and such a difference is also evident in *in situ* fibers (Hernandez-Nicaise et al., 1980). The membrane potential is mainly dependent on the external K^+ concentration within the range 10-100 mM. In this range the relationship between the membrane potential and the logarithm of external K^+ concentration is linear with a slope of 51.2 mV for RMFs and 54 mV for LMFs (Bilbaut et al., 1988a). In *in situ* RMFs, following a treatment of the mesoglea by hyaluronidase, there is a 58 mV change for a tenfold change in external K^+, but there are deviations from the predictions of the Nernst Equation at lower levels of K^+ (Hernandez-Nicaise et al., 1980; Bilbaut et al., 1988a). In isolated muscle cells of *Mnemiopsis*, the resting membrane potential is 59 ±0.7 mV. As in *Beroe*, the relationship between membrane potential and extracellular K^+ follows the Nernst equation only at high K^+. At low K^+, substitution of chloride by propionate provides evidence for a significant resting chloride permeability of the muscle membrane (Anderson, 1984).

4.2. Action Potentials

In *Beroe*, both radial and longitudinal muscle cells produce an overshooting action potential in response to intracellular current injection. In isolated cells the action potential amplitude is 79 mV ±1.5 mV in longitudinal cells and 84.9 ±2 mV in radial cells (Bilbaut et al., 1988a). Although the two types of muscle cells are morphologically similar, the waveforms of their action potentials are significantly different. In LMFs (Fig. 1A), the repolarizing phase is prolonged by a plateau potential of variable duration (between 4.5 and 17.5 ms). In RMFs, action potentials with plateau potentials are never observed (Fig. 1B). Similar differences have been recorded from *in situ* fibers (Hernandez-Nicaise et al., 1980). A second difference is that RMFs can generate a long-lasting train of repetitive firing during prolonged current injection, whereas in LMFs, the response is only brief.

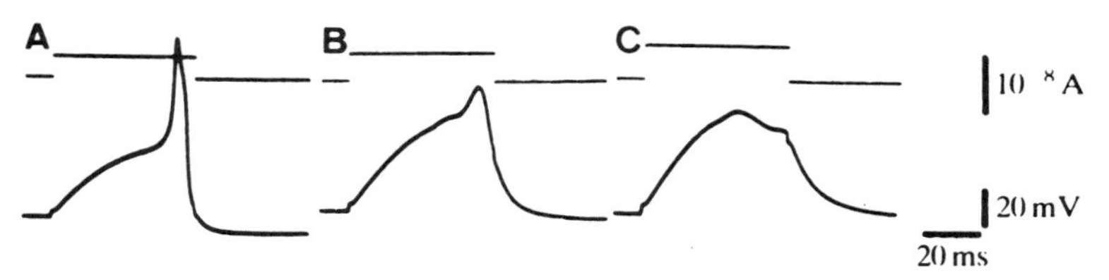

Figure 2. Effect of Co^{++} on the action potential of an LMF isolated from *Beroe*. (A) Action potential in Na^{+}-deficient ASW. (B) and (C) Progressive disappearence of the action potential overshoot in the presence of Co^{++} (20 mM). Upper trace: current monitor and zero potential (from Bilbaut et al., 1988a).

In *Mnemiopsis* (Anderson, 1984), isolated muscle cells produce action potentials that closely ressemble those from LMFs isolated from *Beroe* (Fig 1C). The action potentials are 75 ±1.3 mV in amplitude and are similar in form in all muscle cells studied. They display a characteristic inflexion in the beginning of the repolarizing phase and a marked undershoot. In *Mnemiopsis*, the duration of action potential increases as the membrane potential is made less negative by the injection of a constant depolarizing current.

4.3. Ionic Dependence of Action Potentials

The ionic dependence of the action potential has been examined in the muscle cells of *Beroe* (Hernandez-Nicaise et al., 1980; Bilbaut, et al., 1988a) and *Mnemiopsis* (Anderson, 1984). In both cases the tetrodotoxin (TTX)-resistant depolarizing phase is dependent on both Na^{+} and Ca^{++}. This conclusion is supported by the following observations.

1. Action potentials can be recorded in Ca^{++}-free artificial sea water (Ca^{++}-free ASW), but the overshoot is greatly reduced. In LMFs isolated from *Beroe*, an overshoot of 5-6 mV is recorded in Ca^{++}-free ASW instead of the 19-21 mV overshoot found in normal ASW (Fig. 4B). In *Mnemiopsis*, the addition of EGTA (1-2 mM) to Ca^{++}-free ASW produces a similar reduction in action potential amplitude.

2. Although Na^{+}-free solutions have not been tested, because the absence of Na^{+} (Tris-substitution) causes irreversible contraction in *Beroe* and sprouting in *Mnemiopsis*, lowering the extracellular Na^{+} concentration to 5% (*Beroe*) and 25% (*Mnemiopsis*) has little effect. Fast overshooting action potentials are maintained in LMFs isolated from both *Beroe* and *Mnemiopsis*. In *Beroe*, the overshoot decreases by no more than 5-6 mV (Fig 2A).

3. In RMFs isolated from *Beroe*, action potentials are abolished in Na^{+}-deficient ASW, indicating a significant contribution from Na^{+}, but action potential amplitude is also reduced in Ca^{++}-free ASW. Furthermore, following the addition to normal ASW of Ca^{++}-conductance blockers, such as Co^{++} (15-20 mM) or Cd^{++} (1-2 mM), the amplitude of the action potential is reduced.

4. Co^{++} and Cd^{++} have similar effects on the action potentials of LMFs isolated from both *Beroe* and *Mnemiopsis*. In the presence of either divalent cation, spike amplitude is decreased.

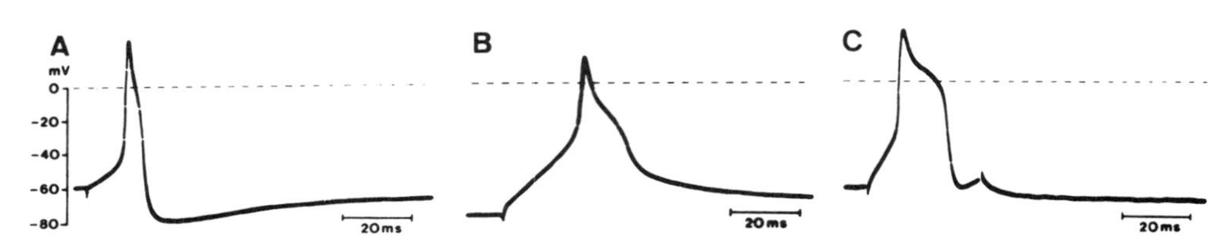

Figure 3. (A) A single action potential from a muscle cell isolated from *Mnemiopsis* (normal ASW). (B) In the presence of D600 in ASW, the spike became significantly longer. (C) An action potential recorded in the presence of TEA (100 mM). Note the pronounced broadening of the spike (from Anderson, 1984).

5. In *Beroe*, LMFs fail to produce action potentials when Co^{++} is added to Na^{+}-deficient ASW (Fig. 2).

6. Other potent Ca^{++} channel blockers have been tested on ctenophore muscle cells. Low concentrations of verapamil (2 mg l^{-1}), have no effect on the action potentials of *in situ* muscle cells of *Beroe* (Hernandez-Nicaise et al., 1980) but at higher concentrations (60 mg l^{-1}), the overshooot is increased and the falling phase prolonged. Verapamil (50 mg l^{-1}) and D600 (10 mg l^{-1}) (Fig. 3B) reduce spike amplitude in the isolated muscle cells of *Mnemiopsis* (Anderson, 1984) in much the same way as Ca^{++}-free ASW.

7. In *Beroe*, Sr^{++} and Ba^{++} added to Ca^{++}-free ASW fully restore the action potential amplitude in both muscle cell types (Fig. 4), although the duration is increased to twice its normal value in the presence of Sr^{++} and can reach 900 ms in the presence of Ba^{++}.

Although there are significant differences between the sensitivity of the different muscle types to Na^{+}-free ASW, it is clear that in each case the depolarizing phase is dependent upon both Na^{+} and Ca^{+}. The repolarizing phase is apparently dependent on K^{+}, although once again, there are differences in pharmacology. In LMFs of *Beroe* and *Mnemiopsis*, tetraethylammonium (TEA) (10-50 mM in *Beroe* and 100 mM in *Mnemiopsis*, Fig. 3C) slightly lengthens the repolarizing phase of the action potential and 4-aminopyridine (4-AP) (2 mM) has a similar effect. In *Beroe*, a long-lasting plateau phase, 400 ms in duration, is observed in the presence of TEA in Ca^{++}-free ASW. In RMFs, TEA (10-50 mM) has very little effect, and the action potentials are

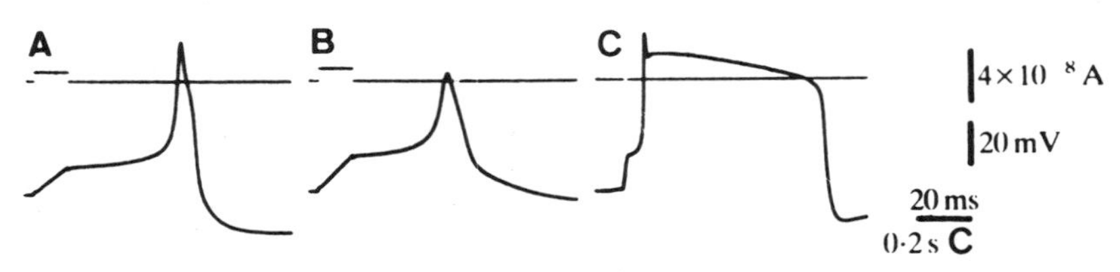

Figure 4. Effect of Ba^{++} (10 mM) on the membrane response (lower trace) of an LMF isolated from *Beroe*. (A) Action potential in ASW. (B) Membrane response after 2 min in Ca^{++} -free ASW. (C) Long-lasting, depolarizing plateau potential recorded after the addition of Ba^{++} to Ca^{++}-free ASW (1 min incubation) (from Bilbaut et al., 1988a).

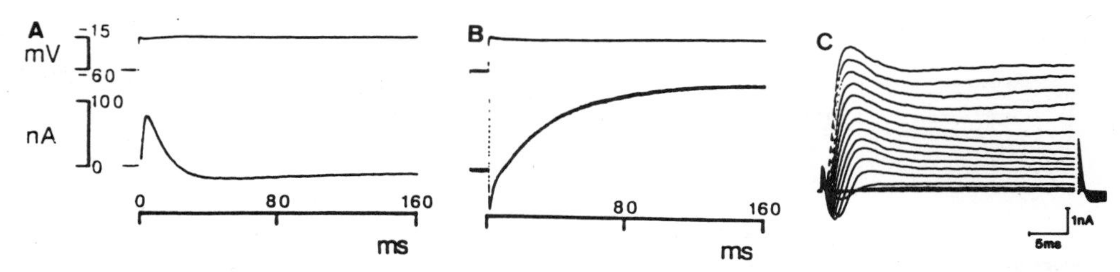

Figure 5. Ion currents in smooth muscle cells isolated from ctenophores. A) and B) Currents recorded from LMF fragment (A) and RMF fragments (B) of *Beroe* during a 45 mV depolarizing step (top trace). Holding potential in each case -60mV (from Bilbaut et al., 1988b). (C) Whole-cell recording of membrane currents of a muscle cell fragment of *Mnemiopsis*. Holding potential -70 mV. First voltage step to -40 mV with 7.5 mV increments thereafter (from Dubas et al., 1988).

also 4-AP-insensitive. In Ca^{++}-free ASW containing TEA or 4-AP, strong current pulses induce oscillating plateau potentials which are not seen in normal ASW.

4.4. Ionic Membrane Currents in Muscle Cells

Voltage clamp experiments have been performed on muscle cells from both *Beroe* and *Mnemiopsis*. In *Beroe*, conventional two-microelectrode techniques have been used on isolated fragments 0.4-0.6 mm in length (Bilbaut et al., 1988b). Calculations from equations derived for a short cable show that current injected midway along a resting fiber (length constant 2 mm) produces a voltage difference between the center and each end of about
1%. Use of a third recording micropipette allowed us to confirm experimentally that, even in our most elongated fibers, the errors are within 1% for pulses of up to 50 mV. Fibers are in a relaxed state when penetrated, but they invariabily shorten during the course of the experiment. When this occurs, the cells form spheres. As the currents recorded from these spheres are identical to those recorded from shorter, relaxed cells, they are used to test the effects of larger amplitude pulses.

In *Mnemiopsis*, Dubas et al. (1988) have taken similar care to check the adequacy of their two-microelectrode voltage-clamp. With a length constant of 1.7 mm at the resting potential, muscle fragments 240 μm in length can be considered to be isopotential. Ionic currents have been also recorded from fragments as small as 20-50 μm using a patch-pipette in the whole-cell configuration.

In order to reduce contraction, muscle cells have been voltage-clamped in low-Ca^{++} bathing solution (2 mM for *Beroe* and 1.7 mM for *Mnemiopsis*).

Total Ionic Membrane Currents

There are significant differences in the ionic currents recorded from the different cell types in *Beroe*. In RMFs, a 45 mV depolarizing step from a holding potential of -60 mV produces an inward current followed by a rapidly-activating, steady, outward current (Fig. 5B). In LMFs under the same conditions, a brief, inward, transient is

Figure 6. Whole-cell recording of membrane currents of a muscle cell fragment isolated from *Mnemiopsis*. (A) Family of inward currents. Holding potential: -70 mV. First voltage step to -40 mV with 5 mV increments thereafter. (B) Family of inward currents recorded from the same cell as (A) in response to the same voltage paradigm but after a 50 ms prepulse to -10 mV. Transient inward current is absent. c) Transient inward currents obtained by digital subtraction of the records presented in (A) and (B). Internal patch pipette solution contained Cs^+ and TEA (from Dubas et al., 1988).

followed by a large outward current that inactivates rapidly to reveal, once again, the long-lasting inward current (Fig. 5A). With larger pulses, a maintained, steady, outward current also develops. In *Mnemiopsis*, the LMF ionic currents resemble those in *Beroe*, except that the steady current appears to develop at lower potentials. There is an early, inward current followed by a transient outward current and a maintained steady-state outward current (Fig. 5C).

Inward Currents

In *Beroe*, analysis of the inward current has been hindered by the presence of overlapping outward currents and the fact that in Na^+-deficient ASW, depolarizing pulses induce an irreversible contraction and loss of excitability. The inward currents are TTX-insensitive but are reduced by the Ca^{++} channel blockers Co^{++} and Cd^{++}. The presence of either Co^{++} or Cd^{++} in normal ASW markedly depresses the inward current but never blocks it completely. Furthermore, in RMFs we have been able to show that the inward current increases with increasing extracellular Ca^{++} (from 2 to 10 mM), and the participation of Ca^{++} in the inward current is clear in both types of cells (Bilbaut et al., 1988b).

Inward currents in LMFs of *Mnemiopsis* have been studied with the whole-cell patch clamp technique (Dubas et al., 1988). By substituting Cs^+ and TEA for K^+ in the pipette patch solution, outward currents are blocked and the inward current is recorded alone. The inward current can be clearly subdivided into two components on the basis of their activating and inactivating properties: a transient inward current which peaks in 5 ms then decays, and a steady-state current. The transient inward current is fully inactivated by depolarizing prepulses which do not affect the steady-state current (Fig. 6). Both ionic currents activate at the same potential (-40 mV) and reverse around +60 mV.

The total inward current is diminished in Ca^{++}-free ASW containing EGTA. For a tenfold increase in extracellular Ca^{++} concentration, in Na^+-deficient ASW, the reversal potential of the inward current changes by 22 mV, close to the 29 mV shift predicted by the Nernst equation. Furthermore, the inward currents are partially blocked by Co^{++} (5 mM), D600 (200 μM) and La $^{+++}$ (2 mM).

Table 1. List of ionic currents in muscle cells isolated from ctenophores *Beroe* and *Mnemiopsis*

Muscle types		*Beroe* radial	*Beroe* longitudinal	*Mnemiopsis* longitudinal
I_{IN}	Ca^{++} current	yes	yes	yes
	Na^{+} curent	?	?	*
I_{OUT}	early transient voltage-activated	yes	yes	yes
	steady-state voltage-activated	yes	(yes)	yes
	Ca^{++}-activated	?	yes	yes
	early transient Na^{+}-Ca^{++}-activated	no	no	yes

*In *Mnemiopsis* muscle cells, Na^{+} inward current flows through Ca^{++} channels (Dubas et al., 1988).

In Na^{+}-deficient ASW, both components of the inward current are slightly reduced. In Ca^{++}-free ASW, a five-fold increase in the extracellular Na^{+} concentration produces a 26 mV shift in reversal potential, instead of the 40 mV change predicted from the Nernst equation. Inward currents are unaffected by TTX (10 μM) or STX (10 μM) added to either low Ca^{++}-ASW or Ca^{++}-free ASW.

Outward Currents

Voltage clamp experiments reveal the existence of a great diversity of outward currents in ctenophore muscle cells (Table 1). They may be classified according to their time course, voltage dependence and gating properties. Two distinct voltage-activated currents and two distinct cation-activated currents have been identified.

In *Beroe* LMFs, a large, early transient, outward current is voltage-activated by stepping the membrane from a -60 mV holding potential. At -15 mV this transient current reaches a peak in 7 ms and inactivates in 30 ms. It is strongly depressed by conditioning prepulses (Fig. 7A). Inactivation is present at the resting potential and is only fully removed when the holding potential is made more negative than -90 mV (Fig. 7B). Although TEA-insensitive, the early outward current is reversibly blocked by 4-AP, and it is probably carried by K^{+} as its tail current reversal potential is at the K^{+} equilibrium potential. When the extracellular K^{+} concentration is changed, the tail

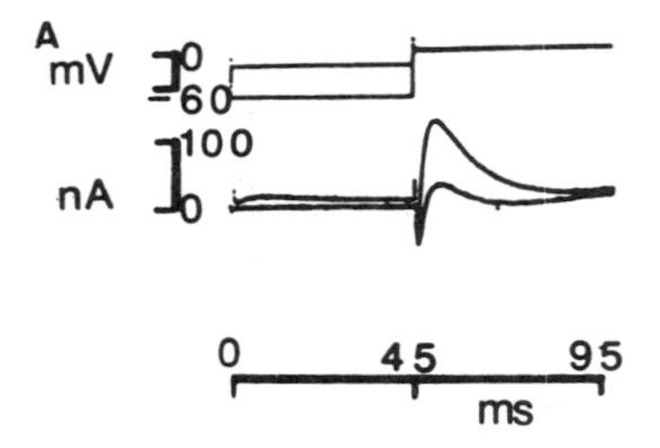

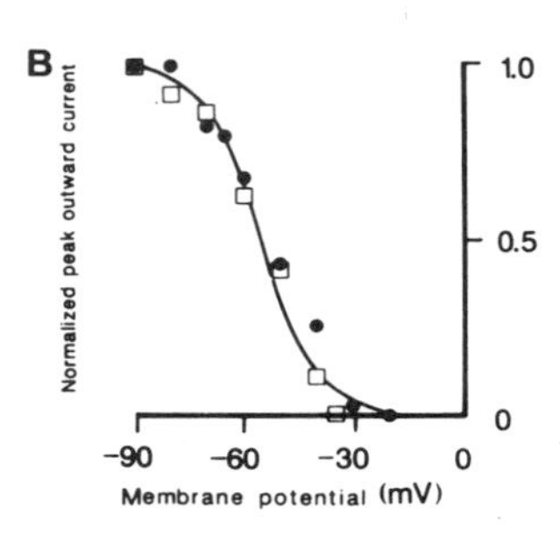

Figure 7. (A) Transient outward currents elicited by a step depolarization to 0 mV in an LMF isolated from *Beroe*. The recording consists of two superimposed traces; in one, the membrane potential was shifted from the holding potential (-60 mV) to -70 mV shortly before the test pulse. In the other, there was an equivalent period of depolarization. The effect of a 45 ms conditioning depolarization to -40 mV was to significantly depress the transient outward current. **(B)** Inactivation of the transient outward current at different membrane potentials. The peak outward current recorded from a typical LMF of *Beroe* was measured and normalized after 50 ms conditioning steps to potentials in the range -20 to -90 mV (●). Holding potential -50 mV, 40 mV test pulse. For comparison, normalized values for the transient outward current from a RMF isolated from *Beroe* are shown for conditioning pulses over the same voltage range (□) (from Bilbaut et al., 1988b).

current reversal potential shifts by an amount predicted by the Nernst equation. In *Mnemiopsis* LMFs, a similar current has been shown to be Cd^{++}-insensitive.

In RMFs of *Beroe*, a similar transient outward current, also reversibly blocked by 4-AP (Fig. 8), appears as a small shoulder at the beginning of the rising phase of a maintained outward current. Its activation and inactivation characteristics are the same

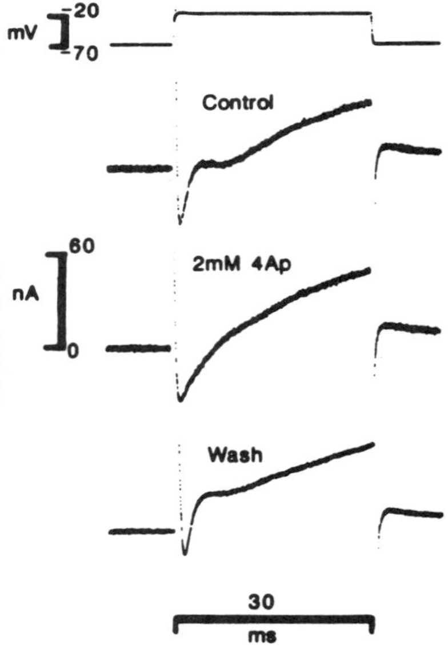

Figure 8. Effects of 4-AP on the outward transient current in a RMF isolated from *Beroe*. Membrane currents in response to a 50 mV depolarization from a holding potential of -70 mV before, and after, exposure to 2 mM 4-AP. The 4-AP reversibly blocked the transient on the rising phase of the outwardly-directed current (from Bilbaut et al., 1988b).

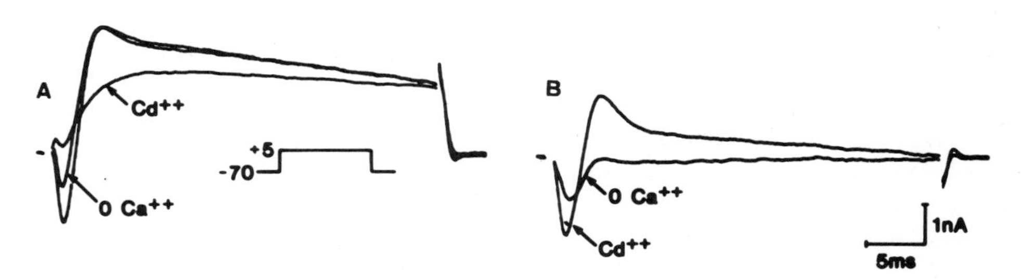

Figure 9. Whole-cell recording from muscle cell fragment of *Mnemiopsis*. (A) Total membrane currents generated by the voltage step shown, in low-Ca^{++} ASW, Ca^{++}-free ASW (zero Ca^{++}) and Ca^{++}-free ASW with 5 mM Cd^{++}. These records were obtained from the same cell fragment and have been superimposed. (B) Fractions of the total membrane current lost by superfusion with zero Ca^{++} and with zero Ca^{++} with 5 mM Cd^{++}. These records were obtained by digital subtraction of the records shown in (A) (from Dubas et al., 1988).

as those observed in the LMF transient current. By analogy with the early transient outward current reported in molluscan neurons, these currents will be termed I_A (Connor and Stevens, 1971).

In LMFs of *Beroe* a Co^{++}-sensitive, delayed, outward current is activated by stepping the membrane to near 0 mV. This delayed current, which reaches a peak within about 100 ms, is strongly depressed in the presence of Co^{++} (20 mM) or TEA (50 mM). Its current-voltage curve is typically N-shaped. Such characteristics are similar to Ca^{++}-activated K^+ currents (I_{KCa}) decribed in other excitable cells (Meech, 1978).

In *Mnemiopsis* LMFs, depolarizing pulses activate a late, Cd^{++}-sensitive, transient outward current that lasts several seconds. It activates between -40 and -50 mV, reaches a peak at +25 mV, and completely inactivates. Its tail current reversal potential is close to E_K. As in LMFs of *Beroe*, the late transient current of *Mnemiopsis* muscle cells is likely to be Ca^{++}- activated.

A second, Ca^{++}-activated, outward current has been identified in muscle cells of *Mnemiopsis*. This is an early transient current which is blocked by Cd^{++} (4-5 mM), but persists in Ca^{++}-free ASW (Fig. 9). Both the transient Cd^{++}-sensitive current and the inward current activate at the same potential (-25 mV) and they both reach a peak at around 0 mV. According to Dubas et al. (1988), the Cd^{++}-sensitive, early transient current is Na^+- or Ca^{++}-activated and had been termed $I_{KCa/Na}$.

In the presence of 5 mM Cd^{++}, isolated *Mnemiopsis* LMFs display two outward currents: the early, transient, voltage-activated current described above, and a steady-state outward K^+ current which slowly activates at holding potential more positive than -40 mV.

In *Beroe* RMFs, outward currents are dominated by a rapid, voltage-activated, steady- state outward current. This current activates at depolarizing potentials more positive than -30 mV. It is depressed by TEA and is insensitive to external Co^{++}.

5. Discussion

This report shows that the ionic currents identified in ctenophore muscle cell membranes are similar to the ionic currents described in muscle cells of higher animals. In both cases, membrane currents are voltage- and cation-activated, their time courses are similar and they are blocked by the same substances.

5.1. Cell Membrane Excitability

The cell membrane of ctenophore muscles is negatively polarized. Although a significant chloride permeability has been identified in *Mnemiopsis*, K^+ appears to dominate the membrane potential. At rest, the membrane potential remains stable and does not display endogenous oscillations.

As Horridge (1966) suggested, ctenophore muscle cells propagate impulses in the form of overshooting action potentials. They, therefore, appear to be electrically excitable in the same way as the muscle cells of higher organisms.

5.2. Ionic Currents in Ctenophore Muscle Cell Membrane

Inward Current

In invertebrate muscle cells, the action potentials are most often Ca^{++} spikes (insect, Washio, 1972; crustacean, Fatt and Ginsborg, 1958; Werman and Grundfest, 1961; mollusc, Kidokoro et al., 1974). Where voltage-clamp experiments have been performed, inward current has been identified as a pure Ca^{++} current (Hagiwara and Byerly, 1981).

In the muscle cells of *Beroe* and *Mnemiopsis*, action potentials persist in either Ca^{++}-free or Na^+-deficient ASW, suggesting that the rising phase is dependent on the presence of both Na^+ and Ca^{++}. Na^+-Ca^{++}-dependent action potentials have been described in other contractile cells, such as the protozoan *Actinocoryne* (Febvre-Chevallier et al., 1986), swimming muscle cells of the hydrozoan medusea *Polyorchis* (Spencer and Satterlie, 1981), protochordate muscles (Hagiwara and Kidokoro, 1971), or vertebrate cardiac muscle cells (for review, see Reuter, 1975). In vertebrate smooth muscle cells, where Ca^{++} appears to be the major component of the depolarizing phase (Brading et al., 1969; Walsh and Singer, 1980; Isenberg and Klöckner, 1985), membrane electrogenesis may be maintained in a solution free of either Na^+ or Ca^{++} (Prosser et al, 1977; Mironneau et al., 1982).

The participation of both Na^+ and Ca^{++} in the action potential of ctenophore muscle cells raises the question of whether or not the ionic current flows through separate membrane channels. In *Mnemiopsis*, inward currents in Ca^{++}-free ASW are blocked by 5 mM external Cd^{++}, and Dubas et al. (1988) have concluded that Na^+ passes through Ca^{++} channels. A similar situation exists in some vertebrate smooth muscle cells. In the absence of extracellular Ca^{++}, membrane electrogenesis is accomplished by a TTX-insensitive Na^{+} current which enters through Ca^{++} channels (Jmari et al., 1987). In general, however, the currents observed in Ca^{++}-free ASW in *Mnemiopsis* are somewhat resistant to Ca^{++} channel blockers; 2 mM La^{+++} gives only a partial block, for example, and the effect of 5 mM Cd^{++} is not readily reversible. It is, therefore, difficult to exclude a contribution from a TTX-resistant Na^{+} channel.

Two types of Ca^{++} channel co-exist in the muscle cell membrane of *Mnemiopsis*. As described in excitable cells of higher phyla (Nowycky et al., 1985), one type of channel supports a transient, inward, Ca^{++} current (I_T), the other supports a steady-state, inward Ca^{++} current (I_L). Furthermore, inactivation of Ca^{++} channels in *Mnemiopsis* muscle cells is voltage-dependent (Dubas et al., 1988). In vertebrate muscle cells, inactivation of Ca^{++} channel may be voltage and/or Ca^{++} -dependent (Jmari et al., 1986).

In *Beroe* muscle cells, the inward current in the presence of both Na^{+} and Ca^{++} is reversibly depressed, but not blocked, by Co^{++} or Cd^{++}. The effects of Na^{+}-free ASW have not been investigated with the voltage-clamp technique but the action potential recorded from RMFs is abolished within 1 minute of exposure to a Na^{+}-deficent solution containing 25 mM Na^{+}. A similar solution had no effect on LMF action potentials. Furthermore, although Co^{++} (20 mM) has little effect on the radial fiber action potential, it fully blocks the action potential recorded from the LMF in Na^{+}-deficient solution. This clearly indicates that the depolarizing phase of action potentials of the RMF type may be dominated by Na^{+} flowing through Co^{++}- and TTX-resistant channels (Bilbaut et al., 1988a).

Outward Current

In excitable membranes, membrane repolarization is achieved most often by one of a large variety of K^{+} conductances. In the muscle cell membranes of the ctenophores *Beroe* and *Mnemiopsis* , many different types of K^{+} channel have been characterized by voltage-clamp (Table 1). Two outward currents, the early transient current and the steady-state current are voltage-activated. The early transient outward current, which resembles that described in some vertebrate smooth musles (Vassort, 1975), was first reported by Hagiwara et al. (1961) and has been well-characterized by Connor and Stevens (1971) in *Aplysia* neurons.

There is also a steady-state outward K^{+} current, which is present in a number of excitable cells, including vertebrate smooth muscle cells (Mironneau and Savineau, 1980) and two other cation-activated outward currents. The first is the well-known Ca^{++}-activated K^{+} current characterized by Meech and Standen (1975) in gastropod neurons. This outward current, which activates as a consequence of the increase in internal Ca^{++} concentration, has also been identified in smooth muscle cells of vertebrates (Berger, et al., 1984; Walsh and Singer, 1983). The second cation-activated outward current is an unusual Na^{+}-Ca^{++}-activated K^{+} current found only in muscle cells of the ctenophore *Mnemiopsis*.

5.3. Diversity of Ionic Currents

The structural organization of ctenophore muscle cells is homogenous. Electron microscopy has revealed no differences between the muscle cells of *Beroe* and *Mnemiopsis* (Hernandez-Nicaise and Amsellem, 1980; Hernandez-Nicaise et al., 1984). However, while the muscle cells of the two ctenophores are structurally similar, there are significant physiological differences. The waveforms of action potentials recorded from the LMFs of *Beroe* and *Mnemiopsis* are similar. The beginning of the repolarizing phase is characterized by an inflexion which is absent in radial fibers. On the other hand, in *Beroe*, RMFs are capable of producing prolonged trains of repetitive pulses.

Voltage-clamp experiments performed on identified muscle cells of both ctenophores reveal a large diversity of ionic channels that co-exist in their cell membranes. Each type of muscle cells (radial in *Beroe*, longitudinal in *Beroe* and *Mnemiopsis*) is equipped with a collection of ionic channels which is different from one cell type to another.

Ca^{++} participates in the inward current in each type of ctenophore muscle cell, but whether Na^{+} is also involved under normal physiological conditions remains unclear.

Early, voltage-activated K^{+} currents have been identified in all muscle cells examined. In *Beroe*, the early outward current is large in LMFs, while in RMFs it is barely visible. In *Mnemiopsis*, the voltage-activated, early current overlaps the Cd^{++}-sensitive, cation-activated transient current. The delayed, outward current of *Beroe* RMFs is a steady-state, voltage-activated current, while in *Beroe* LMFs, the delayed outward current is dominated by a Ca^{++}-activated K^{+} current. In *Mnemiopsis*, the K^{+} current is activated so slowly by Ca^{++} that its contribution to the repolarizing phase of action potentials is unlikely (Dubas et al., 1988).

Although surprising, the diversity of ionic membrane conductances in ctenophore muscle cells may be related to their different patterns of muscle activity. Activation of muscle myofibrills depends on an increase in internal Ca^{++} and this depends on the action potential waveform. Instead of having groups of structurally different muscle fibers or elaborate patterns of innervation as in higher animals, *Beroe* achieves coordinated movements from muscles whose mechanical response to stimulation depends on the ionic conductances present in the cell membrane. The longitudinal muscle systems and radial muscle systems work successively or simultaneously to elaborate elementary behaviors (Bilbaut et al., 1988b). Coordinated body movements arise from different regenerative impulses in structures of identical morphology. Simple effector systems are driven to produce complex behavior by complex membrane events.

Ctenophore smooth muscle fibers differ structurally from vertebrate smooth muscle cells as they lack dense bodies, attachment plates, and surface vesicles (caveolae). Nevertheless, all smooth muscle cell types must solve the same problem in order to contract: impulses propagated over the cell surface must raise the internal Ca^{++} concentration to activate myofibrills in the absence of well-developed sarcoplasmic reticulum and T-tubule systems. As the present report shows, the population of ion channels found in the cell membrane of ctenophore muscle fibers is comparable to that described in many vertebrate smooth muscle cells, and the problem appears to have been solved in much the same way. Unfortunately, we are not yet in a position to be able to say whether or not this is a case of parallel evolution, and we cannot yet compare the structures of the different membrane channels with those of higher animals, however similar they may be in function.

References

Anderson, P. A. V., 1984, The electrophysiology of single smooth muscle cells isolated from the ctenophore *Mnemiopsis*, *J. Comp. Physiol. B* **154**:257-268.

Berger, W., Grygorcyk, R., and Schwarz, W., 1984, Single K^+ channels in membrane evaginations of smooth muscle cells, *Pfülgers Arch.* **402**:18.

Bilbaut, A., Meech, R. W., and Hernandez-Nicaise, M.-L., 1988a, Isolated giant smooth muscle fibres in *Beroe ovata*: ionic dependence of action potentials reveals two distinct types of fibre, *J. exp. Biol.* **135**:343-362.

Bilbaut, A., Hernandez-Nicaise, M.-L., and Meech, R. W., 1988b, Membrane currents that govern smooth muscle contraction in a ctenophore, *Nature* **331**:533-535.

Brading, A., Bulbring, E., and Tomita, T., 1969, The effect of sodium and calcium on the action potential of the smooth muscle of guinea-pig teania coli, *J. Physiol. (Lond.)* **200**:637-654.

Connor, J. A., and Stevens, C. F., 1971, Voltage clamp studies of a transient outward current in gastropod neural somata, *J. Physiol. (Lond.)* **213**:21.

Dubas, F., Stein, P. G., and Anderson, P. A. V., 1988, Ionic currents of smooth muscle cells isolated from the ctenophore *Mnemiopsis*, *Proc. R. Soc. Lond. B* **233**:99-121.

Fatt, P., and Ginsborg, B. L., 1958, The ionic requirements for the production of action potentials in crustacean muscle cells, *J. Physiol. (Lond.)* **142**:516-543.

Febvre-Chevallier, C., Bilbaut, A., Bone, Q., and Febvre, J., 1986, Sodium-calcium action potential associated with contraction in the heliozoan *Actinocoryne contractilis*, *J. exp. Biol.* **122**:177-192.

Hagiwara, S., and Byerly, L., 1981, Calcium channel, *Ann. Rev. Neurosci.* **4**:69-125.

Hagiwara, S., and Kidokoro, Y., 1971, Na and Ca components of action potential in amphioxus muscle cells, *J. Physiol. (Lond.)* **219**:217-232.

Hagiwara, S., Kusano, K., and Saito, N., 1961, Membrane changes on *Onchidium* nerve cell in potassium-rich media, *J. Physiol. (Lond.)* **155**:470-489.

Hernandez-Nicaise, M.-L., and Amsellem, J., 1980, Ultrastrucrure of the giant smooth muscle fiber of the ctenophore *Beroe ovata*, *J. Ultrastr. Res.* **72**:151-168.

Hernandez-Nicaise, M.-L., Mackie, G.O., and Meech, R. W., 1980, Giant smooth muscle cells of *Beroe*. Ultrastructure, innervation, and electrical properties, *J. Gen. Physiol.* **75**:79-105.

Hernandez-Nicaise, M.-L., Nicaise, G., and Anderson, P. A. V., 1981, Isolation of giant smooth muscle cells from the ctenophore *Mnemiopsis*, *Am. Zool.* **21**:1012.

Hernandez-Nicaise, M.-L., Bilbaut, A., Malaval, L., and Nicaise, G, 1982, Isolation of functional giant smooth muscle cells from an invertebrate: structural features of relaxed and contracted fibers, *Proc. Nat. Acad. Sci. USA* **79**:1884-1888.

Hernandez-Nicaise, M.-L., Nicaise, G., and Malaval, L., 1984, Giant smooth muscle fibers of the ctenophore *Mnemiopsis leydii*: ultrastructural study of in situ and isolated cells, *Biol. Bull.* **167**:210-228.

Horridge, G. A., 1966, Pathways of coordination in ctenophores, in: *The Cnidaria and their Evolution* (W. J. Reese, ed.), Academic Press, London.

Isenberg, G., and Klöckner, U., 1985, Calcium currents of smooth muscle cells isolated from the urinary bladder of the guinea-pig: inactivation, conductance and selectivity is controlled by micromolar amounts of $[Ca]_o$, *J. Physiol. (Lond.)* **358**:60P.

Jmari, K., Mironneau, C., and Mironneau, J., 1986, Inactivation of calcium channel current in rat uterine smooth muscle: evidence for calcium- and voltage-mediated mechanisms, *J. Physiol. (Lond.)* **380:**111-126.

Jmari, K., Mironneau, C., and Mironneau, J., 1987, Selectivity of calcium channels in rat uterine smooth muscle: interaction between sodium, calcium and barium ions, *J. Physiol. (Lond.)* **384:**247.

Kidokoro, Y., Hagiwara, S., and Henkart, M. P., 1974, Electrical properties of obliquely striated muscle fiber membrane of *Anodonta glochidium*, *J. Comp. Physiol.* **90:**321.

Meech, R. W., 1978, Calcium-dependent potassium current in nervous tissues, *Ann. Rev. Biophys. Bioeng.* **7:**1-18.

Meech, R. W., and Standen, N. B., 1975, Potassium activation in *Helix aspersa* neurones under voltage-clamp: a component mediated by calcium influx, *J. Physiol. (Lond.)* **249:**211-239.

Mironneau, J., and Savineau, J. P., 1980, Effects of calcium ions on outward membrane currents in rat uterine smooth muscle, *J. Physiol. (Lond.)* **302:**411-425.

Mironneau, J., Eugene, D., and Mironneau, C., 1982, Sodium action potentials induced by calcium chelation in rat uterine smooth muscle, *Pfülgers Arch.* **395:**232-238.

Nowycky, M. C., Fox, A., and Tsien, R. W., 1985, Three types of neuronal calcium channels with different calcium agonist sensitivity, *Nature* **307:**468.

Prosser, C. L., Kreulen, D. L., Weigel, R. J., and Yau, W., 1977, Prolonged action potentials in gastrointestinal muscles induced by calcium chelation, *Amer. J. Physiol.* **233:**C19-C24.

Reuter, H., 1975, Divalent cations as charge carriers in excitable membranes, in: *Calcium Movement in Excitable Cells* (P. F. Baker and H. Reuter, eds.), Pergamon Press.

Stein, P. G., and Anderson, P. A. V., 1984, Maintainance of isolated smooth muscle cells of the ctenophore *Mnemiopsis*, *J. exp. Biol.* **110:**329-334.

Spencer, A. N., and Satterlie, R. A., 1981, The action potential and contraction in subumbrellar swimming muscle of *Polyorchis penicillatus* (Hydromedusae), *J. Comp. Physiol.* **144:**401.

Vassort, G., 1975, Voltage-clamp analysis of transmembrane ionic currents in guinea-pig myometrium: evidence for an initial potassium activation triggered by calcium influx, *J. Physiol. (Lond.)* **252:**713-734.

Walsh, J. W., and Singer, J. J., 1980, Calcium action potential in single freshly isolated smooth muscle cells, *Amer. J. Physiol.* **239:**C162-C174.

Walsh, J. W., and Singer, J. J., 1983, Ca^{2+}-activated K^{+} channels in vertebrate smooth muscle cells, *Cell Calcium* **4:**321-330.

Washio, H., 1972, The ionic requirements for the initiation of action potentials in insect muscle fibers, *J. Gen. Physiol.* **59:**121.

Werman, R., and Grundfest, H., 1961, Graded and all-or-none electrogenesis in arthropod muscle. II. The effects of alkali-earth and onium ions on lobster muscle fibers, *J. Gen. Physiol.* **45:**997-1027.

Chapter 22

Polyclad Neurobiology and the Evolution of Central Nervous Systems

HAROLD KOOPOWITZ

1. Introduction

The rationale for studying flatworm neurobiology has long been the fact that, anatomically, their nervous systems appear to be intermediate between those of the Cnidarians and the more centralized ones of higher metazoans. It has been thought that understanding flatworm nervous systems would lead to an appreciation of, and insight into, the events involved with the early evolution of brains and central nervous systems. One of the problems of this approach is that the early events that occurred during the initial centralization of nervous systems must have happened so long ago in time that present day organisms can only represent a mere shadow of the actual events. We do not know if centralization occurred several times and independently. Certainly one suspects that the centralization in the Cnidaria and in the Hemichordata might have been independent of those of other metazoans. It is not clear if this also implies that the initial centralization of the protostome and deuterostome lines is independent.

In terms of their gross body structure, as well as their protostomous nature, possession of spiral cleavage and trochophore larva, the cotylean polyclad flatworms appear to have affinities with the nemertean and molluscan phyla and, by extension, with the annelids and arthropods. However, nearly all of the electrophysiological studies we have done are based on acotylean species, where there is direct development without a free-swimming larva. Determining which of these two groups is the most primitive, and the position of the polyclads in relation to other Turbellarian taxa, is not

HAROLD KOOPOWITZ • Ecology and Evolutionary Biology, University of California, Irvine, California 92717, USA.

Evolution of the First Nervous Systems
Edited by P.A.V. Anderson
Plenum Press, New York

at all clear (Thomas, 1986). Ehlers (1986) considers the phylum as monophyletic, with the polyclads fairly close to the base of the Platyhelminth cladogram. Smith et al. (1986), on the other hand, see the Platyhelminthes as polyphyletic, being made up from three independent monophyletic groups. One feature that the polyclads share with most of the other turbellarian taxa is a unique kind of sperm which possesses two flagellae; the sperm swims with the flagellae leading and the nucleus trailing, another unique feature. These features tend to withdraw the polyclads and most other flatworms from consideration as an ancestral group (Hendelberg, 1986). Nevertheless, recent and intriguing work on comparisons of 5S rRNA sequences from a widely divergent group of animals by Hori and his co-workers (1988) place the acotylean polyclad, *Planocera reticulata*, close to the chordate lines and diverging earlier than the cnidarians. The data shows a clear divergence between the polyclad and triclad species investigated. We may have to wait for additional work on other conservative genes before a truly satisfactory phylogeny can be reconstructed. Since speculations of physiological evolution must be based on phylogenetic realities and not how we think systems must have evolved, the paucity of hard phylogenetic data makes these considerations extremely difficult, and our speculations probably are no more than a cartoon of reality.

2. Neuroanatomy

2.1. The Nature of the Plexus

The feature of flatworms that is the most reminiscent of that of the Cnidaria is the presence of the various nerve plexuses. Several of these have been described in the polyclads (Koopowitz and Chien, 1974; Koopowitz, 1982) but they are also known from many of the other groups (Ehlers, 1986; Kotikova 1986; Kotikova and Joffe, 1988). There are ventral and dorsal submuscular plexuses, a subepithelial plexus beneath the basement membrane and an infraepithelial plexus. In common with many other phyla, a stomatogastric plexus associated with the gut has also been described in the flatworms. In some other turbellarian taxa, obvious plexuses do not seem to exist. In *Mesostoma ehrenbergii*, a "typhloplanid" rhabdocoel, we have not yet been able to demonstrate a plexiform arrangement, but it may exist. Microscopic turbellarians, particularly those that are interstitial, may show reductions in the plexus and in some, a plexus may not exist. Is the plexus a primitive characteristic, indicative of some phylogenetic relationship with the Cnidaria or is it merely a secondary, but necessary, arrangement needed to innervate flat fields of muscle? If the latter is the case, one might expect the plexus to be concerned solely with the motor side of the system. However, there is some evidence for the occurrence of a sensory nerve-net in *Notoplana* (Koopowitz, 1973). A physiological (i.e. conducting) sensory nerve-net feeding into the brain of the polyclad *Freemania* was demonstrated some years ago (Koopowitz, 1975) and a similar one is also known from *Notoplana* (Koopowitz, unpublished). A perculiarity of these nerve nets is that they function in the presence

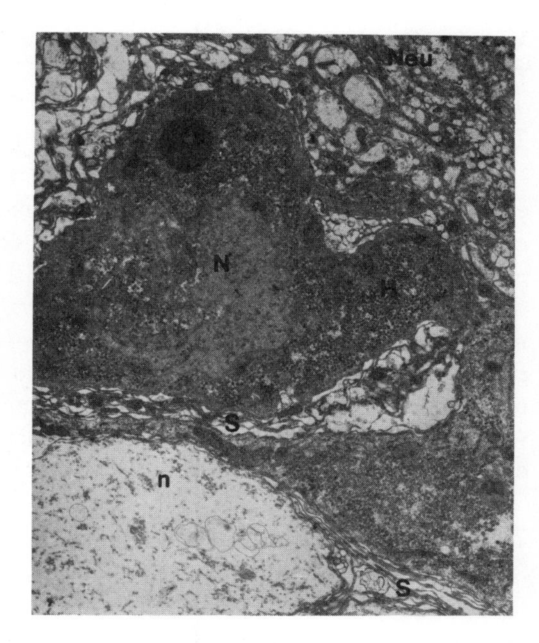

Figure 1. Section through a BRA cell from the brain of *Notoplana acticola*. This cell had been filled with HRP for identification. H = horseradish peroxidase granules; n = neurite of adjacent cell; N = nucleus of BRA cell; Neu = neuropile; S = layers of glial sheath. 5,000X.

of high concentrations of Mg^{++}. One can, however, demonstrate that conduction occurs within this net and not as epithelial conduction. Physiological nerve nets have not been demonstrated in other groups of turbellarians.

2.2. Cell Anatomy

The range of cell types in the flatworms is enormous. Compared to the "higher" invertebrates the polyclads have more bipolar and multipolar cells. The dominant cell type in the submuscular plexus appears to be bipolar, but cell somata are not common within the plexus. Bipolar cells also occur in the cerebral ganglion of polyclads. Bipolar cells seem to be widespread in other orders in the Turbellaria, and several interesting bipolars are now known from the brain of *Mesostoma* (Koopowitz and Elvin, unpublished). Typical invertebrate monopolar neuromorphologies are known from flatworm ganglia, a good example being the VP cell from *Notoplana*, but their distribution is not well understood. One might point out that monopolar neurons are not exclusively invertebrate; the optic tectum in fish has layers of monopolar neurons (Coss and Globus, 1978). However, many of the polyclad neurons which appear to be monopolar using conventional light microscopy, are seen to be multipolar after they have been filled with fluorescent dyes, e.g., BRA cells in *Notoplana*. These cells are heteropolar, multipolar cells and, hence, unlike the classical multipolar cells of cnidarian nerve nets that are isopolar. Polyclad neurons seem more akin to the heteropolar multipolar cells of vertebrates, but there are features of polyclad multipolar cells that are also reminiscent of invertebrate monopolar neurons, particularly the presence of small branches along the axon, that we presume are dendritic in function.

We have found structures in both *Notoplana* and *Mesostoma* that are reminiscent of vertebrate dendritic spines.

2.3. Glia

Glia is present in both the plexus and central nervous system of polyclads and has also been described in several triclads (Golubev, 1988). In *Notoplana*, glia subdivides axons in the nerve trunks of the plexus into discrete bundles, while the largest axons may be individually wrapped by sheets of glial cells (Koopowitz and Chien, 1974). In the brain, Keenan and Koopowitz (unpublished) have found that the individual neuronal somata are sheathed by many layers of glia that may function to isolate the cell bodies from one another. The same is true for major axonal branches (Fig. 1). As many as ten layers of glia have been found interposed between adjacent neurites. The well-developed sheathing around the somata may explain some of the difficulties encountered by workers trying to patch clamp flatworm cerebral neurons.

2.4. Neuromuscular Junctions

The situation in *Notoplana* is similar to that found in the nematodes, cephalochordates and several other animal taxa, inasmuch as the muscle cells send processes to the nerve branches of the ventral plexus (Chien and Koopowitz, 1972). It is possible that this form of innervation is a common plan among many of the lower animal groups.

3. Neurophysiology

Almost nothing is known, in a comparative sense, about the electrophysiology of different flatworm systems. Only a few polyclads have been examined. In part, the lack of comparative data is due to several seemingly intractible problems for making traditional electrophysiological preparations. Not only are the animals small, but they have no hard parts by which they can be anchored during dissection. This, combined with their acoelous nature, makes it difficult to expose the nervous system for manipulation. Furthermore, "abused" preparations are able to "self-destruct," and it has been very difficult to develop an adequate, reversible anesthetic for fresh water flatworms. In polyclads there is a distinctive and tough sheath around the brain and major nerves that facilitates recognition of the nervous tissue during dissection, but in other groups, such as the triclads, the distinctive sheath is not present and the necessary microdissections are even more difficult. Finally, adequate ringers for freshwater turbellarians have still not been worked out.

Polyclads have shown themselves to be the most tractible, but even here problems have been encountered. When electrical responses cannot be recorded, one is never sure if this is because the cells are "silent" or if the preparation has somehow been injured. Despite this, we feel that it must be possible to carry out electrophysiology on other turbellarian taxa and such comparative data is essential for a proper understanding of the status of flatworm nervous systems.

3.1. Ion Channels

A variety of ion channels have been found in the polyclads *Notoplana acticola* and *Alloeoplana californica* (Keenan and Koopowitz, 1981 and 1984; Solon and Koopowitz, 1982). The data has been summarized in figure 2. In all likelihood, this scheme is far from complete. We would like to know if the flatworm membrane systems and their ion channels are intermediate between, or similar to, those of other groups of animals, or if they are totally different. Such comparisons are, however, fraught with difficulties, unless one has a large enough sample. We are still far from that point.

In *Notoplana*, cells have been found that produce all-or-nothing action potentials and graded action potentials; yet others appear to be electrically silent. In most *Notoplana* and *Alloeoplana* cells there are all-or-nothing action potentials which last for only a brief duration; half-durations are less than 0.9 msec. These are considerably shorter than those frequently recorded from neurons in the Cnidaria and Mollusca. In *Notoplana*, there appear to be both fast tetrodotoxin (TTX)-sensitive Na^+ currents and TTX-insensitive Na^+ channels (Keenan and Koopowitz, 1981). These channels may be distributed in different cells but can also occur in the same cell. Most of our ion studies have involved the BRA cells in *Notoplana*. We have no evidence, as yet, for Ca^{++} spikes or a significant contribution by Ca^{++} to the action potential in these animals. Removal of extracellular Ca^{++} does not affect the amplitude of the spike, nor does addition of 180 mM Mg^{++} or 10 mM Cd^{++} (Keenan and Koopowitz, 1984). It should be mentioned, however, that cells bathed in Na^+-free sea water produce action potentials 30% the amplitude of those recorded in the presence of normal Na^+. While measurements show the 9 mM Na^+ still remains after rinsing in Na^+-free medium, this may be too little to account for the residual spike, suggesting that there may be a component of the spike carried by Ca^{++} or another ion. Addition of TTX will not reduce the spike amplitude in those cells further.

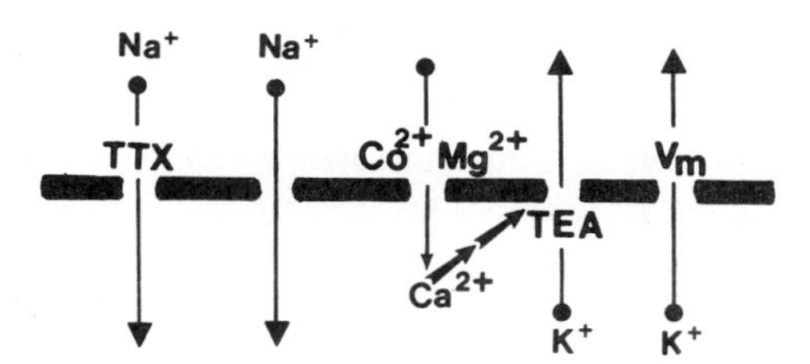

Figure 2. Scheme for ion channels in the membrane of *Notoplana acticola* BRA cells. Outside of the cell is the top of the figure. Agents that block or control the various channels are indicated at the position where they probably work. Note that Ca^{++} works back on the TEA-sensitive K^+ channel. The other K^+ channel is thought to be voltage-dependent. While it is not indicated, the slow, TTX-insensitive Na^+ channel is probably also voltage-activated.

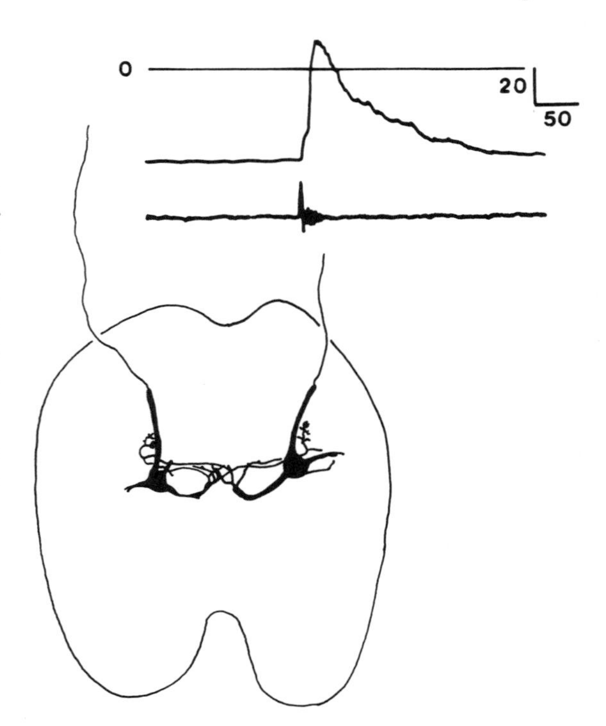

Figure 3. Graded potentials recorded intracellularly from the BUG cells of *Notoplana acticola* brain. The tracing is the response to a small water-borne vibration (Hauser, unpublished). Note the overshooting potential produced once the cell attains the threshold. Lowest trace is the vibration artifact. A camera lucida drawing of a pair of dye-coupled BUG cells is shown below. Dye was injected into one of the cells. Fine axons emerge from the dorsal aspect of the brain. Brain is approximately 600 μm in diameter.

There appear to be two K^+ currents in flatworm neuurons, a voltage-sensitive channel and a Ca^{++}-activated K^+ channel. One interesting point of difference between the polyclads and other taxa lies in the mode of action of tetraethylammonium (TEA). In most preparations TEA blocks K^+ conductances when it is applied in the bath. In both *Notoplana* (Keenan and Koopowitz, 1984) and *Alloeoplana* (Solon and Koopowitz, 1982), TEA must be injected into the cell in order to block K^+ conductances. This must mean a basic difference in the structural configuration of the channels.

3.2. Graded Potentials

One pair of dye-coupled cells that straddle the central dorsal aspect of the brain, the BUG cells, appear to be interneurons involved with photo- and vibration sensitivity. These cells produce graded, positive, overshooting and long-lasting action potentials (Fig. 3). The ionic mechanisms underlying these responses of the BUG cells have yet to be investigated.

3.3. Silent Cells

The Ventral Plexus (VP) cells, another dye-coupled pair in the brain, extend into the ventral plexus, where they send neurites into each major branch of the plexus. These cells appear to be silent. We have been unable to drive them electrically and we have been unable to record either synaptic or spontaneous activity from them. It is possible that the physical cable properties of the cell, such as the profuse branching, may shunt any activity away from the electrodes so that the cells merely appear to be silent. We are reasonably sure from morphological grounds that these cells are, indeed, neurons and not glia.

3.4. Electrotonic Coupling

Dye-coupled pairs of neurons are common in *Notoplana acticola* and *Alloeoplana californica* and they have also been found in the rhabdocoel *Mesostoma ehrenbergii*. These cell pairs are bilaterally arranged and are invariably dye coupled. Sometimes, an entire bilaterally-symmetrical circuit, containing several cells on each side, may be found. In *Notoplana* we have been able to record and stimulate concurrently from both members of the BRA cell pair (Keenan and Koopowitz, in preparation). Electrotonic coupling is variable from preparation to preparation and coupling coefficients from 0.12 to 0.68 have been measured. The animals' ability to function with only one lateral half of the brain (see below) may be tied to the fact that neurons on both sides of the brain may act as a single unit, despite the fact that coupling is not perfect.

The variety of electrical responses of different cells in any one preparation cautions one to be careful before drawing inferences concerning evolutionary trends unless an adequate sample of cell types has been examined in any preparation.

4. Neurochemistry

Electron microscopy of the brain of *Notoplana* has revealed the presence of a variety of synaptic vesicles, ranging in shape from spherical to compressed, and in appearance, from clear to dense-cored (Koopowitz and Chien, 1975). In *Notoplana* and some other polyclads, a variety of putative neurotransmitters have been identified using physiological, fluorescence and immunocytochemical methodologies. Keenan et al. (1979) demonstrated depressive responses to both glycine and GABA in the longitudinal nerve cords of *Notoplana*. The effects of both drugs could be reversed by picrotoxin, bicuculline or strychnine. It is thought that those transmitters might activate chloride conductances in the synaptic regions of the neuron. Glutamate and asparatate act as excitatory transmitters in the same tissue. Gruber and Ewer (1962) demonstrated that adrenergic drugs had an excitatory effect on muscular activity of the polyclad *Planocera*, but, in *Notoplana*, we have been unable to demonstrate the presence of adrenaline or noradrenaline in the central nervous system. We have, however, shown that there is a small population of dopaminergic cells in the brain of *Notoplana* and that these change in activity and number with age (Hauser and Koopowitz, 1987). A similar population has also been found in the rhabdocoel *Mesostoma* (unpublished). The function of this dopaminergic innervation is unknown.

With regard to both acetylcholine and serotonin, neither transmitter, nor their blocking agents, appear to have any effect on the physiology of the *Notoplana* nervous system. Although non-specific esterase stains do stain the plexus in that animal, only in *Enchiridium pharynx* has any physiological response to ACh ever been recorded

(Stone and Koopowitz, 1976). This points out that the absence of a putative transmitter within one member of a taxon does not necessarily mean that it is absent from an entire class or phylum. Typical serotonergic fluorescence cannot be detected in *Notoplana*, and no serotonergic cells have been identified in polyclads. However, 5-HT antiserum does react with nerve cells in animals of other orders (Reuter, 1988; Reuter et al., 1986). Finally, in recent years, a series of small putative neuropeptides of the RF-amide family have been described in a number of turbellarians (Wikgren et al, 1986).

It is clear from the above, that many of the neurotransmitters present in other phyla appear to be present in the flatworms also. Once again, however, we must either acknowledge the early appearance of these substances in phylogeny, or consider that the same molecules evolved independently several times.

5. Behavior

5.1. Natural Behavior

One feature of flatworms that is quite surprising is the pliable nature of their behavior. Feeding behavior, which has received the most attention, demonstrates this succinctly. *Planocera gilchristi*, a polyclad that extracts periwinkles from their shells, is able to interrupt normal feeding when it is disturbed, by inserting an "escape and carry" sequence, whereby it carries the prey to a safe place and then resumes the feeding sequence. In addition, the animals may position the same prey in different ways, each of which requires a somewhat different feeding strategy (Koopowitz, 1970). Recently we (Wrona and Koopowitz, unpublished) have started to investigate the feeding behavior of *Mesostoma ehrenbergii*, a rhabdocoel. These animals will accept *Daphnia*, *Tubifex* and a variety of mosquitos. Each prey item is dealt with somewhat differently and it takes different lengths of time to complete the behavior. Initial prey recognition involves mechanical stimuli and, as far as we are aware, chemical senses play no role in prey recognition. An animal, for example, will not attack a motionless mosquito larvae, although it might crawl directly over it. Moments later, however, in response to movement of the same larva, the worm will strike at and proceed to capture and feed on the prey. *Mesostoma* responds to waves generated in the water with a rapid strike-attack, and will attack a moving glass bead. The length of time spent on the bead depends on the behavior of the bead. If the bead remains stationary after the worm has struck, then the worm will leave the object after a brief period of exploration, but if the bead continues to move, the worm will extend its period of exploration but will not proceed further with the behavior normally elicited in feeding. The behavior, therefore, is not an entire sequence brought into action by a single initial stimulus but, rather like the behavior of polyclads such as *Planocera* and *Notoplana* (Koopowitz, 1970; Koopowitz et al., 1976), feeding behavior is a concatenation of separate behavioral reflexes, each of which must be initiated by its own appropriate stimulus.

5.2. Learning

Since the early work on planarian learning (reviewed by Corning and Kelly, 1973) fell into disrepute, little attention has been focused on learning at this phylogenetic level. In my laboratory we have tried, for many years, to come up with paradigms that demostrate conditioning, but with little success. In fact, there has been, to date, no incontrovertible evidence that flatworms could actually learn. Recently, however, Hauser (in preparation) has been able to demonstrate instrumental conditioning in *Notoplana aciticola*, by pairing a food stimulus (capillary tube with brine shrimp homogenate) with an aversive stimulus (20 μl squirt of 0.75 M KCl on the tail). Learning is rapid and worms learn to avoid the food stimulus after between two and seven trials. Mean time to learn is four trials. Retention is very short; the longest retention time recorded is 10 minutes, with a mean retention interval of 3.8 minutes. We are satisfied that the results demonstrate instrumental conditioning; several different controls seem to rule out pseudo-conditioning or non-associative learning. We have not been able to demonstrate long-term retention in this preparation. Decerebrate *Notoplana* are able to feed and also respond to the noxious stimulus but we have been unable to demonstrate associative learning, conclusively, with this paradigm.

It is possible, however, to demonstrate a kind of long term improvement in righting behavior in decerebrate *Notoplana* (Gallemore et al., in preparation). Righting behavior is a very fast (<2 sec), coordinated activity in normal cerebrate polyclads. Decerebate flatworms are able to right after a fashion; they thrash around until a portion of the margin comes in contact with the substrate and the animals eventually pull themselves over. Mean righting time for decerebrate animals, 24 hours after surgery, is 26 sec. After 20 days there is an improvement in time to right to 16 sec. If we examine the number of times the animals are able to right themselves, we find that initially 47 percent of the animals are unable to right themselves but after 15 days this has dropped to 7 percent. Worms that had not righted within 150 sec were flipped over. The improvement recorded could be due to either recovery from surgical trauma or some kind of learning. If one decerebrates a batch of worms and then waits 15 days to test their righting behavior, one finds that only 20 percent are unable to right. Thus, there is post-surgical improvement without experience. However, when those animals are given 5 trials a day, the ability to turn improves very rapidly, suggesting that there may, in fact, be a learning component of some sort.

5.3. Electrophysiology of Response Decrement

The neurobiological basis of the above types of learning is not known. Habituation is often considered to be the simplest form of learning. Investigations show that response decrement in central interneurons in *Notoplana* has many of the same properties as behavioral habituation. These are intriguing from a phylogenetic point of view, because they seem to involve novel mechanisms. Response decrement appears to be dependent on Ca^{++} ions. This was first seen in the polyclad, *Freemania litoricola*, using extracellular electrodes on the brain (Fig. 4). Later, this effect was confirmed with intracellular recordings from the BRA cells of *Notoplana*, (Hauser et al., in preparation). In *Notoplana* the effect appears, at least in part, to be a post-synaptic phenomenon. If one compares the responses to current injection, interposed between decrementing sensory responses, one finds that the responses to the current injection also decrease, suggesting a post-synaptic effect (Fig. 5). In the absence of decrementing sensory responses, there is no change in response to pulses of injected current. Postsynaptic decrement is also accompanied by a prolonged hyperpolarization of the postsynaptic cell.

6. Redundancy and the Brain

A series of experiments involving lesions and transplantations suggest that considerable redundancy occurs in the brain. In *Notoplana*, if one lateral half of the brain is excised, behaviors requiring brain coordination reappear as soon as the severed

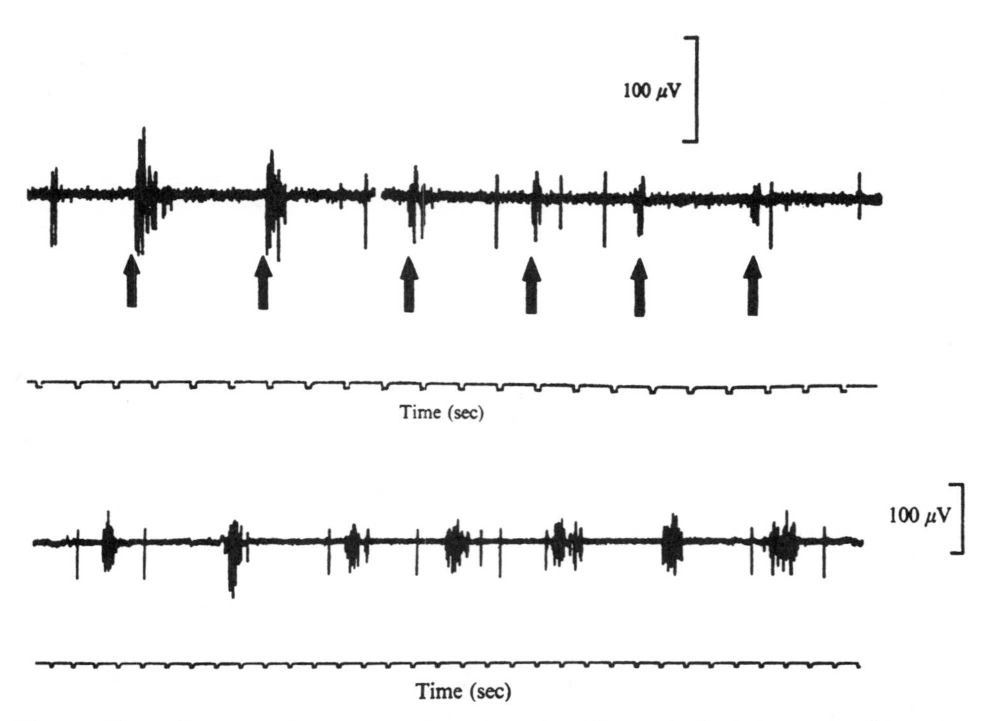

Figure 4. Extracellular potentials measured from the brain of *Freemania litoricola* at the position of the BRA cells. Responses are to water borne vibrations at the anterior edge of the worm. Upper trace: from a preparation in normal sea water. Note that the response diminishes. Lower trace: same preparation anaesthetized in 50 percent sea water and isotonic MgCl. Note that there is no decrease in response to the repetitive stimuli (after Koopowitz, 1975).

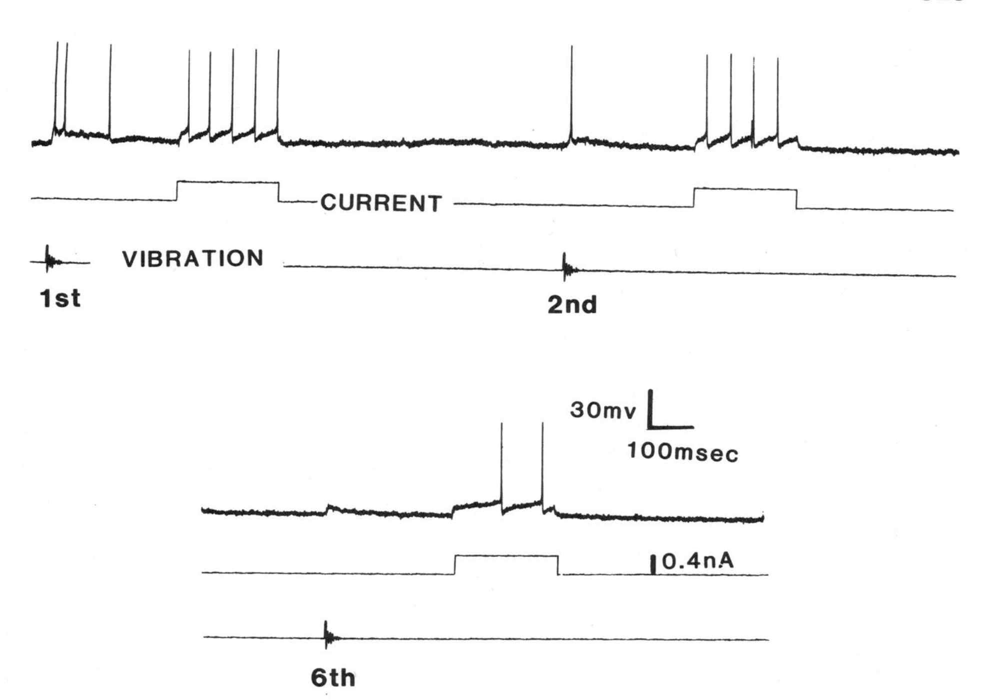

Figure 5. Intracellular recordings from a BRA cell in *Notoplana acticola*, illustrating the response to current injections interposed between responses to vibration. Note that the response to the 6th current injection is a decrease in the number of spikes produced. Cells not subjected to the vibration stimuli will produce unchanging responses for up to 20 current injections (Keenan, unpublished).

nerves have reconnected to the remaining half of the brain (Faisst et al., 1980). Appropriate avoidance-turning requires that the animal possess a "body sense;" a stimulus on the right anterior side necessitates an extension to the left side of the body and vice versa. Animals without half of the brain are able to complete this behavior, irrespective of the side of the animal that has the brain or the side that is stimulated. In another series of experiments, we have shown brain-controlled behavior returns to an animal if the brain is rotated through 180° and/or is inverted (Davies et al., 1985). If, however, the brain is only rotated 90°, then brain-dominated behavior does not reappear. The brain need not be positioned in its anterior location for this; when the brain is transplanted to the rear of the animal, it is able to re-establish behavioral controls, including that of avoidance turning (Gallemore et al., in preparation). Brain transplantations between animals are equally effective.

These behavioral recoveries following brain transplantation are rapid and occur within a few days, suggesting that there may not be time to regrow new neuronal pathways to any great extent. Understanding how the brain is able to reintegrate into the nervous system would also help in understanding how the peripheral nervous system is organized physiologically. The transplant experiments suggest that there might possibly be a system of nerve nets that the brain can feed into, irrespective of its position.

7. Evolution and Perspectives

The most popular and pervasive ideas about the origins of centralization are based on the views that cephalization involved a gathering of sense organs and their interneurons into a single discrete anterior system, which subsequently enlarged to produce the brain. An alternate suggestion is that the central nervous system originated as a devise for coordinating the two sides of the body in a primitive, planuloid ancestor. In modern polyclads, much of the neuronal basis of behavior occurs in peripherally organized reflexes. These peripheral reflexes are coordinated and sequenced by the brain, but can act without it (Gruber and Ewer, 1962; Koopowitz, 1970; Koopowitz et al., 1976). The fact that much of the activity of the brain involves inhibiting peripheral reflexes makes a sensible model for the early evolution of brains. It is difficult to see how the polyclad organization could have been derived from the situation found in many of the microscopic flatworms of other orders, which do not appear to have a plexiform or even peripheral arrangement to the nervous system. The simplification found in many of the microscopic flatworms should probably not be considered primitive, but is perhaps the reduction to be expected when animals loose bulk.

At the cellular level, we find that flatworms possess an array of ion channels and neurotransmitters, but that not all channels or transmitters found in higher animals have been described. However, because there are a variety of cell types between and within species, it would be presumptive to suggest that the flatworms might be lacking in some of these components, before a large enough sample has been studied. We might expect to find that most of the cellular mechanisms evolved early in the evolution of nervous systems. The differences between animal groups probably reflect how cells are arranged in circuits, rather than evolution of new basic mechanisms. Even at this phylogenetic level, we do find complex local circuitry involving complex serial synapses. Obviously, some new mechanisms probably have evolved since the flatworms, but until a complete inventory has been made, the real differences will have to await elucidation.

ACKNOWLEDGEMENTS. I would like to express my thanks to the many students in my laboratory who make the continuing work on flatworm neurobiology possible. This work was written during the tenure of a grant from the Whitehall Foundation.

References

Chien, P., and Koopowitz, H., 1972, The ultrastructure of neuromuscular systems in *Notoplana* acticola, a free-living polyclad flatworm, *Z. Zellforsch.* 133:277-288.

Coss, R. G., and Globus, A., 1978, Spine stems on tectal interneurones in jewel fish are shortened by social stimulation, *Science* **200:**787-790.

Corning, W. C., and Kelly, S., 1975, Platyhelminthes: the Turbellarians, in: *Invertebrate Learning*, (Corning, W. C., Dyal, J. A., and Willows, A. O. D., eds.) 1:171-218, Plenum Press, New York.

Davies, L., Keenan, L., and Koopowitz, H., 1985, Nerve repair and behavioral recovery following brain transplantation in *Notoplana acticola*, a polyclad flatworm, *J. Exp. Zool.* **235:**157-173

Ehlers, U., 1986, Comments on a phylogenetic system of the Platyhelminthes, in: *Advances in the Biology of Turbellarians and Related Platyhelminthes* (S. Tyler, ed.), *Hydrobiologia* **132:**1-12.

Faisst, J. L., Keenan, L. and Koopowitz, H., 1980, Neuronal repair and avoidance behavior in the flatworm, *Notoplana acticola*, *J. Neurobiol.* **11:**483-496.

Gruber, S., and D. W. Ewer, 1962, Observations on the myo-neural physiology of the Polyclad, *Planocera gilchristi*, *J. exp. Biol.* **39:**459-477.

Golubev, A. L., 1988, Glia and neuroglia relationships in the central nervous system of the Turbellaria (Electron microscopic data), in: *Free-living and Symbiotic Platyhelminthes* (P. Ax, U. Ehlers, and B. Sopott-Ehlers, eds.), *Prog. Zool.* **36:**185-190.

Hauser, M., and Koopowitz, H., 1987, Age-dependent changes in fluorescent neurons in the brain of *Notoplana acticola*, a polyclad flatworm, *J. Exp. Zool.* **241:**217-225.

Hendelberg, J., 1986, The phylogenetic significance of sperm morphology in the Platyhelminthes, in: *Advances in the Biology of Turbellarians and Related Platyhelminthes* (S. Tyler, ed.), *Hydrobiologia* **132:**53-58.

Hori, H., Muto, A., Osawa, S., Takai, M., Lue, K., and Kawakatsu, M., 1988, Evolution of *Turbellaria* as deduced from 5S ribosmal RNA sequences, in: *Free-living and Symbiotic Platyhelminthes* (P. Ax, U. Ehlers, and B. Sopott-Ehlers, eds.), *Prog. Zool.* **36:**163-167.

Keenan, C. L., and Koopowitz, H., 1981, Tetrodotoxin-sensitive action potentials from the brain of the polyclad flatworm *Notoplana acticola*, *J. Exp. Zool.* **215:**209-213.

Keenan, L., and Koopowitz, H., 1984, Ionic bases of action potentials in identified flatworm neurones, *J. Comp. Physiol.* A **155:**197-208.

Keenan, L., Koopowitz, H., and Bernardo, K., 1979, Primitve nervous systems: action of aminergic drugs and blocking agents on activity in the ventral nerve cord of the flatworm *Notoplana acticola*, *J. Neurobiol.* **10:**397-407.

Koopowitz, H., 1970, Feeding behavior and the role of the brain in the polyclad flatworm, *Planocera gilchristi*, *Anim. Behav.* **18:**31-35.

Koopowitz, H., 1973, Primitive nervous systems. A sensory nerve net in the polyclad flatworm *Notoplana acticola*, *Biol. Bull.* **145:**352-359.

Koopowitz, H., 1975, Electrophysiology of the peripheral nerve net in the polyclad flatworm *Freemania litoricola*, *J. exp. Biol.* **62:**469-479.

Koopowitz, H., 1982, The neurobiology of free-living flatworms, in: *Electrical Conduction and Behavior in Simple Invertebrates* (G. Shelton, ed.), Clarendon Press, Oxford.

Koopowitz, H., and Chien, P., 1974, Ultrastructure of nerve plexus in flatworms. I. Peripheral organization, *Cell. Tiss. Res.* **155:**337-351.

Koopowitz, H., and Chien, P., 1975, Ultrastructure of nerve plexus in flatworms. II. Sites of synaptic activity, *Cell. Tiss. Res.* **157:**207-216.

Koopowitz, H., Silver, D., and Rose, G., 1976, Primitive nervous systems. Control and recovery of feeding behavior in the polyclad flatworm, *Notoplana acticola*, *Biol. Bull.* **150:**411-425.

Kotikova, E. A., 1986, Comparative characterization of the nervous system of the Turbellaria, in: *Advances in the Biology of the Turbellarians and Related Platyhelminthes* (S. Tyler, ed.), *Hydrobiologia* **132:**82-92.

Kotikova, E. A., and Joffe, B. I., 1988, On the nervous system of the dalyellioid turbellarians, in: *Free-living and Symbiotic Platyhelminthes* (P. Ax, U. Ehlers, and B. Sopott-Ehlers, eds.) **36:**191-194.

Reuter, M., 1988, Development and organization of nervous system visualized by immunocytochemistry in three flatworm species, in: *Free-living and Symbiotic Platyhelminthes* (P. Ax, U. Ehlers, and B. Sopott-Ehlers, eds.), *Prog. Zool.* **36:**181-184.

Reuter, M., Lehtonen, M., and Wikgren, M., 1986, Immunocytochemical demonstration of 5-HT-like and FMRF-amide-like substances in whole mounts of *Micostomum lineare* (Turbellaria), *Cell Tiss. Res.* **246:**7-12.

Smith, J., Teyler, S., and Rieger, R. M., 1986, Is the Turbellaria polyphyletic? in: *Advances in the Biology of Turbellarians and Related Platyhelminthes* (S. Tyler, ed.), *Hydrobiologia* **132**:13-21.

Solon, M., and Koopowitz, H., 1982, Multimodal interneurones in the polyclad flatworm, *Alloeoplana californica*, *J. Comp. Physiol.* **147**:171-178.

Stone, G., and Koopowitz, H., 1976, Primitive nervous systems: electrophysiology of the pharynx of the polyclad flatworm *Enchiridium punctatum*, *J. exp. Biol.* **65**:627-642.

Thomas, M. B., 1986, Embryology of the Turbellaria and its phylogenetic significance, in: *Advances in the Biology of Turbellarians and Related Platyhelminthes* (S. Tyler, ed.), *Hydrobiologia* **132**:105-115.

Wikgren, M., Reuter, M., and Gustafsson, M., 1986, Neuropeptides in free-living and parasitic flatworms (Platyhelminthes). An immunocytochemical study, in: *Advances in the Biology of Turbellarians and Related Platyhelminthes* (S. Tyler, ed.), *Hydrobiologia* **132**:93-99.

Chapter 23

Enigmas of Echinoderm Nervous Systems

JAMES L. S. COBB

1. Introduction

There are three reasons why it is difficult to place echinoderm nervous systems in an evolutionary context. In the first place, although the phylum is clearly deuterostome, there is very little beyond speculation to relate the origins of it to earlier metazoan groups. This argument applies equally, of course, to the chordates themselves since there is evidence that they shared a common ancestor with echinoderms. The second reason is that the primative echinoderms were more likely related to the sea-lilies (Crinoidea) than to the other 5 extant classes (Ophiuroidea-brittlestars, Echinodea-sea urchins, etc., including the new class Concentricycloidea). There is a fundamental difference between the crinoids and the rest, in that the former live mouth up and the rest, effectively, mouth down. A consequence of this is that the role of the nervous systems are radically different, with the ectoneural nervous system being much less significant in the crinoids and the aboral being dominant. The problem here, of course, is that almost all we know about echinoderm nervous systems is derived from more advanced non-crinoids. The third reason is the most enigmatic. Echinoderms start off as perfectly respectable bilaterally symmetrical larvae from which a star is born. There are no really definitive studies which show for certain how the larval nervous system is related to that of the adult. Furthermore, there has never been a satisfactory suggestion for the advantages of the pentameric symmetry of the phylum, and there has been no effective escape from this body form for 500 million years. It must, thus, be possible that the radially symmetrical nervous system of the adult is

JAMES L. S. COBB • Gatty Marine Laboratory, University of St. Andrews, St. Andrews, Fife KY16 8LB, United Kingdom.

basically a secondarily evolved system. Although radial symmetry is considered a restrictive and primitive body form, as in the cnidaria, this cannot be the case in echinoderms, if their ancestors were bilateral.

2. Separate Nervous Systems

The Crinoidea, considered the primitive extant group, have a poorly developed ectoneural system, and a more extensive mesodermal hyponeural system, but the main central and radial nervous system is the aboral one. In the other extant classes, perhaps because they live the other way up, the ectoneural is the extensive and dominant nervous system and the hyponeural, which is undoubtedly mesodermal, is purely motor. It is now clear that the distribution of the hyponeural system is related to a developmental failure. The ectoneural system does not penetrate the basement membrane underlying all ectodermally-derived tissues. This means that ectoneural nerve endings cannot directly innervate muscle and do not directly connect with the hyponeural system. In both cases, varicose vesicle-filled axons of the ectoneural nervous system are assumed to release transmitter which diffuses across the basement membrane. This can be seen in the radial nerve cord connections of asteroids and ophiuroids (Cobb, 1970) and in the innervation of the muscles of the ampullae (Cobb, 1970) and, as extensively described and discussed by Florey and Cahill (1980), for the innervation of the tubefeet. In situations where the ectoneural system was thought to innervate muscles, such as in the spines and pedicellariae, it has recently been shown by Cobb (1987), for spines, and M. Ghyoot (unpublished) for pedicellariae, that there are discrete nodes of hyponeural tissue that directly innervate the muscles, but which are themselves postsynaptic to ectoneural nerves across a basement membrane.

Under some circumstances where there are no hyponeural motor nerves, long processes run from the muscles, as narrow specialized axon-like processes, to form a synaptic contact across a basement membrane (described in Cobb, 1967 and original description corrected in Cobb, 1987). In the arms of brittlestars, there are hyponeural motor nerves, but the bundles of muscle tails are also innervated across the basement membrane that lies aboral to the ectoneural radial nerve cord (Cobb and Stubbs, 1981). The reasons for the presence of both forms of innervation is not clear, but it is consistent.

3. Ultrastructure

When prepared using standard transmission EM protocols, echinoderm nerves look very similar to those of other phyla. They contain filaments which, presumably, are microfilaments and intermediate filaments. Microtubules are also present, but their distribution is puzzling in that they are often absent. This may, however, be a fixation effect, since the phylum is notoriously difficult to fix. Nerve endings, both neuroneuronal and neuromuscular, are simple varicose endings which are filled with a variety of vesicle types, and do not show morphological specializations of the pre- and post-synaptic membranes and cytoplasm. They, thus, resemble those found in much of the vertebrate autonomic system (Cobb and Pentreath, 1977). There is, however, one recent exception to this and that is the innervation of juxta-ligamental cells located underneath the arm plates. Here, "classical" chemical synapses have recently been found (Cobb, 1985b). The area under the arm plates would be subject to particular mechanical stress and the presence of more conventional chemical synapses in this region may support the idea that much of chemical synapse morphology is related to sticking cells together, rather than the transmission process (Cobb and Pentreath, 1977). There are no structural or functional reports of electrical synapses between any cell types in adult echinoderms. This is not to say that they do not exist but that extensive searches have not identified them, although again poor fixation is often a serious handicap.

4. Lack of Glial Cells

The lack of glial cells in all parts of the echinoderm nervous system has been discussed by Pentreath (1987). There are also no blood spaces between the tissue and the axon bundles are very tightly packed (see Cobb, 1970 and Cobb and Stubbs, 1981). Given our increasing understanding of the functional significance of glial cells, this is a major enigma. Equally puzzling is the great reduction in rough endoplasmic reticulum, which is particularly noticeable in the cell bodies of the giant neurons in brittlestars. In these cells, it can be seen, even at the light microscope level, that the cell bodies are 70% or more full of an amorphous material (Cobb and Stubbs, 1981) which, on the basis of fixation reactions, is glycogen. This material is also present, but unevenly distributed, in many of the axons. The energy demands of nerve cells may be met by the presence of large amounts of glycogen, but this relatively enormous volume leaves little room for the other cellular components required by the cell. For instance, how are these cells able to manufacture the many complex proteins which need to be continually synthesized and transported by the endoplasmic reticulum system? We know nothing of axon transport systems in echinoderms.

5. Mutability of Connective Tissue

The property of sea urchin spines to lock solid upon stimulation has long been known, as has the ability of holothurians to stiffen the body wall. Although the locking structure in echinoderms was initially thought to be a catch muscle, it was shown by

Takahashi (1967) to be a connective tissue ligament. It has recently become clear that the change in the viscosity of the connective tissue is not only a very rapid phenomenon, but may be brought about by the nervous system. Wilkie (see review, 1988) has described a class of nerve-like cells, called juxta-ligamental cells, that send processes containing large variably-sized, granular vesicles between the bundles of collagen. The connective tissue matrix can reversibly change state, apparently allowing the collagen bundles to slide past each other, and this is controlled by the nervous system. The whole phenomenon has been extensively reviewed recently by Wilkie (1988) and Motokawa (1988), but no aspects of the biochemistry of the process are understood.

6. Ionic Basis of Action Potentials

There have been many studies on the pharmacology and physiology of the echinoderm nervous system, based on crude organ bath studies and, in the case of neurophysiology, very crude recording techniques. This work has recently been reviewed by Cobb and Moore (1989) but, truthfully, it does little to enlighten us about the functioning of the nervous system. An important advance came when Brehm (1977) showed that there were classes of giant fibers in the radial nerve cords of brittlestars and that these could be recorded from. His studies hinted that a Ca^{++}-based action potential occurred. Tuft and Gilly (1984), again using a brittlestar preparation, but recording compound potentials, described two classes of neurons, one with a Na^{+} spike, and another with a Ca^{++} spike. Berrios et al. (1985) suggested that a Ca^{++} spike was present in the long spines of the sea-urchin *Diadema*, though again, this study relied on extracellular recordings of compound action potentials. Cobb and Moore (1988) have re-examined this question using intracellular recording techniques and can show no involvement of Na^{+} ions in the action potentials of any neurons. Furthermore, this work demonstrated that choline, the Na^{+} ion substitute used by Tuft and Gilly (1984), produces anomalous results which might be interpreted, incorrectly, as demonstrating the presence of Na^{+}-dependent action potentials. Using the criteria employed by Baccaglini (1978) for dorsal root ganglion neurons of *Xenopus*, it appears that both the ectoneural and hyponeural nerves produce Ca^{++} spikes. The current evidence is only tentative and, until voltage clamp analysis of the various currents carried by particular ions has been achieved (and this is not technically very easy), absolute certainty about the ionic basis of the action potential will not be possible.

7. Centralization, Receptors, and Giant Fibers

The classical idea of the echinoderm nervous system envisages the circumoral nerve ring and immediately adjacent nerve cords acting as a sort of "brain" (Smith,

1966). A long time ago (Cobb, 1970) challenged this idea and proposed that the radial nerve cords consisted of a series of linked segmental ganglia and that all were of equal status. The author has developed this theme (Cobb and Moore, 1989) and hypothesizes that any single ganglion of any arm can control the behavior of the entire animal and that the circumoral ring merely connects the nerve cords together. The echinoderms, therefore, by any reasonable definition, do not have a central nervous system, although it is still sometimes forcefully argued that the whole of the radial nervous system and circumoral nerve ring constitutes the CNS. There is almost no evidence to support the ideas of Smith (1966), although it is a very logical hypothesis (the golden rule in echinoderm biology is that the obvious explanation is always the wrong one!). The evidence for the hypothesis of the author is growing, but it must be admitted that it is a long way from conclusive.

The peripheral part of the nervous system receives input from countless sensory receptors that are almost never organized into receptor organs, and we now know that echinoderms are exquisitely sensitive to environmental change at levels far below those anticipated from behavioral observations (Moore and Cobb, 1985a,b; 1986). Based on physiological studies, it is clear that they also have internal proprioreceptors which monitor body and appendage position. There is no nerve net, but there are numerous peripheral ganglia which are associated with various surface structures, such as spines, pedicellaria and tubefeet. We have no direct evidence, but it seems likely that the peripheral input to the radial nerve cords in each segment is by interneurons and not by the sensory neurons themselves.

Brehm's (1977) description of giant fibers opened a new era in echinoderm biology, in that it showed that a functional description of the nervous system might be possible, based on recordings from single neurons. What the giant fibers do is still far from clear. In the case of the giant motor fibers of the hyponeural nerve cord, intracellular recordings from single cells have allowed us to examine in some detail the input by the ectoneural system onto these neurons, and have shown that when arm movements occur, cells on opposite sides of the same arm receive mirror image inhibitory and excitatory synaptic potentials that summate to produce contraction on one side of the arm and relaxation on the other (Cobb, 1988 and in preparation). The cell bodies of the hyponeural neurons, as well as those of all the ectoneural cells examined are non-excitable.

We know less about the ectoneural giant fibers, which exist in two size classes and which conduct at different volocities (the slower at 35 cm sec^{-1} and the faster at 80 cm sec^{-1}). The slower-conducting fibers relay information from the site of stimulation, giving rise to excitatory and inhibitory input to hyponeural motor neurons. This, in turn, creates various arm movements. Repetitive stimulation causes the information to be relayed over greater and greater distances within the radial nervous system (Cobb and Moore, 1989). The intriguing giant fibers are the larger ones. These fire to any stimulus that is threatening, and the same individual fiber will fire to any stimulus modality (chemical, noxious, shadow, etc). Spikes in these giant fibers can be recorded extracellularly as activity is conducted throughout the nervous system, but simultaneous intracellular recordings from the motor neurons shows no consistent, correlated, synaptic activity.

Morphologically, ectoneural giant fibers appear to originate in each segment and then run centrally for at least six segments. At either end they break up into a fine varicose plexus and they do not appear to receive input or make output at any segment they pass through. This implies that there are overlapping populations of neurons in each segment. Tuft and Gilly (1984) have suggested that single giant fibers run the entire length of the nerve cord. However, although we still cannot rule out this idea on morphological grounds, no single neuron has been filled completely with lucifer yellow, and there is much evidence against such an idea.

Functionally, these neurons conduct information throughout the nervous system whenever a peripheral stimulus reaches threshold (stimulus thresholds and responses are discussed in Cobb and Moore, 1989), and they appear to conduct in both directions. However, although recordings from such neurons in whole animal preparations show this, it has not been confirmed by subsequent lucifer yellow fills that the impaled cells were, indeed, longitudinal neurons.

We know still less about other neurons, except that intracellular records have been obtained from neurons that are postsynaptic to both size classes of longitudinal neurons, and that lucifer fills show neurons of varying morphology within a single ganglion (see Cobb, 1985a). Many of these neurons are extremely small.

8. Speculation

The echinoderms are a very significant marine phylum, but we know very little about function at the cellular level and, as far as the nervous system is concerned, too much of what information there is has come from a single laboratory. In the context of the origins of nervous systems they have some relevance in that they re-invented the wheel (radial symmetry), since this is considered a primitive feature. They are interesting in that they are a major offshoot in the deuterstome line of evolution towards the ultimate nervous system, as found in humans. The apparent presence of Ca^{++} spikes is an anomaly, if Na^{+} spikes evolved to avoid the difficulties a massive influx of Ca^{++} ions is likely to cause (Hille, 1984). Nevertheless, such is the paucity of knowledge that what follows must be mere speculation.

Why did radial symmetry evolve in the echinoderms, if it is disadvantageous to advanced metazoa, and why has a secondary bilateralism never been re-imposed, since one arm might have been expected to become the dominant one? Why can ectodermal nerves not penetrate into mesodermal tissues, and why is this such an unbreakable rule that presumably the hyponeural system evolved from a mesodermal cell type to work in combination with muscle tails to overcome this developmental failure? The function of the hyponeural system itself is interesting, since in all respects it resembles other nervous systems except that there do not appear to be any synaptic interaction between the individual neurons. The presence of the Ca^{++} spike might be more easily explained if it were confined to the hyponeural system, since one could then hypothesize that the hyponeural tissue evolved from muscle cells, as an advance over the muscle tail system itself. However, Ca^{++} spikes also occur in the ectoneural system. It would be very valuable to know the ionic basis of the action potential in the larval nervous system. However, the echinoderm genome does include the information required for producing Na^{+} channels, since Na^{+} and Ca^{++} ions are involved in the production of action potentials in starfish oocytes and the respective currents have been studied using voltage and current clamp techniques (Miyazaki et al., 1975a,b) (see also Cobb, 1982 for review). This appears to indicate that the Ca^{++} spike is not the retention of a primitive feature.

The lack of glial cells is puzzling and, combined with the reduction of rough endoplasmic reticulum in the neurons, must place a large question mark over the processes of respiration and metabolism of echinoderm nerve cells. It is just possible that the total failure of all attempts to use uptake systems to trace axons (cobalt, HRP, etc.), despite substantial efforts, might be a consequence of these cells using a mechanism other than the axoplasmic transport systems we understand in other nervous systems, for supplying their distant parts with metabolites and proteins.

Mutability is fascinating, although neither of the two basic questions have been answered. We know little of the pharmacology of the neural involvement, save that the phenomenon is Ca^{++}-dependent (but it would be surprising if it were not, if a Ca^{++} spike is involved) and we have no understanding of the biochemistry of the changes in structure of the glyco-proteins and proteoglycans of the embedding matrix of the collagen bundles. The same cell processes as are found in connective tissue and which originate axon-like from the innervated justa-ligamental cells, occur in the muscle of the gut and water-vascular system. Does this imply that muscle cells can slip past each other as easily as collagen bundles? There is no doubt that the mutable connective tissue is the basic factor for the evolutionary success of the phylum, since heavy appendages can be moved using muscles, but then efficiently fixed in new positions by the mutability of the connective tissue. Stiff body walls and radiating spines without energy consumption! The ultimate speculation may be that the disadvantages of radial symmetry and the need for two nervous systems may be outweighed by the advantages of mutability. To me at least, this requires some quite extraordinary evolutionary mechanism involving loss of developmental potential of one sort and replacement by another. Perhaps, one day, an examination of the genetic basis of development will clarify this phenomenon.

Finally, how much can the giant fiber system tell us generally about the neural basis of behavior, which is, after all, the prime reason for the author's involvement in this group of animals? It is not a straightforward giant fiber system where speed is of the essence (though the relatively high speed of locomotion of brittlestars is, of course, a highly significant correlate). If speed alone were significant, one might expect to find electrical synapses between units, but instead, the neurons break up into very small varicose endings. Furthermore, available evidence is against the physical continuity of single neurons over very long distances. We can produce a partial answer to the question, are brittlestars typical of the other groups? The general layout appears to be the same. However, even in brittlestars, where there are giant fibers, the nervous system is technically difficult to work with at the cellular level, and neurons are all much smaller in the other classes. Confirmation may, therefore, be a long time coming.

We still know almost nothing about integration in the segmental ganglia of the radial nerve cords. We do know something about sensory input, a little about conduction through the nervous system by the different size classes of inteneurons, and rather more about motor output by the hyponeural nervous system. It is enough, however, to show us that all ganglia are equal, and that there are no specialized central ganglia that coordinate behavior. The conclusion from this is that complex whole-animal behavior can be initiated from any segmental ganglion in the radial nerve cords, however peripheral it may be, and that control of behavior shifts to other ganglia according to circumstances. The best guess at present must be that the largest class of giant fiber that occurs in the ectoneural system (15 μm cell bodies and 10 to 12 μm diameter axons) is the basis of a system that allows any part of the nervous system to subsequently command the behavior of the whole animal by way of other classes of neuron, and also allows for that command to shift from one part of the nervous system to another. This is reviewed in more detail in Cobb (1988) and Cobb and Moore (1989).

The echinoderm nervous systems may be secondarily evolved, rather than primitive, for reasons that we do not understand. However, echinoderms show so many unusual phenomena that, in a comparative sense, they are very valuable in understanding the evolution of nervous systems.

References

Baccaglini, P., 1978, Action potentials of embryonic dorsal root ganglion neurones in *Xenopus* tadpoles, *J. Physiol. (Lond.)*, **283**:585-604.

Berrios, A., Brink, D., del Castillo, J., and Smith, D. S., 1985, Some properties of the action potentials conducted in the spines of the sea-urchin *Diadema antillarum*, *Comp. Biochem. Physiol.* **81A**:15-23.

Brehm, P., 1977, Electrophysiology and luminescence of an ophiuroid radial nerve cord, *J. exp. Biol.* **71**:213-227.

Cobb, J. L. S., 1967, The innervation in the ampulla of the starfish *Astropecten*, *Proc. Roy. Soc. B* **168**:91-99.

Cobb, J. L. S., 1970, The significance of the radial nerve cord in asteroids and echinoids, *Z. Zellforsch* **108**:457-474.

Cobb, J. L. S., 1982, Membrane physiology of echinoderms, in: *Membrane Physiology of Invertebrates* (R. B. Podesta, ed.), Marcel Dekker, New York.

Cobb, J. L. S., 1985a, The neurobiology of the ectoneural/hyponeural synaptic connection of an echinoderm, *Biol. Bull.* **168**:432-446.

Cobb, J. L. S., 1985b, The motor innervation of the oral arm plate ligament in the brittlestar *Ophiura*, *Cell Tissue Res.* **242**:685-688.

Cobb, J. L. S., 1987, Neurobiology of the Echinodermata, in: *Invertebrate Nervous Systems*, pp. 483-527 (M. A. Ali, ed.), Plenum Press, New York.

Cobb, J. L. S., 1988, A preliminary hypothesis to account for the neural basis of behaviour in echinoderms, in: *Proc. 6th Int. Echinoderms Conf.*, pp. 565-575 (R. D. Burke, ed.), Balkema, Rotterdam.

Cobb, J. L. S., and Moore, A., 1988, Studies on the ionic basis of the action potential of the brittlestar *Ophiura ophiura*, *Comp. Biochem. Physiol.* **91A:**821-827.

Cobb, J. L. S., and Moore, A., 1989, Studies on the integration of sensory information by the nervous system of the brittlestar *Ophiura ophiura*, *Mar. Behav. Physiol.* **14:**211-222.

Cobb, J. L. S., and Pentreath, V. W., 1977, Anatomical studies on simple invertebrate synapses using stage rotation electron microscopy and densitometry, *Tiss. Cell* **9:**125-135.

Cobb, J. L. S., and Stubbs, T., 1981, The giant neurone system in ophiuroids. I. The general morphology of the radial nerve cords and circumoral ring, *Cell Tissue Res.* **219:**197-207.

Florey, E., and Cahill, M. A., 1980, Cholinergic motor control of sea urchin tubefeet: Evidence for transmission without synapses, *J. exp. Biol.* **88:**281-292.

Hille, B., 1984, *Ionic Channels of Excitable Membranes*, Sinauer, Sunderland, Mass.

Miyazaki, S., Ohmori, H., and Sasaki, S., 1975a, Action potential and non-linear current voltage relation in starfish oocytes, *J. Physiol. (Lond.)* **246:**37-54.

Miyazaki, S., Ohmori, H., and Sasaki, S., 1975b, Potassium rectifications of the starfish oocyte membrane and their changes during maturation, *J. Physiol. (Lond.)* **246:**55-78.

Moore, A., and Cobb, J. L. S., 1985a, Neurophysiological studies on photic responses in *Ophiura*, *Comp. Biochem. Physiol.* **80A:**11-16.

Moore, A., and Cobb, J. L. S., 1985b, Neurophysiological studies on the detection of amino acids by *Ophiura ophiura*, *Comp. Biochem. Physiol.* **82A:**395-399.

Moore, A., and Cobb, J. L. S., 1986, Neurophysiological studies on the detection of mechanical stimuli by *Ophiura ophiura*, *J. Exp. Mar. Biol. Ecol.* **104:**125-141.

Motokawa, T., 1988, Catch connective tissue: A key character for echinoderm's success, in: *Echinoderm Biology*, pp. 39-55 (R. D. Burke, ed.), Balkema, Rotterdam.

Smith, J. E., 1966, The form and functions of the nervous system, in: *Echinoderm Physiology*, pp. 503-512 (R. A. Boolootian, ed.), Wiley, New York.

Pentreath, V. W., 1987, Functions of invertebrate glia, in: *Invertebrate Nervous Systems*,pp. 61-104 (M. A. Ali, ed.), Plenum Press, New York.

Takahashi, K., 1967, The catch apparatus of the sea-urchin spine. II. Response to stimuli, *J. Fac. Sci. Tokyo Univ. Sec. IV* **11:**109-120.

Tuff, P. J., and Gilly, W. F., 1984, Ionic basis of action potential propagation along two classes of "giant axons" in the ophiuroid *Ophiopteris papillosa*. *J. exp. Biol.* **113:**337-350.

Wilkie, I. C., 1988, Design for disaster, The ophiuroid inter-vertebral ligament, in: *Echinoderm Biology*, pp. 25-38 (R. D. Burke, ed.), Balkema, Rotterdam.

Chapter 24

Summary of Session and Discussion on Electrical Excitability

BERTIL HILLE

The papers of the day showed that modern protists have many of the features needed for making neurons. In particular, they have a wide variety of voltage-gated Ca^{++} and K^{+} channels that are functionally very similar to those of neurons and they have several kinds of mechanoreceptive channels. Protists respond sensitively to chemicals and to touch. They can conduct Ca^{++} spikes and coordinate motor outputs with the underlying Ca^{++} fluxes. *Actinocoryne* makes a Na^{+}/Ca^{++} spike. They have receptors coupled to GTP-binding proteins. However, no channel that gates open within milliseconds of binding a ligand (such as is found in fast chemical synapses) has been reported in protists.

One topic of discussion was the phylogenetic relationship of cnidaria and ctenophores to other animal phyla. The conference was, in a sense, predicated on the assumpton that coelenterate nervous systems offer the earliest glimpse at steps in the evolution of animal nervous systems. This idea depends on the traditional placement of coelenterates, as descending from a multicellular animal that is also ancestral to a flatworm form that gives rise to virtually all higher animal phyla:

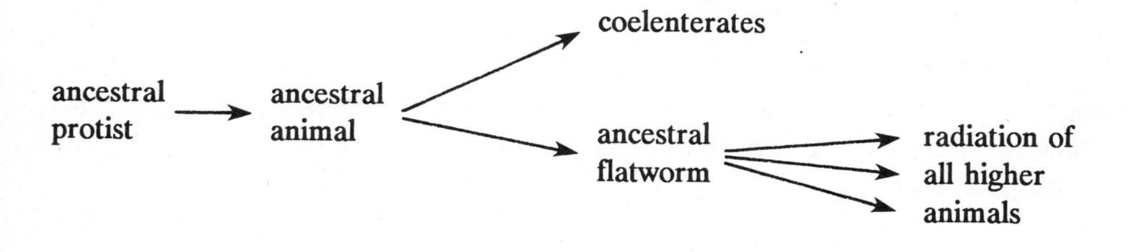

BERTIL HILLE • Dept. of Physiology, University of Washington, Seattle, Washington 98195, USA.

Nervous systems of the higher animal phyla seem to share homologous and highly differentiated cellular mechanisms and signal molecules. They have excitatory nicotinic acetylcholine receptors, inhibitory $GABA_A$ receptors opening Cl^- channels, tetrodotoxin-sensitive Na^+ channels, fast, Ca^{++}-dependent vesicular release of neurotransmitters from synapses, gap junctions at electrical synapses, pathfinding that permits specific wiring of complex, polarized neural circuits, and so forth. Thus, such mechanisms, which seem to be absent in protists, must have achieved their basic form at least in the nearest common ancestor of the higher phyla, the ancestral flatworm. The papers of the conference would then help to define how many of these characteristics were achieved already by the animals that were ancestral both to coelenterates and to the ancestral flatworm. This includes differentiation of an extended neuronal cell type, which is, however, multipolar and neither very directed nor supported by the specific glial cell type. New features apparently also include the evolution of a mechanism for rapid Ca^{++}-dependent chemical synaptic transmission as judged by identifiable synapses with synaptic delays of less than 1 ms in cnidaria and ctenophores, the use of voltage-gated Na^+ channels for propagation in axons, the development of morphologically and electrophysiologically recognized gap junctions and of proteins that cross-react with anti-gap junction antibodies, and, perhaps, the recruitment of many neurotransmitter molecules resembling those of higher animals.

An alternative phylogeny (known in the literature) was discussed in which coelenterates and flatworms arose as two independent multicellular experiments from unicellular protists.

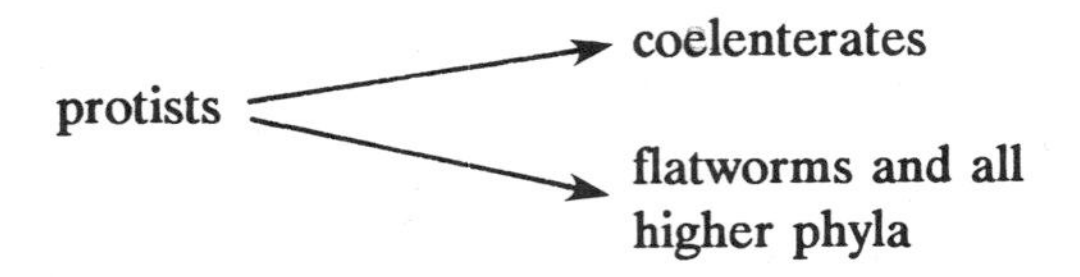

Within this scheme the coelenterate nervous system would illustrate the degree to which solving similar problems leads to similar solutions through convergent evolution since it would be unrelated to the nervous system of other animals. The participants had a wide variety of reactions to such phylogenetic questions and did not unanimously reject the possibility of independent origins. Opinions were expressed that convergent and parallel evolution could be so powerful as to be able even to give proteins with much sequence identity, that we may never be able to distinguish homology from analogy, that structure is diagnostic and after many more amino-acid sequences are available we will have a clearer idea. Furthermore, it was felt that many of the key mechanisms may already be present in protists and we simply have not studied them enough, that all phyla including protists have evolved so much since these early branchpoints that the important clues may be erased, and that flatworms may have arisen several times and may not be ancestral to higher animals.

Part 3

SENSORY MECHANISMS

Chapter 25

Chemoreception in Unicellular Eukaryotes

JUDITH VAN HOUTEN

1. Introduction

Every organism is surrounded by chemicals and each has the means to extract information from this chemical environment and to respond appropriately. This process, chemoreception, underlies the sensory processes of taste and smell in vertebrates and invertebrates, and of chemotaxis and chemokinesis in unicellular organisms. There is no unifying theme among the types of chemical stimuli that affect chemoreceptors, but within this confusion of compounds there is order, because each stimulus fits into the context of the life of the organism: folic acid attracts the paramecia and slime-mold amoebae that feed on bacteria, the stimulus source; amino acids attract lobsters and catfish that prey on fresh or decaying muscle; pheromones are commonly used by insects or protozoans to attract mates. There are, however, aspects common to the various chemosensory transduction systems: they all appear to be initiated at the membrane surface of a receptor cell by the interaction of a stimulus with a receptor molecule (or perhaps in some cases, the membrane directly), and, subsequently, this interaction is transduced into intracellular messengers. These second and third messengers are limited in number and, for the most part, consist of cyclic nucleotides, permeant ions, phosphoinositides, diacyl glycerol, arachidonic acid and internal pH levels.

JUDITH VAN HOUTEN • Department of Zoology, University of Vermont, Burlington, Vermont 05405, USA.

Evolution of the First Nervous Systems
Edited by P.A.V. Anderson
Plenum Press, New York

Table 1. Eukaryotic External Chemoreceptors

Dictyostelium cAMP chemoreceptor
yeast mating type pheromone receptors
Arbacia resact receptor
Paramecium cAMP chemoreceptor
Tetrahymena enkephalin/opiate receptor

By examining chemosensory transduction pathways in organisms ranging from unicells to metazoa, common threads of an evolutionary pathway could, perhaps, become evident. However, it is not clear that any thread of evolution can be followed from unicells to the nervous systems of higher organisms. For example, it may not be possible to separate the higher chemosensory schemes that are directly descended from those in unicells from successful schemes that have appeared independently, over and over again, across phyla. Therefore, the immediate importance of examining unicellular organisms here is not to trace chemosensory pathways from ciliates to vertebrates, but to examine unicellular examples of successful chemosensory transduction pathways and to begin to determine why their means of dealing with external stimuli are successful.

An additional reason for studying unicellular organisms for evidence of prototype chemosensory mechanisms is that they are proving to be more amenable than metazoans to the biochemistry necessary to identify and purify receptors. Our first glimpses of external chemoreceptors are coming from eukaryotic unicells (Table 1) and they do, indeed, have counterparts in internal vertebrate receptors (see chapter by Carr in this book) and may well have counterparts in the external receptors of vertebrates also.

Some receptors in unicells undoubtedly function to detect local, environmental stimuli that indicate the presence of food or of harmful conditions and that are beyond the control of the organism. With the evolution of sex, mating pheromones and their receptors could participate in chemical signalling between unicellular organisms of the same species, thus facilitating mating. This review will deal exclusively with *external* chemosensory signalling as represented by olfaction, gustation and unicellular chemoresponse, and highlight parallels between these various external chemoreceptor systems. The intriguing parallels between the external receptors of unicellular eukaryotes and those for neurotransmitters and other *internal* receptor systems has been discussed by Carr (Chapter 6, this book).

The common themes of chemoreception to be taken up in this chapter are receptor processes, perireceptor events, and second messengers. This survey will begin with an introduction to olfaction and gustation and then move to similar chemosensory transduction pathways in the unicells *Dictyostelium*, *Chlamydomonas*, yeast, *Arbacia* spermatozoa, *Paramecium*, and *Tetrahymena*.

2. Olfaction

In vertebrate olfactory systems, the cilia of primary receptor neurons are exposed to odorants that dissolve into the overlying mucus layer of the olfactory epithelium. It is thought that, in general, these cilia contain the olfactory receptors and serve as the initiation site for the sensory transduction process (Adamek et al., 1984). Recently, olfactory binding proteins (OBP) have been identified in several vertebrate species and they exist in the mucous layer and to bind a wide range of odorant molecules (Snyder et al., 1988). Therefore, it is possible that receptors detect not the odorant alone, but an OBP-stimulus complex. Additionally, the OBP may function to protect odorants from degradation and to facilitate partitioning of hydrophobic molecules into the mucous layer.

The interaction between the stimulus and the olfactory receptor cells results in the initiation of, or a change in the frequency, of action potentials (Fig. 1). Each olfactory receptor cell sends its axon directly to the olfactory bulb, and it is likely to be the pattern of firing of receptor cells over the entire olfactory epithelium that is interpreted by higher centers for quantity and quality of odorants (see Kauer, 1987 and Getchell and Getchell, 1987 for review).

The mechanisms by which the ion channels in receptor cells are opened by receptor-ligand interaction are yet to be determined, although several possibilities are being tested. The adenylate cyclase activity of olfactory cilia is high, G protein-dependent (Lancet and Pace, 1987) and activated by odorants (Lancet and Pace, 1987; Sklar et al., 1986). Therefore, the receptor may interact with a G protein, perhaps the one that is unique to olfactory receptor neurons (G_{olf}; Jones and Reed, 1989), which, in turn, will activate adenylate cyclase. The resultant cAMP would then interact with channels and initiate action potentials (Pace and Lancet, 1987). Indeed, there are reports of nucleotide-sensitive conductances in olfactory receptor neurons and of a correlation between receptor-cell population activity and cyclase activation (Gold and Nakamura, 1987; Lowe et al., 1988). Alternatively, there are reports of direct stimulus gating of channels (Labarca et al., 1988), in which case, the activation of adenylate cyclase by the same stimuli might function in a slower process of adaptation. It is possible that the ligand binding sites function here as ligand-gated ion channels, akin to the nicotinic cholinergic receptor, or alternatively, interaction between stimulus and

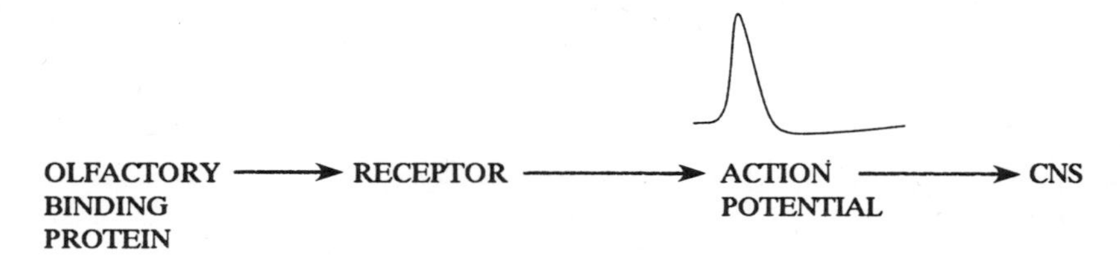

Figure 1. Schematic of the pathway for olfactory transduction.

membrane lipid might perturb the membrane structure sufficiently to open channels. These two pathways, nucleotide vs. ligand gating, are not mutually exclusive; with time we may discover a diversity of olfactory mechanisms that all result in the production of action potentials in receptor cells.

As mentioned above, some odorants stimulate adenylate cyclase and cAMP may function as a second messenger either in the transduction process, to open ion channels, or in desensitization. However, there are other odorants, particularly putrid ones, that do not stimulate adenylate cyclase (Sklar et al., 1986) and are not reported to directly gate channels. These stimuli may utilize different second messengers such as IP_3 and/or Ca^{++}. Indeed, in catfish, there is evidence for rapid changes in inositol phospholipids as a result of olfactory stimulation (Bruch et al., 1987a; Bruch et al., 1987b).

Invertebrate olfactory systems also have been examined in detail and the lobster and insect olfactory sensilla have been particularly well studied. Here, the receptor cells are enclosed in a porous cuticle. Stimuli bind to receptors on the dendrites of these cells and affect action potential production. A pattern of firing emerges across the broadly- and narrowly-tuned receptor cells, and the concentration and qualities of an odorant blend are, thus, encoded (Derby and Atema, 1987; Ache, 1987). While receptor molecules have been difficult to locate and second messengers not yet well characterized, the invertebrate systems have been particularly fruitful for the study of perireceptor events, that is, the processes that occur in the sensillar lymph in the vicinity of the receptor, to protect, carry, and finally degrade the stimulus (Getchell et al., 1984). Unlike light and touch, chemical stimulation outlasts its transient application and must be destroyed or removed from the receptor in order for the cell to resensitize and respond to the next wave of stimulus (Atema, 1987). Hydrophobic stimuli must cross the aqueous lymph to reach receptors in the insect antennae and must be protected from potent degrading enzymes (Vogt et al., 1987). Insect binding proteins that serve this function and are the counterparts of the mammalian OBP have been known for some time (Vogt et al., 1987). The aesthetasc sensillae of lobsters contain dendrites of chemosensory cells that respond to the nucleotides ATP, ADP and AMP, but not the nucleoside adenosine, through M1- and M2-type purinergic receptors. These sensillae also possess potent ectonucleotidase activity that can remove nucleotide stimulants from the vicinity of the receptors in the brief interval between flicks of the antenulles (Carr et al., 1987, 1989).

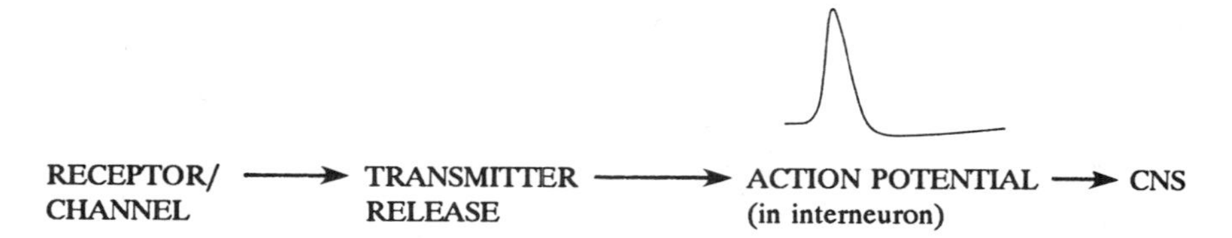

Figure 2. Schematic for the gustatory transduction pathway.

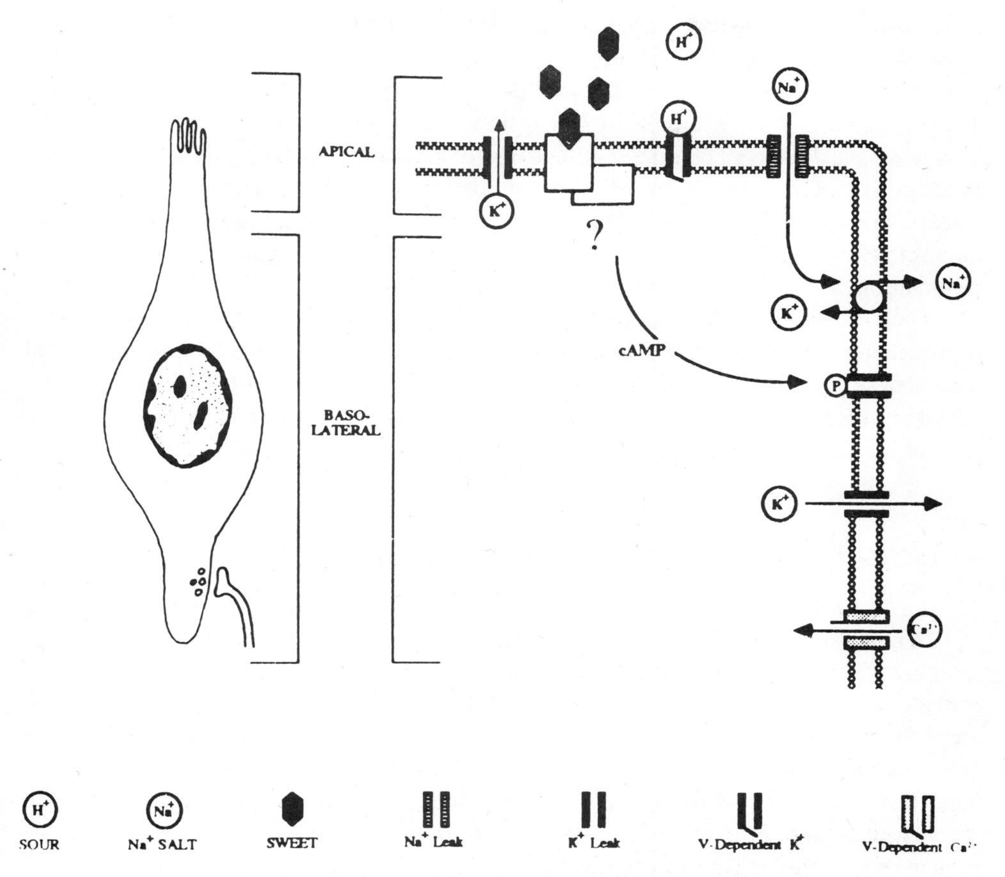

Figure 3. Schematic of taste transduction mechanisms (from Kinnamon, 1988).

To date, no vertebrate or invertebrate olfactory receptors have been identified and purified, although there are some hopeful candidates from insect (Vogt et al., 1988) and dog (Price and Willey, 1987, 1988). It seems clear that receptors account for the specificity of olfaction, particularly as observed in fish, but it is possible that non-receptor mechanisms may also play a role, particularly for volatile, lipid-soluble stimuli (Dionne, 1988; Lerner et al., 1988).

3. Gustation

Taste receptor cells are modified neurons that synapse with interneurons, whose axons traverse to the CNS. Generally, taste stimuli elicit a calcium-dependent release of neurotransmitter by depolarizing the cell, thereby opening voltage- sensitive Ca^{++} channels (Kinnamon, 1988; Roper, 1989) (Fig. 2).

The mechanisms by which tastants depolarize cells are diverse and include the direct inhibition of K^+ channels by sour stimuli (H^+); passage of Na^+ through passive, amiloride-sensitive Na^+ channels; generation of cAMP by sugars binding to receptors, followed by indirect inhibition of K^+ channels by cAMP-dependent phosphorylation; release of intracellular Ca^{++} by second messengers generated after bitter tastants interact with receptors; and direct opening of channels by arginine in catfish taste (Kinnamon, 1988) (Figs. 2 and 3).

There is no doubt that future work will identify a variety of mechanisms in taste transduction ranging from receptor interactions that generate one of several second messengers, to direct ligand gating (Kinnamon, 1988; Roper, 1989). To date, no gustatory receptor has been isolated, but strong possibilities can be found in catfish (Bryant et al., 1987) and in mammalian sweet taste (Persaud et al., 1988).

4. Unicellular Eukaryotes

In general, unicellular eukaryotes detect and respond to external chemicals through a highly schematized pathway (Fig. 4). The most obvious difference between this and transduction in olfactory and taste receptor cells is in the response. In unicells, chemical stimulation results in a change in motility and not, ultimately, in synaptic transmission. However, unicells can be electrically excitable and changes in their electrical activity are involved in at least some unicellular sensory pathways below. (See Van Houten and Preston, 1987 and Van Houten, 1990 for a more comprehensive review of unicellular chemoreception.)

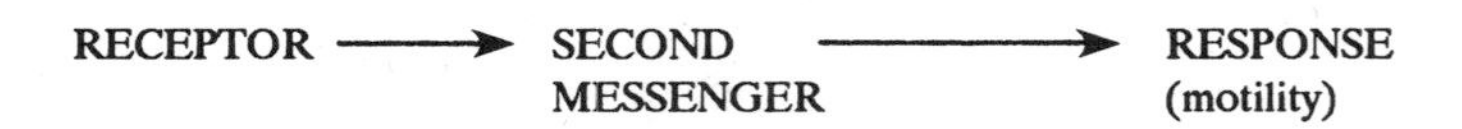

Figure 4. Sensory transduction pathway of a unicellular eukaryote.

4.1. *Dictyostelium*

When amoebae of *Dictyostelium* run out of the bacteria they feed on, they begin the process of developing into a multicellular slug by responding to pulses of cAMP that emanate from focal cells. The cells migrate up the gradient of cAMP until they congregate and aggregate into a slug. When a cell is stimulated with cAMP, it, in turn, releases a pulse of cAMP (the relay), while, at the same time, orienting and transiently moving toward the origin of the stimulus. The chemoresponse can, therefore, be divided into two components, orientation of the cytoskeletal motile apparatus required for chemotaxis, and activation of adenylate cyclase to produce cAMP for the relay.

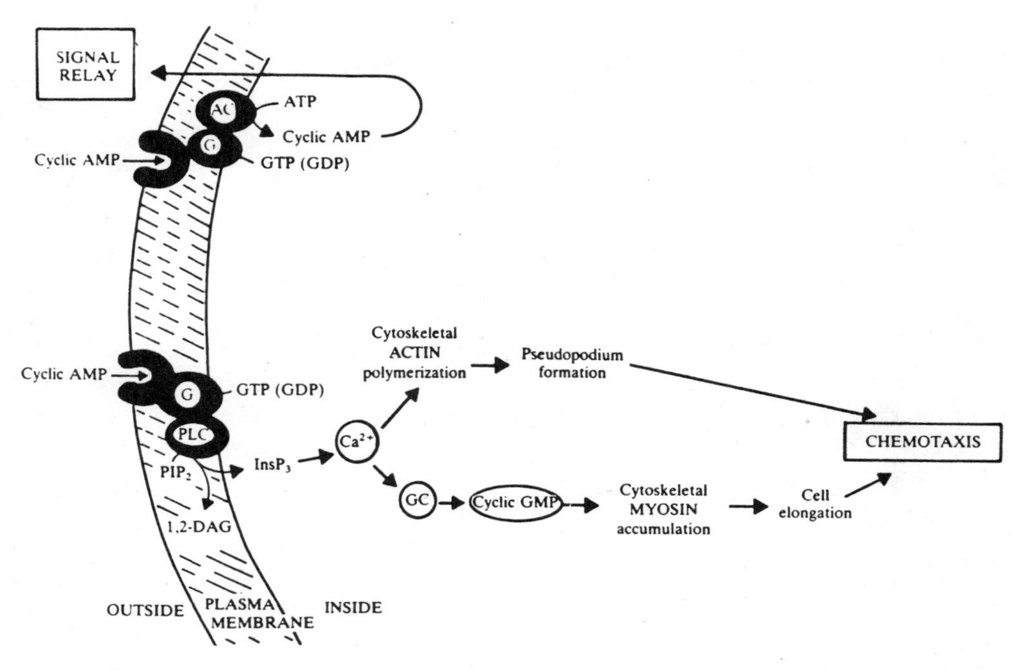

Figure 5. Pathways of sensory transduction in *Dictyostelium* (from Newell et al., 1988).

These two processes can be separated and studied independently (see Newell et al., 1988 for review).

These two chemosensory transduction pathways are initiated at two types of cAMP receptors, with different affinities (Janssens and Van Haastert, 1987), and both sets of receptors interact with G proteins (Kumagai et al., 1989). In the relay system, the adenylate cyclase is activated by a receptor-activated G protein and cAMP is produced for the relay. In the chemotaxis system, phospholipase C is activated, resulting in an increase in IP_3 and the liberation of Ca^{++} from non-mitochondrial stores (Small et al., 1987; Europe-Finner and Newell, 1987). The Ca^{++}, in turn, activates the guanylate cyclase to increase levels of cGMP, a long-acknowledged internal messenger for *Dictyostelium* chemosensory transduction (Devreotes and Zigmond, 1988). The increase in intracellular Ca^{++} and cGMP affects the cell's cytoskeletal motor apparatus and the cell becomes motile and oriented (Newell, 1986; Newell et al., 1988). There are also potent ecto- and exo-enzymes that function to clear the signal surrounding the cell (see chapter by Carr in this book), thus enabling it to respond to the next wave of cAMP.

A *Dictyostelium* membrane cAMP receptor has been identified (Table 1) and its gene cloned (Klein et al., 1988). The protein exists as a doublet on gels, with an apparent molecular weight of 40-43 kD, depending upon its phosphorylation state. As the cell desensitizes in the presence of stimulus, the receptor becomes covalently modified. The cDNA sequence of the receptor indicates that this receptor could be part of the G-mediated superfamily of receptors that includes the β-adrenergic and muscarinic receptors, since the inferred amino acid sequence has motifs for the characteristic seven membrane-spanning regions, and a G protein interaction site.

4.2. *Chlamydomonas* gametes

Chlamydomonas gametes begin mating by sticking to cells of the complementary mating type. There are species-specific agglutinin molecules that mediate this process in *C. eugametos* and *C. reinhardtii* (Musgrave and van den Ende, 1987; Goodenough et al., 1985). These are large glycoproteins ($>10^3$ kD) with hooks that anchor them into the membrane. Agglutination triggers a series of events, but it is not clear how the interaction of molecules at the tips of the flagella is communicated to the body of the cell. Nonetheless, the adenylate cyclase is activated early in the process and an increase in internal cAMP is sufficient to trigger the cascade of mating responses (Pasquale and Goodenough, 1987).

Ligand binding has been shown to change the levels of intracellular cAMP or cGMP during *Chlamydomonas* agglutination (Pasquale and Goodenough, 1987; Musgrave and Van der Ende, 1987), *D. discoideum* aggregation (Janssens and Van Haastert, 1987), sperm chemotaxis (Garbers et al., 1986), neutrophil chemotaxis (Sha'afi and Molski, 1988), vertebrate olfaction and gustation (see Teeter and Gold, 1988; Lancet and Pace, 1987; Anholt, 1987 for reviews). However, only in *Chlamydomonas* and *D. discoideum* are there clear indications that cyclic nucleotides are the internal messengers that function as links in the sensory transduction pathways.

4.3. *Saccharomyces*

There are two mating types of the haploid yeast *Saccharomyces cerevisisae*, 'a' and α. Each mating type produces a pheromone that arrests cells of the complementary mating type in G1 and induces changes in the cell in preparation to mating. The mating factors bind to specific receptors on the complementary cell's surface (Table 1). The receptor for α factor is coded for by the STE2 gene and that for 'a' factor by the STE3 gene (Burkholder and Hartwell, 1985; Nakayama et al., 1988; Hagen et al., 1986). The amino acid sequences that are inferred from the DNA sequences give pictures of two receptors that are similar in structure. However, it is curious that while their hydropathy plots are virtually superimposable with seven potential membrane spanning regions, the proteins are utterly different in primary amino acid sequence (Hagen et al., 1986). Their deduced structures resemble those of members of the class of receptors that interact with G proteins (Herskowitz and Marsh, 1987; Marsh and Herskowitz, 1988). The carboxy-terminus is implicated in endocytosis and down regulation of the receptor (Reneke et al., 1989).

There is mounting evidence that yeast mating pheromone receptors interact with G proteins (Whiteway et al., 1989; Blinder et al., 1989; Marsh and Herskowitz, 1988). The interesting twist to the G protein function in yeast is that the α subunit dissociates from the β and γ units upon ligand binding and the free $\beta\gamma$-complex may initiate the hormone response (Whiteway et al., 1989). Alternatively, the genetic evidence would also be compatible with the inhibition of the sensory pathway by the G protein trimer and release from inhibition by the binding of the ligand to the receptor and dissocation of the trimer (Dietzel and Kurjan, 1987; Miyajima et al., 1987). Although G proteins may function in the pheromone sensory pathway, there is no evidence for the activation of adenylate cyclase by G proteins or for the role of cAMP as second messenger.

4.4. Sea urchin spermatozoa

Secretions from the eggs of the sea urchins *Stronglyocentrotus purpuratus* and *Arbacia punctulata* stimulate spermatozoan motility and metabolism, thereby facilitating fertilization (Trimmer and Vacquier, 1986). Upon stimulation, levels of cAMP and cGMP increase in the sperm and there is a net H^+ and K^+ efflux, a Ca^{++} influx, and guanylate cyclase is dephosphorylated. The stimulus for these changes comes from two peptides, speract and resact, which consist of 10 and 14 amino acids, respectively (Dangott et al., 1984).

Speract and resact stimulate spermatozoa in a species-specific manner through receptor proteins in spermatozoan membranes (Table 1). Speract and resact binding proteins have been identified as proteins of 77 kD and 160 kD molecular mass, respectively (Dangott et al., 1984; Shimomura et al., 1986). The resact receptor is particularly interesting because it is the guanylate cyclase itself (Singh et al., 1988). The atrial natiuretic factor receptor of vertebrates, likewise, is the guanylate cyclase (Lowe et al., 1989), and may be the second example of an important, and previously unrecognized mechanism of signal transduction.

4.5. *Paramecium*

Paramecium tetraurelia responds to chemicals in its aqueous environment. In particular, fermentation and other bacterial products are attractants, probably signifying the presence of food (Van Houten, 1978). Stimuli, such as folic acid, acetate, and cAMP hyperpolarize the cells (Van Houten, 1979). This hyperpolarization causes changes in ciliary beating: the cells move more smoothly and turn less frequently as a consequence. This, in turn, causes populations of cells to accumulate, by biased random walk (Van Houten, 1978; Van Houten and Van Houten, 1982). The stimuli are thought to interact with the cell at specific receptor sites (Schulz et al., 1984; Smith et al., 1987; Sasner and Van Houten, 1989) and single-site mutations eliminate both binding and chemoresponse (DiNallo et al., 1982; Schulz et al., 1984; Smith et al., 1987). Cilia are not essential for chemoreception: deciliated cells show the characteristic hyperpolarization in the presence of attractants (Preston and Van Houten, 1987a,b) and only a minority of binding sites are on the cilia (Schulz et al., 1984; Smith et al., 1987).

The ionic basis of the hyperpolarization is not completely clear (Preston and Van Houten, 1987 a,b), but appears to be due to the activity of an electrogenic ion pump, perhaps carrying Ca^{++} (Wright and Van Houten, 1989). For stimuli such as NH_4Cl, however, there is indirect evidence that internal pH levels participate in the transduction pathway. Recently, a Ca-ATPase activity, that has all the hallmarks of being the pump involved in the chemoresponse, has been identified (Wright and Van Houten, 1989). Additionally, a 115 kD molecular weight protein has been identified as a Ca-ATPase phospho-enzyme intermediate, and is likely to be the pump implicated in homeostatsis and/or chemoresponse (Wright and Van Houten, 1989). There is no evidence for internal second messengers other than Ca^{++} functioning in the *Paramecium* chemosensory pathway. Cyclic AMP levels and phosphoinositol lipids do not appear to change with stimulation.

The cAMP receptor of *Paramecium* (Table 1) is the most recent of the eukaryotic external chemoreceptors to be identified (Van Houten et al., 1990; Schlichtherle et al., 1989). The receptor is a doublet, 48 kD molecular weight cell, body membrane glycoprotein that elutes from cAMP affinity columns. Although a doublet of proteins in the 40 kD range is reminiscent of the *Dictyostelium* cAMP receptor, it is a not immunologically crossreactive with antibodies against the *Dictyostelium* cAMP receptor. Additionally, it has no antigenic similarity to the regulatory subunit of the cAMP-dependent protein kinase or the immobilization-antigen related proteins of similar size from *Paramecium*. Antibodies against this protein inhibit chemoresponse to cAMP, but to no other stimulus tested (Schlichtherle et al., 1989). The N terminal sequence has been determined in preparation for sequencing the gene for this protein.

Recently, a membrane protein from the ciliate *Tetrahymena* has been identified as the analog of the vertebrate opiate or enkephalin binding protein (O'Neill et al., 1987; Table 1). This protein is conserved across phyla, as judged by its immuno-crossreactivity, size, pharmacology, and binding properties. Its function in *Tetrahymena* is not clear, but cells accumulate in response to enkephakin, and this response is blocked by naloxone (O'Neill et al., 1987; Zipser et al., 1988). *Tetrahymena* cells are attracted to amino acids and peptides, including platelet-derived growth factor (PDGF) (Leick and Hellung-Larsen, 1985; Levandowsky et al., 1984), suggesting that enkephalin peptides may be acting through peptide receptors that serve primarily to detect food cues.

5. Summary

Eukaryotes, ranging from slime molds to amoebae, have many common themes in their chemosensory transduction pathways. Receptors, G proteins, second messengers, stimulus-binding proteins, and degradative enzymes are found to participate in these pathways up and down the phylogenetic scale. Diversity within an organism and even within a receptor cell is also a hallmark of chemosensory transduction. Taste and olfactory receptor cells do not limit themselves to one transduction pathway, and neither do *Dictyostelium*, *Arbacia* sperm, or *Paramecium*.

Therefore, in considering the evolution of chemical senses, it would be important to recognize that there are several successful approaches used by "primitive" animals and, if these reflect the raw material from which vertebrates evolved their chemosensory systems, we can expect variations on many basic, successful themes.

It is also important to remember in our search for an ancestral nervous system that the processes of receptor-mediated signal transduction are not unique to neurons. From fibroblasts to neurons to *Dictyostelium*, a diverse range of receptors, some mediated by G-proteins and others not, affect ion channels, enzyme activity, gene transcription, cell division, and so on, either directly or through second messengers. Therefore, among the genes that characterize neurons and that are to be sequenced and analyzed by the molecular geneticists for cladistic analysis could be genes expressed uniquely in neurons, as well as genes expressed broadly. Many of the gene products that play a role in chemosensory transduction will be in the latter class.

References

Ache, B., Chemoreception in invertebrates, in: *Neurobiology of Taste and Smell* (T. Finger and W. Silver, eds.), J. Wiley, 1987.

Adamek, G. D., Gesteland, R. C., Mair, R. G., and Oakley, B., 1984, Transduction physiology of olfactory receptor cilia, *Brain Res.* **310**:87-97.

Anholt, R. R. H., 1987, Primary events in olfactory reception, *Trends in Biochem. Sci.* **12**:58-62.

Atema, J., 1987, Aquatic and terrestrial chemoreceptor organs: morphological and physiological designs for interfacing with chemical stimuli, in: *Comparative Physiology: Life in Water and on Land* (P. Dejours, L. Bolis, C. Taylor, and E. Weibel, eds.), Fidia Research Series.

Blinder, D., Bouvier, S., and Jenness, D. D., 1989, Constitutive mutants in the yeast pheromone response: ordered function of the gene products, *Cell* **56**:4799-4786.

Bruch, R., Kalinoski, D. L., and Huque, T., 1987, Role of GTP-binding regulatory proteins in receptor-mediated phosphoinositide turnover in olfactory cilia, *Chem. sen.* **12**:173.

Bruch, R., Rull, R. D., and Boyle, A. G., 1987, Olfactory L-amino acid receptor specificity and stimulation of potential second messengers, *Chem. Sen.* **12**:642-643.

Bryant, B., Brand, J. G., Kalinoski, D. L., Bruch, R. C., and Cagan, R. H., 1987, Use of monoclonal antibodies to characterize amino acid taste receptors in catfish: effects on binding and neural responses, in: *Olfaction and Taste IX* (S. Roper and J. Atema, eds.), NY Acad. Sci.

Burkholder, A. C., and Hartwell, L. H., 1985, The yeast alpha-factor receptor, *Nucl. Acid Res.* **13**:8463-8473.

Carr, W. E. S., Ache, B. W., and Gleeson, R. A., 1987, Chemoreceptors of crustaceans: similarities to receptors for neuroactive substances in internal tissues, *Environ. Health Perspec.* **71**:31-46.

Carr, W. E. S., and Gleeson, R. A., 1989, Stimulants of feeding behavior in marine organisms: receptor and perireceptor evolution provide insight into mechanisms of mixture interactions, in: *Perception of Complex Smells and Tastes* (D. Laing, W. Caine, B. Ache, and R. McBride, eds.), Academic Press, Australia.

Dangott, L. J., and Garbers, D. L., 1984, Identification and partial purification of the receptor for speract, *J. Biol. Chem.* **259**:13712-13716.

Derby, C. D., and Atema, J, 1988, Chemoreceptor cells in aquatic invertebrates, in: *Sensory biology of aquatic animals,* (J. Atema, R. Fay, A. Popper, and W. Tavolga, eds.), Springer-Verlag.

Devreotes, P. and Zigmond, S., 1988, Chemotaxis in eukaryotic cells, *Ann. Rec. Cell Biol.* **4**:649-686.

Dietzel, C., and Kurjan, J., 1987, The yeast SCG1 gene: a G_a-like protein implicated in the a- and a-factor response pathway, *Cell* **50**:1001-1010.

DiNallo, M. C., Wohlford, M., and Van Houten, J., 1982, Mutants of *Paramecium* defective in chemokinesis to folate, *Gen.* **102:**149-158..

Dionne, V., 1988, How do you smell? Principle in question, *TINS* **11:**188-189.

Europe-Finner, G. N., and Newell, P. C., 1987, GTP analogues stimulate inositol trisphosphate formation transiently in *Dictyostelium*, *J. Cell Sci.* **87:**513-518.

Garbers, D. L., Noland, T. D., Dangott, L. J., Ramarao, C. S., Bentley, J. K., 1986, The interaction of egg peptides with spermatozoa, in: *Molecular and Cellular Aspects of Reproduction* (D. Dhindsa and Om. P. Bahl, eds.).

Getchell, T. V., and Getchell, M. L., 1987, Peripheral mechanisms of olfaction: biochemistry and neurophysiology, in: *Neurobiology of Taste and Smell* (T. Finger and W. Silver, eds.), J. Wiley and Sons.

Getchell, T. V., Margolis, F. L., and Getchell, M. L., 1984, Perireceptor and receptor events in vertebrate olfaction, *Prog. Neurobiol.* **23:**317-345.

Gold, G., and Nakamura, T., 1987, Cyclic nucleotide-gated conductances: a new class of ion channels mediates visual and olfactory transduction, *TIPS* **8:**312-316.

Goodenough, U., Adair, W. S., Collin-Osdoby, P., and Heuser, J. E., 1985, Structure of the *Chlamydomonas* agglutinin and related flagellar surface proteins in vitro and in situ, *J. Cell Biol.* **101:**924-941.

Hagen, D. C., McCaffrey, G. and Sprague, G. F., 1986, Evidence the yeast STE3 gene encodes a receptor for the peptide pheromone a factor: gene sequence and implications for the structure of the presumed receptor, *Proc. Natl. Acad. Sci. (USA)* **83:**1418-1422.

Herskowitz, I., and Marsh, L., 1987, Conservation of a receptor/signal transduction system, *Cell* **50:**995-996.

Janssens, P. M. W. and Van Haastert, P. J. M., 1987, Molecular basis of transmembrane signal transduction in *Dictyostelium discoideum*, *Microbiol. Rev.* **51:**396-418.

Jones, D. T., and Reed, R. R., 1989, G_{olf} : an olfactory neuron specific-G protein involved in odorant signal transduction, *Science* **244:**790-795.

Kauer, J. S., 1987, Coding in the olfactory system, in: *Neurobiology of Taste and Smell* (T. Finger and W. Silver, eds.), J. Wiley and Sons.

Kinnamon, S. C., 1988, Taste transduction: a diversity of mechanisms, *TINS* **11:**491-496.

Klein, P. S., Sun, T. J., Saxe, C. L., Kimmel, A. R., Johnson, R. L., and Devreotes, P. N., 1988, A chemoattractant receptor controls development in *Dictyostelium discoideum*, *Science* **241:**1467-1472.

Kumagai, A., Pupillo, M., Gundersen, R., Miake-Lye, R., Devreotes, P., and Firtel, R. A., 1989, Regulation and function of G_a protein subunits in *Dictyostelium*, *Cell* **57:**265-267.

Labarcha, P., Simon, S., and Anholt, R. H., 1988, Activation of odorants of multistate cation channel from olfactory cilia, *Proc. Nat. Acad. Sci. USA* **85:**944-947.

Lancet, D., and Pace, U., 1987, The molecular basis of odor recognition, *TIBS* **12:**63-66.

Leick, V., and Hellung-Larsen, P., 1985, Chemosensory responses in *Tetrahymena*: the involvement of peptides and other signal substances, *J. Protozool.* **32:**550-553.

Lerner, M., Reagan, J., Gyorgi, T., and Roby, A., 1988, Olfaction by melanophores: what does it mean?, *Proc. Natl. Acad. Sci. (USA)* **85:**261-264.

Levandowsky, M. L., Cheng, M., Kehr, A., Kim, J., Gardner, L., Silvern, L., Tsang, L., Lai, G., Chung, C., and Prakash, E., 1984, Chemosensory responses to amino acids and certain amines by the ciliate *Tetrahymena* in a flat capillary assay, *Biol. Bull.* **167:**322-330.

Lowe, D., Chang, M-S., Hellmess, R., Chin, E., Singh, S., Garbers, D., and Goeddel, D., 1989, Human atrial natriuretic peptide receptor defines a new paradigm for second messenger signal transduction, *EMBO J.* **8:**1377-1384.

Lowe, G., Nakamura, T., and Gold, G. H., 1988, EOG amplitude is correlated with odor-stimulated adenylate cyclase activity in the bullfrog olfactory epithelium, *Chem. Sen.* **13:**710.

Marsh, L., and Herskowitz, I., 1988, STE2 protein of *Saccharomyces cerevisisae* is a member of the rhodopsin/ß- adrenergic family and is responsible for recognition of the peptide ligand alpha factor, *Proc. Natl. Acad. Sci.(USA)* **85:**3844-3859.

Miyajima, I., Nakafuku, M., Nakayama, N., Brenner, C., Miyajima, A., Kaibuchi, K., Arai, K., Kaziro, Y., and Matsumoto, K., 1987, GPA1, a haploid-specific essential gene, encodes a yeast homolog of mammalian G protein which may be involved in mating factor signal transduction, *Cell* **50:**1011-1019.

Musgrave, A., and van den Ende, H., 1987, How *Chlamydomonas* court their partners, *TIBS* **12:**469-473.

Nakamura, N., Kaziro, Y., Arai, K-I., and Matsumoto, K., 1988, Role of STE genes in the mating factor signaling pathway mediated by GPA1 in *Saccharomyces cerevisisae, Molec. Cell. Biol.* **8:**3777-3783.

Newell, P. C., Europe-Finner, G. N., Small, N. V., and Liu, G., 1988, Inositol phosphates, G-proteins and ras genes involved in chemotactic signal transduction of *Dictyostelium, J. Cell Sci.* **89:**123-127.

O'Neill, J. B., Pert, C. B., Ruff, M. S., Smith, C. C., Higgins, W. J., and Zipser, B., 1987, Identification and characterization of the opiate receptor in the ciliated protozoan, *Tetrahymena, Brain Res.*, in press.

Pasquale, S.M., and Goodenough, U., 1987, Cyclic AMP functions as a primary sexual signal in gametes of *Chlamydomonas reinhardtii, J. Cell Biol.* **105:**2279-2292.

Persaud, K. C., Chiavacci, L., and Pelosi, P., 1988, Binding proteins for sweet compounds from gustatory papillae of the cow, pig and rat, *Biochim. Biophys. Acta* **967:**65-75.

Preston, R. R., and Van Houten, J. L., 1987a, Chemoreception in *Paramecium tetraurelia*: acetate and folate-induced membrane hyperpolarization, *J. Comp. Physiol.* **160:**525-535.

Preston, R. R., and Van Houten, J. L., 1987b, Localization of chemoreception properties of the surface membrane of *Paramecium, J. Comp. Physiol.* **160:**537-541.

Price, S., and Willey, A., 1987, Benzaldehyde binding protein from dog olfactory epithelium, *Ann. N.Y. Acad. Sci.* **510:**561-564.

Price, S., and Willey, A., 1988, Effects of antibodies against odorant binding proteins on electrophysiological responses to odorants, *Biochim. Biophys. Acta* **965:**127-129.

Reneke, J. E., Blumer, K. J., Courchesne, W. E., and Thorner, J., 1988, The carboxy-terminal segment of the yeast α-factor receptor is a regulatory domain, *Cell* **55**:221-234.

Roper, S., 1989, The cell biology of vertebrate taste receptors, *Ann. Rev. Neurosci.* **12:**329-353.

Sasner, J. M., and Van Houten, J. L., 1989, Evidence for a *Paramecium* folate chemoreceptor, *Chem. Sen.*, in press. Schlichtherle, I. M., Cote, B., Zhang, J., and Van Houten, J. L., 1989, The cyclic AMP chemoreceptor of *Paramecium, Chem. Sen.*, abstract, in press.

Schulz, S., Denaro, M., Xypolyta-Bulloch, A., and Van Houten, J., 1984, Relationship of folate binding to chemoreception in *Paramecium, J. Comp. Physiol.* **155:**113-119.

Sha'afi, R. I., and Molski, R. F. P., 1987, Signalling for incresed cytoskeletal actin in neutrophils, *Biochem. Biophys. Res. Comm.* **145:**934-941.

Sha'afi, R. I., and Molski, R. F. P., 1988, Activation of the neutrophil, *Prog. in Allergy*, in press.

Shimomura, H., Dangott, L. J., and Garbers, D. L., 1986, Covalent coupling of a resact analogue to quanylate cyclase, *J. Biol. Chem.* **259**:10983-10988.

Singh, S., Lowe, D. G., Thorpe, D. S., Rodriguez, H., Kuang, W.-J., Daggott, L., Chinkers, M., Goeddel, D. V., and Garbers, D. L., 1988, Membrane guanylate cyclase is a cell-surface receptor with homology to protein kinases, *Nature* **334:**708-712.

Sklar, P. B., Anholt, R., and Snyder, S. H., 1986, The odorant-sensitive adenylate cyclase of olfactory receptor cells, *J. Biol. Chem.* **261:**15538-15543.

Small, N. V., Europe-Finner, G. N., and Newell, P. C., 1987, Adaptation to chemotactic cyclic AMP signals in *Dictyostelium* involves the G-protein, *J. Cell Sci.* **88:**537-545.

Smith, R. A., Preston, R. R., Schulz, S., and Van Houten, J. L., 1987, Correlation of cyclic adenosine monophosphate binding and chemoresponse in *Paramecium, Biochim. Biophys. Acta* **928:**171-178.

Snyder, S. H., Sklar, P. B., and Pevsner, J., 1988, Molecular mechanisms of olfaction, *J. Biol. Chem.* **263:**13971-13974. Teeter, J. H., and Gold, G. H., 1988, A taste of things to come, *Nature* **331:**298-299.

Trimmer, J. S., and Vacquier, V. D., 1986, Activation of sea urchin gametes, *Ann. Rev. Cell Biol.* **2:**1-26.

Van Houten, J., 1978, Two mechanisms of chemotaxis in *Paramecium, J. Comp. Physiol.* **127:**167-174.

Van Houten, J., 1979, Membrane potential changes during chemokinesis in *Paramecium, Sci.* **204:**1100-1103.

Van Houten, J., 1990, Signal Transduction in chemoreception, in: *Modern Cell Biology* (J. Spudich, ed.), in press.

Van Houten, J., and Preston, R., 1987, Chemoreception in single-celled organisms, in: *Neurobiology of Taste and Smell* (T. Finger and W. Silver, eds.), John Wiley and Sons.

Van Houten, J., and Van Houten, J. L., 1982, Computer analysis of *Paramecium* chemokinesis behavior, *J. Theor. Biol.* **98:**453-468.

Vogt, R. G., 1987, The molecular basis of pheromone reception: its influence on behavior, in: *Pheromone Biochemistry* (G. Prestwich and G. Blomquist, eds.), Academic Press.

Vogt, R. G., Prestwich, G. D., and Riddiford, L. M., 1988, Sex pheromone receptor proteins, *J. Biol. Chem.* **263**:3952-3959.

Whiteway, M., Hougan, L., Digard, D., Thomas, D., Bell, L., Saari, G., Grant, F., O'Harra, P., and MacKay, V. L., 1989, The STE4 and STE18 genes of yeast encode potential ß and γ subunits of the mating factor receptor-coupled G protein, *Cell* **56**:467-477.

Wright, M. V., and Van Houten, J. L., 1989, Calcium ATPase as part of the transduction pathway in *Paramecium* chemoreception, *Chem. Sen.*, abstract, in press.

Zipser, B., Rutt, M. R., O'Neill, J. B., Smith, C. C., Higgins, W. J., and Pert, C. B., 1988, The opiate receptor: a single 110 K Da recognition molecule appears to be conserved in *Tetrahymena*, leech and rat, *Brain Res.* **463**:296-304.

Chapter 26

The Functional Significance of Evolutionary Modifications found in the Ciliate, *Stentor*

DAVID C. WOOD

1. Introduction

The ciliate protozoan, *Stentor coeruleus*, is notable for the range of its sensory capabilities, the variety of its behavioral responses, and its ability to assume different morphologies. This versatility is well illustrated in H. S. Jennings's (1906) account of the response of sessile *Stentor* to carmine particles introduced into the medium surrounding their frontal field. These extended and trumpet-shaped *Stentor* were reported to respond initially to this irritant by bending to one side. If carmine particles continued to be present, the animals reversed the direction of their ciliary beat, thereby driving the carmine particle-containing fluid away from them. If this also did not remove the carmine particles, the *Stentor* contracted into a ball. They then went through several cycles of gradual re-extension and all-or-none contraction until finally they detached, assumed a pear-shaped configuration, and swam away. While more recent investigators have found that this behavioral sequence is not reproducible (Reynierse and Walsh, 1967), all the behavioral responses and cellular morphologies described in it are easily observable. In addition, sessile *Stentor* contract when mechanically stimulated or in response to certain chemicals, and motile *Stentor* swim away from light sources, as a result of their blue-green pigmentation.

The variety and specialized nature of some of this animal's behavioral and sensory capabilities indicate that a number of biochemical and morphological changes occurred during its evolution from its ciliate ancestors, since few other extant living ciliates

DAVID C. WOOD • Department of Behavioral Neurosciences, University of Pittsburgh, Pittsburgh, Pennsylvania 15260, USA.

Evolution of the First Nervous Systems
Edited by P.A.V. Anderson
Plenum Press, New York

display such versatility. In agreement with this conclusion, ciliate taxonomists place *Stentor* among the most "highly evolved" ciliates, basing this on either the complex structure of *Stentor*'s buccal cavity and membranellar band (Corliss, 1979) or the pattern of fibers present in their hexagonally-packed polykinetids (Small and Lynn, 1985). Unfortunately, a more definitive statement about the evolutionary relationship between ciliates and other phyla, or among classes of ciliates, must await extensive RNA and DNA sequence analysis, and only a few such studies have been made (Clark and Cross, 1988).

The fact that many of the sensory and behavioral capabilities of *Stentor* differ from those of the majority of ciliates, suggests that these functional capabilities evolved during ciliate evolution. Two such capabilities will be described. The apparent selective advantage of these characteristics will then be discussed briefly with the intent of pointing out some factors which normally are not considered important in the evolution of neural or quasi-neural function.

In nature *Stentor* are most commonly found in the extended, trumpet-shaped form. When extended, they are generally attached at their narrow end to a piece of vegetation or debris. Abrupt movement of the fluid surrounding an animal or movement of its point of attachment often causes the animal to contract, a response requiring only 10 ms. Intracellular recordings reveal that contractions are preceded by a fixed sequence of transmembrane potentials. Mechanical stimuli elicit receptor potentials, which rise in 5 to 50 ms to a peak and then decay exponentially, with a time constant between 150 and 200 ms. Receptor potentials, which reach amplitudes of 15-25 mV, trigger all-or-none action potentials, with amplitudes in the range of 65-75 mV. The onset of an action potential precedes the onset of a contraction by 1.5 - 2.0 ms and, hence, appears to trigger the contraction, as will be discussed in more detail later. When *Stentor* contract, they simultaneously undergo a reversal in the direction of their ciliary beat. Thus, the sequence:

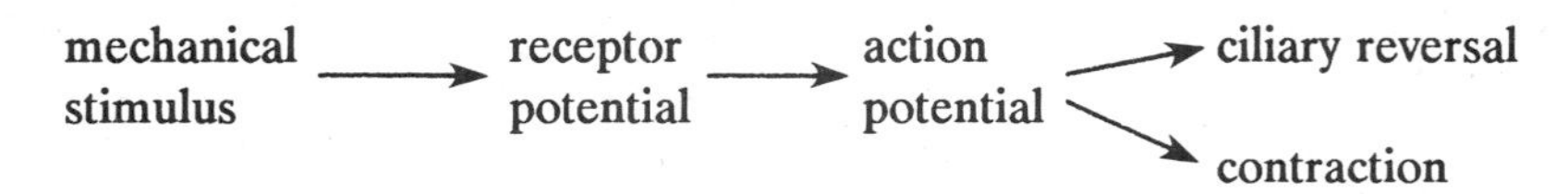

provides a simple model on which to make elaborations (Wood, 1970b).

2. A Voltage-Dependent Mechanoreceptor Channel

In cells voltage-clamped to their resting potential (-48 to -56mV), mechanical stimuli elicit brief (<50 ms) inward currents. Ionic substitution experiments indicate that this receptor current is carried predominantly by Ca^{++} ions (Wood, 1982). Unlike most receptor currents, this current is also highly voltage-dependent. When the membrane potential is clamped to -60 to -70 mV at the time of mechanical stimulation, the receptor current is either very small or completely blocked. Conversely, the receptor current elicited by a constant intensity mechanical stimulus increases in amplitude sigmodally as the transmembrane potential is depolarized from resting level to values approaching and including -20 mV. For depolarizations above -20 mV, receptor currents decrease linearly, until a reversal potential near +21 mV is reached. The voltage-dependent segment of this I-V relation is well fit, if the mechanoreceptor channel conductance is assumed to have a Boltzmann distribution function with an e-fold change in channel proportions being produced by a 12.4 - 12.6 mV change in transmembrane potential. This analysis assumes that individual channel molecules can be shifted by membrane depolarization from a "U" (Unresponsive) conformation to an "R" (Responsive) conformation, where the "U" form does not increase its conductance when mechanically stimulated, while the "R" form does increase its conductance. Hyperpolarization shifts "R" conformation channels to the "U" conformation.

A similar analysis of the voltage dependence of the depolarizing mechanoreceptor currents of *Paramecium* or *Stylonychia* has not been reported, but the published data on these species suggest that their depolarizing mechanoreceptor channels are not so highly voltage-dependent (de Peyer and Machemer, 1978; Satow et al., 1983). Deitmer (1981) has reported a limited degree of voltage dependence for the hyperpolarizing mechanoreceptor of *Stylonychia*. However, in the main, it appears that the voltage-dependent mechanoreceptor channel present in *Stentor* probably evolved from voltage-independent mechanoreceptor channels present in other ciliates. Similarly, voltage-dependent receptor channels have been observed for a variety of neurotransmitters in a diverse array of species, without any evolutionary progression between them being evident (Pellmar, 1981). The question, therefore, arises: "What selective advantage does a protozoan with voltage-dependent mechanoreceptor channels have over a similar protozoan with only voltage-independent mechanoreceptor channels?"

At first glance, the selective advantage would appear to be with those protozoa which possess voltage-independent mechanoreceptor channels. Since the conductance of the voltage-dependent channel at resting potential is only 30-35% of its maximal value, *Stentor* are less sensitive to mechanical stimuli than they might be. This factor undoubtedly contributes to the substantial difference between the mechanical stimulus sensitivity of *Paramecium* (Eckert et al., 1972) or *Stylonychia* (de Peyer and Machemer, 1978), which produce a 10 mV receptor potential to a punctate stimulus elicited by a 5-10 μm styllus movement, and *Stentor*, which require a 15-25 μm movement of the entire animal to produce an equivalent receptor potential.

On the other hand, voltage-dependent mechanoreceptor channels appear to be advantageous to *Stentor* because they are the basis of its ability to habituate. When sessile, extended *Stentor* are stimulated mechanically, they contract. During the 30 to 45 sec following a contraction, they gradually re-extend. A second mechanical stimulus, applied 1 min after the first, will, generally, elicit another contraction, but when the mechanical stimuli are repeated at the rate of 1/min the probability of eliciting a contraction decreases progressively. This decrease in response probability has been studied parametrically and has been found to exhibit the characteristics of habituation

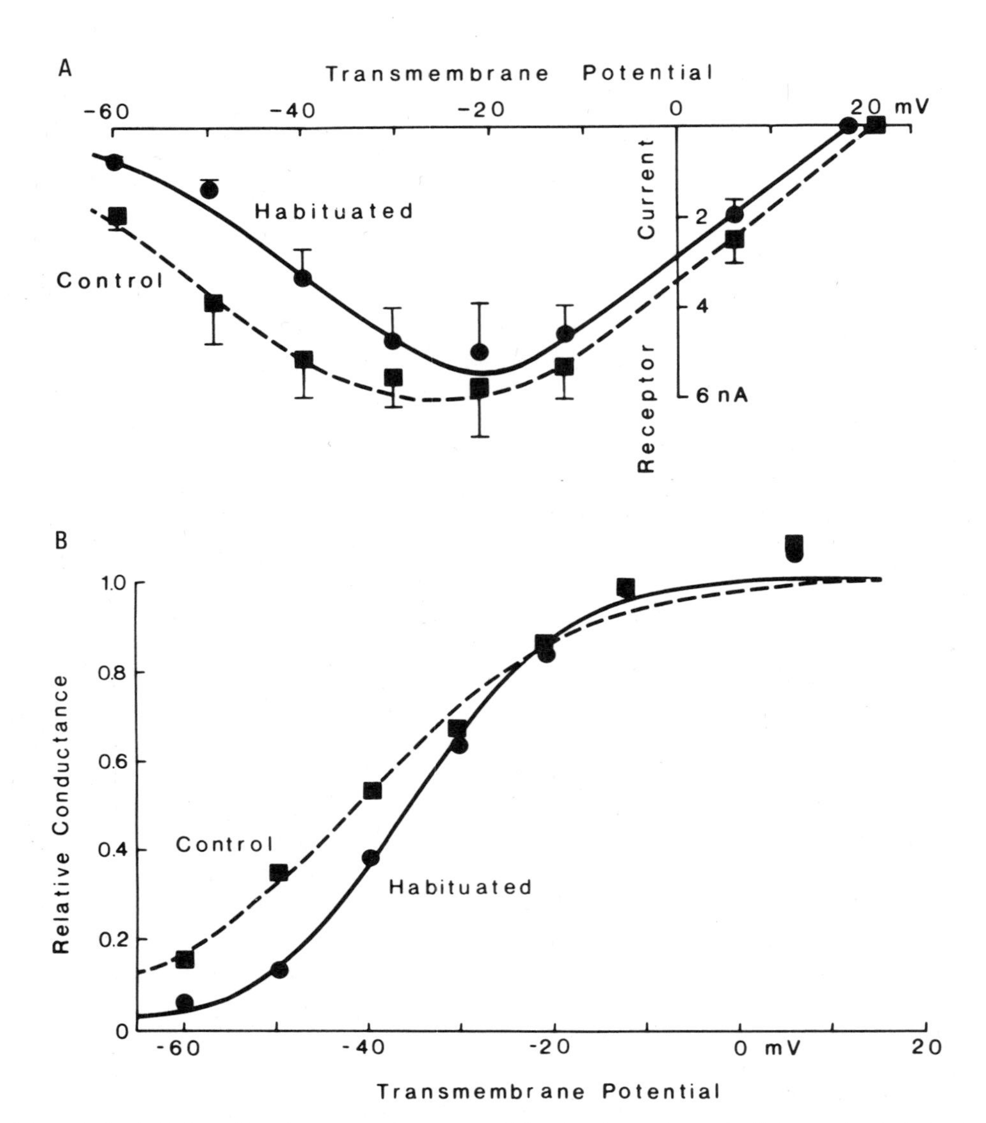

Figure 1. (A). I-V plots of mechanoreceptor currents elicited in voltage-clamped cells by mechanical stimuli applied 800 ms after stepping the transmembrane potential to the values indicated along the abscissa. An initial set of data was generated using widely spaced stimuli, the animals were then habituated, and a second set of data was collected from the same animals, while they were in the habituated state. **(B).** Voltage dependence of the mechanoreceptor conductance, as determined from the data for control and habituated animals shown in A. The curves are the best least-squares fit of the Boltzmann equation to the data.

(Wood, 1970a), as observed in metazoa (Thompson and Spencer, 1966). Habituation to repetitive mechanical stimuli appears advantageous to *Stentor* because this process allows the animal to adapt to repetitive innocuous stimuli, as first suggested by Jennings (1906). *Stentor* is generally attached to vegetation near the surface of ponds and slow-moving streams, where the innocuous repetitive stimuli they experience are probably waves. Since *Stentor* can feed only when extended, habituating to waves increases the amount of time the animal can feed. Habituating to repetitive, innocuous stimuli still leaves *Stentor* sensitive to more intense mechanical stimuli which might be produced by the movement of predators or larger, potentially dangerous waves.

For the amplitude of mechanical stimulus we normally use, the probability of response in a habituated animal is about 30%. Likewise, the receptor potential and receptor current amplitudes in a habituated animal decrease to 31% and 34%, respectively, of their control values (Wood, 1988a). When cells are voltage-clamped to -20 mV or more positive potentials, the maximal receptor current was observed to be the same before and after habituation (Fig. 1A)(Woods, 1988b). Similarly, the reversal potential of the mechanoreceptor current remained at 20-21 mV after habituation. These data indicate that the total number of potentially functional mechanoreceptor channels is the same before and after habituation. On the other hand, the slope of the I-V relation between -60 and -20 mV changes significantly after habituation. The voltage dependence of the mechanoreceptor conductance shifts from 12.4 mV/e-fold change in the control state to 9.6 mV/e fold change in the habituated animal (Fig. 1B). This change in voltage dependence accounts for the reduced receptor current seen in habituated animals at resting potential and, hence, is the physiological basis of habituation. Consequently, *Stentor* can habituate to repetitive mechanical stimuli exactly because their mechanoreceptor channels are voltage-dependent. Or, when viewed from the opposite perspective, the ability to habituate makes it selectively advantageous for *Stentor* to possess a voltage-dependent mechanoreceptor channel.

It is noteworthy that modifications of voltage-dependent ionic channels are also observed in metazoa during learning. Presently, known examples of such modifications include: 1) the Ca^{++}channel of *Aplysia* sensory neurons during habituation (Klein et al., 1980), 2) the serotonin-sensitive K^+ channel of *Aplysia* sensory neurons during sensitization (Klein and Kandel, 1980; Siegelbaum et al., 1982), and 3) the A-channels of *Hermissenda* type B photoreceptors during conditioning (Alkon et al., 1982).

3. The Function of Action Potentials in *Stentor*

As a second issue, I would like to consider the functional significance of voltage-dependent ion channels and, hence, action potentials, for protozoa in general, and *Stentor* in particular. In metazoa, action potentials are needed to propagate electrical signals along axons which are long relative to the axon's length constant. In protozoa, action potentials appear unnecessary for this function. For *Paramecium caudatum*,

Eckert and Naitoh (1970) calculated a length constant of 1350 μm which is 5.4 times the body length of a 250 μm *Paramecium*. For *Stentor*, the length constant is also large relative to the body length (Wood, 1982). Thus, in both species the cytoplasm is approximately isopotential, and there is no need for the cell to generate action potentials to propagate electrical signals across the cell surface.

Nevertheless, action potentials are physiologically significant because they amplify transmembrane potentials and currents. For instance, a 15-25 mV receptor potential in *Stentor* is converted to a 65-75 mV action potential with a concomitant increase in Ca^{++} influx. A 20 mV receptor potential is produced by a Ca^{++} influx of about 100 pC but triggers voltage-dependent Ca^{++} channels to produce a Ca^{++} influx of over 500 pC, thereby increasing Ca^{++} influx 5-fold. An amplification of this magnitude is physiologically significant. For instance, a Ca^{++} concentration greater than 1 μm is necessary to induce ciliary reversal in *Paramecium* models (Naitoh and Kaneko, 1972; Kung and Naitoh, 1973). Since their voltage-dependent Ca^{++} channels are localized exclusively in the plasma membrane of the cilia (Dunlap, 1977; Machemer and Ogura, 1979), the Ca^{++} influx producing an action potential enters directly into the intraciliary space where it has been calculated to raise the Ca^{++} concentration to 10 μm or more, thereby eliciting ciliary reversal (Eckert, 1972; Eckert et al., 1976). On the other hand, the Ca^{++} current producing a mechanoreceptor potential is both insufficient and inappropriately localized to elicit this response. Action potentials are also significant in ciliates because they coordinate the Ca^{++} influx of all the motile cilia and, thereby, produce a nearly synchronous ciliary reversal response.

All-or-none contractions can be elicited in a number of ciliates by electrical, mechanical, and chemical stimuli and, in *Stentor*, by light. The electrical threshold for elicitation of these contractions adhers to the conventional strength-duration curve (Fabczak et al., 1973; Hawkes and Holbertson, 1974) and shows accommodation (Wood, unpublished results), thereby exhibiting properties characteristic of action potentials. Indeed, an action potential was produced concomitantly with every contraction when the recordings were obtained from *Stentor* maintained at room temperature (Wood, 1970b, 1971, 1988a). This correlation is most striking when, as a result of illumination, extended cells gradually depolarize for as long as 60 sec before producing an action potential and simultaneously contracting (Wood, 1973, 1976). Infrequently, extended cells produce an action potential and simultaneously contract, without apparent external stimulation. These observations leave no doubt that action potentials and contractions are functionally related. Microelectrode recordings made while monitoring the penetrated cell's length with a photomultiplier reveal that action potential onset precedes the onset of the contraction by 0.8 to 2.6 ms at room temperature. Thus, action potentials appear to trigger contractions in *Stentor* as well as elicit ciliary reversal, as in other protozoa. Action potentials also appear to trigger contractions in other protozoa: *Vorticella* (Shiono et al., 1980), *Actinocoryne* (Febvre-Chevalier et al., 1986), and *Zoothamnium* (Moreton and Amos, 1979). However, action potentials have not been recorded in association with contractions in *Spirostomum*

(Ettienne, 1970; Sleigh, 1970), though ciliary reversal, a behavior produced by an action potential, generally does occur during a contraction (Clark, 1946).

Contractions in protozoa, like those in metazoa, are produced by an increase in intracellular free Ca^{++}, as evidenced by the findings that intracellular injection of Ca^{++} solutions into *Spirostomum* induces a total and maintained contraction (Hawkes and Holbertson, 1974), and the chelation of extracellular Ca^{++} with EGTA, produces *Stentor* which are extended but paralyzed (Huang and Pitelka, 1973). However, the most convincing evidence for Ca^{++} dependence comes from the many studies of glycerated *Vorticella* and *Carchesium* stalks, which can be induced to contract by the addition of 0.2 - 0.5 μM Ca^{++} solutions (Levine, 1956; Hoffmann-Berling, 1958; Amos, 1971; Hawkes and Rahat, 1976; Townes, 1978; Ochiai et al., 1979). The proteins contained in these stalks, called spasmins, have been isolated and found to bind Ca^{++} at concentrations as low as 0.1 μM (Amos et al., 1975; Yamada and Asai, 1982; Yamada-Horiuchi and Asai, 1985). Antibodies to *Carchesium* spasmins cross-react with proteins from other contractile ciliates, including *Stentor*, suggesting that spasmins are the contractile components in these other ciliates (Ochiai et al., 1988). On the other hand, these antibodies do not cross-react with metazoan muscle proteins, and so appear to be unique to protozoans.

While there are several longitudinal fiber systems in *Stentor* which might contain spasmins, most evidence suggests that the M-bands or myonemes are the contractile components. M-bands are located at approximately 6 μm intervals around the surface of *Stentor*, at a depth of 1-3 μm beneath each of the longitudinally oriented rows of kineties, and hence, are 1-3 μm beneath the surface membrane. M-bands are relatively large structures with cross-sectional dimensions on the order of 2 and 6-10 μm. In contracted animals the M-bands are both shorter and thicker than in extended animals (Bannister and Tatchell, 1968). They also re-extend more rapidly than the animal as a whole and are, therefore, thrown into convolutions during re-extension of the animal. In extended *Stentor*, the M-bands contain primarily a longitudinally-oriented array of 3-5 nm filaments (Huang and Pitelka, 1973; Kristensen et al., 1974), while the M-bands of contracted animals contain more 8-12 nm tubular structures, which may represent coiled 3-5 nm filaments. This structural change is accompanied by a pronounced decrease in M-band birefringence during contraction (Kristensen et al., 1974), supporting the conclusion that the M-bands are the contractile components.

The submembrane position and large size of the M-bands make it impossible for the Ca^{++}, which enters the ciliary membrane during an action potential, to be sufficient to directly activate the biochemical machinery which produces contractions. This conclusion is supported by the occasional observation that an extended *Stentor*, chilled to 8-10 °C, produces action potentials but does not contract. Thus, a mechanism for excitation-contraction coupling had to evolve in protozoa, as it did in metazoa.

The presence of such a coupling mechanism is suggested by the observation that extended *Stentor*, chilled to 8-10 °C, sometimes partially or fully contract during the passage of hyperpolarizing current pulses applied during the early stages of a recording session, but do not produce action potentials. In these cases, 5-50 mV depolarizations are superimposed on the induced membrane hyperpolarization (Fig. 2A). Since these transient depolarizations have a consistent waveform, a voltage threshold, and reliably signal the occurence of a contraction, they appear to be a reflection of an all-or-none intracellular event which normally produces contractions. Occasionally, one of these depolarizations is sufficiently large to elicit a ciliary action potential (Fig. 3). In extreme cases, this sequence produces a bizarre voltage record, wherein a full-blown action potential appears to be triggered directly by a hyperpolarization (Fig. 4A).

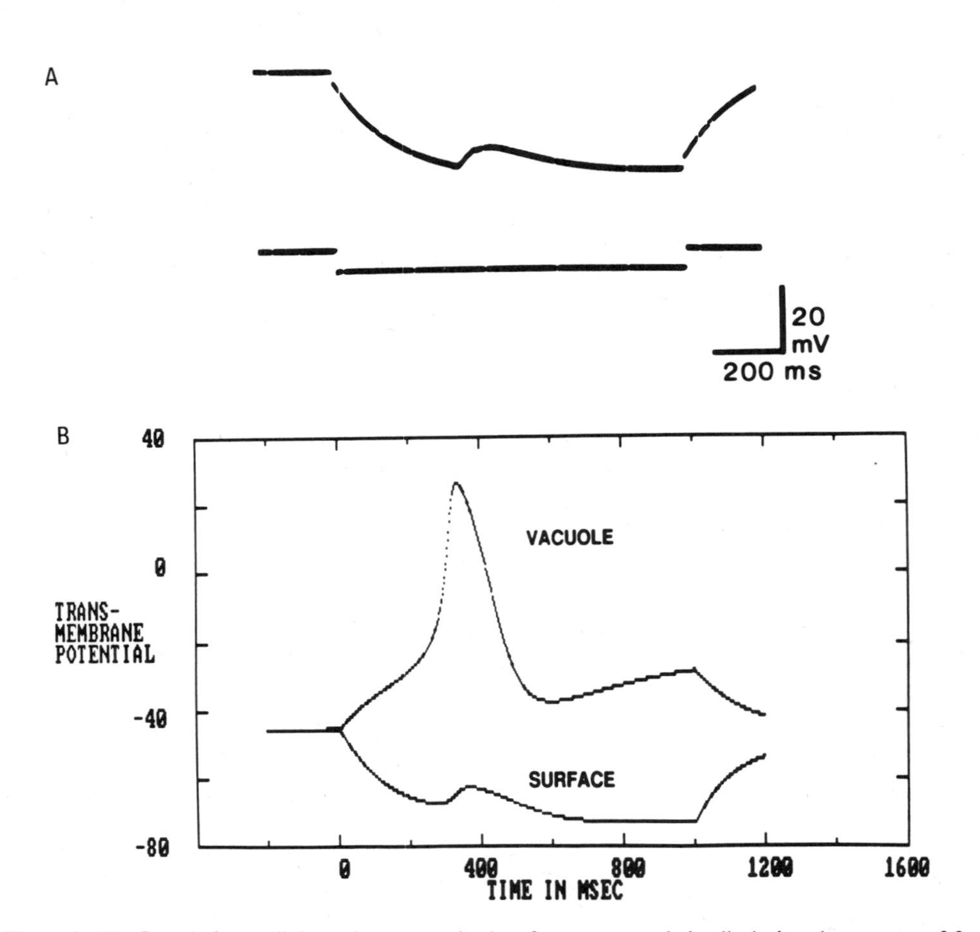

Figure 2. (A). Sample intracellular voltage record taken from an extended cell, during the passage of 2 nA of inward current. The animal contracted when the small depolarization that is superimposed on the exponential hyperpolarization occurred. **(B).** Output of the mathematical model, assuming 2 nA of current was being passed by the current electrode into the vacuole. The calculated values for the potential across the surface membrane and the vacuole membrane are both plotted.

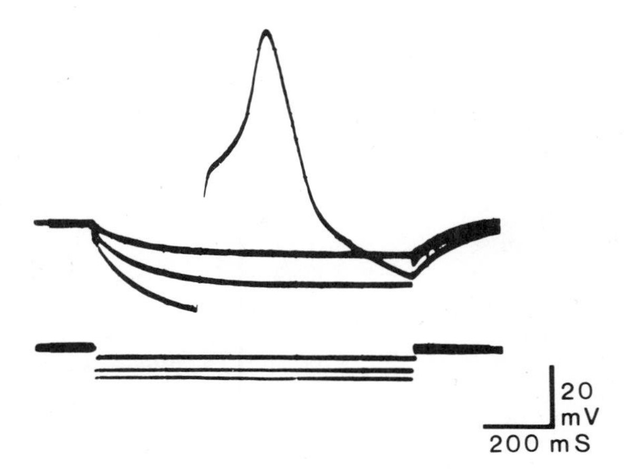

Figure 3. Sample intracellular voltage records taken from an extended cell during the passage of 1, 2, and 3 nA of inward current. In response to the 3 nA current pulse, the cell produced a 50 mV depolarization, which was just above threshold for eliciting a ciliary action potential. The animal contracted during the 3 nA current pulse.

Given the correlation between these depolarizations and contractions, the site of their generation is of considerable interest. However, we have not yet succeeded in recording from such a site for a long enough period to inject sufficient dye to determine its exact locus. Nevertheless, we have engaged in a certain amount of speculation about their anatomical and physiological basis. Published electron micrographs of M-bands show large vacuoles bounding the sides of the M-band (Bannister and Tatchell, 1968, 1972; Huang and Pitelka, 1973). Our own electron micrographs reveal that a portion of these vacuolar membranes frequently comes into direct apposition with the surface membrane structure.

On this anatomical basis, I hypothesize that the vacuolar membrane acts as a pathway to conduct electrical signals from the surface membranes to the M-bands. To test this hypothesis, we have constructed both electronic (Roy, 1972) and mathematical models which incorporate the essential anatomical and physiological elements of the presumed system. The mathematical model uses the Hodgkin-Huxley equations with parameters modified to fit *Stentor* action potentials. These models make 3 basic assumptions: 1) that the vacuolar membrane contains voltage-dependent Ca^{++} channels and is capable of producing action potentials, like the surface membrane, 2) that the fluid in the vacuole contains a high concentration of Ca^{++}, like the extracellular medium, and 3) that the site of surface membrane - vacuolar membrane apposition is a low resistance site, specialized for the passage of electrical current.

Using either type of model, it has been possible to produce voltage records in which hyperpolarizing current pulses elicit intracellularly recorded depolarizations similar to those observed, assuming that the current electrode was actually in the vacuole, while the voltage electrode was recording from the cytoplasm (Fig. 2B). In this case, the stimulating current must be flowing inward across the surface membrane and, hence, hyperpolarizing it. However, the same current is passing from the cytoplasm to the vacuolar fluid and, thus, is an "outward", depolarizing current for the vacuolar membrane. In these models, current-induced depolarizations of the vacuolar membrane

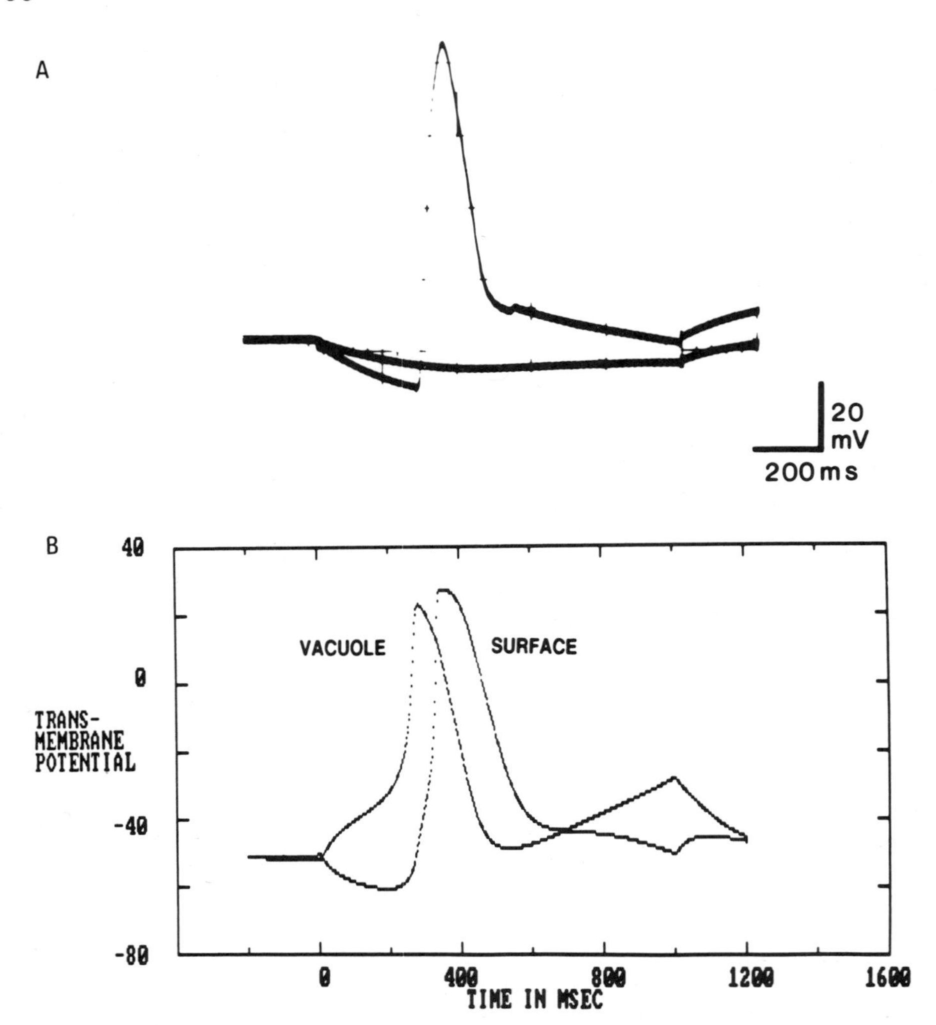

Figure 4. (A). Sample intracellular voltage record as in figure 2A, except that a surface or ciliary membrane action potential was elicited. **(B).** Output of the mathematical model as in figure 2B, showing the calculated values for the voltage across both surface and vacuole membranes. The resistance across the vacuole membrane - surface membrane apposition was assumed to be lower in this case than in the case shown in figure 2B.

membrane can be sufficient to elicit action potentials in that membrane. When this occurs, some of the current produced by the vacuole action potential flows outward through the surface membrane, producing the observed depolarization, and back into the vacuole through the site of low-resistance membrane apposition. If the resistance assumed for this area of vacuolar membrane - surface membrane apposition is reduced sufficiently, the depolarization induced in the surface membrane by current from the

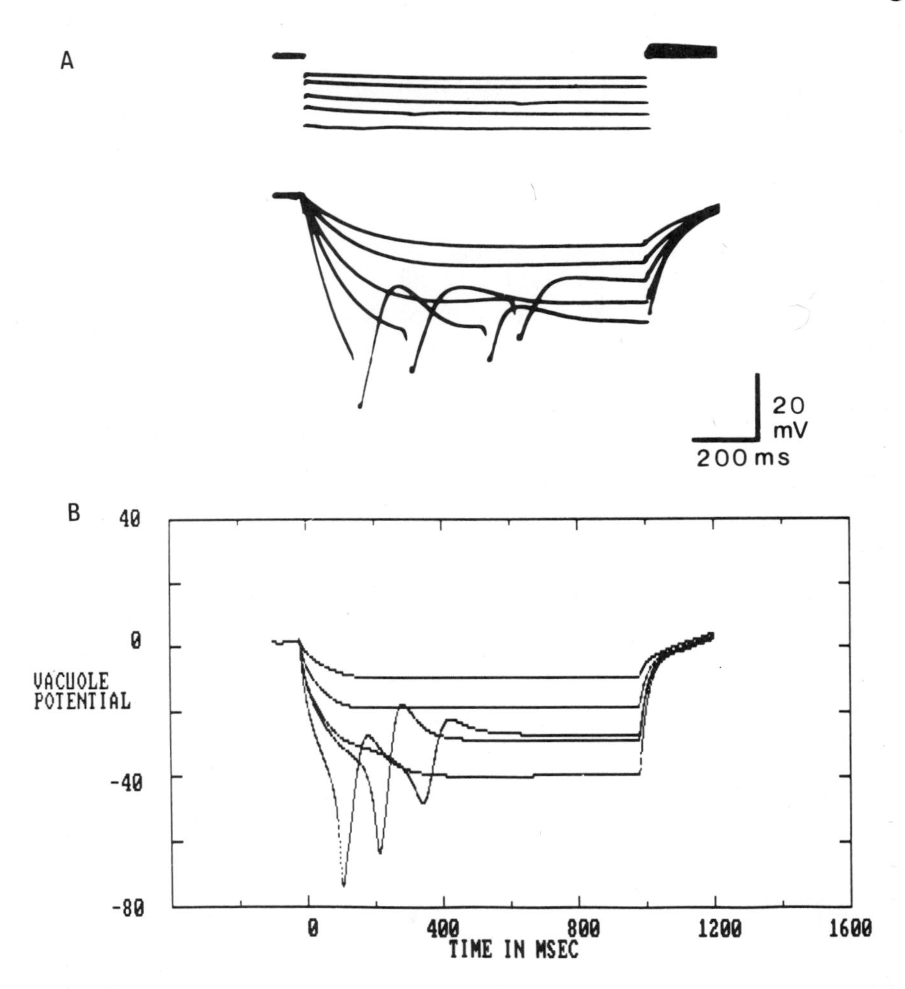

Figure 5. (A). Sample voltage records obtained from an intracellular site that had a steady-state potential near 0 mV, and presumed to be a vacuole. The negative-going spikes were correlated with contractions. **(B).** Output of the mathematical model showing calculated values for the voltage difference between the vacuole and the extracellular medium. The parameters in the model were the same as used to generate the output shown in figure 4B.

vacuolar membrane action potential becomes large enough to trigger an action potential in the surface membrane, analogous to those we have observed (Figs. 4A).

As an unexpected bonus from this model building, we discovered that, if both the voltage and current electrodes are presumed to be in the vacuole, then current pulses which are inward across the surface membrane and "outward" across the vacuole membrane, elicit negative-going action potentials, similar to those recorded and published nearly 20 years ago (Wood, 1970b; Chen, 1972) (Fig. 5A). These records are similar to the hyperpolarizing action potentials recorded from the sap vacuole of *Noctiluca* (Hisada, 1957; Chang, 1960; Eckert, 1965a). Such action potentials are negative-going because the action potential elicited in the vacuolar membrane allows positive charge, presumably in the form of Ca^{++} ions, to flow through the vacuolar membrane into the M-bands and cytoplasm leaving the vacuole more negatively charged.

These models suggest a functional significance for action potentials in contractile protozoa, in addition to the production of ciliary reversal. In these models, action potentials elicited in the plasma membrane are sufficient to elicit action potentials in the vacuolar membrane and, thereby, to produce an efflux of Ca^{++} from the vacuole into the adjacent M-band. This Ca^{++} current is calculated to raise the Ca^{++} concentration throughout the M-band to 3-5 μM, thus exceeding the Ca^{++} concentration which activates spasmins. If a receptor potential, but not an action potential, is produced in the plasma membrane, then a very small potential is recorded across the vacuolar membrane and essentially no Ca^{++} flux occurs. Therefore, these models suggest that action potentials in *Stentor* are necessary to couple electrical events produced across the surface membrane with the intracellular release of Ca^{++} needed to elicit contractions. That is, action potentials and the voltage-dependent channels producing them appear necessary for excitation-contraction coupling in *Stentor*.

These models also suggest a solution to an apparent paradox. At the beginning of this section it was noted that *Stentor* have a long length constant, relative to their body length and, therefore, all electrical signals and the responses produced by them, i.e., contractions, should occur simultaneously over the entire cell surface. However, recordings from electrodes encapsulated by the cell surface indicate there is, in fact, a propagated electrical signal (Wood, 1972). Likewise, the bodily contraction is propagated along the cell length (Jones et al., 1970; Nielsen, 1975). These observations can be explained if it is assumed that the action potentials which directly produce myonemal contraction occur in the vacuolar membrane. These vacuoles have cross sectional dimensions of only a few micrometers and presumably contain a fluid of lower ionic strength than that of the cytoplasm; therefore, they should have a short length constant. As a consequence, the inside of the vacuole need not be isopoential and action potentials could be propagated along them at measurable velocities, producing contraction in the neighboring myoneme as they pass. It is noteworthy that action potentials are propagated along the vacuolar membrane of *Noctiluca* (Eckert, 1965b).

It must be pointed out that these models are based on the high affinity of spasmin for Ca^{++}, and morphological features unique to *Stentor*. Therefore, the models would require modification to account for excitation-contraction coupling in *Spirostomum* and contractile peritrichs. Likewise, they represent a morphological and biochemical mechanism for increasing intracellular free Ca^{++}, which is different from that employed by vertebrate muscle.

References

Alkon, D. L., Lederhendler, I., and Shoukimas, J. J., 1982, Primary changes of membrane currents during retention of associative learning, *Science* **215:**693-695.

Amos, W. B., 1971, Reversible mechanochemical cycle in the contraction of *Vorticella*, *Nature* **229:**127-128.

Amos, W. B., Routledge, L. M., and Yew, F. F., 1975, Calcium binding proteins in a vorticellid organelle, *J. Cell Sci.* **19:**203-211.

Bannister, L. H., and Tatchell, E. C., 1968, Contractility and the fibre systems of *Stentor coeruleus*, *J. Cell Sci.* **3:**295-308.

Bannister, L. H., and Tatchell, E. C., 1972, Fine structure of the M fibres in *Stentor* before and after shortening, *Exptl. Cell Res.* **73:**221-225.

Chang, J. J., 1960, Electrophysiological studies of a non-luminescent form of the dinoflagellate *Noctiluca miliaris*, *J. Cell. Comp. Physiol.* **56:**33-42.

Chen, V. K., 1972, The electrophysiology of *Stentor polymorphus*: An approach to the study of behavior, University Microfilms: Ann Arbor, MI.

Clark, A. M., 1946, The reactions of isolated parts of *Spirostomum*, *J. exp. Biol.* **22:**88-94.

Clark, C. G., and Cross, G. A. M., 1988, Small-subunit ribosomal RNA sequence from *Naegleria gruberi* supports the polyphyletic origin of amoebas, *Mol. Biol. Evol.* **5:**512-518.

Corliss, J. O., 1979, *The Ciliate Protozoa Characterization, Classification and Guide to Literature*, 2nd ed., Pergamon Press, New York.

de Peyer, J., and Machemer, H., 1978, Hyperpolarizing and depolarizing mechanoreceptor potentials in *Stylonychia*. *J. Comp. Physiol.* **127:**255-266.

Deitmer, J. W., 1981, Voltage and time dependent characterization of the potassium mechanoreceptor current in the ciliate, *Stylonychia*, *J. Comp. Physiol.* **141:**173-182.

Dunlap, K., 1977, Localization of calcium channels in *Paramecium caudatum*, *J. Physiol. (Lond.)* **271:**119-133.

Eckert, R., 1965a, Bioelectric control of bioluminescence in the dinoflagellate *Noctiluca*, *Science* **147:**1140-1142.

Eckert, R., 1965b, Asynchronous flash initiation by a propagated triggering potential, *Science* **147:**1142-1145.

Eckert, R., 1972, Bioelectric control of ciliary activity, *Science* **176:**473-481.

Eckert, R., and Naitoh, Y., 1970, Passive electrical properties of *Paramecium* and problems of ciliary coordination, *J. Gen. Physiol.* **55:**467-483.

Eckert, R., Naitoh, Y., and Friedman, K., 1972, Sensory mechanisms in *Paramecium*. I. Two components of the electric response to mechanical stimulation of the anterior surface. *J. exp. Biol.* **56:**683-694.

Eckert, R., Naitoh, Y., and Machemer, H., 1976, Calcium in the bioelectric and motor functions of *Paramecium*, *Symp. Soc. Exp. Biol.* **30:**233-255.

Ettienne, E. M., 1970, Control of contractility in *Spirostomum* by dissociated Ca^{++} ions. *J. Gen. Physiol.* **56:**168-179.

Fabczak, S.,Korohoda, W., and Walczak, T., 1973, Studies on the electrical stimulation of contraction in *Spirostomum*. I. Conditions of the quantitative measurements, *Cytobiol.* **7:**152-163.

Febvre-Chevalier, C., Bilbaut, A., Bone, Q., and Febvre, J., 1986, Sodium-calcium action potential associated with contraction in the heliozoan *Actinocoryne contractilis*, *J. exp. Biol.* **122:**177-192.

Hawkes, R. B., and Holberton, D. V., 1974, Myonemal contraction of *Spirostomum*. I. Kinetics of contraction and relaxation, *J. Cell Physiol.* **84:**225-236.

Hawkes, R. B., and Rahat, M., 1976, Contraction and volume reduction of the glycerolated *Carchesium* spasmoneme: Effects of alkali earth ions, *Experientia* **32:**160-162.

Hisada, M., 1957, Membrane resting and action potentials from a protozoan, *Noctiluca scintillans*, *J. Cell. Comp. Physiol.* **50:**57-71.

Hoffmann-Berling, H., 1958, Der Mechanismus eines neuen, von der Muskelkontraktion verschiedenen Kontraktionszyklus, *Biochim. Biophys. Acta* **27:**247-255.

Huang, B., and Pitelka, D. R., 1973, The contractile process in the ciliate, *Stentor coeruleus*, *J. Cell Biol.* **57**:704-728.

Jennings, H. S., 1906, *Behavior of the Lower Organisms*, Columbia University Press, New York.

Jones, A. R., Jahn, T. L., and Fonseca, J. R., 1970, Contraction of protoplasm III. Cinematographic analysis of the contraction of some heterotrichs, *J. Cell. Physiol.* **75**:1-8.

Klein, M., and Kandel, E. R., 1980, Mechanism of calcium current modulation underlying presynsptic facilitation and behavioral sensitization in *Aplysia*, *Proc. Natl. Acad. Sci. USA* **77**:6912-6916.

Klein, M., Shapiro, E., and Kandel, E. R., 1980, Synaptic plasticity and the modulation of the Ca^{++} current, *J. exp. Biol.* **89**:117-157.

Kristensen, B. I., Nielsen, L. E., and Rostgaard, J., 1974, Variations in myoneme birefringence in relation to length changes in *Stentor coerulus*, *Exptl. Cell Res.* **85**:127-135.

Kung, C., and Naitoh, Y., 1973, Calcium-induced ciliary reversal in the extracted models of "pawn", a behavioral mutant of *Paramecium*, *Science* **179**:195-196.

Levine, L., 1956, Contractility of glycerated *Vorticellae*, *Biol. Bull.* **111**:319.

Machemer, H., and Ogura, A., 1979, Ionic conductances of membranes of ciliated and deciliated *Paramecium*. *J. Physiol. (Lond.)* **296**:49-60.

Moreton, R. B., and Amos, W. B., 1979, Electrical recording from the contractile ciliate *Zoothamnium geniculatum* Ayrton, *J. exp. Biol.* **83**:159-167.

Naitoh, Y., and Kaneko, H., 1972, Reactivated triton-extracted models of *Paramecium*: Modification of ciliary movement by calcium ions, *Science* **176**:523-524.

Nielsen, L. E., 1975, High-speed cinematographic analysis of electrically-induced contraction in *Stentor coeruleus*, *Comp. Biochem. Physiol.* **50**:463-468.

Ochiai, T., Asai, H., and Fukui, K., 1979, Hystersis of contraction-extension cycle of glycerinated *Vorticella*, *J. Protozool.* **26**:420-425.

Pellmar, T. C., 1981, Transmitter control of voltage-dependent currents, *Life Sci.* **28**:2199-2205.

Reynierse, J. H., and Walsh, G. L., 1967, Behavior modification in the protozoan *Stentor* re-examined, *Psychol. Rec.* **17**:161-165.

Roy, G., 1972, A simple electronic analog of the squid axon membrane: The NEUROFET, *IEEE Trans. Biomed. Eng.* **19**:60-63.

Satow, Y., Murphy, A. D., and Kung, C., 1983, The ionic basis of the depolarizing mechanoreceptor potential of *Paramecium tetraurelia*, *J. exp. Biol.* **103**:253-264.

Shiono, H., Hara, R., and Asai, H., 1980, Spontaneous membrane potential changes associated with the zooid and vacuolar contractions in *Vorticella convallaria*, *J. Protozool.* **27**:83-87.

Siegelbaum, S. A., Camardo, J. S., and Kandel, E. R., 1982, Serotonin and cyclic AMP close single K^{+} channels in *Aplysia* sensory neurons, *Nature* **299**:413-417.

Sleigh, M. A., 1970, Some factors affecting the excitation of contraction in *Spirostomum*, *Acta Protozool.* **7**:335-352.

Small, E. B., and Lynn, D. H., 1985, Phylum ciliophora Doflein, 1901, in: *An Illustrated Guide to Protozoa* (J. J. Lee, S. H. Hulner, and E. C. Bovee, eds.), Soc. of Protozool., Lawrence, Kansas.

Thompson, R. F., and Spencer, W. A., 1966, Habituation: A model phenomenon for the study of neuronal substrates of behavior, *Psychol. Rev.* **73**:16-43.

Townes, M. M., 1978, The participation of ionic strength and pH in the contraction induced in *Vorticella* by calcium and magnesium, *Comp. Biochem. Physiol.* **61A**:555-558.

Wood, D. C., 1970a, Parametric studies of the response decrement produced by mechanical stimuli in the protozoan, *Stentor coeruleus*, *J. Neurobiol.* **1**:345-360.

Wood, D. C., 1970b, Electrophysiological studies of the protozoan, *Stentor coeruleus*, *J. Neurobiol.* **1**:363-377.

Wood, D. C., 1971, Electrophysiological correlates of the response decrement produced by repeated mechanical stimuli in the protozoan, *Stentor coeruleus*, *J. Neurobiol.* **2**:1-11.

Wood, D. C., 1972, Generalization of habituation between different receptor surfaces of *Stentor*. *Physiol. Behav.* **9**:161-165.

Wood, D. C., 1973, Stimulus specific habituation in a protozoan, *Physiol. Behav.* **11**:349-354.

Wood, D. C., 1976, Action spectrum and electrophysiological responses correlated with the photophobic response of *Stentor coeruleus*, *Photochem. Photobiol.* **24**:261-266.

Wood, D. C., 1982, Membrane permeabilities determining resting, action, and mechanoreceptor potentials in *Stentor coeruleus*, *J. Comp. Physiol.* **146**:537-550.
Wood, D. C., 1988a, Habituation in *Stentor*: A response-dependent process, *J. Neurosci.* **8**:2248-2253.
Wood, D. C., 1988b, Habituation in *Stentor*: Produced by mechanoreceptor channel modification, *J. Neurosci.* **8**:2254-2258.
Yamada, K., and Asai, H., 1982, Extraction and some properties of the proteins, Spastin B, from the spasmoneme of *Carchesium polypinium*, *J. Biochem.* **91**:1187-1195.
Yamada-Horiuchi, K., and Asai, H., 1985, Partial purification of the Ca^{++}-binding proteins from the spasmoneme of *Carchesium*, *Comp. Biochem. Physiol.* **81B**:409-413.

Chapter 27

Hydromedusan Photophysiology: An Evolutionary Perspective

STUART A. ARKETT

"Jellyfish have no brains and they do not look like they are able to integrate [light] information. Yet, medusae must integrate information, even if we do not understand how they do it." W. M. Hamner, 1985.

1. Introduction

Although there has been a long standing debate on the number of phyletic lines of photoreceptor structure evolution (Eakin, 1982; Salvini-Plawen, 1982; Vanfleteren, 1982), there seems to be a growing consensus that photoreceptor organelles may have arisen anew several times from photosensitive ciliated ectodermal cells (Salvini-Plawen and Mayr, 1977; Burr 1984). Modifications to a simple photosensitive cell leading to a more complex "eye" appear to be a "...polyphyletic reaction to respective selective pressures..." (Salvini-Plawen, 1982). This process has led to an extraordinarily diverse array of photoreceptor types. To generate this diversity of photoreceptors, modifications were presumably functionally adaptive to a particular group of animals. This basic premise implies that there is a positive correlation between the degree of structural complexity and photoreceptor function (Salvini-Plawen and Mayr, 1977), revealed either by its photophysiology or by the behavior of the animal. Yet, for many groups of simple invertebrates, and particularly cnidarians, there has been insufficient information to begin to examine this correlation.

Hydromedusae exhibit a wide array of photoreceptors, ranging from a presumptive photoreceptive "l'organe lamellaire" in *Phialidium* (Bouillon and Nielsen, 1974), to a

STUART A. ARKETT • Department of Physiology, The Medical School, University of Bristol, Bristol BS8 1TD, United Kingdom.

simple epithelial eye spot in *Leuckartiara* (Singla, 1974), to a complex, pigmented cup with a lens-like structure in *Cladonema* (Bouillon and Nielsen, 1974). The diversity of hydromedusan photoreceptors has often been used to illustrate a presumptive evolution of photoreceptive structures from simple, photosensitive ciliated cells to a more complex photoreceptor. If the above correlation is to hold, then we might expect those medusae with complex photoreceptors to show a correspondingly complex photophysiology and behavior.

The purpose of this paper is: 1) to examine the relationship between the structural diversity of hydromedusan photoreceptor organs and behavioral photophysiology and 2) to examine how hydromedusan photophysiology compares with that of higher metazoans. These comparisons are possible only because there have been many morphological, behavioral, and electrophysiological studies over the years. Data surveyed from a variety of studies, from whole animal to cellular, indicate that medusae with more complex photoreceptors tend to show a greater complexity of behavior, and presumably, physiology. Furthermore, even though cnidarians have long been separated from other metazoans, hydromedusan photophysiology exhibits several features that are commonly observed in higher metazoans, suggesting a convergence of hydromedusan and other metazoan photoreceptor structure and photophysiology.

Table 1 presents a summary of a survey of several aspects of hydromedusan photic behavior. I have divided this behavior into three catagories: diel vertical migration, shadow responses, and spawning. Diel vertical migration is typified by an upward movement in the water column at dusk and a downward movement at dawn. Shadow responses are behaviors that are exhibited, typically within a few seconds, in response to rapid changes (either increase or decrease) in light intensity. Spawning is not usually a behavior of the whole animal, but is merely the release of gametes. I have also separated the medusae into two major divisions, those with discrete ocelli and those without ocelli. The hydromedusan ocellus is a simple pigment cup with the cilium-based receptor cell bodies forming the base. Although the second group lacks discrete ocelli, in at least one case (e.g., *Phialidium*), a cilium-based structure which is purported to be photosensitive has been described.

This list is not exhaustive because I have selected examples where there is information on at least two of these behaviors. The information is collected from a variety of sources including behavioral, electrophysiological, and field studies. In some cases, the information is anecdotal, which precludes precise determination of distances migrated or timing of responses. I have also made value judgments on the strength or weakness of diel migrations, shadow responses, and spawning times based on the information available, and I appreciate that other workers may interpret the information differently. Where gaps in the literature existed, and wherever possible, I have made personal observations and sought information from the personal observations of others. Yet, even with these limitations, I think this survey serves to illustrate a positive correlation between photoreceptor structure and behavioral physiology, and to form a framework for future studies.

2. Trends in Table

1. Most medusae make diel vertical migrations to some extent regardless of whether they have discrete ocelli or not.

Diel migrations may sometimes be distinct with an upward movement within an hour of sunset and a marked downward movement near sunrise. Several medusae, notably *Bougainvillia*, *Polyorchis*, *Gonionemus*, *Stomotoca*, typify this pattern. It is illustrative to note that neither *Gonionemus* nor *Stomotoca* have ocelli. Indeed, some of the best migrators, in terms of the distances travelled and the match between the timing of their movements and the time at which light is changing most rapidly (just before dawn and just after sunset), are medusae lacking ocelli. Field studies have shown that *Proboscidactyla* migrates at least 45 m (Moreira, 1973) and the narcomedusan *Solmissus* migrates over 500 m (Benovic, 1973; Mills and Goy, 1988). The behavior of *Phialidium* illustrates one of the few cases of reversed diel vertical migration, that is, up in the water column during the day with a well-defined downward movement at night (Russell, 1925; Mills, 1983). *Mitrocoma* also shows sharply-defined migrations that coincide with rapidly changing light levels, despite the absence of ocelli.

Several medusae, both with and without ocelli, exhibit somewhat weak migrations (e.g., *Aequoria*, *Aegina*, *Sarsia*, *Spirocodon*). However, one must be careful in this generalization since for some of these medusae, good quantitative information on their diel distribution is lacking. The problem of sampling plankters to detect temporal and spatial positional changes in the water column is well known (Pearre, 1979; Omori and Hamner, 1982) and it may be especially difficult for medusae. However, Mills' "tank" studies have provided valuable information on the timing of migrations, while submersible work by Mackie and Mills (1983) has given us an insight into the behavior and movements of rarer, deep-water medusae. Anecdotal accounts are very useful, if not entertaining, as evidenced by the description by Murbach (1909) of *Gonionemus* medusae which "... become restless in their native haunts as darkness approaches."

2. Some hydromedusae both with and without ocelli exhibit, in addition to diel vertical migrations, a distinct response to rapid changes in light intensity. These medusae tend to, although not necessarily, show more clearly-defined diel migration and spawning times.

Several hydromedusae respond to rapid changes in light intensity. A common response to rapid decreases in light intensity is a burst of swimming contractions and often simultaneous tentacle contactions. This behavior has been called an "off response" (Singla, 1974), a "shadow reflex" (Kikuchi, 1947), or a "shadow response" (Tamasige and Yamaguchi, 1967). This behavior is commonly thought to function as an escape mechanism, although for one medusa at least, it has been shown that a more likely function is in initiating diel migrations (Arkett, 1985). Table 1 shows several species that exhibit a typical shadow response. *Bougainvillia*, *Polyorchis*, and *Gonionemus* all respond to a rapid decrease in light intensity with a few rapid swimming contractions. These medusae tend to show distinct diel migrations. A notable

Table 1. Comparison of photic behavior of selected hydromedusae

	DVM	Shadow Response	Spawning
Medusae with ocelli			
Bougainvillia principis	+[18]	I - no reponse[31] D - swim[31]	SS, SR +[18]
Polyorchis penicillatus	+[3, 18]	I - crumple[4], ERG+[34] D - swim[1,4,5,6,] ERG+[34]	<1 hr OFF +[15] SS[18]
P. karafutoensis	?	D - swim[32]	1-1.5 hr ON[22]
Spirocodon saltatrix	+ (weak)[11]	I - ERG+ UV inhibits swim D - ERG+[24] VIS swim[9,23,25]	1-1.5 hr OFF+[36,37]
Sarsia tubulosa	0[17]	I - swim[26], ERG+[35] D - ERG+[35], swim stops[26]	2 hr ON[15]
Tiaropsis multicirrata	+[38]	I - swim[38] D - "quiet"[38]	<0.5 hr ON+[15]
Medusae without ocelli			
Gonionemus vertens	++[18]	I - swim, turning[21,38] D - swim[21]	1 hr OFF+[15,18,28]
Proboscidactyla ornata	++[20]	?	?
P. flavicirrata	?	0[po]	<0.5 hr OFF+[15]
Stomotoca atra	+[18]	I - inhibits TP & swim D - accel. TP & RP[12] no swim[po]	SR+[18] <0.5 hr ON & evening[15]
Phialidium gregarium	0 - Rev[16,17]	0[po]	<0.5 hr ON[15] 1 hr before SR[27]
P. hemisphaericum	0 - weak Rev[29]	?	1 hr ON[10]
Mitrocoma cellularia	Active, SR & SS[18]	0[po]	SR+[18]
Solmissus albescens	+[7,19]	0[19]	?
Euphysa spp.	+(weak)[13]	0[po]	0.5 hr ON[15]

Table 1 (continued)

	DVM	Shadow Response	Spawning
Aequoria victoria	0, mid-morning rise[18]	0[30]	mid to late morning[18]
Obelia spp.	+[20], +(weak)[29]	0[po]	<1 hr ON[15]
Aegina citrea	0 - (weak+)[13,14,33]	0[po]	?
Rathkea octopunctata	0[17]	active swim in light[38]	1 hr ON[15]
Aglantha digitale	+(weak)[2,14,29,33]	0[po]	2 hr ON[15]
Tesserogastria musculosa	0[8]	0[8]	after ON[8]
Mitrocomella spp.	?	0[po]	1 hr OFF[15]

Abbreviations. DVM - diel vertical migration, ranging from no detectable migration (0) to distinct pattern (++). Responses to rapid changes in light intensity are shown under "shadow responses". I - rapid increase or D - rapid decrease in light intensity. 0[po] indicates that I have observed no response to either rapid (1 sec) increase or decrease in visible light. UV - ultraviolet (350 nm) light. VIS - visible (500 nm) light. ERG+ indicates depolarization of electroretinogram. TP - tentacle pulse, RP - ring pulse. Spawning information is given as the time after light ON or OFF that gametes begin to be shed. + indicates spawning times are sharply defined. SS - sunset, SR - sunrise

1. Anderson PAV, Mackie GO (1977)
2. Arai MN, Fulton J (1973)
3. Arkett SA (1984)
4. Arkett SA (1985)
5. Arkett SA, Spencer AN (1986a)
6. Arkett SA, Spencer AN (1986b)
7. Benovic A (1973)
8. Hesthagen IH (1971)
9. Hisada M (1956)
10. Honegger T, et al. (1980)
11. Kikuchi K (1947) and Ohtsu, pers. com.
12. Mackie GO (1975)
13. Mackie GO (1985)
14. Mackie GO, Mills CE (1983)
15. Miller RL (1980)
16. Mills CE (1981)
17. Mills CE (1982)
18. Mills CE (1983)
19. Mills CE , Goy J (1988)
20. Moreira GS (1973)
21. Murbach L (1909)
22. Nagao Z (1963)
23. Ohtsu K (1983a)
24. Ohtsu K (1983b)
25. Ohtsu K, Yoshida M (1973)
26. Passano LM (1973)
27. Roosen-Runge EC (1962)
28. Rugh R (1929)
29. Russell FS (1925)
30. Satterlie RA (1985a)
31. Singla CL (1974)
32. Tamasige M, Yamaguchi T (1967)
33. Thurston MH (1977)
34. Weber C (1982a)
35. Weber C (1982b)
36. Yoshida M (1959)
37. Yoshida M, et al. (1980)
38. Zelikman EA (1969)

exception to this is *Sarsia*. Despite having well developed ocelli (Singla and Weber, 1982b) and being very responsive to rapid changes in light intensity (complex ERG, Weber, 1982b; stops swimming upon light increases, Passano, 1973), *Sarsia* does not appear to be a good migrator (Mills, 1982). Some medusae that lack ocelli and show no distinct shadow responses (e.g., *Phialidium*, *Mitrocoma*, *Solmissus*, *Proboscidactyla*) also make distinct diel migrations. In particular, *Mitrocoma* shows rapid up and down movements at dusk and dawn (Mills, 1983), but lacks a response to shadows. *Stomotoca* may also be included in this group, because, despite an acceleration of activity from a system that controls tentacle contractions (tentacle pulse - TP system, Mackie, 1975), this medusa does not show a burst of swimming activity in response to rapid decreases in light intensity. There is another group of medusae that does not show a distinct response to rapid changes in light intensity and does not show very distinct diel migrations. This pattern may be typified by *Rathkea*, *Aegina*, *Euphysa*, *Aequoria*, and *Aglantha*. Again, I must emphasize here that the temporal and spatial resolution of studies used to gain migration information may preclude determination of their true migration patterns. These medusae may be better migrators than the available data would suggest.

Most medusae release gametes within a few hours of the most rapid changes in light conditions, that is, around dusk and dawn. It is at this time when medusae are most active in their migrations. *Gonionemus* and *Polyorchis*, for example, begin spawning within an hour after dark and finish within 15 minutes (Mills, 1983). Both *Stomotoca* and *Bougainvillia* spawn in a short period of time at dawn and dusk. Examples of medusae that lack shadow responses but still show precise spawning periods are *Mitrocoma* and *Euphysa*. Although *Aequoria* and *Phialidium* spawn when they are most active, the duration of the spawning period tends to be poorly defined.

Clearly, spawning is closely tied with diel migrations, but it is not understood how light triggers gamete maturation and release. Gametes can mature and be released from isolated gonads (Ikegami et al., 1978), suggesting that input from marginal photoreceptors or the central nervous system is unnecessary. Experiments by Yoshida et al. (1980) suggest that a conducting system may spread light information around gonadal tissue. Gap junctions connect gonadal epithelial cells and electrical activity may spread via these junctions. Gonadal epithelial tissue is also often pigmented and this pigmentation may play a role in the photosensitive mechanism (Roosen-Runge, 1962).

3. Hydromedusan Versus Other Metazoan Photophysiology

I present here the available information for various aspects of hydromedusan photophysiology to demonstrate the similarities with other metazoan photoreceptors. Most of what is known comes from studies on medusae with ocelli. Based on the relatively little information we have about their photophysiology, hydromedusan photoreceptors appear to share many features with those of higher metazoans.

3.1. Morphology

There have been many studies on the morphology of hydromedusan ocelli and I have already mentioned several of these above. This work has stimulated, in part, a great deal of speculation on how simple hydromedusan photoreceptors, exemplified by the simple eye spot in *Leuckartiara*, may be a starting point for the derivation of "eyes" in higher metazoans (see Westfall, 1982). I do not wish to add to the voluminous literature on this subject, but it is clear that photoreceptor cells of hydromedusan ocelli do posses the basic features (e.g., increased surface area, cilium and related structures, mitochondria, microtubules, pigment granules in associated pigment cells) of higher metazoan photoreceptors (see Salvini-Plawen and Mayr, 1977; Burr, 1984).

Extraocular photoreceptor morphology is one area that we know little about and additional information would be useful in comparisons with higher metazoans. A careful examination of those medusae which lack obvious ocelli, but exhibit distinct photic behavior (e.g., *Gonionemus*, *Mitrocoma*,) may provide us with more information about the cilium-based lamellar bodies, which have been found in the distal portion of sensory cells in the outer nerve-ring of *Phialidium*, (Bouillon and Neilsen, 1974) and *Polyorchis* (Satterlie, 1985b). The axons of these cells merge with the extensive neuronal networks of the outer nerve-ring. These bodies have been found in other animals (e.g., ctenophores, Horridge, 1964; Aronova, 1979; molluscs, Wiederhold et al., 1973, Henkart, 1975; insects, Arikawa et al., 1980) where extraocular photosensitvity either has been demonstrated or highly suspected.

3.2. Extraocular Photosensitivity

Examples of extraocular photosensitivity, both neuronal and epithelial, appear throughout the metazoa (see Yasuda, 1979). This type of photosensitivity appears to be well represented amongst hydromedusae as well, as evidenced by the photic behavior of medusae that lack ocelli. For most of these medusae, the location of the photosensitivity is unknown, although there are indications that it is intraneuronal (see above). Physiological evidence for extraocular photosensitivity was first presented by Anderson and Mackie (1977). They showed that motor neurons, which control swimming in the hydromedusan *Polyorchis*, appeared to be directly photosensitive. This conclusion has been challenged by additional work on *Polyorchis* by Arkett and Spencer (1986a), who provided evidence suggesting that it is not the swimming motor neurons that are photosensitive, but rather an identifiable neuronal system ("O" system) in the outer nerve-ring. This system extends projections toward each ocellus and appears to be presynaptic to the swimming motor neurons. Furthermore, swimming motor neurons from *Polyorchis* isolated in primary culture do not show any detectable photosensitivity (A. N. Spencer, personal communication). The apparent photosensitivity of the swimming motor neurons in *Spirocodon* (Ohtsu, 1983a) may also be due to input from the outer nerve-ring, but this has not been examined. Electrophysiological studies on those medusae that show distinct photic responses, but lack ocelli (e.g., *Gonionemus*, *Stomotoca*, *Mitrocoma*), should give us a better idea of the nature of extraocular photosensitivity.

3.3. Photopigment

Several studies on a few selected hydromedusae indicate that rhodopsin is the functional visual pigment as it is for vertebrate and most invertebrate photoreceptors (Anderson and Andrews, 1982; Goldsmith and Bernard, 1985). The action spectra for the shadow response of *Polyorchis* is around 550 nm (Arkett, 1985) and the maximum electroretinogram (ERG)-like response is 530 nm (Weber, 1982a). *Sarsia* shows a maximum ERG response at 540 nm (Weber, 1982b). *Spirocodon* shows a peak shadow response between 480 and 500 nm (Yoshida, 1969). These responses are to be expected if a rhodopsin (maximal absorbance 498 nm) or rhodopsin-like photopigment were operating in the photoreceptor.

Photosensitivity appears to be localized in the extensive ciliary membrane of ocellar receptor cells. Intramembranous particles (IMPs) with diameters 8-9 nm and at densities of 5000-6000/μm^2 have been found associated with the microvillar and ciliary membranes of receptor cells from *Spirocodon* ocelli (Takasu and Yoshida, 1984). These particles are similiar in size and density to IMPs found in rod and cone outer segment plasma membrane from *Xenopus* (Besharse and Pfenninger, 1980), which are thought to contain rhodopsin (Jan and Revel, 1974). Furthermore, Yoshida (1972) has provided evidence suggesting that the chromophore in *Spirocodon* ocelli is a "retinol-like substance".

There is also evidence that hydromedusan photoreceptors have two peaks of sensitivity, one in visible light and one in ultraviolet (UV) light. Ohtsu (1983a) showed that intracellular recordings from giant swimming motor neurons hyperpolarize at the onset of 350 nm light. Swimming is thus inhibited. Subsequent exposure to visible light (500 nm) disinhibits the motor neurons, the membrane potential returns to pre-stimulus levels, and swimming resumes. It is not clear from this study, however, whether the motor neurons themselves, or another neuronal or non-neuronal system presynaptic to them, are directly responsive to UV light.

3.4. Receptor Potential

There is almost no cellular information available on the response of hydromedusan photoreceptors, yet there are indications from a few studies that these receptors function in a manner similiar to other photoreceptors. Electroretinogram (ERG) -like recordings from the ocelli of *Sarsia* and *Polyorchis* show graded, positive potential changes in response to the onset of varying light intensities (Weber, 1982a,b). Ocellar recordings from *Spirocodon* also show a positive, transient potential change in response to the onset of 500 nm light (Ohtsu, 1983b). Weber concluded, by comparing his recordings with vertebrate ERGs, that the hydromedusan photoreceptor hyperpolarizes in response to light. It must be borne in mind that ocellar recordings are difficult to interpret because, like ERG recordings from vertebrate eyes (Tomita, 1972), they represent complex potentials arising from a number of potential sources. Ocellar recordings may include electrical responses from photoreceptors, second-order neurons (if present), tentacle motor neurons, and tentacle myoepithelial effectors. Weber's conclusion of a hyperpolarizing photoreceptor may be premature, since in none of these studies were intracellular electrode recordings from receptor cells made to corroborate this conclusion.

Intracellular electrode recordings from the identified neuronal system in the outer nerve-ring of *Polyorchis* ("O" system) show a graded response to increases and decreases in light intensity (Arkett and Spencer, 1986b). Lucifer Yellow fills of this system shows that portions of it extend up to each ocellus. Typical "O" system recordings can also be made from the ocellus. Based on these findings, Arkett and Spencer (1986a) have proposed that the electrically-coupled "O" system is the primary photoreceptor system in *Polyorchis*. If this is so, then in *Polyorchis* at least, the photoreceptors respond to increases in light intensity with graded depolarizations and to decreases in light intensity with graded hyperpolarizations. If hydromedusan photoreceptors do depolarize in response to light, then they would seem to fit into the frequently-cited generalization that invertebrate photoreceptors depolarize (although with known exceptions, see McReynolds, 1976), while vertebrate photoreceptors hyperpolarize in response to light. Yet, there is much more we need to discover about hydromedusan photophysiology before we can make any definitive conclusions. For example, we know nothing about transduction mechanisms or light-sensitive conductance changes in hydromedusan photoreceptors and whether these changes are mediated by second messengers, such as cyclic GMP.

3.5. Electrical Coupling

There is some evidence that hydromedusan photoreceptors, like the receptors of both vertebrates and other invertebrates (see Fain et al., 1976; Jarvilehto, 1979; Laughlin, 1981), are electrically coupled to each other through extensive gap junctions. This is not unexpected since most (but not all) neuronal systems identified in hydromedusae thus far show electrical coupling between units within each system. The "O" system, which appears to connect the numerous ocelli and may function as the photoreceptor system in *Polyorchis*, is electrically coupled (Spencer and Arkett, 1984). Arkett and Spencer (1986b) have proposed that the electrical coupling properties of the "O" system and the radially arranged ocelli may be important in the integration of photic information in a way analogous to that of photoreceptors in other animals. Morphological evidence for electrical coupling is given by Singla and Weber (1982b) who have found gap junctions between receptor cell bodies in the ocelli of *Sarsia*, although none have been observed in ocelli of *Polyorchis* (Singla and Weber, 1982a), or *Spirocodon* (Toh, Yoshida, and Tateda, 1979). It seems likely that if photosensitive neuronal systems are identified in medusae, such as *Gonionemus* or *Phialidium*, they will also be electrically coupled.

3.6. Transmitters

As indicated by this meeting, we are just beginning to find out what transmitters are functioning in the cnidarian nervous system. There are only a few studies that have given us clues as to what transmitters might function at photoreceptor/second order neuron synapses. At the ultrastructural level, both electron-lucent and electron-dense synaptic vesicles, ranging in size from 60-110 nm diameter, have been found within receptor cells (Singla and Weber, 1982a,b; Toh et al., 1979; Yamamoto and Yoshida, 1980). Predicting what transmitters these vesicles may contain is difficult without further information. There was some suggestion several years ago that a neuropeptide might function in the photoreceptor system of *Polyorchis*. This suggestion was based on the remarkable similarity between the morphology of the "O" system, from Lucifer Yellow fills (Spencer and Arkett, 1984), and the FMRFamide immunoreactive neuronal system which sends processes up to the ocelli (Grimmelikhuijzen and Spencer, 1984). However, by combining immunostaining and Lucifer Yellow injections, Spencer (1988) has concluded that the "O" system does not contain antigens to FMRFamide antisera. I am aware of only one other coelenterate example of a transmitter possibly involved in a photoreceptor system. Aronova (1979) has demonstrated the presence of cholinesterase activity at neuronal synapses near "presumptive photoreceptor cells" in the aboral organ of the ctenophore *Beroe*, but this has not, to my knowledge, been followed up. Characterization of cnidarian neurotransmitters is an active area of research and, as yet, there is little information on the ones that are involved in the photic system. However, if photoreceptor properties are conserved throughout the metazoa, then we should be looking for transmitters that appear to function in vertebrate retina and invertebrate photoreceptors (e.g., L-glutamate, ACh, GABA, dopamine; see Lam et al., 1982; Daw et al., 1989).

4. Conclusions

The primary purpose of this paper was to examine the relationship, if any, between the diversity of hydromedusan photoreceptors and behavioral photophysiology. In this way, I had hoped to illustrate a possible scenario for the evolution of hydromedusan photoreceptors through functionally adaptive modifications. With what information is available, there would appear, at first glance, to be little indication of a positive correlation between the functional (at least as indicated by behavior and some physiology) and structural complexity of hydromedusan photoreceptors. Ocelli-bearing medusae do not necessarily exhibit more clearly defined diel migrations than medusae without ocelli. Nor are medusae with ocelli the only ones to exhibit a more complex photic behavior, namely shadow responses. However, I suggest that there is, in fact, such a positive correlation, but it will be revealed only by a more detailed examination of several types of photoreceptors exhibited by medusae.

As I have already mentioned, the presumptive evolution of the hydromedusan ocellus has received a great deal of attention. However, there has been little regard for the role of neuronal and non-neuronal photosensitivity in the evolution of hydromedusan photoreceptors. Salvini-Plawen (1982) has advanced the idea that a photosensitive, monociliate ectodermal cell acts as a stem cell from which other photoreceptors can be derived. I have used this model as a guide to propose how hydromedusan photoreceptors may have split into two lines, which generate the diversity of photoreceptors and physiology we see in extant medusae.

The precursor for photoreceptors in general, and hydromedusae in particular, is a photosensitive, ciliated, ectodermal cell. The hydrozoan conducting epithelium (ectoderm) has been shown to have a sensory function (e.g., mechanoreception; Mackie and Passano, 1968) and it is possible that it could also function as a light receptor. Mackie (1970) suggested that this type of "neuroid" epithelia may function in more general responses. Such a response may be a light-induced change in the pacemaker activity of neurons controlling swimming and, thus, have a direct effect on diel migrations. There is evidence for this type of neuro-epithelial connection. Depolarization of conducting epithelia does lead to an inhibition of swimming motor neurons in *Stomotoca* (Mackie, 1975). It is unknown whether this type of photoreceptor is represented in extant hydromedusae. *Solmissus*, which appears to swim constantly, may use this type of photoreceptor, but Mills and Goy (1988) report no evidence for diel differences in swimming rate. Yet, even slight diel changes in swimming frequency may result in significant migrations. Several other medusae, which show somewhat weak diel migrations (Table 1), may also be included in this group, but it is difficult to do so with the available information about their physiology and morphology.

From this simple photoreceptor, I propose that two lines of hydromedusan photoreceptors may be derived, one leading to the more commonly known ocelli and the other leading to photosensitive neurons. In one line, it seems possible that the photosensitive, ciliated ectodermal cell became internalized, but retained the basic cilium structure. This structure may have been modified and now has more neuron-like properties. An example of this may be seen in the lamellar bodies of the sensory cells within the outer nerve-ring of *Phialidium* (Bouillon and Nielsen, 1974). With this modification, the photoreceptor became localized within the nervous system. This would seem to be an important modification because photic information could now be received more directly by the central nervous system. This information may, thus, be integrated to coordinate specific or localized behavior. However, it remains to be conclusively demonstrated that these lamellar bodies in hydromedusae do impart some photosensitivity to the neurons.

It would seem possible that a next step in this line is the loss of the cilium and related structures, while the neurons retain the photosensitive property. We may see this condition in *Gonionemus*. No lamellar bodies have been observed in the nerve rings (Schnorr von Carolsfeld, 1984), yet *Gonionemus* shows a complex photic behavioral repertoire including a distinct diel migration, a pronounced shadow response, and a light directional sense (Murbach, 1909). Again, because the photoreceptor is located within the nervous system, light information may be integrated to produce localized responses, such as the turning behavior of *Gonionemus*. Another example of a local response may be the shadow-induced acceleration of the tentacle pulse (TP) system in *Stomotoca*, which leads to tentacle contractions (Mackie, 1975). A detailed electrophysiological study on the photic behavior of these medusae is greatly needed.

The other line of hydromedusan photoreceptor evolution may lead to the highly-structured ocellus. The ectodermal cell again becomes more neuron-like, with the distal portion expanding in surface area by folding of the ciliary membrane. An axon emerges from the cell body and connects with the outer nerve-ring. The photoreceptors are localized, and together with associated pigment cells, are arranged radially. The simple ocellus of *Leuckartiara* has been used to illustrate a possible first stage in this line (Singla, 1974; Salvini-Plawen and Mayr, 1977). Unfortunately, we know little about the photic behavior of *Leuckartiara*. It does not appear to have a shadow response, but it does show a temporary increase in swimming frequency as it passes through a light beam (personal observations). Further modifications to photoreceptor cells are apparent in medusae like *Bougainvillia* and *Polyorchis*, (Eakin and Westfall, 1962; Singla, 1974) which both show pronounced shadow responses and distinct diel migrations. The ocellar photoreceptors of *Cladonema* appear to have undergone the greatest modifications, forming a lens-like structure (Bouillon and Nielsen, 1974), although it is unknown what, if any, function this structure plays. Unfortunately, we know almost nothing about the photophysiology of this medusa aside from an anecdotal account of *Cladonema ucidai*, which abruptly starts swimming after illumination with UV followed by visible light (Ohtsu, personal communication).

It is difficult to place any one medusa into a stage of this scheme of proposed photoreceptor evolution. Indeed, there may be features that are present in both lines. For example, *Polyorchis*, which has well-differentiated ocelli, also has lamellar bodies in the outer nerve-ring (Satterlie, 1985b). This may indicate that the proposed lines are not direct. Furthermore, one must remember that this proposed scheme has been constructed with the information available from a small fraction of the great diversity of extant hydromedusae. However, I do think that there is now a good indication of a positive correlation between the structural and functional complexity of hydromedusan photoreceptors. Those medusae with the most highly modified photoreceptors exhibit the most complex photic behavior. Future studies on some of the key medusae will, hopefully, strengthen this correlation.

From what we know about their photophysiology and morphology, hydromedusan photoreceptors appear to have several features in common with photoreceptors of higher metazoans. These findings, together with the knowledge from the fossil record (see Glaessner, 1962) and, more recently, from ribosomal RNA sequence data (Field et al., 1988) that suggests that cnidarians have been separated from other metazoans for a long time, provide strong evidence for an independent, and possibly convergent, evolution of hydromedusan photoreceptors. Clearly, other metazaons have more complex photoreceptors, which provide more detailed information for more complex behavior. However, for hydromedusae, the most basic function of a photoreceptor, that is the detection of changes in light intensity, is fulfilled by their relatively simple "eyes".

ACKNOWLEDGEMENTS. I thank Drs. J. Costello, C. Mills, K. Ohtsu for their observations on various hydromedusae.

References

Anderson, P. A. V., and Mackie, G. O., 1977, Electrically coupled, photosensitive neurons control swimming in a jellyfish, *Science* **197**:186-188.

Anderson, R. E., and Andrews, L. D., 1982, Biochemistry of retinal photoreceptor membranes in vertebrates and invertebrates, in: *Visual Cells in Evolution* (J. A. Westfall, ed.), pp. 1-22, Raven Press, New York.

Arai, M. N., and Fulton, J., 1973, Diel migration and breeding cycle of *Aglantha digitale* from two locations in the N. E. Pacific, *J. Fish. Res. Bd. Can.* **30**:551-553.

Arikawa, K., Eguchi, E., Yoshida, A., and Aoki, K., 1980, Multiple extraocular photoreceptive areas on genitalia of butterfly, *Papilio xuthus*, *Nature* **288**:700-702.

Arkett, S. A., 1984, Diel vertical migration and feeding behavior of a demersal hydromedusan (*Polyorchis penicillatus*), *Can. J. Fish. Aquat. Sci.* **41**:1837-1843.

Arkett, S. A., 1985, The shadow response of a hydromedusan (*Polyorchis penicillatus*): behavioral mechanisms controlling diel and ontogenic vertical migration, *Biol. Bull.* **169**:297-312.

Arkett, S. A., and Spencer, A. N., 1986a, Neuronal mechanisms of a hydromedusan shadow reflex. I. Identified reflex components and sequence of events, *J. Comp. Physiol.* **159**:201-213.

Arkett, S. A., and Spencer, A. N., 1986b, Neuronal mechanisms of a hydromedusan shadow reflex. II. Graded response of reflex components, possible mechanisms of photic integration, and functional significance, *J. Comp. Physiol.* **159**:215-225.

Aronova, M. Z., 1979, Electron microscopic investigation of the presumptive photoreceptive cells in the aboral organ of the ctenophore *Beroe cucumis*, *Zhurnal Evolyutsionnoi Biokhimii i Fiziologii* **15**:59-601 (in Russian).

Benovic, A., 1973, Diurnal vertical migration of *Solmissus albescens* in the southern Adriatic, *Mar. Biol.* **18**:298-30.

Besharse, J. C., and Pfenninger, K. H., 1980, Membrane assembly in retinal photoreceptors. 1. Freeze-fracture analysis of cytoplasmic vesicles in relationship to disc assembly, *J. Cell Biol.* **87**:451-463.

Bouillon, J., and Nielsen, M., 1974, Étude de quelques organes sensoriels de cnidaires, *Arch. Biol.* **85**:307-328.

Burr, A. H., 1984, Evolution of eyes and photoreceptor organelles in the lower phyla, in: *Photoreception and Vision in Invertebrates* (M. A. Ali, ed.), pp. 131-178, Plenum Press, New York.

Daw, N. W., Brunken, W.J., and Parkinson, D., 1989, The function of the synaptic transmitters in the retina, *Ann. Rev. Neurosci.* **12**:205-225.

Eakin, R. M., 1982, Continuity and diversity in photoreceptors, In:*Visual Cells in Evolution* (J. A. Westfall, ed.) pp 91-105, Raven Press, New York.

Eakin, R. M., and Westfall, J. A., 1962, Fine structure of photoreceptors in the hydromedusan, *Polyorchis penicillatus*, *Proc. Natl. Acad. Sci. U.S.* **48**:826-833.

Fain, G. L., Gold, G. H., and Dowling, J. E., 1976, Receptor coupling in the toad retina, *Cold Spring Harbor Symp. Quant. Biol.* **40**:547-561.

Field, K. G., Olsen, G. J., Lane, D. J., Giovannoni, S. J., Ghiselin, M. T., Raff, E. C., Pace, N. R., and Raff, R. A., 1988, Molecular phylogeny of the animal kingdom, *Science* **239:**748-753.

Glaessner, M. F., 1962. Pre-Cambrian Fossils, *Biol. Rev.* **37:**467-494.

Goldsmith, T. H., and Bernard, G. D., 1985, Visual pigments of invertebrates, *Photochemistry and Photobiology* **42:**805-809.

Grimmelikhuijzen, C. J. P., and Spencer, A. N., 1984, FMRF-amide immunoreactivity in the nervous system of the medusa *Polyorchis penicillatus*, *J. Comp. Neurol.* **230:**361-371.

Hamner, W. M., 1985, The importance of ethology for investigations of marine zooplankton, *Bull. Mar. Sci.* **40:**414-424.

Henkart, M., 1975, Light-induced changes in the structure of pigmented granules in Aplysia neurons, *Science* **188:**155-157.

Hesthagen, I. H., 1971, On the biology of the bottom-dwelling trachymedusae *Tesserogastria musculosa*, *Norwegian J. Zoology* **19:**1-19.

Hisada, M., 1956, A study on the photoreceptor of a medusa, *Spirocodon saltatrix*, *J. Fac. Sci. Hokkaido Univ. Ser. VI, Zool.* **12:**529-533.

Honegger, T., Achermann, J., Stidwell, R., Littlefield, L., Baenninger, R., and Tardent, P., 1980, Light-controlled spawning in *Phialidium hemisphaericum* (Leptomedusae), in: *Development and Cellular Biology of Coelenterates* (P. Tardent and R. Tardent, eds.), pp. 83-88, Elsevier/North Holland Biomedical Press, Amsterdam.

Horridge, G. A., 1964, Presumed photoreceptive cilia in a ctenophore, *Quarterly Journal Microsc. Sci.* **105:**311-317.

Ikegami, S., 1977, The occurrence and some properties of a spawning inducing substance in the testis of the hydromedusan *Spirocodon saltatrix*, *Agric. Biol. Chem.* **41:**2311-2312.

Ikegami, S., Honji, N., and Yoshida, M., 1978, Light-controlled production of spawning-inducing substance in jellyfish ovary, *Nature* **272:**611-612.

Jan, L. Y., and Revel, J. P., 1974, Ultrastructural localization of rhodopsin in the vertebrate retina, *J. Cell Biol.* **62:**257-273.

Jarvilehto, M., 1979, Receptor potentials in invertebrate visual cells, in: *Handbook of Sensory Physiology*, Volume VII/6A, *Comparative Physiology and Evolution of Vision in Invertebrates, A: Invertebrate Photoreceptors* (H. Autrum, ed.), pp. 315-356, Springer-Verlag, Berlin.

Kikuchi, K., 1947, On the shadow reflex of *Spirocodon saltatrix* and their vertical distribution in the sea, *Zool. Mag.* **57:**144-146 (in Japanese).

Lam, D. M-K., Frederick, J. M., Hollyfield, J. G., Sarthy, P. V., and Marc, R. W., 1982, Identification of neurotransmitter candidates in invertebrate and vertebrate photoreceptors, in: *Visual Cells in Evolution* (J. A. Westfall, ed.), pp. 65-80, Raven Press, New York.

Mackie, G. O., 1970, Neuroid conduction and the evolution of conducting tissues, *Quart. Rev. Biol.* **45:**319-332.

Mackie, G. O., 1975, Neurobiology of *Stomotoca*. II. Pacemaker and conduction pathways, *J. Neurobiol.* **6:**339-378.

Mackie, G. O., 1985, Midwater macroplankton of British Columbia studied by submersible PISCES IV, *J. Plankton Res.* **7:**753-777.

Mackie, G. O., and Mills, C. E., 1983, Use of the PISCES IV submersible for zooplankton studies in coastal waters of British Columbia, *Can. J. Fish. Aquat. Sci.* **40:**763-776.

Mackie. G. O., and Passano, L. M., 1968, Epithelial conduction in hydromedusae, *J. Gen. Physiol.* **52:**600-621.

McReynolds, J. S., 1976, Hyperpolarizaing photoreceptors in invertebrates, in: *Neural Principles in Vision* (F. Zettler and R. Weiler, eds.), pp. 394-409, Springer-Verlag, Berlin.

Miller, R. L., 1980, Species-specificity of sperm chemotaxis in the hydromedusae, in: *Developmental and Cellular Biology of Coelenterates* (P. Tardent and R. Tardent, eds.), pp. 89-94, Elsevier/North Holland Biomedical Press, Amsterdam.

Mills, C. E., 1981, Diversity of swimming behaviors in hydromedusae as related to feeding and utilization of space, *Mar. Biol.* **64:**185-189.

Mills, C. E., 1982, *Patterns and mechanisms of vertical distribution of medusae and ctenophores,* Ph.D. dissertation, 384 pp., University of Victoria, Victoria, B.C.

Mills, C. E., 1983, Vertical migration and diel activity patterns of hydromedusae:studies in a large tank, *J. Plankton Res.* **5**:619-635.

Mills, C. E., and Goy, J., 1988, In situ observations of the behavior of mesopelagic *Solmissus narcomedusae* (Cnidaria, Hydrozoa), *Bull. Mar. Sci.* **43**:739-751.

Moreira, G. S., 1973. On the diurnal migration of hydromedusae off Santos, Brazil, *Pub. Seto Mar. Bio. Lab.* **20**:537-566.

Murbach, L., 1909, Some light reactions of the medusa *Gonionemus*, *Biol. Bull.* **17**:354-368.

Nagao, Z., 1963, The early development of the anthomedusa, *Polyorchis karafutoensis* Kishinouye, *Annot. Zool. Jpn.* **36**:187-193.

Ohtsu, K., 1983a, UV-visible antagonism in extraocular photosensitive neurons of the anthomedusa *Spirocodon saltatrix* (Tilesius), *J. Neurobiol.* **14**:145-155.

Ohtsu, K., 1983b, Antagonizing effect of ultraviolet and visible light on the ERG from the ocellus of *Spirocodon saltatrix* (Coelenterata:Hydrozoa), *J. exp. Biol.* **105**:417-420.

Ohtsu, K., and Yoshida, M., 1973, Electrical activities of the anthomeudsan, *Spirocodon saltatrix* (Tilesius), *Biol. Bull.* **145**:532-547.

Omori, M., and Hamner, W. M., 1982, Patchy distribution of zooplankton: behavior, population assessment, and sampling problems, *Mar. Biol.* **72**:193-200.

Passano, L. M., 1973, Behavioral control systems in medusae: a comparison between hydro- and scyphomedusae, *Pub. Seto Mar. Bio. Lab.* **20**:615-645.

Pearre, S., 1979, Problems of detection and interpretation of vertical migration, *J. Plankton Res.* **1**:29-44.

Roosen-Runge, E. C., 1962, On the biology of sexual reproduction of hydromedusae, Genus Phialidium Leuckhart, *Pacific Science* **16**:15-24.

Romanes, G.J., 1885, *Jellyfish, starfish, and sea urchins*, 323 pp., D. Appleton & Co., New York.

Rugh, R., 1929, Egg laying habits of *Gonionemus* murbachii in relation to light, *Biol. Bull.* **57**:261-266.

Russell, F. S., 1925, The vertical distribution of marine macroplankton. An observation on diurnal changes, *J. Mar. Biol. Assoc. UK* **13**:769-809.

Salvini-Plawen, L. v., 1982, On the polyphyletic origin of photoreceptors, in: *Visual Cells in Evolution* (J. A. Westfall, ed.), pp. 137-154, Raven Press, New York.

Salvini-Plawen, L. v., and Mayr, E., 1977, On the evolution of photoreceptors and eyes, in: *Evoutionary Biology*, Volume 10 (M.K. Hecht, W.C. Steere, B. and Wallace, eds.), pp. 207-263, Plenum Press, New York.

Satterlie, R. A., 1985a, Central generation of swimming activity in the hydrozoan jellyfish *Aequoria aequoria*, *J. Neurobiol.* **16**:41-55.

Satterlie, R. A., 1985b, Putative extraocular photoreceptors in the outer nerve ring of *Polyorchis penicillatus*, *J. exp. Zool.* **233**:133-137.

Schnorr von Carolsfeld, J., 1984, *Contributions to Hydromedusan Neuroethology from a Study on Two Olindiads:* Gonionemus vertens *(Agassiz 1865)* Eperetmus typus *(Bigelow 1915)*, 283 pp., Victoria, British Columbia, Canada, University of Victoria, Master's Thesis.

Singla, C. L., 1974, Ocelli of hydromedusae, *Cell. Tissue Res.* **149**:413-429.

Singla, C. L., and Weber, C., 1982a, Fine structure studies of the ocelli of *Polyorchis* penicillatus (Hydrozoa, Anthomedusae) and their connections with the nerve ring, *Zoomorphology* **99**:117-129.

Singla, C. L., and Weber, C., 1982b, Fine structure of the ocellus of *Sarsia tubulosa* (Hydrozoa, Anthomedusae), *Zoomorphology* **100**:11-22.

Spencer, A. N., 1988, Effects of Arg-Phe-amide peptides on identified motor neurons in the hydromedusa *Polyorchis penicillatus*, *Can. J. Zool.* **66**:639-645.

Spencer, A. N., and Arkett, S. A., 1984, Radial symmetry and the organization of central neurones in a hydrozoan jellyfish, *J. exp. Biol.* **110**:69-90.

Takasu, N., and Yoshida, M., 1984, Freeze-fracture and histofluorescence studies on photoreceptive membranes of medusan ocelli, *Zoological Science* **1**:367-374.

Tamasige, M., and Yamaguchi, T., 1967, Equilibrium orientation controlled by ocelli in an anthomedusa, *Polyorchis karafutoensis*, *Zool. Mag.* **76**:35-36.

Thurston, M. H., 1977, Depth distribution of *Hyperia spinigera* Bovallius 1889 (Crustacea:Amphipoda) and medusae in the North Atlantic Ocean, with notes on the associations between Hyperia and coelenterates, in: *A Voyage of discovery: George Deacon 70th anniversary volume* (M. Angel, ed.), pp. 499-536, Pergamon Press Ltd., Oxford.

Toh, Y., Yoshida, M., and Tateda, H., 1979, Fine structure of the ocellus of the hydromedusan, *Spirocodon saltatrix*, I. Receptor cells, *J. Ultrastructure Res.* **68:**341-352.

Tomita, T., 1972, The electroretinogram, as analyzed by microelectrode studies, in: *Handbook of Sensory Physiology*, Volume VII/2, *Physiology of photoreceptor organs* (M.G.F. Fuortes, ed.), pp. 635-666, Springer-Verlag, Berlin.

Vanfleteren, J. R., 1982, A monophyletic line of evolution? Ciliary induced photoreceptor membranes, in:*Visual Cells in Evolution* (J. A. Westfall, ed.) pp 107-136, Raven Press, New York.

Weber, C., 1982a, Electrical activities of a type of electroretinogram recorded from the ocellus of a jellfish, *Polyorchis penicillatus* (Hydromedusa), *J. Exp. Zool.* **223:**231-243.

Weber, C., 1982b, Electrical activity in response to light of the ocellus of the hydromedusan *Sarsia tubulosa*, *Biol. Bull.* **162:**413-422.

Westfall, J. A., 1982, *Visual cells in evolution*, 161 pp., Raven Press, New York.

Wiederhold, M. L., MacNichol, E. F. Jr., and Bell, A. L., 1973, Photoreceptor spike responses in the hardshell clam, *Mercenaria mercenaria*, *J. Gen. Physiol.* **61:**24-55.

Yamamoto, M., and Yoshida, M., 1980, Fine structure of ocelli of an anthomedusan, *Nemopsis dofleini*, with special reference to synaptic organization, *Zoomorphology* **96:**169-181.

Yasuda, M., 1979, Extraocular photoreception, in *Handbook of Sensory Physiology*, Volume VII/6A, *Comparative Physiology and Evolution of Vision in Invertebrates, A: Invertebrate Photoreceptors* (H. Autrum, ed.), pp. 581-640, Springer-Verlag.

Yoshida, M., 1959, Spawning in coelenterates, *Experientia* **15:**11-13.

Yoshida, M., 1969, The ocellar pigment of the anthomedusa *Spirocodon saltatrix*: Does its photoreduction bear any physiological significance? *Bull. Mar. Biol. St. Asamushi* **13:**215-219.

Yoshida, M., 1972, Detection of a retinol-like substance and the relative abundance of carotenoids in different tissues of the anthomedusa, *Spirocodon saltatrix*, *Vision Res.* **12:**169-182.

Yoshida, M., Honji, N., and Ikegami, S., 1980, Darkness induced maturation and spawning in *Spirocodon saltatrix*, in: *Developmental and Cellular Biology of Coelenterates* (P. Tardent and R. Tardent, eds.), pp. 75-82, Elsevier/North Holland Biomedical Press.

Zelikman, E. A., 1969, Structural features of mass aggregations of jellyfish, *Oceanology* **9:**558-564.

Chapter 28

Summary of Session and Discussion on Sensory Mechanisms

M. S. LAVERACK

Because this was the last session of the workshop, the discussion encompassed a variety of different aspects of nervous systems. Nevertheless, sensory mechanisms form a useful basis for the final discussion, since it is quite easy to envisage their role in the evolution of a nervous system.

The first nervous systems may have originated millions of years ago amongst organisms that were multicellular, and which demonstrated feeding movements. Such animals may have been essentially static since they did not search for food, but reacted instead to chemicals in the environment that promoted increased water circulation. Later, perhaps, some bending towards a particular point might be possible as it became possible to coordinate muscles (the movement of whole multicellular bodies would presumably come later when larger prey became available). Within this framework, and that put forward by Carr (Chapter 6), it is not surprising to find that certain amino acids (e.g., glutamic acid and glycine) and purines are stimulants to nerve cells but, at the same time, one must question why other naturally occurring feeding stimulants are not utilized as neurotransmitters, and why small chain peptides are so significant.

The natural world is a continuum in which all parts reside, interact and influence all others. Amongst organisms of all kinds there are apparently advantages for coordination; bacteria, for instance, show concerted and directional responses to outside influence in the form of, for example, chemical and magnetic stimuli. In the case of eukaryotes, the presence of directed responses often, but not always, implies the existence of a conduction system amongst the component parts. The parts in question may be contained within a single cell or be distributed amongst many. It further indicates that whilst ion (e.g., Ca^{++}) exchange across membranes may be

M. S. LAVERACK • Gatty Marine Laboratory, University of St. Andrews, St. Andrews KY16 8LB, Scotland, U.K.

Evolution of the First Nervous Systems
Edited by P.A.V. Anderson
Plenum Press, New York

similar in principle amongst bacteria, protists, coelenterates, echinoderms and vertebrates the end result should be a modification of behavior. This behavior may be simple, resulting in no more than ciliary reversal in the case of a unicellular organism, or complex, but the transmission of information occurs over a distance. There may be clear signs of a common origin, especially in the biochemistry of ion uptake and exchange, or there may be convergence, arriving at a similar outcome from different directions.

If the surface of an individual cell or organism were sensitive to chemical stimuli, it is possible that transduction of that stimulus would give rise to an influx of ions across the cell membrane, i.e., a current. If the cells in question were electrically coupled by way of gap junctions or their functional equivalents, then information could spread throughout a multicellular organism. This is the basis of epithelial conduction; the ability of epithelial cells to produce and propagate action potentials. Epithelial conduction is known amongst cnidarians, urochordates and the embryos of some amphibians and fish, and may be predicted in others, such as cephalopods in which embryonic behavior is noted intracapsularly. The proposition that epithelial conduction might represent the first form of nervous system, including behavioral consequences on musculature and swimming, was, however, not well received by the participants. Simple radiation of an electrical signal from a single point, indiscriminately across an organism, as in epithelial conduction, is not a directed response. In the case of known sensory epithelial conduction systems, they inform the animal that it has been stimulated but provide no information as to the location of that stimulus. Similarly, on the motor side, the diffuse nature of epithelial conduction is not easily adapted to produce much coordinated behavior; all effectors within a coupled array are invariably activated.

On the other hand, protozoa have developmental stages, with functional modifications from one stage to the next, and they show food selection, but are these alterations and actions just a question of the distribution of different ion channels? Can behavior occur without a nervous system? In sponges, limited local behavior appears to be a function of single cells. Older descriptions of behavior (e.g., Jennings, 1923, *The Behavior of the Lower Organisms*, Columbia Univ. Press, New York) suggest that protozoa carry out a considerable repertoire of behavior, but the most sophisticated modern measurements of behavior seem to suggest that Protozoa do very few things, like moving backwards or forwards, or contracting. Is behavior (or the modification of locomotion) simply only a question of ionic channels and the number of such? Are food selection, attachment, detachment and ingestion all variations on the same mechanism? It is generally agreed that sponges do not possess nervous systems.

It is arguable, instead, that a major impetus to nervous system development was the appearance of the bilaterality found in higher organisms; such a development may have given rise to the need to coordinate the behavior of the two sides of the animal, and to the development of discrete local nervous systems. However, nervous systems, as we recognize them, are first noted amongst Cnidaria, which are radially organized.

A reasonable conclusion from this line of thinking is that polyphyletic origins for the nervous system are probable, as is the likely development of component parts such as giant fibers. Stem groups may not now exist and to look for beginnings in Cnidaria or sponges may be in error.

An important step in any discussion of the evolution of the first nervous system is to first define what is meant by a nervous system. This is not easy. One can argue in terms of properties such as polarity of communication and directed information transfer. For instance, perhaps the nervous system is first of all polarized to allow rapid conduction in particular directions in an animal body, implying a requirement for improved communication for certain parts of behavior (e.g., escape and feeding) whilst other slower activities such as gametogenesis may be a slower event controlled through endocrine mechanisms. However, endocrine systems use the same structures (polarized axons and synapse-like release sites) as nerve cells, weakening this argument.

A second basic feature of the nervous system may be the property of inhibition, which would surely be important for the coordination of the two sides of a bilaterally symmetrical animal. However, true inhibitory synapses are not known with certainty in any member of the first group of animals to possess a nervous system, the Cnidaria, nor in Ctenophora. Examples of what would appear to be IPSP- mediated processes occur in the Hydrozoa, but in most cases, there is still insufficient evidence to point to the inhibitory process having a synaptic basis. The crumpling response of a hydromedusa is associated with inhibition of swimming, but this effect is believed to be a field effect, similar to that which occurs in the Mauthner neurons in vertebrates. In addition, evidence was presented by Spencer (Chapter 3) for an inhibitory effect of dopamine in the hydromedusa *Polyorchis*, but it is not clear whether this effect is mediated via inhibitory synapses or by alternative pathways, since dopamine is known to modulate synaptic effects without being involved in synaptic transmission. Perhaps the best possibility for inhibitory synapses is to be found in *Hydra*. In nerve-free *Hydra*, nematocysts discharge under all conditions, but this does not happen in complete *Hydra*, implying that the synapses onto nematocytes (see Chapter by Hobmayer and David) are inhibitory.

This workshop has concentrated on data from adult organisms; larval stages have been essentially ignored. It is not clear that larval stages are any more primitive than juvenile and adult forms, being part of the same life cycle; nonetheless some larvae show distinct differences in the way in which sensory information is processed by the nervous system. For example, direct interaction between larval ciliary sensors (e.g., apical organ in polychaete trochophores) and ciliary locomotory organs (prototroch) is not seemingly repeated or continued in later life. Larval systems seemingly deserve more attention in the future.

Molecular biological analysis of certain proteins (e.g., globins, etc.) has shown no reliable relationships. It was suggested that a checklist of the fundamental properties of a nervous system should be prepared to enable molecular biologists and gene technologists to establish the presence or absence of such features specifically. This would be an improvement on claims that the presence of a particular protein or polypeptide in widely diverse groups, implies some relationship or functional significance. However, an attempt to compile a list of the essential building blocks with which to build a nervous system, foundered on the many individual views of what comprised essential bits that could be arranged into a significant whole. No answers were offered to the question "What is a nervous system and how did it begin?"

Part 4

PLENARY LECTURE

Chapter 29

Evolution of Cnidarian Giant Axons

G. O. MACKIE

1. Introduction

Speculations about evolution have an irresistable fascination, partly because they are hard to prove wrong, partly because they make us think about origins and look for clues in development. Cnidarians may have been the first metazoans to evolve, although this seems rather unlikely in view of the fact that they are all carnivores, preying on other metazoans. Given, however, that of the surviving phyla they alone retain the presumably ancestral (diploblastic) body plan, it becomes especially interesting to look at their nervous systems for clues as to how nerves evolved. This is not just a selling point for use in grant applications. There really is no better group in which to look for clues. At the same time, given the 700 million years which have elapsed since cnidarians appeared in the fossil record, any clues about nervous origins can only be of the most general nature. Cnidarian nerves are not obviously primitive in functional terms, only in the way they are laid out as center-less nets. Of course the layout can be far more complex than this. Some of them do have centres. At the last count, the jellyfish *Aglantha* had six, possibly seven, physiologically distinct neuronal subsets running in the marginal nerve rings. My efforts to understand their interactions have aged me prematurely. Other speakers at this meeting have looked at ionic channels, morphogenetic factors and neurochemicals for clues. Molecular biology may (probably will) eventually help us reconstruct phylogeny, and place the cnidarians in their proper place on the many-branched tree of neural evolution, but we are still far from this point. As of now, we do not know in what precursor cell line neurons first arose, through what stages they passed to assume their now familiar form, or what

G. O. MACKIE • Biology Department, University of Victoria, Victoria, British Columbia V8W 2Y2, Canada.

Evolution of the First Nervous Systems
Edited by P.A.V. Anderson
Plenum Press, New York

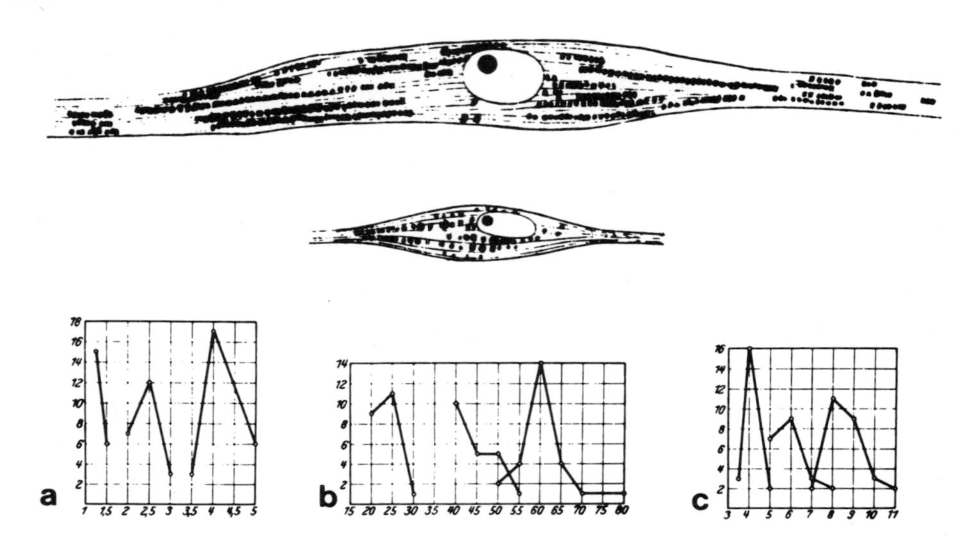

Figure 1. Size relationships of bipolar neurons in *Rhizostoma* (from Bozler, 1927). The neurons fall into three size classes. Representatives of the largest and smallest are shown above, to the same magnification. Graphs below plot numbers of neurons on the ordinate against neurite diameter (**a**), cell length (**b**) and cell body diameter (**c**) measured in eyepiece divisions on the abscissa.

functions they originally served. Electrical signalling may have been a secondary, or late development. Rather than rehashing the many theories which have been advanced on these questions, I propose to restrict myself here to considering specifically those cnidarian neurons which, in one way or another, for one reason or another, have assumed giant proportions relative to the majority of neurons in the same animal, and to speculating about how they evolved. They probably evolved after the main patterns and connection of neurons were already established, in order to provide rapid conduction pathways, and they certainly evolved independently in many different groups. A general reference covering giant axons and escape behavior in a wide range of animal groups is the recent book by Eaton (1984).

2. Scyphomedusan Giant Fiber Nerve Net (GFNN)

GFNN neurons in *Cyanea* are bipolar motor neurons innervating the swimming muscles. Lucifer Yellow injections (Anderson and Schwab, 1981) show no dye coupling between the neurons, confirming evidence from electron microscopy that gap junctions are absent and, incidently, confirming Bozler's (1927) rejection of the older, syncytial theory. They vary considerably in size, with cell bodies ranging from a few microns to 20 μm and axon diameters from 1-5 μm (Anderson and Schwab, 1981). Bozler (1927) identified three size classes of bipolar neurons within this range (Fig. 1).

It would be interesting to have this observation confirmed, as the three size classes might represent a polyploid series. Somatic polyploidy has been demonstrated in the myoepithelium of *Physalia*, where 2n, 4n and 8n cells (possibly even higher) occur (Mackie, unpublished). If the process involves some form of endopolyploidy (Geitler, 1953) there need be only one nucleus. Most of the neurons in *Rhizostoma*, including the largest bipolars have a single nucleus, but, rarely, there are two (Bozler, 1927). Of course, size variations could be achieved without polyploidy, given the possession of a set of genes controlling size or rate of growth, with different alleles expressed in different neuroblasts (see Bennett, 1984), but Bozler's three size classes look suspiciously like polyploidy.

3. *Velella* Closed System

Schneider's (1892) belief that the neurons in this network are syncytially connected was supported by Mackie (1960), on the basis of silver preparations showing what appear to be continuous stretches of uninterrupted axoplasm between cell bodies in parts of the net. A second nerve net (open system) in the same animal looks quite different. Carré (1978), however, found gap junctions between closed system neurons and concluded that the system was, therefore, not a syncytium. Mackie et al. (1988) have responded that the presence of gap junctions between some neurites in the net does not preclude the possibility of syncytial fusion in other parts. They envisage a progression of steps whereby neurons joined by gap junctions engulf one another (Fig. 2). Excess cell membrane would be generated by the fusion process. It is significant therefore that what appear to be internalized cell membrane fragments still held together by gap junctions are often seen in the cytoplasm.

If this interpretation is correct, a giant fiber net might arise in evolution, as it appears to form during larval or post-larval development, by the fusion of smaller elements connected by gap junctions and we should look at other systems, including the following two cases, with the same possibility in mind.

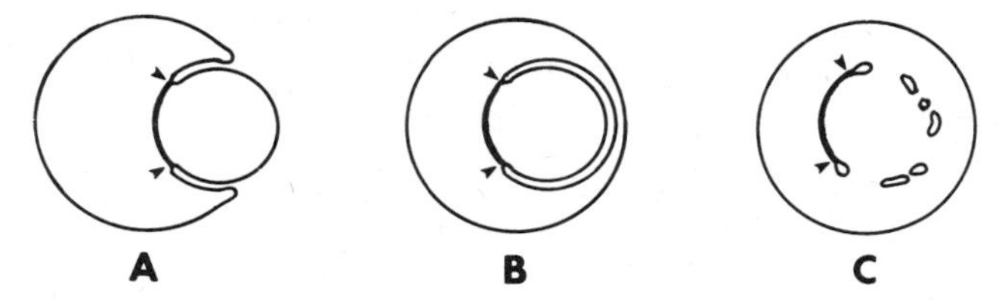

Figure 2. One way in which fusion seems to be occurring in the closed system of *Velella* (from Mackie et al, 1988). Neurites adhere by an extensive gap junction (A). Larger neurite creeps around (B) and engulfs (C) the smaller one. Fragments of internalized cell membrane with gap junctions remain in cytoplasm.

4. *Polyorchis* Swimming Motor Neuron Network (SMN)

Rapid conduction around the margin of the impulses that trigger swimming is a feature of all hydromedusae. The SMN system has been best studied in *Polyorchis*, but the findings apply to *Aequorea* and at least some other species (review by Satterlie and Spencer, 1987). When injected with Lucifer Yellow, the network is seen to consist of short, chunky strands rarely exceeding 100 μm in length, up to 25 μm wide, and interconnected by strap-like bridges some of which are almost as wide as the primary strands. The nucleus makes little or no bulge in the strand it lies in. Gap junctions have been observed between adjacent neurons, but these do not seem to be very abundant and rarely exceed 0.5 μm in width, according to Satterlie's (1985) data for *Aequorea*, whereas the bridges displayed by Lucifer injection can be more than an order of magnitude larger. Spencer (1979) tellingly remarks that "it may be that the giants are in cytoplasmic continuity with one another since the membranes separating axons are often incomplete. If fusion does occur, then the swimming motor neurons form a syncytium." Like closed system neurons in *Velella*, *Polyorchis* SMNs often show what appear to be internalized cell membrane relics containing appositions indistinguishable from gap junctions. Physiologically, the neurons show tight electrical coupling. This would be expected whether the connections are made by extensive gap junctional complexes, by open cytoplasmic bridges, or by a mixture of the two. For a conclusive answer to the contact/continuity question, HRP should be injected into these axons, as this marker does not cross gap junctions.

5. Siphonophore Stem "Giant Syncytium"

Several species of physonectid and diphyid siphonophores have a pair of giant axons running longitudinally in the dorsal ectoderm of the stem. Physiological evidence suggests that each giant is associated with its own diffuse nerve net, but this needs to be verified at the histological level, e.g., by injecting different dyes into the two giants. Physiologically, the two systems appear to function independently, acting synergistically to cause contractions of the myoepithelium (discussion in Mackie, 1984). A. N. Spencer has injected Lucifer Yellow into the giant axons of *Nanomia* (reported in Mackie, 1984) and of *Halistemma* (Grimmelikhuijzen et al., 1986) and has shown that the giants are dye coupled to diffusely distributed nerve nets in both species. (My own carboxyfluorescein injections confirm these findings in *Nanomia*). There is no question that the giants are syncytial structures, as many nuclei can be discerned within continuous stretches of cytoplasm. Whether or not the plexuses associated with the giant fibers are also syncytia remains to be determined. The neurons composing them could be joined by gap junctions. However, the stem plexus in *Halistemma* resembles both the *Velella* closed system and the *Polyorchis* SMN system, in being composed

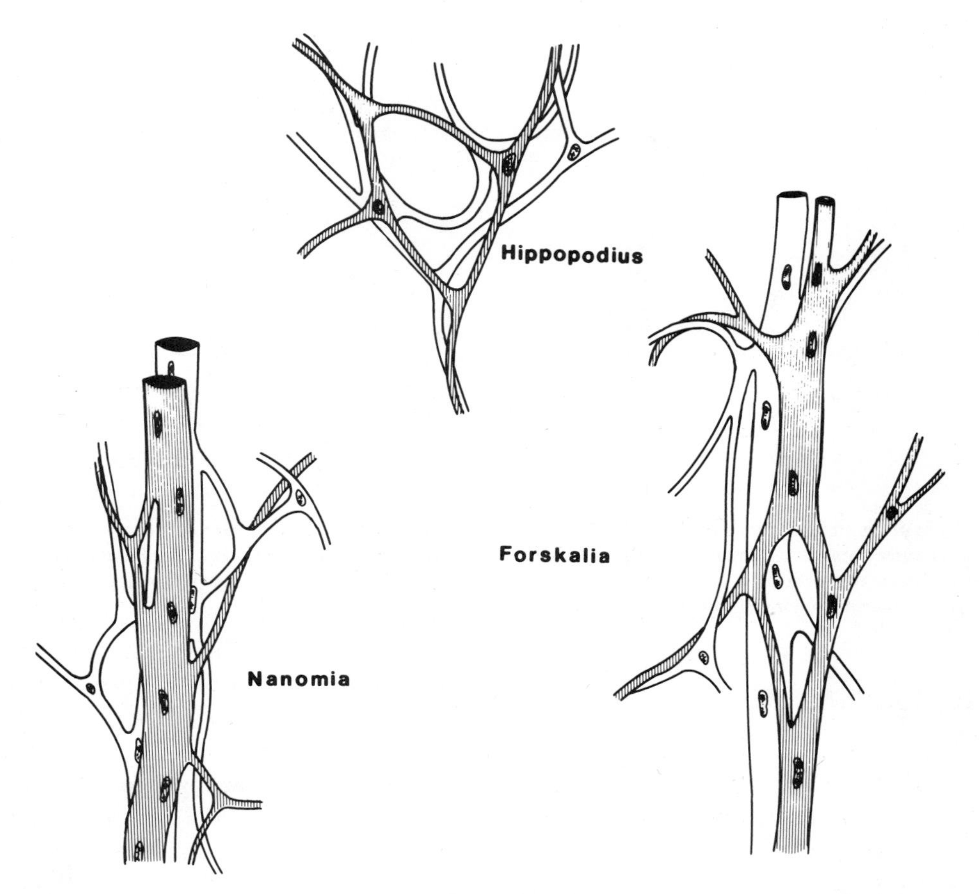

Figure 3. Possible evolution of siphonophore stem giant axons. In all three species, physiological evidence suggests that a double innervation is present. In *Hippopodius* there are no giant axons and the two systems are therefore presumed to exist in the form of diffuse nets. In *Nanomia*, a giant axon has "condensed" within each nerve net. *Forskalia* is similar to *Nanomia*, but the giant axons are less clearly differentiated from the associated nets. The nets are shown as syncytia, but the actual nature of the intercellular connections has not yet been established (from Mackie, 1984).

of composed of relatively short, chunky strands interconnected by thick bridges, favoring the syncytial interpretation and lending support to Schneider's (1892) contention that the giant axon along with its associated nerve net constitutes a single "giant syncytium".

The rather scanty evidence from other siphonophores suggests an evolutionary progression of stages leading up to the *Nanomia* condition (Fig. 3). Thus, in evolution, giant axons might have "condensed" out of coupled or fused small neuron networks, in situations where a high conduction velocity offered a selective advantage. There is no information on the ontogeny of this system.

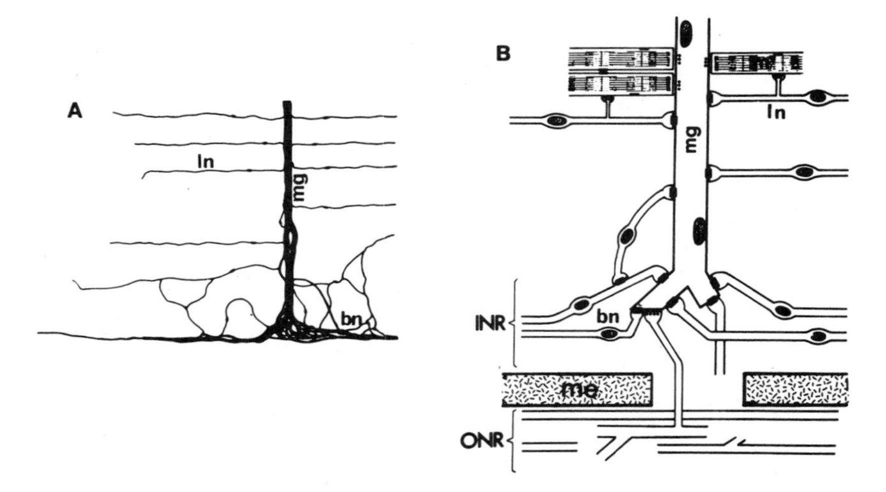

Figure 4. The motor giant axon complex of *Aglantha*. (A). Drawing of a Lucifer Yellow preparation. (B). Interconnections as determined by optical and electron microscopy. Abbreviations: bn, basal neurons; INR, inner nerve ring; ln, lateral neurons; m, swimming muscle; me, mesogloea; mg, motor giant axon; ONR, outer nerve ring (from Weber et al., 1982).

6. *Aglantha* Motor Giant Axon (MOG)

This axon is exceptionally interesting physiologically (see chapter by Meech, in this book), but is also interesting histologically. The axon itself is about 1.5 cm long and 40 μm in diameter in a large specimen, with nuclei scattered all through its axoplasm. As in a striated muscle fiber, the spindle-shaped nuclei are dwarfed by the huge tube they lie in, and cause no bulges. The MOG is dye coupled along much of its length to smaller "lateral" neurons which run out across the muscle sheet. At its base, it is dye coupled to a network of "basal" conventional neurons which form a loose bundle running in the inner nerve ring (Fig. 4). Gap junctions have been located by electron microscopy at the points where basal neurons contact the MOG. There is no indication of cytoplasmic confluence in these areas. Horseradish peroxidase, which does not cross gap junctions, fails to pass from the MOG to the marginal network (Weber et al., 1982).

I have tried to raise *Aglantha* larvae to see how the giants form, but without success. The larvae invariably die while still in the actinula stage. Thanks to Claudia Mills, we have been able to examine a small number of early post-larvae from the plankton. Details of our findings will be presented elsewhere (Mackie and Singla, in preparation). Despite indifferent fixation, well-formed MOGs can be recognized in animals larger than about 1.0 mm bell diameter. The axons make synapses with the swimming muscles and these little larvae perform swimming pulsations. In an earlier stage (0.6 mm) there are no MOGs although the fine radial nerve fibers which will run next to them are present. Some isolated cells, which could be pre-MOG neuroblasts, are located along the line where the MOG will form and it is tempting to infer that the MOG is formed from them, presumably by fusion, but this still needs to be demonstrated. Stages intermediate between the 0.6 and 1.0 mm stages are needed to fill out the picture. An observation consistent with the fusion theory is that MOGs in mature animals not infrequently show bifurcations or give off side branches which flow back into the main axon after wandering separately for some distance. If the axon arose from within a network of smaller units, such persistent net-like configurations might be expected.

Regardless of how the giants originate in ontogeny, it would seem likely that they arose in evolution by some process involving cytoplasmic pooling within a network of conventional, coupled neurons and, of course, they remain functionally associated with networks of conventional neurons (the lateral and basal neurons) to which they are coupled by gap junctions.

7. *Aglantha* Ring Giant Axon (RG)

The ring giant runs around the margin of *Aglantha* in the outer nerve ring. It appears to have no side branches, no beginning and no end, but is a perfect annulus, perhaps the only such neuron in the animal kingdom. It is also peculiar in containing a large fluid-filled vacuole which, in a fully-fledged RG, occupies 85% of the volume of the whole cell, confining the cytoplasm to a narrow cortex (Fig. 5). These unusual features not unnaturally led earlier workers to view it as something other than a neuron. Roberts and Mackie (1980) showed that when excited, it conducts all-or-none action potentials at high velocities (up to 2.6 m. s^{-1}) leading to excitation of the motor giant neurons and escape behavior.

The RG is a multinucleate syncytium. The cytoplasm forms a thin cortical layer around the enormous vacuole (Fig. 6B), and the nuclei lie in this layer (Fig. 6C). No membrane partitions are seen in the cytoplasm. Dyes injected into the vacuole spread through long stretches in which many nuclei can be seen. The nuclei fluoresce brightly, showing that dye has crossed the membrane between the cytoplasm and the vacuole. Ultrastructurally the RG resembles an axon in receiving synapses from adjacent neurons, and in making symmetrical ("two-way") synapses with other neurons (Fig. 6B). It does not appear to make synapses with epithelial cells. It makes gap junctions, identified in lanthanum preparations by their connexon-like particles (65-75 Å diameter with 15-20 Å central depression), with adjacent epithelial cells, and with small-diameter neurons in the outer nerve ring. In many fixations, the axoplasm has a dense, grainy appearance, similar to that seen in surrounding epithelial cells (Fig. 6C).

Regarding these connections with other cells, electrophysiological evidence (to be presented more fully elsewhere) suggests that the RG is electrically coupled, though weakly, to the adjacent epithelial cells. RG spikes cause <12 mV depolarizations in these cells. This would fit with the observations of gap junctions between the two. The occurrence of two-way synapses with local small-diameter neurons fits with observations on interactions between the RG and a slowly conducting neuron system I call the pre-slow wave (PSW) system which runs in parallel with the RG. PSW spikes depolarize the RG sufficiently (by about 10 mV) to raise it to spike threshold. Regarding conduction in the opposite direction, RG spikes cause an increase in the normally slow conduction velocity of PSW events initiated at the same time. (A similar effect in *Nanomia* has been termed the "piggyback effect" (discussed in Mackie, 1984) as it is assumed that the faster system is depolarizing the slower and carrying its spike along with it on its back, as it were.) The finding of gap junctions between the RG and small-diameter axons fits with the observation that EPSP input into the RG from one system at least, the mechanoreceptor hair cells (Arkett et al., 1988), is blocked in 2 mM octanol. These junctions continue to transmit in 10-20 mM Co^{++} .

The RG can be detected histologically in post-actinula larvae with bell diameters of only 0.2 mm. At this stage, the RG is about 3.0 μm in diameter and resembles a conventional axon except in having vacuoles forming in the cytoplasm, apparently in association with Golgi complexes (Fig. 6A). Measured with a digitizer, vacuolar area is typically about 10-20% of total cross sectional area at this stage. The same RG may

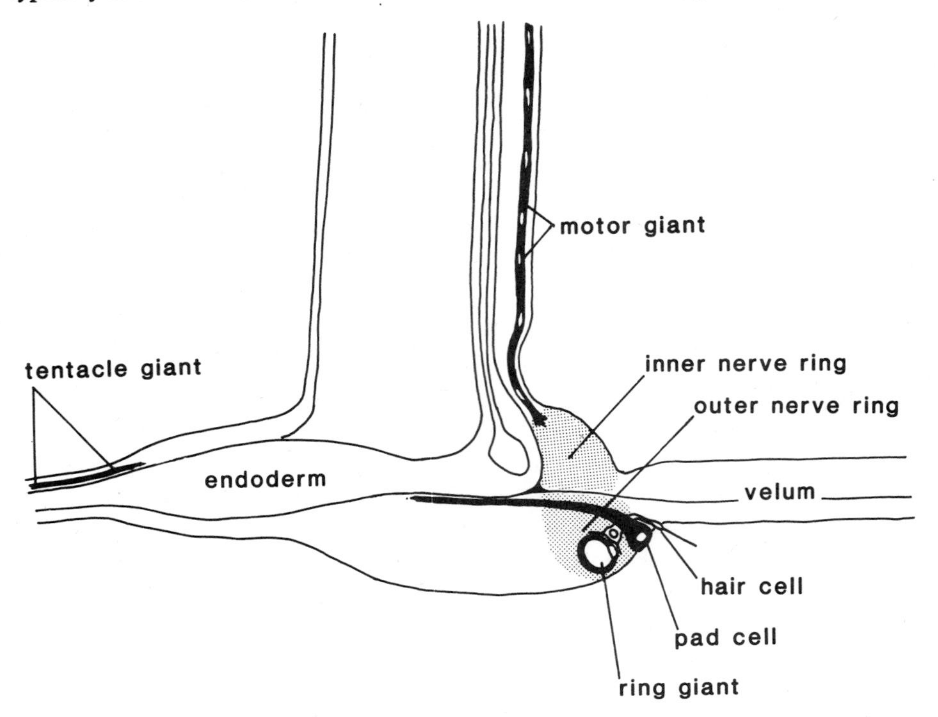

Figure 5. Radial section through the margin of *Aglantha*, showing positions of cells discussed.

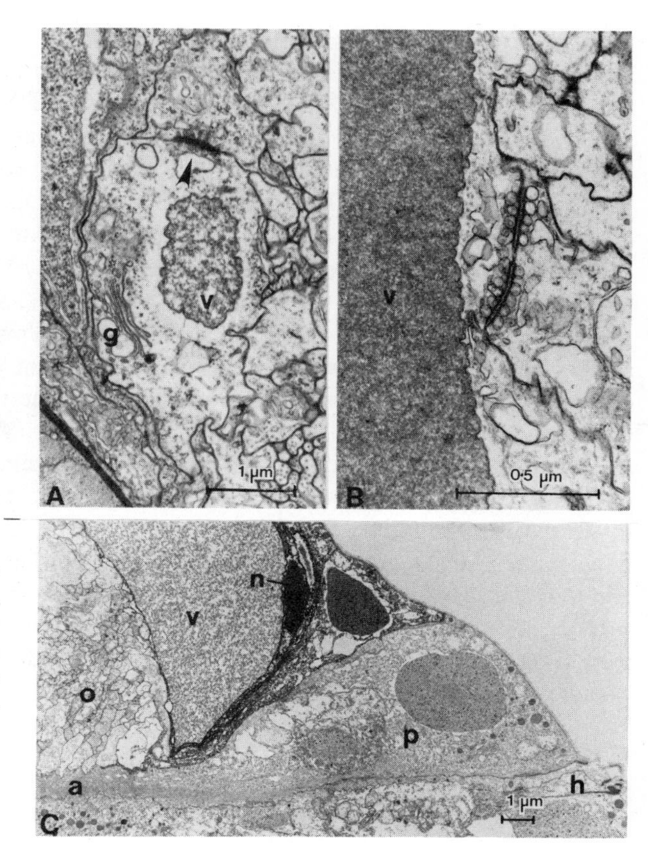

Figure 6. Electron micrographs of the ring giant and pad cell (contributed by Dr. C. L. Singla, University of Victoria). (**A**). section through an early stage of the RG axon from a 2.5 mm diameter larva. Note prominent Golgi complex (g) and vacuole (v) surrounded by microtubules. Arrowhead shows two-way synapse with an adjacent neuron. (**B**). Cortical cytoplasm of mature ring giant with two-way synapse to adjacent axon. RG vacuole (v) on left. C. Radial section showing pad cell (p) and its axon-like process (a). Outer nerve ring (o) is seen to left of the RG, whose nucleus (n) and vacuole (v) are labelled. Note grainy, electron-dense cytoplasm of RG which makes it resemble the adjacent epithelial cell. A hair cell (h) is separated from the pad cell by an epithelial sheathing process.

show a single large vacuole in some sections and several small ones in others, as if the vacuoles were coalescing. Each vacuole is surrounded by a ring of microtubules. (This is also true in mature cells, Fig. 6B). The adult pattern of synapses is present. By the 1.0 mm stage the vacuole has expanded to fill about 80% of the cell, and the RG is now 6 μm in diameter. We have no evidence regarding the very earliest stages in the formation of the RG. Before the vacuoles appear, there might be nothing to distinguish the RG from a conventional small axon. It is not known how the multinucleate condition arises.

Is the RG a "true neuron" or a transformed epithelial cell? The development suggests that it starts as a true neuron, becomes vacuolated and (somehow) multinucleate. It preserves the essential features of a neuron, the ability to produce sodium spikes, and the ability to make and receive synapses. The peculiar dark, grainy appearance of the cytoplasm sometimes seen in both the RG and local epithelial cells is not a consistent feature and may be an artefact. The presence of a vacuole filling most of the interior makes it look very unlike a conventional neuron, but this feature would make any cell look odd. The presence of gap junctions with other neurons is not a surprising feature, as many hydrozoan neurons are so joined. The gap junctions and evidence of coupling with epithelial cells is unusual, something not normally to be expected in a conventional nerve cell. However, *Aglantha* is unusually well endowed with gap junctions. They have been found between hair cells (which are essentially neurons with a sensory process) and epithelial cells as well. All in all, it seems best to assume that the RG evolved from within the nervous system, not from an epithelial cell ancestor.

The vacuole may be a mechanical adaptation allowing the axon to shorten and swell symmetrically, and so prevent it from buckling or kinking when the margin contracts violently in escape swimming behavior. The vacuole may also be useful in increasing the conduction velocity of RG impulses. Action potentials and EPSPs recorded from the vacuole are not inverted, but resemble those recorded from the cytoplasm. The resting potential of the vacuole is -15 to -20 mV, compared with -60 to -65 mV for the cortical cytoplasm. All the inputs and outputs are made with the outer membrane, and it seems likely that this membrane generates the action potential and that the inner membrane is leaky, allowing current to flow longitudinally in the vacuole, as it would in the axoplasm of a conventional giant axon. A leaky inner membrane is also needed to explain the apparent ability of dyes to cross from the vacuole into the cytoplasm, staining the nuclei.

8. Other Giant Axons in *Aglantha*

The tentacle giant is a giant axon running along the dorsal side of each tentacle (Fig. 5) (see Roberts and Mackie, 1980). This axon synapses with tentacle muscles and is implicated in rapid muscular contractions and ciliary arrests that accompany escape behavior.

Our preliminary electron microscopic examination raised the possibility that the giant axon is formed from the fused processes of the pair of peculiar "pad cells" (Fig. 5), which are epithelial cells lying close to the hair cell clusters (tactile combs) on the velum, near the tentacle base. We are in the middle of trying to verify this relationship. The pad cell does indeed put out a long process resembling a giant axon which runs toward the tentacle (Figs. 5, 6C), but a direct link between this process and the tentacle giant is proving hard to demonstrate (Louise Page, University of Victoria, personal communication). Further, recordings from the pad cell soma fail to show action potentials when the tentacle giant is stimulated. Thus, at the time of writing, it is not at all certain that the tentacle giant stems from the pad cell process. If, however, this proves to be the case, we would have a uniquely interesting example of an axon of epithelial origin, the long-sought-for missing link between excitable epithelia and true nerves (see discussion following Horridge, 1966).

9. Discussion and Conclusions

We have considered three possible ways in which giant axons could have evolved in cnidarians.

1. The first is by endomitotic polyploidy within a population of diploid neurons. The giant bipolar neurons of scyphomedusae should be examined with this possibility in mind. Polyploid neurons formed in this way need not be multinucleate.

2. The second is by modification of an epithelial cell to form an axon-like structure. It is known that epithelia conduct action potentials in many hydrozoa, and some of them could have become specialized for rapid conduction by forming long, thick structures like giant axons in the nervous system. Unless, however, the *Aglantha* pad cell proves to be such a case, there is no good evidence to support this possibility. The ring giant axon shows some features that suggest an epithelial derivation, but the bulk of the evidence, especially from development, suggests that it is a true neuron.

3. The third way is by fusion within a neuronal network, a process which may have occurred several times in the Hydrozoa. The examples from *Velella*, *Polyorchis* and the siphonophore stem all show a certain family resemblance, consisting of relatively short thick processes connected by what appear, at least superficially, to be continuous stretches or bridges of cytoplasm through which (in the last two cases) dyes can be shown to pass freely. In the siphonophore stem the net is dye coupled to a multinucleate giant axon, which is certainly a true syncytium (but this does not prove that the network itself is syncytial). HRP injections should help here.

Assuming, however, that all these systems are true syncytia in whole or part, we can imagine how they may have evolved from networks of conventional neurons joined

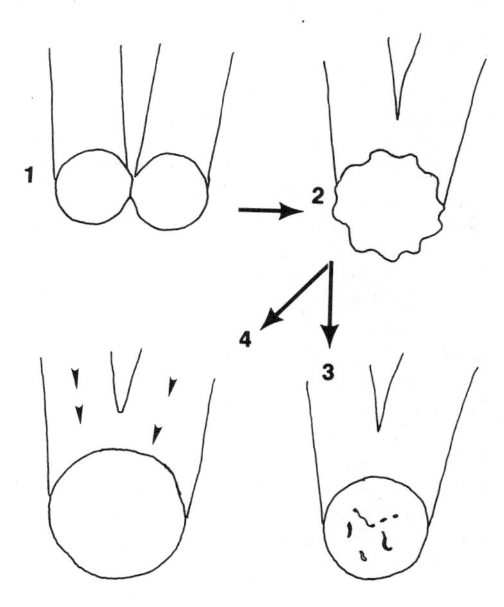

Figure 7. Two axons fuse at a gap junction (1) producing excess cell membrane at fusion site (2) because, if volume doubles, surface area only increases by a factor of 1.4. Internalization of excess membrane (3) could occur or axoplasm could flow in (arrowheads) from other parts of the net to fill out the membrane, leading to progressive giantization (4).

by gap junctions. A tendency for the membrane to break down at junctional sites would result in axoplasmic confluence and the production of thicker strands in the net, and sometimes in the formation of single, distinctive giant axons.

Is there something about extensive gap-junctional membrane appositions which predisposes cells to take the final step and fuse, becoming true syncytia? A sufficiently dense and extensive array of connexon particles might constitute a weak point, at which the membrane tended to lose lateral cohesion and break. Regarding the tendency for the fusion product to have few, thick strands, we can imagine (Fig. 7) that at points where the membrane has broken down, the axons would tend to flow together into larger-diameter units, altering the surface area/volume ratio and resulting in a local excess of cell membrane. If this membrane were not disposed of, axoplasm would continue to flow in from neighboring areas to fill out the slack. This would eventually lead to the production of giant axons or nets composed of short chunky strands. This process of coalescence or condensation could be regulated in various ways, e.g., by secretion of new axoplasm or by internalization and disposal of the excess membrane produced during fusion; thus, the system could be stabilized at any point between a diffuse net and a single giant axon. Another way of limiting the coalescence process would be by strengthening the membrane at gap junctions so that they were retained intact. This would stop the syncytial part from taking over these regions. The *Aglantha* motor giant axon provides a case in point, being a multinucleate syncytium which is connected by gap junctions to conventional neurons at its base and sides.

References

Anderson, P. A. V., and Schwab, W. E., 1981, The organization and structure of nerve and muscle in the jellyfish *Cyanea capillata* (Coelenterata; Scyphozoa), *J. Morph.* **170:**383-399.

Arkett, S. A., Mackie, G. O., and Meech, R. W., 1988, Hair cell mechanoreception in the jellyfish *Aglantha digitale*, *J. exp. Biol.* **135:**329-342.

Bennett, M. V. L., 1984, Escapism: some startling revelations, in: *Neural Mechanisms of Startle Behavior* (R. C. Eaton, ed.), Plenum Press, New York.

Bozler, E., 1927, Untersuchungen über das Nervensystem der Coelenteraten, *Zeit. Zellforsch.* **5:**244-262.

Carré, D., 1978, Étude ultrastructurale du système nerveux de *Velella velella* (Cnidaria: Chondrophoride), *Vie et Milieu* **28:**221-235.

Eaton, R. C. (ed.), 1984, *Neural Mechanisms of Startle Behavior*, Plenum, New York.

Geitler, L., 1953, Protoplasmatologia (Heilbrunn and Weber, eds.), Bd.6c, Lange, Maxwell and Springer, London.

Grimmelikhuijzen, C. J. P., Spencer, A. N., and Carré, D., 1986, Organization of the nervous system of physonectid siphonophores, *Cell Tiss. Res.* **246:**463-479.

Horridge, G. A., 1966, Pathways of coordination in ctenophores, in: *The Cnidaria and their Evolution* (W. J. Rees, ed.), Academic Press, London.

Mackie, G. O., 1960, The structure of the nervous system in *Velella*, *Quart. J. Micr. Sci.* **101:**119-131.

Mackie, G. O., 1984, Fast pathways and escape behavior in Cnidaria, in: *Neural Mechanisms of Startle Behavior* (R. C. Eaton, ed.), Plenum Press, New York.

Mackie, G. O., Singla, S. L., and Arkett, S. A., 1988, On the nervous system of *Velella* (Hydrozoa: Chondrophora), *J. Morph.* **198:**15-23.

Roberts, A., and Mackie, G. O., 1980, The giant axon escape system of a hydrozoan medusa, *Aglantha digitale*, *J. exp. Biol.* **84**:303-318.
Satterlie, R. A., 1985, Central generation of swimming activity in the hydrozoan jellyfish *Aequorea aequorea*, *J. Neurobiol.* **16**:41-55.
Satterlie, R. A., and Spencer, A. N., 1983, Neuronal control of locomotion in hydrozoan medusae. A comparative study, *J. Comp. Physiol.* **150**:195-206.
Satterlie, R. A., and Spencer, A. N., 1987, Organization of conducting systems in "simple" invertebrates: Porifera, Cnidaria and Ctenophora, in: *Nervous Systems in Invertebrates* (M. A. Ali, ed.), Plenum, New York.
Schneider, K. C., 1892, Einige histologische Befünde an Coelenteraten, *Jena. Zeit. Naturwiss.* **27**:387-461.
Spencer, A. N., 1979, Neurobiology of *Polyorchis*. II. Structure of effector systems, *J. Neurobiol.* **10**:95-117.
Weber, C., Singla, C. L., and Kerfoot, P. A. H., 1982, Microanatomy of the subumbrellar motor innervation in *Alantha digitale* (Hydromedusae: Trachylina), *Cell Tiss. Res.* **223**:305-312.

Chapter 30

Concluding Remarks

PETER A. V. ANDERSON

A major aim of this workshop was to try to arrive at a consensus about how the first nervous systems might have evolved. To this end, we examined data about many of the cell biological functions that underlie nervous function, in an array of animals ranging from protists through lower invertebrates.

Attempting to achieve this in this manner relies on the validity of an important premise; by examining certain cell biological processes in extant organisms, we can gain insight into the evolutionary process in question. It is clear that the various protists and lower invertebrates currently being examined are certainly not primitive; they have had as much opportunity for change, in terms of evolutionary time, as those groups typically referred to as advanced. At the same time, however, the apparent conservation of certain of the mechanisms by which organisms at all phylogenetic levels function suggests that there may be some validity to this approach.

Protists are capable of many of the functions we normally attribute to neurons; they produce action potentials and other electrical events, they communicate with one another (intercellular communication), and transduce sensory stimuli into meaningful responses. However, the remarkable thing is that in many cases, the mechanisms underlying these activities are so similar to those found in neurons in higher animals. For instance, as Carr shows (Chapter 6), the signalling molecules, receptors and transducing mechanisms used by fungi and slime molds for intercellular communication are remarkably similar to those used by nerve cells for the same purpose. Similarly, the basic mechanisms used by protozoans to transduce chemical stimuli are essentially the same as those used by higher invertebrates and vertebrates for olfactory and gustatory transduction (Van Houten, Chapter 25). Clearly, in interpreting such details, one is

PETER A. V. ANDERSON • Whitney Laboratory and Departments of Physiology and Neuroscience, University of Florida, 9505 Ocean Shore Blvd., St. Augustine, Florida 32086, USA.

Evolution of the First Nervous Systems
Edited by P.A.V. Anderson
Plenum Press, New York

faced with the dilemma of whether or not such similarities reflect conservation or convergence. However, as Carr points out, the peptide receptor of fungi is a seven membrane-span protein, and as such, falls within the rhodopsin family of receptor molecules; it is not a six-span protein as are the family of voltage-activated ion channels, or any other of the numerous configurations undoubtedly possible. Furthermore, the sequence of subsequent events involving G-proteins, etc. are, once again, the same as those in "higher" animals. They do not, for instance, use methylation, as do bacteria, but instead rely on phosphorylation in the same manner as higher animals. Thus we see remarkable similarities at all stages in a multistep process. I would argue, therefore, that given that a truly primitive fungi would have a variety of mechanisms available to it for each stage in such a transduction process, the fact that the entire sequence is so similar in such widely diverse groups has to reflect conservation rather than convergence. Whether or not this thesis can be extended to include other cellular properties remains to be seen, but it will be interesting to see whether ion channels in protozoans fit into the six-span family for ion channels in "higher" organisms.

One problem with our trying to define how the first nervous systems might have evolved is the fact that the first nervous systems, as we recognize them, already possess most, if not all, of the features of the complete and complex nervous system of higher animals. Neurons in cnidarians, the "first" phylum to possess a recognizable nervous system, produce fast overshooting, Na^{+}-dependent action potentials (see Chapters 3, 19 and 20, by Spencer, Anderson and Meech), interact by way of chemical and electrical synapses (see Chapter 3, by Spencer), and are important in development both for the substances they secrete (Hoffmeister and Dübel, Chapter 4) and for their role in cell-cell interactions (Hofmayer and David, Chapter 5). It is unfortunate that no one has yet found neuron-like interactions in a colonial protozoans, for instance or that the substrate by which signals propagate through certain sponges is not known. However, within the phylum Cnidaria, there are hints of what could be considered to be evolutionary intermediates. Gap junctions, the structural correlate of electrical coupling, are widespread in the most advanced members of this phylum (the Hydrozoa), but have not yet been identified in the other classes. This may be a technical problem since if the connexons are widely distributed, perhaps in arrays similar to the large loops of particles found in ctenophores (Hernandez-Nicaise et al., Chapter 2), then they might have escaped notice in conventional thin section microscopy. We have essentially no information on functional electrical coupling in those classes that lack recognizable gap junctions, simply because only one nerve net, the motor nerve net of the scyphozoan jellyfish *Cyanea*, has been examined with this question in mind. However, it is curious that this very same nerve net serves no integrative function, yet relies solely on chemical synaptic transmission to transmit activity over quite large distances (1 m in a large animal). Rapid, coordinated conduction through a two-dimensional plexus could, surely, be achieved more efficiently by way of electrical synapses, as has been shown to occur in hydromedusae (Spencer, Chapter 3), than by chemical synapses. The fact that chemical synapses are used in

Cyanea might indicate that gap junctions are not available to the animal. Clearly, future molecular and, perhaps immunocytochemical studies are required to resolve this point.

The cnidarian Na^+ channel may be another example of an evolutionary intermediate. The Na^+ channel of the scyphomedusa *Cyanea* is functionally a Na^+ channel (Anderson, Chapter 19), but pharmacologically a Ca^{++} channel and, thus, might be considered an intermediate in the evolution of Na^+ channels from Ca^{++} channels (Hille, 1984). In contrast, the Na^+ channel of the more advanced hydrozoa, as typified by the hydromedusa *Aglantha* (Chapter 20) is insensitive to Ca^{++} channel blockers. Na^+ channels have not yet been identified in the most "primitive" class, the Anthozoa. Thus, in the Cnidaria, we may be seeing the various steps in the evolution of the Na^+ channel, as hypothesized by Hille (1984). That thesis was based on the distribution of ion channels; specifically the apparent absence of Na^+ channels in the protozoa but their presence in the Cnidaria. In that light, however, it is very curious to find that the marine protozoan *Actinocoryne* produces Na^+-dependent action potentials (Febvre-Chevalier, et al., Chapter 17). One wonders whether the fact that Na^+ currents have not previously been described in protozoans is largely a consequence of the fact that little attention has been paid to marine species. Nevertheless, the presence of a Na^+ spike in the protozoa does not rule out Hille's thesis, since evolution is an on-going process.

One area that is proving to be enormously frustrating is the question of neurotransmitters in the Cnidaria. One would think that by now we would have a handle on such a basic question but, with the exception of some neuropeptides, the question is still unresolved. In the case of peptides, there is good evidence that neurons in various cnidarians synthesize and secrete a variety of peptides (Chapters 4 and 8). In the case of head activator peptide (Hoffmeister and Dübel, Chapter 4) and to a lesser extent the RFamide family of peptides (Grimmelikhuijzen et al., Chapter 9), applications of exogenous peptides have been shown to have physiological effects. However, if, as seems reasonable, neuropeptides function in cnidarians in the same manner as they do in higher animals, then we would expect them to act by way of second messengers. If so, then these compounds cannot be responsible for the very fast synaptic transmission demonstrated at synapses in *Cyanea* (Chapters 3 and 19), *Polyorchis* (Chapter 3) and *Aglantha* (Chapter 20). The very brief synaptic delay of these synapses must surely reflect the action of ligand-gated postsynaptic channels, yet, in the case of *Cyanea*, where putative transmitters can be applied directly to the exposed synapse, none of the recognized neurotransmitters have any action. One cannot, of course, rule out the possibility that peptide-activated ligand-gated channels exist in these animals, but such a process would surely be remarkable.

The whole question of non-peptidergic transmitters in cnidarians is still unresolved. There is increasing evidence that a variety of non-peptidergic transmitters are present in these animals (Chapters 9-11) in both neuronal and non-neuronal locations, but clear demonstration of a synaptic action of any of them is lacking, primarily because cnidarians are notoriously bad preparations for conventional pharmacology. Clearly, resolution of this question awaits the development of new preparations of single isolated postsynaptic cells. However, investigators should be cautious in their interpretation of results obtained in this manner and realize that several criteria must be fulfilled for clear demonstration of the presence of a transmitter. Just as demonstration of immunoreactivity to a particular antibody is only one stage in the confirmatory process, demonstration of a physiological effect of applications of the putative transmitters is also but one stage in that process, as illustrated by the fact that in crayfish, a variety of transmitters apparently activate the same post-synaptic receptor (see Chapter 12, by Hatt and Franke).

An important point in any discussion of nervous system evolution is the position of the Cnidaria in this process. This subject was addressed on several occasions and remains unresolved. As other authors have pointed out, cladograms based on the conserved RNA sequences put the Cnidaria in a variety of positions. It was the general consensus of the participants that the whole question of the evolutionary position of the Cnidaria will be resolved only when the genes for a variety of other conserved molecules are determined, but whatever the outcome, it is clear that the nervous system of the cnidarians is essentially *fait accompli* in terms of nervous system evolution. Regardless of the position of the cnidaria in this process, it was eminently clear that platyhelminthes occupy an important position in the evolution of the nervous system. Not only are they the "first" phylum with a rostral brain (see Koopowitz, Chapter 22) but they are invariably found at the base of most cladograms depicting evolution of the animal kingdom. Clearly, this group requires far more attention that it is currently receiving.

Another commonly expressed sentiment was that it is unfortunate that developmental biologists have devoted so much effort to *Hydra* at the expense of all other cnidarians. While *Hydra* has clear advantages for developmental studies, it is a very difficult preparation for neurophysiology; its neurons are too small to permit intracellular recording, and the fact that it lives in very low ionic strength medium poses major problems for patch clamp recordings from single isolated cells. The result is that we know essentially nothing about the neurophysiology of *Hydra*. Such information is sorely needed to complement the developmental data. Future work by Kass-Simon and others may correct this, but we are still in the unfortunate situation where we know a great deal about a species which is physiologically problematic, and very little about the development of those species which are amenable to physiological examination. This discrepancy should be corrected.

One potentially useful approach to identifying pathways of nervous system evolution would be to identify the particular processes, and ultimately the molecules concerned, that define nervous system function. However, to do this, one must first address the question, what is a nervous system? The most obvious answer, a functional aggregation of neurons, is of little help since one is then faced with the challenging task of assembling a list of the criteria with which to identify a neuron, without, in the case of something like a cnidarian or flatworm, recourse to the developmental origin of the cells. The participants attempted, unsuccessfully, to compile such a list. The most obvious descriptors, electrical excitability and synaptic transmission or other targeting of information, were of little use since non-spiking neurons are common in many animals, epithelial cells spike in a variety of invertebrates and vertebrates, and many endocrine cells target information to other cells by way of released chemicals.

Despite this list of problems and negative results, I think certain conclusions can be reached. First, the title of the workshop, *Evolution of the First Nervous Systems*, was fortuitously correct; there is no such thing as a first nervous system, they probably arose several times. Furthermore, we are correct in calling them the first nervous systems, rather than primitive nervous systems, since the nervous system of a cnidarian or platyhelminth may have evolved just as much as those of higher animals, such as ourselves. I also feel that we are justified making inferences about how the nervous systems might have evolved by studying extant lower forms, since the cellular processes used by a wide range of lower animals are so similar at the molecular level as to argue against convergence, at least for many of the processes in question. We must, however, remember that such organisms have themselves evolved since the cellular process in question first appeared. Clearly this is rather a subjective argument, but as the genes of relevant molecules in both higher and lower forms are identified and cloned, and as other groups (e.g., platyhelminthes) recieve more attention, a more objective evaluation will be possible.

Index